Exploration and Production of Oceanic Natural Gas Hydrate

Michael D. Max · Arthur H. Johnson

Exploration and Production of Oceanic Natural Gas Hydrate

Critical Factors for Commercialization

Second Edition

Springer

Dr. Michael D. Max
Marine and Petroleum Geology Research Group (Natural Gas Hydrate)
iCRAG, School of Earth Sciences
University College Dublin
Belfield, Dublin, Ireland

Arthur H. Johnson (Deceased)

ISBN 978-3-030-13110-4 ISBN 978-3-030-00401-9 (eBook)
https://doi.org/10.1007/978-3-030-00401-9

Cover illustration: Recovering oceanic natural gas hydrate at in situ temperatures and pressures is essential to preserve this ephemeral diagenetic material in its natural geological form. Geotek Coring has developed engineering systems for acquiring, handling and analysing natural gas-hydrate-bearing formations for many years. The picture shows multiple pressurised storage chambers containing rich gas-hydrate-bearing cores that can be shipped to specialist laboratories for advanced geological and geotechnical analysis whilst being maintained at in situ temperatures and pressures (Courtesy of Geotek Ltd.).

This Springer imprint is published by the registered company Springer Nature Switzerland AG
The registered company address is: Gewerbestrasse 11, 6330 Cham, Switzerland

Bill Dillon and I would like to dedicate this second edition to our great friend and colleague Art Johnson. We corresponded on natural gas hydrate in our separate institutions and jobs since the late 1980s until we brought our separate activities and interests (Art, teaching), Max, research including establishing and operating an advanced chemical laboratory) (Bill, USGS emeritus activities and editing) under the umbrella of HEI. In 1999, the three of us established Hydrate Energy International (HEI) when Bill retired from the U.S. Geological Survey and Art retired from Chevron, while I left the Naval Research Laboratory to establish a separate small R&D company, Marine Desalination Systems that investigated using gas hydrate for industrial separation purposes (seawater desalination, industrial water treatment & gas separation) until 2011. Art also was a Lay Methodist Minister, which allowed him to hone his lecture technique. He was also active in the Northern Ireland Peace initiative and through a consortium of New Orleans churches did great work during the hurricane

Katrina recovery. He was always eager to chat to young geologists or to anyone about geology and gas hydrate and to have a beer while doing so. He was an unforgettable partner.

Preface to the Second Edition

The first edition of ***Exploration and Production of Oceanic Natural Gas Hydrate*** was being been written while the hydrocarbon energy industry was undergoing substantial downsizing and the prices of both gas and oil were dropping. Since then industry has largely learned to live with lower prices although prices appear to be slowly rising. Also, production of shale oil and gas in the United States has reached record levels and is still increasing, making the country one of the top producers in the world. The energy picture in the world and especially the United States has changed considerably in the last two and a half years. There appears to be sufficient supply for current and near-term demand, but this does not mean that the NGH resource development is not needed. We feel that the suggestions that we make for lowering the cost of exploration and production for the environmentally very low risk and pure produced gas and water from NGH may be very competitive with conventional hydrocarbon energy. The NGH resource is also available to the deep importing countries (Chap. 3) to whom availability may be important than strict price competitiveness.

This second edition more deeply explores the development of the potentially huge natural gas hydrate (NGH) resource. The term NGH is synonymous with "methane hydrate" and "gas hydrate" and "clathrate" when used to reference the resource. The term NGH also includes compound hydrate which can be composed of two or more hydrocarbons, which can produce natural gas liquids and other gases in addition to methane. Hydrocarbon gases produced from the hydrate resource are essentially the same as the natural gas used by consumers the world over. The NGH resource is a potentially very large gas resource that is approaching the tipping point of commercialization.

This book is written as a resource for deepwater (includes ultra-deepwater) NGH exploration and production activities. It is intended for scientists, students, engineers, company administrators, regulators, and policy makers, but specifically for those interested in developing new technology, responding to opportunities arising from the special attributes of NGH. This book builds on three earlier books. This second edition adds considerable information from drilling expeditions, production tests, and new equipment development. It is assumed that the reader has a background in NGH, marine technology, and at least a general familiarity in the energy industry, including conventional and unconventional fuel sources.

We propose that of all the hydrocarbon fuels, natural gas provides the best option for a high-quality combustible energy resource for base load and peak power demands in an otherwise carbon-free energy supply scenario, while replacing coal and oil-based power in the nearer term. Natural gas is an ideal base load and peak (spiker) on-demand energy source in an otherwise largely renewable energy future. Our focus is on natural gas, and particularly the potentially great oceanic NGH resource. We anticipate that natural gas will be the backbone on-demand fuel of the future and that as the most environmentally friendly of all natural gas resources, NGH has the potential to provide natural gas far into the future.

This book also considers implications of the development of the renewable energy paradigm wherein increasing amounts of renewable, non-combustible energy will eventually dominate energy production. Replacement of oil and especially coal by gas-fired power generation will have the effect of lowering CO_2 emissions.

Many of the recent references that are relevant to potential commercialization of the NGH resource are not published in traditional books and journals because it is too new and the technology is only in development. We do reference industry standard publications, which are almost entirely website references that may have a limited longevity compared with hardcopy and formal electronic publications. These include, but are not limited to, American Association of Petroleum Geologists *Chemical & Engineering News*, Currents (Marine Technology Society), *DW Monday* (Douglas Westwood), *Drilling Contractor, EOS* (American Geophysical Union), *Fuel Fix, GEOExPro, (The) Fuze (*http://energyfuse.org*), InnovOil/Newsbase (*http://www.innovoil.co.uk*), Marine Technology Society, Maritime Reporter and Engineering News, OilOnline, OilVoice, OilPro (went offline 2 Aug 17 but was very useful), Oil & Gas Journal & Oil, Gas & PetroChem Equipment (PennWell Corp.), Offshore Magazine* (PennWell Corp.), *PennEnergy Daily Petroleum Update* (PennWell Corp), *Petro Global News, Power (Official publication of 'Electric Power), Science (American Association for the Advancement of Science), Sea Technology* (Compass Publications), DutilityDive (https://www.utilitydive.com), *Xinhua* (China) daily news, and also other blogs and other energy-related publications of professional organizations and news sources worldwide.

Disclaimer: Throughout this book, we have referenced specific companies, processes, equipment, and developments, among other things to provide examples to assist readers without familiarity of all the subjects and as a way to help them further research topics. Neither the authors, the publisher, nor anyone associated with this book infers any particular support or promotion of any commercial or other entity over any other that may have similar products or services of any type. Dates are American usage (MM/DD/YR) rather than European (DD/MM/YR).

A reference date is commonly (YR/MM/DD) so that references with dates are in order with the youngest first, similar to the alphabetization system.

Book Chapters and Organization: The book is divided into 11 chapters with topical subsections. Some reorganization has been done so to consolidate subject information in individual chapters, in particular, Chaps. 3 and 4. Some chapters have been modified substantially from the first edition, while other chapters are less modified.

Chapter 1 **Energy Overview: Energy Options and Prospects for Natural Gas** This chapter concerns all the major sources of energy, especially the developments in microgrids, widening of the horizontal business model for energy, and renewable energy. It has been modified more than the others with inclusion of the newest references that reflect rapidly developing energy solutions.

Key Topics: Renewable Energy, Fossil Fuels, Gross Domestic Product, Human Development Index, Quality of Life, Energy Mix, Climate

Chapter 2 **Economic Characteristics of Deepwater Natural Gas Hydrate** Basic elements of the NGH system are outlined. What, where, why, and when are considered, although "when" is more thoroughly discussed in Chap. 11.

Key Topics: Gas Hydrate Stability Zone, Biogenic, Thermogenic, Crystallization, Dissociation, Migration, Stability Zone, Sulfate/Methane Transition, Reservoirs, Resource Potential, Environmental Risk

Chapter 3 **Exploration for Deepwater Natural Gas Hydrate** This chapter focuses on NGH as part of a petroleum system that makes exploration amenable to a systematic approach and the impact of geophysical analytical techniques that have been ground-truthed by drilling.

Key Topics: Petroleum System, Stability Zone, Paragenesis, Migration, Reservoir, Exploration, Basin Modeling, Seismic, Bottom Simulating Reflector, Mediterranean, National Programs

Chapter 4 **Potential High-Quality Reservoir Sediments in the Gas Hydrate Stability Zone** The present first-order exploration targets for NGH are marine turbidite sands. On some continental margins, however, complex rift inversions may have placed some high-quality reservoirs into the gas hydrate stability zone. In addition, marine reworking of sediments may have formed large bodies of sandy sediment.

Key Topics: Exploration, Reservoir, Rift-Related Sediments, Paralic Depositional Environments, Aeolian Systems, Sequence Stratigraphy, Mediterranean Sea, Black Sea

Chapter 5 **Valuation of NGH Deposits** NGH is an economic mineral deposit similar in many ways to low-temperature strata-bound mineral deposits. Mineralization by NGH fills porosity in a host bed while displacing pore water, rather than filling all available space in a reservoir volume as is the case for gas or liquid conventional hydrocarbons.

Key Topics: Gas-in-Place, Petrogenesis, Permeability, Mineralization Grade, Cell Valuation, Seismic Response, Creaming Curve

Chapter 6 **Deepwater Natural Gas Hydrate Innovation Opportunities** The natural characteristics of NGH and its reservoir system, physical environment, and relatively low pressures and temperatures allow for new approaches to many aspects of exploration and production.

Key Topics: Depressurization, Exploration, Drilling, CAPEX, Geotechnical Attributes, Reservoir Stability, Production, Stability Zone, Environmental Security, Seafloor Operations, Lightweight Paradigm

Chapter 7 **Leveraging Technology for NGH Development and Production** Deepwater technology development has reached the point where a very large part of the technology necessary to open the NGH resource exists or can be modified to suit natural gas hydrate conditions. The technology deals with production (including processing, power, automated control systems, and remote control of seawater industrial sites).

Key Topics: Robotics, Seafloor Installations, Lightweight Paradigm, Technology Innovation, ROV, AUV, Power Systems, Data Acquisition, Data Management, Drilling, Coiled Tubing, Processing, Sand Control, Flow Assurance

Chapter 8 **New Technology for NGH Development and Production** The specific properties of NGH and its host sediments provide a broad range of opportunities for introducing new approaches for resource development.

Key Topics: Technology Innovation, Exploration, Production, Drilling, Coiled Tubing, Casing, Autonomous Bottom Hole Assemblies, Active Tethered Drilling, Active Wellbore Lining, Sand Control, Gas/Water Separation, Active Reservoir Control, Technology Readiness Level

Chapter 9 **Offshore Operations and Logistics** The main issues concern how to deal with stranded gas and long-term operation of seafloor sites.

Key Topics: Exploration, Production, CAPEX, O&M Costs, Access, Search and Rescue, Spill Response

Chapter 10 **Energy Resource Risk Factors** Risk in general is discussed followed by discussion of risk factors of the different natural gas resources, other fossil fuels, and unconventional energy resources.

Key Topics: Environmental Risk, Reservoir Performance, Access, Exploration, Appraisal, Development, Production, Abandonment, Fracking, Climate, Dependability, Energy Prices, Technology Risk

Chapter 11 **Commercial Potential of Natural Gas Hydrate** Commerciality of a product such as natural gas is decided by a number of factors. These include exploration and production costs, transport to market entry point, and other factors.

Key Topics: Economics, Valuation, EROI, World Gas Market, LNG, CNG, Dissociation, Production Rate, Production Profile, Pressure Management, Infrastructure, Stranded Gas

M.D. Max would like to thank chapter reviewers of the second edition. Chapter 1—Christopher Carstens, B.Sc. Mechanical Engineering, UC Berkeley, Founder, Hydrate Dynamics LLC; Chaps. 2 and 3—William P. Dillon, Ph.D., U.S. Geological Survey, Emeritus; Chaps. 4, 6—Srikumar Roy, Post Doctorate Fellow, University College Dublin, Ireland; Chap. 5—William H. Bird, Ph.D., PGeo, Vancouver, Canada; CAPT Dennis M. Egan PE, US Coast Guard (Retired); Chaps. 7, 8, 10, 11—Jeffrey Eppink, President & Founder, Engeis, LLC.

Dublin, Ireland

Michael D. Max
Arthur H. Johnson

Preface to the First Edition

Throughout human history, technology has changed reality. In the modern world, the industrial revolution ushered in the greatest changes to the existing realities and the current age of electronic automation is again changing reality for most people as economic, political, and social systems struggle to adapt. This book is about one facet of change: Energy and development of technology to underpin the transition from the fossil fuel age to a renewable energy future.

"*Exploration and Production of Oceanic Natural Gas Hydrate*" is our fourth book on the topic of natural gas hydrate (NGH). The term NGH is synonymous with "methane hydrate," "gas hydrate," and "clathrate" when used to reference the resource. The term NGH also includes compound hydrate which can be composed of two or more hydrocarbons, including natural gas liquids and other gases. Hydrocarbon gases produced from the hydrate resource are essentially the same as the natural gas used by consumers the world over. The NGH resource is a potentially very large gas resource that is approaching the tipping point of commercialization.

Our second book concerned the application of the NGH physical–chemical system to the economic geology of NGH. The third mainly concerned the potential for NGH resources in the Arctic Ocean and included resource estimates and procedures along with an outline of the NGH petroleum system analysis.

This book is written as a resource for deepwater (includes ultra-deepwater) NGH exploration and production activities. It is intended for scientists, students, engineers, company administrators, regulators, and policy makers, but specifically for those interested in developing new technology, responding to opportunities arising from the special attributes of NGH. This book builds on our previous three NGH books. The first (2000, 2003) summarized the understanding of NGH issues. The second (2006) concerned physical chemistry applied to the formation of NGH concentrations and its economic geology. The third focused on NGH resource potential in the Arctic Ocean. Besides our books, there exists an expanding NGH literature on topics including climate, seafloor morphology and sedimentation, drilling and other seafloor interventions, safety, biosystem interaction, and the primary driver of NGH research and development funding, which is energy. We assume that the reader has a background in NGH, marine technology, and a familiarity the energy industry, including conventional and unconventional fuel sources. The book is intended to inform current industrial and technology trends including the future possibilities for NGH. Summaries for other power sources are readily available and are widely understood by the environmental and industrial communities. We refer to nonrenewable energy sources, particularly fossil fuels, when necessary in discussion.

Although this book has been written during the post-2014 oil and natural gas price crash, market rebalancing of supply and demand should once again raise interest in oil and gas exploration and production. From a supply-chain perspective, this point in the hydrocarbon economic cycle is a period for innovation to bring through new approaches and technology for deepwater developments to improve efficiency and lower cost. Deepwater will be a cost-competitive source of world-class hydrocarbon reserves. With this in mind, we stress two strategic opportunities. First, natural gas is the best fuel to backstop the wide use of renewable energy and second, NGH has a relatively low environmental risk to the biosphere. We intend to show how existing technology can be leveraged and new technology developed so that cost of NGH development can be reduced so that it has a competitive advantage on a cost basis.

Of hydrocarbon fuels, natural gas provides the best option for a high-quality combustible energy resource for base load and peak power demands in an otherwise carbon-free energy supply scenario, while replacing coal- and oil-based power in the nearer term. Natural gas is an ideal base load and peak (spiker) on-demand energy source in an otherwise largely renewable energy future. Our focus is on natural gas, and particularly the potentially great oceanic NGH resource. We anticipate that natural gas will be the backbone on-demand fuel of the future and that as the most environmentally friendly of all natural gas resources, NGH has the potential to provide natural gas far into the future.

This book also considers implications of the development of the renewable energy paradigm wherein increasing amounts of renewable, non-combustible energy will eventually dominate energy production. Natural gas will almost certainly emerge as the preferable fossil fuel backstop for renewable energy and the transition to it. The content of this book strongly supports the position of COP21 (Conference of Parties under the United Nations Framework Convention on Climate Change), in which natural gas is also considered to be the combustible backbone fuel of the future. Replacement of oil and especially coal by gas-fired power generation will have the effect of lowering CO_2 emissions.

The physical attributes of the NGH are very different from conventional hydrocarbon resources in a number of important aspects. Springing from these differences, new approaches to exploration and production offer options for development and utilization of new technology and methods that have the potential to dramatically lower the cost of exploration and production. These opportunities could dramatically lower the cost of commercialization of the resource and make NGH competitive with other natural gas resources on a produced basis.

Many of the recent references that are relevant to potential commercialization of the NGH resource are not published in traditional books and journals because it is too new and the technology is only in development. We do reference industry standard publications, which are almost entirely website references that may have a limited longevity compared with hardcopy and formal electronic publications. These include, but are not limited to, the *Oil & Gas Journal* (PennWell Corp.), *Offshore Magazine* (PennWell Corp.), *PennEnergy Daily Petroleum Update* (PennWell Corp), *Petro Global News, DW Monday* (Douglas Westwood), *GEOExPro, OilOnline, OilVoice, Drilling Contractor, Fuel Fix* and other blogs, *Xinhua* (China) daily news and also major energy-related news sources in other countries, *Sea Technology* (Compass Publications), *Chemical & Engineering News*, publications of the Marine Technology Society, *EOS* (American Geophysical Union), and the American Association of Petroleum Geologists. Further, many of our suggestions for development of technology and practices that we regard as having the potential to lower the cost of NGH exploration and production are new, although they may be consistent with some elements of current technology development. We have made and retain PDFs of all materials referenced from the Web.

Disclaimer: Throughout this book, we have referenced specific companies, processes, equipment, and developments, among other things to provide examples to assist readers without familiarity of all the subjects and as a way to help them further research topics. Neither the authors, the publisher, nor anyone associated with this book infers any particular support or promotion of any commercial or other entity over any other that may have similar products or services of any type.

Book Chapters and Organization: The book is divided into 11 chapters with topical sections. The organization and discussion of topics stress the place of natural gas in a long-term energy future, as well as the attributes of the NGH system that pertain to the exploration and production. Chapters are broadly referenced so that readers will have a head start to deeper research in each topic. Key references are often used in more than one chapter.

Chapter 1 Energy Overview: Prospects for Natural Gas Availability of energy to fuel motors, widespread electricity generation and distribution, and industrialize food and water production and distribution is the basis of modern civilization and the socioeconomic well-being of nations. Energy is the economy. Nations that have access to abundant energy have many options that energy-poor nations do not.

The change from vertically organized power companies having traditional fossil fuel electrical generation plant(s) serving relatively small areas to new horizontally organized electric power systems serving much larger areas is necessary to draw on renewable as well as fossil fuel energy. This allows companies to meet electric load demands over much larger regions while using as much renewable energy as is available at any time. The importance of natural gas as an on-demand backstop to renewable energy is correlated with the development of the low environmental risk NGH resource for the renewable energy paradigm.

Key Topics: Renewable Energy, Fossil Fuels, Gross Domestic Product, Human Development Index, Quality of Life, Energy Mix, Climate

Chapter 2 Economic Characteristics of Deepwater Natural Gas Hydrate Basic elements of the NGH system are outlined that we regard as the most important to use of existing and development of new technology that could have the potential to dramatically lower the cost of exploration and production. This chapter captures the basic elements of NGH that are important for exploration and production: How NGH sequesters gas from the natural gas flux, controls of crystallization and growth, and the NGH prospect zone. What, where, why, and when are considered, although "when" is more thoroughly discussed in Chap. 11.

Key Topics: Gas Hydrate Stability Zone, Biogenic, Thermogenic, Crystallization, Dissociation, Migration, Stability Zone, Sulfate/Methane Transition, Reservoirs, Resource Potential, Environmental Risk

Chapter 3 Exploration for Deepwater Natural Gas Hydrate This chapter focuses on NGH as part of a petroleum system that makes exploration amenable to a systematic approach and the impact of geophysical analytical techniques that have been ground-truthed by drilling. The chapter describes seismic and electrical approaches to prospect analysis. A general case for exploration for NGH concentrations in the principal geological target for a marine turbidite sand reservoir host is discussed. Special attention is given to NGH potential in the Mediterranean and Black seas, in which little NGH exploration has been carried out. New original work by the authors is included. The NGH exploration programs of nations throughout the world are summarized.

Key Topics: Petroleum System, Stability Zone, Paragenesis, Migration, Reservoir, Exploration, Basin Modeling, Seismic, Bottom Simulating Reflector, Mediterranean, National Programs

Chapter 4 Potential High-Quality Reservoir Sediments in the Gas Hydrate Stability Zone This type of reservoir is not currently regarded as an NGH exploration target but under certain conditions, they could be. Because of the much higher quality of the reservoirs, we include it for completeness. Although the first-order exploration target for NGH is marine turbidite sands, on some continental margins complex early tectonic, sedimentary depositional situations may have existed during transition to conditions of fully marine sedimentation. In particular, rift clastics associated with the separation of continental crust masses, and shallow marine sediments, would normally underlie the marine sequences that developed as rifting transitioned into passive continental margins of new ocean basins. In addition, the formation of paralic sediments in that may now be in the upper marine succession associated with oceanic flooding. The Black and Mediterranean seas and the northern Gulf of Mexico are discussed in some detail because their younger history of these basins could have yielded high-quality reservoirs of the type that are generally much more deeply buried by sediment in open ocean continental margins. Other rifted margins of interest are more lightly discussed.

Key Topics: Exploration, Reservoir, Rift-Related Sediments, Paralic Depositional Environments, Eolian Systems, Sequence Stratigraphy, Mediterranean Sea, Black Sea

Chapter 5 Valuation of NGH Deposits NGH is an economic mineral deposit similar in many ways to low-temperature strata-bound mineral deposits. Mineralization by NGH fills porosity in a host bed while displacing pore water, rather than filling all available space in a reservoir volume as is the case for gas or liquid conventional hydrocarbons. Filling over 80 % of host sediment porosity may occur in high-grade deposits. The valuation of an NGH concentration will more closely resemble the volume or cell approach to metallic mineral deposits. This discussion suggests other methods than classical drilling that takes forward the current ability to predict closely NGH in a reservoir host and its concentration using a geotechnical approach to seismic interpretation alone.

Key Topics: Gas-in-Place, Petrogenesis, Permeability, Mineralization Grade, Cell Valuation, Seismic Response, Creaming Curve

Chapter 6 Deepwater Natural Gas Hydrate Innovation Opportunities In this chapter, we elaborate on physical/chemical and petroleum system aspects of NGH and highlight those specific elements of the NGH system that offer potential for new technology development. The natural characteristics of NGH and its reservoir system, physical environment, and relatively low pressures and temperatures allow for new approaches to many aspects of exploration and production.

Key Topics: Depressurization, Exploration, Drilling, CAPEX, Geotechnical Attributes, Reservoir Stability, Production, Stability Zone, Environmental Security, Seafloor Operations, Lightweight Paradigm

Chapter 7 Leveraging Technology for NGH Development and Production Deepwater technology development has reached the point where a very large part of the technology necessary to open the NGH resource exists. The technology that deals with production (including processing, power, and automated control systems and remote control of seawater industrial sites) may be used mainly as they are. Other technology can be optimized with minor redesign for NGH conditions. Industry is already installing special hardware for exploration and processing to the seafloor; the surface handling capabilities for establishing and maintaining seafloor work sites appear to largely exist. Seafloor processing will largely follow current trends in conventional production and processing.

Key Topics: Robotics, Seafloor Installations, Lightweight Paradigm, Technology Innovation, ROV, AUV, Power Systems, Data Acquisition, Data Management, Drilling, Coiled Tubing, Processing, Sand Control, Flow Assurance

Chapter 8 New Technology for NGH Development and Production The specific properties of NGH and its host sediments provide a broad range of opportunities for introducing new approaches for resource development. In addition to robotization, we outline new drilling and wellbore lining processes that are derived from a combination of drilling and tunneling practices.

Key Topics: Technology Innovation, Exploration, Production, Drilling, Coiled Tubing, Casing, Active Bottom Hole Assemblies, Active Tethered Drilling, Active Wellbore Lining, Sand Control, Gas/Water Separation, Active Reservoir Control, Technology Readiness Level

Chapter 9 Offshore Operations and Logistics The main issues concern how to deal with stranded gas and long-term operation of seafloor sites. Transfer of gas from wellhead/processing systems to transport and the trade-offs of pipelines and ships carrying the gas in different compressed formats may follow current practices and developments because once the gas has been produced to the seafloor, it is no different in general from conventional natural gas. Remote and difficult operational areas such as the Arctic are discussed.

Key Topics: Exploration, Production, CAPEX, O&M Costs, Access, Search and Rescue, Spill Response

Chapter 10 Energy Resource Risk Factors Risk in general is discussed first followed by discussion of risk factors of the different natural gas resources, other fossil fuels, and unconventional energy resources. Discussion of specific risk factors deals with overdependence on a single fuel, environmental risk, geohazards, and the risks of other energy sources, followed by business issues including regulations, leasing, price, and business cycles. Finally, exploration, new technology, and cost–benefit analysis are discussed in a renewable energy context. The existing transition to natural gas from other fossil fuels will be accelerated by both cost factors and regulations that will, in part, relate to climate change mitigation.

Key Topics: Environmental Risk, Reservoir Performance, Access, Exploration, Appraisal, Development, Production, Abandonment, Fracking, Climate, Dependability, Energy Prices, Technology Risk

Chapter 11 Commercial Potential of Natural Gas Hydrate Commerciality of a product such as natural gas is decided by a number of factors. These include exploration and production costs, transport to market entry point, regulations that may introduce additional costs (or prove prohibited for other reasons), subsidies for existing competition, and other factors. Climate issues are taken into account, but mainly from exploration and production viewpoints.

Key Topics: Economics, Valuation, EROI, World Gas Market, LNG, CNG, Dissociation, Production Rate, Production Profile, Pressure Management, Infrastructure, Stranded Gas

Notes to Chapters

1. The first use of natural gas hydrate introduces the term NGH, which is used throughout the remainder of each chapter. Where the phrase occurs in the abstract as well as the text, the term is also defined.
2. Dates are in American notation (MM/DD/YR).

State of the Industry and Timing of this Book: Although this book has been written while the hydrocarbon energy industry was undergoing substantial downsizing bordering on collapse of the higher cost segments of exploration and production, publication of this book in late 2016 should see a slowing if not a bottom to the deterioration of the business environment, if not the development of some positive inflections for the industry as a whole. A considerable part of the industry will have downsized, and operational and business innovation should be recognized as key to the energy future. We feel that the suggestions that we make for lowering the cost of NGH exploration and production may be more relevant than if publication took place in an environment dominated by traditional conventional hydrocarbon energy.

Fields of Innovation: The bulk of the new technology suggestions in this book follows a program of innovation undertaken by HEI that culminated in the filing of a number of patents, which may be summarized in this book but are not discussed in detail and are not referenced in text or in the reference lists.

Dublin, Ireland

Michael D. Max
Arthur H. Johnson

Contents

About the Authors

Michael D. Max has a broad background including geology, geophysics, chemistry, acoustics, and information technology. He has a BSc from the University of Wisconsin, Madison, an MSc from the University of Wyoming, and a PhD from Trinity College, Dublin, Ireland. He has worked as a geologist / geophysicist for the Geological Survey of Ireland, the Naval Research Laboratory, Washington, DC, and the NATO Undersea Research Center, La Spezia, Italy. From 1999 to 2011 Michael Max was CEO and Head of Research for Marine Desalination Systems LLC, which established a hydrate research laboratory and explored industrial applications of gas hydrate. He is the author of many scientific publications and four textbooks, and holds over 40 patents. He assisted in the writing of the US Gas Hydrate Research and Development Act of 2000. Michael is a member of the Methane Hydrate Advisory Committee of the Department of Energy 2014-2019, and is Co-Chair, Diving Committee of the Marine Technology Society. He is an Adjunct Professor at the School of Geological Sciences of University College, Dublin, Ireland. Hydrate Energy International, LLC (HEI) has been closed. Michael Max is now carrying on his R&D activities through MaxSystems LLC and University College, Dublin, Ireland.

Art Johnson was a founding partner of Hydrate Energy International, LLC (HEI). Prior to forming HEI in 2002, Art had been a geologist with Chevron for 25 years, where his career included most aspects of hydrocarbon exploration and development. Art was instrumental in initiating Chevron's Gulf of Mexico program for gas hydrate studies in 1995. He advised Congress and the White House on energy issues starting in 1997, and chaired advisory committees for several Secretaries on Energy. He had a longstanding role coordinating the research efforts of industry, universities, and government agencies. Art served as the Gas Hydrate Lead Analyst for the "Global Energy Assessment," an international project undertaken by the International Institute for Applied Systems Analysis (IIASA) of Vienna, Austria and supported by the World Bank, UN organizations, and national governments that evaluated the energy resource bases of the entire planet with a view to addressing energy needs in the decades to come. He was Chair of the Gas Hydrate

Committee of the Energy Minerals Division of the American Association of Petroleum Geologists (AAPG) and was also very active in his Methodist Church and in helping with hurricane relief and peacemaking activities. Much to the sorrow of his good friend and co-author, and of countless other friends, Art unexpectedly passed away on August 9, 2017.

Other Books published by Springer:
Max, M.D. (Ed). Natural Gas Hydrate In Oceanic and Permafrost Environments (2000, second edition 2003)
Max, M.D., Johnson, AH. & Dillon, W.P. Economic Geology of Natural Gas Hydrate (2006)
Max, M.D., Johnson, AH. & Dillon, W.P. Natural Gas Hydrate - Arctic Ocean Deepwater Resource Potential (2013)

Chapter 1
Energy Overview: Energy Options and Prospects for Natural Gas

Abstract Availability of energy is key to wealth, political and military power, and living standards. Energy availability and consumption may be the most reliable measure of an economy. There is a direct relationship between energy consumption and countries' gross domestic product (GDP) and the Human Development Index (HDI). Energy security, which is the relative certainty that energy supplies for a country will be available, constitutes a primary security concern for countries with high energy demands and countries with increasing energy use. Even though there appears to be no immediate shortage of hydrocarbons, this non-renewable resource is being supplemented by renewable energy. The Renewable Energy Era has already begun. Reduction in CO_2 emissions is underway because of government regulations and market forces. A completely renewable energy future may be in our future, but its timing is very uncertain as renewable energy presently contributes less than 10% of energy, and that supply is highly concentrated geographically. The potentially largest natural gas resource remaining on Earth, oceanic natural gas hydrate (NGH), may substantially supplement the natural gas supply far into the future. An additional benefit is that natural gas produces less CO_2 per Btu and also has a much lower pollution potential than any other combustion fuel. Natural gas is the clean hydrocarbon fuel that will reach into the renewable energy future. Its continued availability at affordable prices becomes increasingly important as coal and oil power plants are retired and energy demand becomes increasingly filled by development of renewable or intermittent power sources.

1.1 Energy, GDP, and Society

Energy is the Economy. The fossil fuel energy industry dates back only to the early 18th century Industrial Revolution, (Hall et al. 2003). Prior to that, humans, animals, water power (usually water wheels) installations, and wind provided very limited amounts of energy for work. The transition to electric energy took place in the late 1800s, but widespread transmission did not make it available widely until the mid-1900s.

M. D. Max and A. H. Johnson, *Exploration and Production of Oceanic Natural Gas Hydrate*, https://doi.org/10.1007/978-3-030-00401-9_1

Modern, energy intensive economies have traditionally depended on fossil fuels for energy. Widespread use of energy has transformed human existence over a very short time, allowing lifestyles that could only be dreamed of earlier. We now live in an eclectic socio-economic paradigm (Cleveland et al. 2008), surrounded by mechanical and electronic machines that require staggering amounts of energy. A further impact of energy availability is its price. Lower energy cost has a direct impact upon economic activity and manufacturing, a relationship clearly demonstrated by the low price of natural gas in the United States from 2008 to present, even though the energy industry itself in the U.S. is directly responsible for no more than 1.5% of GDP (Cardarelli and Lusinyan (2016).

Modern society and economies are based on the availability of energy and may be the single most important factor in a nation's standard of living and potential for geopolitical and economic power. The sure availability of energy is also the foundation of national security for the United States (CAN 2017). The wealth of a country is commonly represented by its gross domestic product (GDP), which shows a direct or almost linear relationship with energy use (Fig. 1.1). Today's major powers all have relatively high GDP and energy consumption, although many countries rich in energy resources have not created broad economies or political power. Countries with early, continuing, and increasing energy consumption are now generally referred to as 'developed' or first-world; countries, while the least energy consuming countries are referred to as undeveloped or under-developed or third-world. Those countries with high GDPs consume more

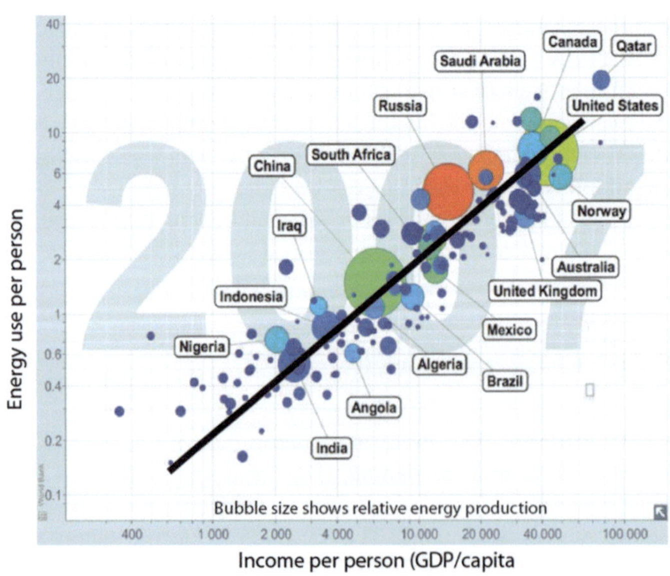

Fig. 1.1 General relationship between per capita GDP and energy use. Modified from World Bank figures http://www.worldbank.org/en/research. Black line represents approximate mean of all data points

energy and it is widely understood that for a country to increase its standard of living and influence in the world, GDP must be raised (Lane 2014). Technology drives standard of living metrics and requires increased energy supplies.

Countries wishing to increase their GDP, their international standing, and improve their living standards, require their own—or secure access to—energy supplies. Such countries as the United States are secure in both conventional hydrocarbons (coal, oil, natural gas) and unconventional fuels (coalbed methane, shale oil and gas, and NGH (NGH)) while some countries such as Japan, is an extreme example of having only a NGH resource potential.

The attributes of modern life taken for granted in developed countries, such as ubiquitous access to labor saving devices, transport, energy-intensive cities, products of our new innovation economy such as computers and mobile communications devices, may only be objects of desire in lower energy consumption and GDP countries. In order to modernize, countries must increase their energy consumption/ GDP. No industry influences the shape of the world more than oil and gas (Nicholls et al. 2014). But simply increasing energy through fossil fuel consumption may not have the desired result. Fossil-fuel economies have been imposed on many societies in Africa, Asia and elsewhere but most have failed to develop the desired 'fossil fuel values' and other societal factors of advanced countries (Turchin 2015).

More sophisticated representations of the relationship between energy use and the 'modernism' of a country relates GDP to a variety of environmental and societal parameters. GDP and energy consumption of countries are closely linked but simple or per capita GDP may not accurately represent the standard of living or quality of life within a country, particularly in developing or energy-rich countries where GDP may be very high but the standard of living for citizens may be lower than would be expected, and in all countries where income inequality is high.

Availability and level of energy use is also strongly related to standard of living parameters that add perspective to how the societal components of GDP are apportioned within each country. The Human Development Index (HDI) of the United Nations Development Program (UNDP) incorporates factors such as life expectancy, education, income inequality, poverty rates, GDP per capita and environmental parameters into a normalized measure on a scale of 0–1 (UNDP 2014a, b). Countries with an HDI approaching 1 have a very high standard of living and countries with lower HDI have lower standards. A detailed analysis of the data by UNDP (2014a) yields somewhat different of view than simply using GDP as the principle measure.

Three distinct trends or data clusters of Quality of Life (QoL) groups (very low, average, and very high) emerge when each country's HDI is plotted versus energy consumption (Fig. 1.2). The general relationship of high energy consuming/GDP countries generally having higher QoL than lower energy consuming/GDP countries, but there is a noticeable reordering, which mainly reflects countries having broader social programs having higher HDI. There appears to be a direct relationship between quality of life and consumption of electrical energy (Alam et al. 1991).

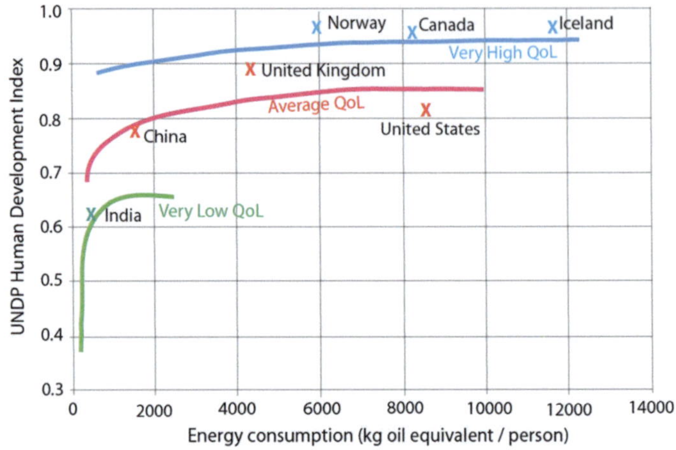

Fig. 1.2 Quality of Life (QoL) trends as a function of energy consumption. Original figure based on UNDP HDI country data plotted as points (Ben 2008), with the positions of a few countries shown. Estimated mean regression of the data points through the data clusters. A mean trend line through all data can also be drawn as a single trend line that would follow the lower part of the very low QoL and arc upward and to the right. Similar trends also occur for carbon dioxide emissions with some countries being in different positions within their groups because the energy mix varies from country to country, with, for instance, Canada and Norway having substantial CO_2-free hydrokinetic power supplies

In order to raise the HDI of the world's 2008 population to 0.9, representing 2400 kg of oil equivalent (5.8 MM Btu per Bbl oil) per person, the world energy consumption would have to rise from its 2004 level of about 450 quadrillion Btu (or quads) to about 630 quads (Ben 2008). Reaching an HDI of 0.9 today (2015) would require about 200 additional quads because of increased population and systematic issues such as increased demand. This is probably not an achievable objective. An increase of this order would require the increased use of hydrocarbon fuels and a dramatic increase in the number of nuclear reactors, even with an equally dramatic increase in renewable energy (Table 1.1). Countries cannot be idle in securing energy supplies with which to fuel improvements in standard of living. The expectation of being used to see the lights coming on when a switch is thrown is going to grow.

Non-renewable energy consists of fossil fuels that have sequestered vast amounts of carbon from the biosphere over hundreds of millions of years by natural processes, and in the case of gas and oil, concentrated by petroleum systems (Max and Johnson 2014). Burning of fossil fuels to provide heat energy with which to generate direct power and electrical energy exhausts combustion products into the atmosphere, recycling carbon to the biosphere. As greater amounts of energy are now required, the amount of carbon, largely in the form of carbon dioxide (CO_2) that is exhausted into the atmosphere, adds to the heat-retaining character of the atmosphere and may be driving changes in climate (IPCC 2013). Renewable energy

Table 1.1 (a & b) U.S. Electricity net generation (Other gases small consumption not shown)

	(a) *Fossil fuel*				
	1. Coal	2. Oil	3. Natural gas	4. Nuclear electric	5. Hydrokinetic pump storage[a]
1950	154,520	33,734	44,559	N/A	N/A
1960	403,467	47,987	157,970	518	N/A
1970	704,394	184,183	372,890	21,804	N/A
1980	1,161,562	245,994	346,240	251116	N/A
1990	1,594,011	126,460	372,765	576,862	−3,508
2000	1,966,285	111,221	601,038	753,893	−5,539
2010	1,847,290	37,061	987,697	806,968	−5,501
2014	1,585,697	30,489	1,121,928	797,067	−6,209

(b) *Renewable energy*

	Conventional	Unconventional					12. Total[b]
	6. Hydrokinetic electric	7. Wood	8. Waste	9. Geothermal	10. Solar/ PV	11. Wind	
1950	100,885	390	N/A	N/A	N/A	N/A	334,088
1960	149,440	140	N/A	33	N/A	N/A	759,156
1970	303,153	18	174	3,246	N/A	N/A	1,535,111
1980	279,182	275	158	5,073	N/A	N/A	2,289,600
1990	292,866	32,522	13,260	15,434	367	2,789	3,037,827
2000	275,573	37,595	23,131	14,093	493	5,593	3,802,105
2010	260,203	37,172	18,917	15,219	1,212	94,652	4,125,060
2014	258,749	43,050	21,269	16,628	18,321	181,791	4,092,935

Electricity in Millions of kWh. Compiled by M. D. Max from U.S. Energy Information Administration deta (eia.gov)

a. Major energy types. #5. Minus number is the energy consumed for each pumping cycle, produced energy added to #6

b. Totals include minor additional fuel sources not included in table

sources are those that naturally exist around us such as solar and moving wind and water whose use does not exhaust carbon into the atmosphere.

Countries with existing high-energy consumption, and countries striving to improve their GDP, must have continued and ever-increasing energy supplies. But economic growth (Harris et al. 2013) such as was responsible for rising standards of living in the 20th century, when wealth was produced mainly as a result of the unrestricted use of hydrocarbon energy, will be more difficult in the 21st century and in the future because the Earth's oil fields are non-renewable in less than a geologic time-frame, and only contain a limited amount of the resource at any price, even with secondary and tertiary recovery techniques. Once consumed, the demand must be replaced by other energy sources to maintain GDP and everything based on it. Today's biggest economies have been built on a backbone of hydrocarbon energy that is simply unsustainable. Transition to hydrocarbon replacement energy sources eventually will be a matter of national survival, regardless of the environmental impetus for the development of renewable energy sources.

Although this book is being written during the possible trough of the post-mid 2014 oil oversupply and price crash, decadal energy demand should increase and fossil fuels, and particularly natural gas, will continue to provide an energy resource. Fossil fuels will probably be limited by environmental concerns long before fossil fuel depletion is an issue. The part increasing population, and the emergence of a growing middle class that can purchase energy, particularly in China, India, and SE Asia will play is likely to threaten world societal stability. This is because energy-intensive lifestyles are the foundation of GDP and they are very attractive. There is some question as to whether the whole energy-climate issue could be considered to be a kind of intelligence test for us.

Hydrocarbon fuels are non-renewable and will eventually become more expensive and less available, as well as being less sociologically acceptable because of carbon emissions. It's not a question of if, but when the use of fossil fuels has to be scaled back. Therefore, the sooner alternative energy sources can be implemented, the easier will be the passage to the new energy paradigm in which renewable energy will be dominant. There is likely to be a dramatic increase energy efficiency, which will have an effect on the link between GDP/quality of life as well as a comparably dramatic increase in renewable energy supplies, not only to increase the overall fuel supply but to replace existing fuel supplies.

This process has already begun with a change from hydrocarbon-dominant energy to a mix of traditional fossil fuels, whose relative percentage of use is being subjected to a type of economic-environmental triage, and renewable energy, for which no definite schedule of implementation is currently possible. An energy revolution that almost certainly spells the endgame for fossil fuels as the main energy source, however, is currently underway (Helm 2017). Between 2007 and 2016 about 56,000 MW of coal fired electrical generation has been lost. Energy costs and prices are down by about 3% and CO2 emissions are also down (Silverstein 2018). There is a wave of market and technology changes in which hydrocarbon (especially coal) power generation is being replaced by principally wind (currently 8% of U.S. national energy at 84,900+ MW) and solar (1+ % at 27,000+ MW). The rate of solar installation is greater than wind (in 2017–2018). Most important and largely overlooked by the political Luddites writhing to preserve coal and coal mining is that over 1.2 million people are currently employed in the wind and solar industries, with opportunities for further employment as other renewable energy opportunities, such as tidal and geothermal, open up.

1.2 The Energy Mix

Energy from more than one source. The renewable energy era has already begun in that the stalwart of energy supply, fossil fuels, and its modern nuclear energy equivalent that produces the high temperature steam that drives the same type of turbine-electric generators, are being supplemented by renewable energy sources (El-Ashry 2015). Clean or renewable energy now accounts for more than half of all new energy supply. The development of a renewable energy future is likely to herald a 4th industrial revolution that will be driven by new renewable energy technology (WEF 2016).

This transition may also be seen as the beginning of the end of the fossil fuel industry (Kaufman 2015), except possibly for natural gas. The value and nature of energy sources (Thakur and Rajput 2011) mean different things to different communities of interest. Energy sources fall into non-renewable and renewable categories (Table 1.1), with environmentalists strongly concerned that renewable energy replace existing reliable energy sources that have powered the industrial revolution and dramatic increases in country GDP and standard of living. The 'old reliables' are fossil fuels, with the 'new reliable' of nuclear energy and to a much lesser extent, leveraging excess energy by storing it. For instance, pump storage, which optimizes continuous power generation from fossil fuel power plants that have excess power during period of low demand can be used to pump water to ponds at elevation. The stored potential energy can then be used to generate electricity through water head, similar to conventional hydro energy, but with energy loss from pumping (Table 1.1). Pump storage can store non-renewable energy to be used in a renewable energy manner. We use the term 'hydrokinetic' to refer specifically to high pressure water systems powered by gravity-driven water head that can drive very large electricity generators usually having fully contained turbines, rather than the more generic term 'water power', which includes all water driven power flowing water or water falling only a short distance. These have been used since at least Roman times to provide direct mechanical power.

Unconventional renewable energy is only unconventional in the sense that it has not been used to provide large amounts of energy for distribution. Wood and wind energy, for instance, have been used since humans discovered fire and sails and windmills, but these were only ever used locally. Today, the world is on the road to a renewable energy paradigm, driven by new, ever more affordable technology and manufacturing.

Although metallurgical coal will continue to be used to manufacture steel, particularly in India, the U.S., and China, although this market is only a small part of what used to be coal production. Natural gas is replacing rapidly both coal and oil as the primary fossil fuel for power generation (Table 1.1) in part because of simple economics; the price of gas is very low and promises to stay low for many years. Petroleum coke produced from oil can also be used in steel manufacture. Coal-to-gas switching (Slingsby 2015) has accelerated since the onset of low prices post-2014 and natural gas is on course to become the major fuel for power generation. Maloney (2018) notes that in 2016, first time gas-fired power generation exceeded that of coal for the first time. He also expects low gas prices to drive down the use of coal and points out that the market force of lower fuel cost will drive fuel swapping. He expects that EIA's predictions that coal's share of the generation market will fall from 30 to 28% by 2019 may be understated.

Coal use has significantly decreased. Almost no one alive today remembers when coal was the major fuel that was once used by ships and trains as well as for electrical power. The fuel switch has been particularly dramatic in Britain, in which the use of coal fell by 52% in 2016 (BBC 2017) and is now 12 times below the peak of 221 million tonne burnt in 1956. Long-standing perceptions have changed in the energy industry, even though the aphorism of 'oil is where the money is' is still correct in energy density terms. But demand controls supply and demand changes depend on cost and technology and a dramatic paradigm shift in the demand for hydrocarbon products could significantly reduce the price of these fossil fuels and essentially destroy hydrocarbon energy industries (Arbib and Siba 2017). Coal use in China, which has underpinned the energy required for China to become the second largest industrialized economy in the world, appears to have peaked (Ye and Lu 2018). Further economic expansion will almost certainly be based on renewable energy, particularly solar, and natural gas.

Coal is also now being displaced by natural gas in electric power generation. Current conventional wisdom is that oil demand may not diminish much any time soon because of its very widespread use in vehicles, despite a number of countries and companies such as Volvo stressing transition to electric vehicles. But because demand for producing petroleum is dominated by its use in vehicles, and particularly in privately owned and operated vehicles, any dramatic shift in the traditional liquid fuel paradigm could dramatically reduce the demand for liquid petroleum fuels, and possibly natural gas.

In addition to fuel displacement by natural gas, the dramatic fall in the cost of renewable energy indicates that in many areas existing coal fired power plants are more expensive than new renewable energy installations (although this may not include battery backup that would yield more dependable long-term power). Smyth (2018) notes that on the basis of competitive costs, accelerated closing of coal fired power plants could actually improve the cash flow of existing energy companies.

Three principle factors are combining to affect ten trillion-dollar revenues of the existing vehicle and petroleum production through use as a combustible fuel economy.

- Battery development will probably improve mileage on a single charge so that range will be directly comparable with existing petroleum fueled vehicles, and faster charging will allow for mid-trip recharging. In addition, adding photovoltaics to individual vehicles will provide for low background recharging, which could be useful especially where cars are used for limited commuting (within vehicle renewable recharging). Battery charging will become easier and more widespread, for instance while parked in some areas of the European Union (Nice, France, personal observation) where the cost of parking and

recharging can be combined for convenience. Experience with improving battery performance in vehicles suggests that life total of 600,000 to a million miles is possible on a single vehicle battery array. In addition, many of the high costs of maintaining a reciprocating diesel or petrol fired motor do not pertain to electric vehicles (Miller 2017). In particular, relatively less expensive electric motors can be changed out much more easily.
- Autonomous vehicles may form the backbone of 'vehicles available on demand'. These have the potential to dramatically alter personal ownership of vehicles. Because vehicles are most efficient when they are in motion and transporting people and goods, it makes sense for some sort of 'Uber' driverless transport paradigm to cut deeply into personally owned vehicles. The main resistance to public transport is not that it is inefficient in terms of time but that it does not take people from exactly where they are to where they want to go. Transport available on demand will fill this need and could dramatically cut personal vehicle ownership.
- Reduction of carbon dioxide (CO_2) emissions that is being infused into the atmosphere by combusting hydrocarbon fossil fuels is required to minimize the negative effects of global warming. (whether or not certain governments understand its importance). Although National enthusiasm for reducing CO_2 in the atmosphere varies considerably, there is a growing awareness in the global public that cutting the amount effused to the atmosphere will also have an effect over time of reducing demand for hydrocarbon fossil fuels.

Some other opportunities that may present themselves:

- Trucks: One of the largest classes of automotive fuel use consumption are the ubiquitous articulated and smaller trucks that seem to dominate the interstate highways of the US and the major roads of many other countries. Regardless of range, switching to electric could have major cost savings per mile, especially when maintenance is taken into consideration and when within-vehicle renewable recharging (above).
- Electric cars and trucks combined have more potential for electric storage than any 'powerwall'. Electric vehicles are essentially moving energy storage devices. When they are not moving they could be part of a distributed electric grid that would utilize the vehicle charging points. In other words, integrate vehicles into local power grids so that they can both charge or optionally provide at least emergency power.
- All other fuel types such as CNG, H_2, fuel cells, etc.

The shift to "cleaner" gas fuel is a response by governments and non-governmental organizations for action to be taken on reducing CO_2 exhaust into the atmosphere to reduce the impact of man-made climate change, which is increasingly being regarded as a security issue (Rüttinger et al. 2015). Major oil and gas companies including BG Group plc, BP plc, Eni S.p.A., Royal Dutch Shell plc, Equinor ASA and Total SA, announced on 5 July 2015 that governments around the world that the United Nations Framework Convention on Climate Change

(UNFCCC) should create clear, stable, carbon pricing systems (PennEnergy 6-1 2015). These would reduce uncertainty and encourage the most cost-effective ways of widely reducing carbon emissions (OECD/EIA 2013). Without a dependable and widespread carbon fee, the world could be faced with reemergence of a coal and renewable energy as an energy mix. Although the situation is not so simple, "Coal is cheap… and renewables are heavily subsidized. The result will be higher emissions, not less" (Reed 2015).

In May 2015 Ben van Beurden, following Patrick Pouyanne, CEO of Total, when he reiterated a policy change of their companies. Shell would alter "oil-and-gas company to a gas-and-oil company", and "Total is gas, and gas is good". ExxonMobil CEO Rex Tillerson noted that "Natural gas and shale have been instrumental in reducing CO_2 emissions to levels not seen since the 1990s". Chevron's John Watson noted that, "Natural gas and LNG will be essential to the energy mix. The challenge is where new supplies will come from in order to meet the rapidly rising demand" (Reed 2015).

Thus, as natural gas is increasingly seen as the cleanest fuel by the energy industry, the type of gas resource, its potential volume, and its environmental factors, which are discussed in Chap. 10, also may become critical to resource development decision-making. With the dramatically increased activity in the LNG industry, gas is beginning to become a widely traded world fuel and new resources beyond conventional gas, coalbed methane, and tight gas become critical to the increased demand.

The conventional renewable energy source, hydrokinetic electrical generation, has held at about the same level because new capacity has been offset by diminished water supply and on-going dam removal. Nuclear energy has held about steady since 2000. The unconventional energy sources (Table 1.1) are rapidly increasing installed capacity. As of December 2017, monthly renewable electricity generation surpassed nuclear (EIA 12/27/17). Unconventional renewable energy, which will almost certainly include tidal power and possibly wave power, is projected to become even more important in the future. Tidal energy has been produced in other countries, particularly France at the Rance 240 MW barrage (de Lieu, 2009) and the 254 MW Sihwa Lake Tidal Power Plant in South Korea. Several smaller facilities are operational including Race Rocks Tidal Power Demonstration Project in Canada and Kislaya Guba on the Barents Sea in Russia. Wave and tidal energy are not yet generated in the United States.

Non-renewable energy sources have been produced by long-term geological and chemical natural processes that have converted dead organic matter into concentrated materials. Non-renewable energy is depleted or removed from available resources when used. These almost entirely consist of fossil fuels such as natural gas, oil, and solid fossil fuels such as coals and lignite. The most environmentally efficient use of waste plant matter is to produce biofuels, particularly natural gas that can be used as CNG or LNG. When it is combusted its CO_2/Btu ratios are similar. Non-renewable energy includes energy derived from fossil (naturally processed organic matter), biomass and organic waste (TREIA 2015). Non-renewable energy usually involves a one-way chemical reaction of combustion to produce electrical or direct motive energy.

Hydrogen is the only combustible fuel (either burning in a flame or in fuel cells) that does not produce CO_2 as a by-product, but it must first be created (manufactured) in some way because it does not exist as free gas except in trace amounts. Producing hydrogen from coal and oil by a reformulation process, for instance the Sassol Process of South Africa, which currently commonly results in production of about the same amount of CO_2 byproduct as if it were burned, although it may be chemically more pure and easy to capture and store. Reformation of methane, however, has the potential to produce very pure H_2 and elemental carbon, which should be pure enough to use in carbon fiber production.

Carbon-free production of hydrogen can be achieved using electrolysis so long as the electrical current required is produced without a CO_2 byproduct. The main competition with CH_4—based H_2 production is electrolysis, especially where renewable energy can be used. Electrolysis is well matched with variable energy input because available energy and the amount of O_2 and H_2 produced are essentially directly related to power. No elaborate power conditioning is required. Low power produces less electrolytic gas and increasingly higher power produces increasing amounts of gas. Research into more efficient electrolysis has the potential to make H_2 the primary high energy power on demand power source because of its carbon-free exhaust (Simonsen 2017). Nuclear energy (uranium or other fission options, fusion energy not yet being available) is a special case of a non-renewable resource main because it does not produce CO_2 exhaust as a byproduct but has the character of baseload power. The nuclear reaction is used to produce heat to generate high temperature steam that produces electricity by powering turbine electric generators. In addition, relatively small volumes of fuel are consumed. Reactors based on other radioactive fuels such as thorium have been proven to be possible, but currently there is no commercial activity using anything but uranium as a fuel for commercial electric power generation.

As part of the fission reaction, some waste products such as plutonium are produced that can be used to further generate electricity. Radioactive waste products are produced during the fission reaction that must be processed and stored long-term. Existing distributed storage of nuclear waste is an unsolved problem, in addition to potential dangers of nuclear accidents such as have occurred at Chernobyl or Fukushima. Unlike fossil fuels, no CO_2 is directly emitted. Fusion energy would be more efficient and safe, but the technology has not yet been mastered. In any case, it can take decades to build and bring a well-understood fission energy power plant into service. Thus, nuclear energy in general is not responsive to the increasingly short-term energy requirements of the emerging horizontally organized energy industry.

Nuclear energy, at least in its present expression as a relatively small number of very large power stations, has a low incident risk but an extremely high impact risk (Sect. 10.7). Nuclear power is becoming less of a contributor to the national power grid as less than 100 nuclear power plants currently operate in the US, five have shut down in the last several years, and dozens more of the old plants on the verge of shutting down because of safety and economic reasons. In addition, two new nuclear plants were cancelled in July of 2017, with the only other two plants licensed to build also likely to be cancelled (Wasserman 2017). The risk and cost of nuclear is definitely beginning to affect the industry as a whole. An expression of this is the early retirement of older nuclear power generation (Laforge 2018). While new reactors are being cancelled because of very high costs older reactors are being closed and companies that operate them are often going bankrupt if public subsidies cannot be found.

Distributed nuclear energy, especially if low radioactive waste reactors, such a s thorium fueled or other small reactors are an option for providing distributed base load. These reactors produce about the 50 MW level, rather than the huge, and hugely expensive traditional boiling water reactors, are getting a second look. A number of small reactors can also be installed at one site, which could raise the output from the site arithmetically. This would provide a high degree of flexibility and cost matching to demand changes. The most recent designs are small enough to be fabricated using a small number of manufacturing facilities and distributed by truck or train and assembled in a short time on site (Schneider 2018). The inherent difficulties of securing a large number of smaller reactors in an era of foreign and domestic terrorism is an unavoidable risk, but one that could be dealt with systematically as part of site selection and preparation.

Fuel impact on CO_2 and other emissions: Renewable energy consists of natural elements in our environment such as wind, sun, moving water, gravity, and heat sources from which energy can be captured to produce electricity, which is increasingly the cornerstone of our modern economy. Whereas non-renewable energy such as fossil fuels are finite and if completely consumed are gone forever, renewable energy does not. Solar energy resources are both indirect and direct (e.g., thermal, photochemical, and photovoltaic). Electricity is also generated by mechanical means using moving water (hydrokinetic driven by gravity, tidal, wave), geothermal (heat provided by warm rocks at depth, such as in active volcanic areas such as in Iceland) (Table 1.1). Renewable energy is commonly generated where the resource exists, such as in windy places or where dams can be built to use the head of falling water to turn electric generators, in contrast to fossil fuels that are normally transported from their traps to points of use.

Renewable energy does not produce CO_2 emissions directly. From environmental and climate standpoints, a future of increased use of renewable energy will be beneficial to reducing carbon dioxide (CO_2) in the atmosphere. In the near term, the amount of CO_2 produced per BTU by the major fuels favors gas as the most environmentally-friendly fuel because it has the highest power to CO_2 ratio (Fig. 1.3) and generally emits fewer chemical and metal pollutants and less soot than either liquid or solid fuels (Table 1.2). Note that all ranks of coal produce about the same CO_2/Btu; the main difference between the 'coal' fuels is that the degree of pollution varies, especially with regard to metals.

Currently, 8 out of 10 of the largest and most productive and economic coal mines in the U.S. are located in the Powder River Basin in NE Wyoming and SE Montana, which is the new center of U.S. coal mining (BSNF 2013; MISO 2015), replacing traditional underground and strip-mined coal from the Appalachians. The

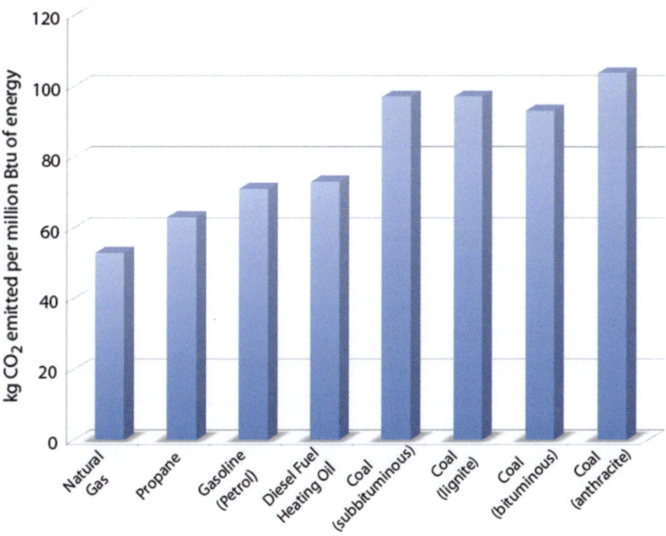

Fig. 1.3 General relationship of CO_2 produced by common combustion fuels. Original figure using U.S. Energy Information Administration data http://www.eia.gov/tools/faqs/faq.cfm?id=73&t=11. Detailed breakdown of international CO_2 production available in IEA (2014)

Table 1.2 Combustion products of common fossil fuels

Fuel type	CO_2 lb/MWH	Common pollutants
Natural gas	1135	Up to 1.3% NOx, 1% SOx, little metal or non-CO_2 carbon-based complex chemical pollution
Oil	1672	SO_2, CH_4, some Hg and minor metals, some hydrocarbon volatiles and chemicals
Coal	2249	SiO_2, Al_2O_3, HCl, HF, SOx, Dioxins, Ba, Sr, B, Mn, Zn, V, Cr. As, Pb, Ni., Cu, Mo, Ti, Be, U, Se, Sb, Cd, Hg, Ra., Volatile & Polycyclic Aromatic hydrocarbons, other radionuclides, formaldehyde, toluene

Natural gas refers to natural gas from all sources. From EPA-1 (2015), ALA (2011) and Ritter (2016). 1 lb = 0.454 kg. Zhang et al. (2014), for instance, found that human activity has increased mercury in the oceans by a factor of from 5 to 6. Some pollutants are present in both exhaust and fly ash

image of the traditional underground coal miner no longer exists as even underground mining is largely mechanized. Although the main reason for the transition to western coal is commercial mainly because it is disposed in nearly parallel thick beds near the surface and can be open-pit mined inexpensively, has a very low sulfur content and is also relatively low in pollutants (DOE 1993), although its CO_2 emissions are about the same as other coals (Fig. 1.3). The Powder River produced coal price of about \$10 per ton is almost 1/4 the cost of coal from Appalachia (Warrick 2015). Market forces or price competition, has caused natural gas rather than renewable energy to displace coal and nuclear in electricity generation (Wiser et al. 2017). The long-term threat of regulatory action to lower atmospheric pollution from coal is also beginning to be a factor in planning for new coal fired power plants. Even with increased energy production, the amount of CO_2 produced since 2005 has decreased significantly and further decreases are anticipated based only on gas substitution for coal (and a small amount of oil-fired production (Mellin 2018; Walton 2018a).

About 74 million short ton of coal was exported from the U.S. in 2015, down from about 97 million short ton in 2014 (EIA 2016). Although coal makes up about 40% of global power generation (Table 1.3), coal-burning power plants produce about 72% of all CO_2 emissions from power plants. Experiments with carbon capture from coal-fired power plant chemically complex exhaust have failed to work reliably, exorbitantly expensive geological sequestration experiments have failed to provide a safe repository for CO_2, and a large scale coal gasification plant that would produce a hydrogen-based combustible gas at Kemper, Mississippi (U.S.), which has been fueled by natural gas for over a year with no CCS may only demonstrate that this option is also too expensive to be practical (Biello 2016). Simply replacing coal fired power plants with cheaper gas also could have a major impact upon emissions quality (purity or level of pollutants other than CO_2) and quantity. In particular, replacement of coal and oil-fired power stations by gas is already having an impact on air quality (Shepstone 2017b). This argues for more exploration for gas reserves, including NGH.

Hydrocarbon-sourced energy from combustion is directly related to CO_2 emissions, whereas renewable energy has an indirect relationship. Although CO_2 may be produced in order to fabricate apparatus or to make the cement for hydrokinetic power (e.g., dam and electric generating infrastructure), CO_2 emissions are a one-time manufacturing and transport cost charge for renewable energy apparatus that is virtually completely depreciated over the life of installations. According to Sandalow et al. (2014), the energy produced from 100 Bcm of natural gas and combusted in a combined cycle gas turbine generator operating at 60% efficiency produces the same energy as combusting 2.1822×10^8 metric tonne of coal while producing 311.2 million tonne less CO_2 than would have been produced from combusting coal. Combined-cycle units offer greater efficiency than traditional combustion turbines, although simple cycle service gas turbine generators are extremely portable and efficient (Siemens 2016). Operational flexibility of

Table 1.3 The mix of fuels that power the world

Fuel	% of fuel
Coal	40.4
Natural gas	22.5
Oil	5
Nuclear	10.9
Renewables	21.2

Data from Warrick (2015). *Note* Wind and solar reached 10% of U.S. energy production in March, 2017 (EIA 2017)

turbine-electric power is vital in supporting intermediate and spiker load demand (DE 2016). The most efficient combined cycle gas fired power plant as of 1 September 2017 is the Bouchain plant in France, which was installed to be a reliable backup for renewable energy (Larson 2017). It has achieved a long-term efficiency of 62.22%. It is thought that 65% efficiency is an achievable objective. This is far better than coal or oil-fired power stations.

Carbon capture from combustion of fossil fuels produces CO_2. Although natural gas produces less than either coal or oil for the same heat production, another factor in its favor is that gas normally has fewer pollutants. Under most circumstances this makes the exhaust from completely combusted natural gas a better candidate exhaust from which to extract the CO_2 so that it can be sequestered or used for other purposes, such as enhanced oil recovery. We do not discuss CO_2 sequestration here because it is a very broad topic onto itself and would possibly be longer than this book. Nevertheless, and interesting innovation, whose mass balances are not known to us suggests that the exhaust gas could be put to work to generate further energy without combustion, the product of the energy from combustion and post-capture work being additive.

Willis (2017a) suggests that because exhaust from natural gas combustion could be passed through an air separation unit to isolate the CO_2. Although this reduced energy efficiency by about 10%, compressing the CO_2 to a supercritical fluid that has a much more efficient energy conversion character, which results in a downstream turbine generator that can be less than 1/10th the size of a standard combustion turbine (Willis 2017a). Data from a pilot supercritical combined cycle pilot plant suggested that power plants of this type could be essentially zero CO_2 emission. CO_2 that would not be required for the combined cycle plant could be sequestered (somehow). Although a zero-emission combined cycle natural gas power plant would be only a little more energy efficient overall than a conventional combined cycle plant, it is proposed that zero emissions could be achieved with about the same capital and operating costs.

The future of natural gas-generated electricity production may lie in methane fuel cells in which methane and air produce relatively pure CO_2 and water in a low-temperature reformatting process for methane at the anode, with pure water the only product on the cathode side of the cell (Joglekar et al. 2015). The cathodic separation of water from CO_2 may offer a more elegant solution to downstream carbon capture and sequestration (CCS) than separation of these materials from a high temperature exhaust stream.

Solar Technology Solar energy generally refers to the direct production of electricity from sunlight. Although the Earth receives about 120,000 trillion W, or only one-billionth of the sun's energy in one hour, it is equivalent to about the total energy consumed by humans in an entire year (Jacoby 2016a). It is generally dependable in that long-term planning for its availability can be forecast with some precision on an annual and diurnal basis, although the intensity of the light may be diminished by cloud cover or other light absorbing and diffusing materials in the atmosphere such as soot or water vapor. Solar photo-voltaic panels are semiconductor devices that have the principle attribute of converting sunlight directly to electricity, producing power without any mechanical components. Solar also has an extremely low environmental impact when deployed (in contrast to wind, for instance, which is a known bird-killer).

Photovoltaic (PV) cells are extremely attractive for producing electricity because they contain no moving parts and have proven to be very durable and dependable and are generally pollution-free and with recoverable materials upon decommissioning. Whereas wind energy technology is relatively mature and has strong perceived negatives, such as noise from huge moving propellers, danger to birds, and interference with broadcast signals, amongst others, the field of solar technology is more commonly perceived as having fewer negatives and offers substantial promise of improvement.

The backbone of solar panels is silicon photovoltaic (SPV) technology. SPV is a solid, rigid, crystalline material that is physically and chemically very stable. The SPV was first deployed over 35 years ago and has yielded performance data showing its durability and performance over time. This provides a knowledge-based confidence in its use commercially (where dependability is a key commercial factor), and allows for SPV to be conservatively costed over times that are becoming comparable with conventional power stations. SPV absorbs infrared as well as light in the visible spectrum and converts about 15–20% of the sun's incident light energy to electricity at an output of about 0.75 V in bright sunlight. This is the PV technology that is now widely used both on Earth and in space. It took about 40 years of product improvement to get from a few % to reach today's SPV characteristics.

SPV can be considered to be a mature technology in which further improvements are liable to be incremental. The main improvements that can be made in a mature technology are in manufacturing and mass manufacture to bring down costs. Chile, for instance, is installing mass solar energy power plants. In other words, instead of continually making small improvements in technology, produce good-enough technology in very large amounts to reduce unit costs. In August 2016, Chile contracted with Spanish developer SOlarpack Corp. Technologica for 120 MW at the cost of 2.91 US cents per kWh (Sanders 2016). Haegel et al. 2017) estimate that from a base of PV shipments in 2015 of about 57 GW, it is feasible that by 2030 installed solar will be producing from 3 to 10 TW.

The most prominent current competitors in the solar technology field are dye-sensitized or thin-film solar (TPV), which has a more complex crystalline structure that is generally referred to as 'perovskite' (Lee et al. 2012; Yang et al. 2015), which obtains a considerable proportion of its energy from higher frequency light in the visible spectrum than do SPV. Although commercialization of perovskite is presently hindered by manufacturing and stability issues, it is envisaged that hybrid or tandem cells fabricated by affixing a thin transparent perovskite PV layer, for instance configured to absorb energy from the blue spectrum of light, over SPV, which absorbs energy from the lower end of the light spectrum. In this dual-PV mode, efficiencies of about 30% should be reached (Sivaram et al. 2015).

TPV not only has a potentially much greater potential efficiency than SPV, but output line voltage of 1.2 V, or about 30% higher than conventional silicon PV.

These have been improved from outputs of less than 4% to over 20% in about 5 years and have a theoretical efficiency of 30–40%. de Quilettes et al. (2015) believe that it is likely that the performance of perovskite devices will approach the thermodynamic limits for solar cells and other light emitting devices. Chen et al. (2015) have produced efficient and stable large area perovskite solar cells that Sessolo and Bolink (2015) regard as a major step on the R&D ladder for development of highly efficient thin-film solar cells. Lead and other metals may occupy crystal lattice sites in TPV cells, but these are not perceived as a pollution hazard because of the way they are strongly bound into the crystal lattice. Research and improvement is mainly concerned with achieving the best components and crystal structure for development of stable TPV, while at the same time exploratory work is proceeding on other crystalline thin film systems, such as Cd/Te (18.3% efficient), CIGS, copper indium gallium (di)selenide, (20.4% efficient), CIS, copper-indium-selenium (19.75% efficient), organic tandem cells (12.0% efficient) (Montgomery 2013), which have achieved somewhat higher efficiencies in the last two years. PPV can be made nearly transparent so that compound PV using PPV over SPV is possible.

SPV is relatively expensive to make because temperatures of about 1000 °C have to be reached in silica melts in special clean rooms to produce high purity silicon wafers. The organic–inorganic PPV materials are manufactured using temperatures of about 100 °C using solvents and vapor deposition on substrates, which can be flexible or of irregular shape (such as the exterior surfaces of automobiles). The manufacturing costs of thin film should be less expensive than silicon but thin film solar uses a variety of more expensive materials with different life cost structures. Multi-crystalline silicon solar wafers cost about $0.82–$0.62/W, which provides a benchmark for production costs and a comparison with CeTe thin film (Shiao 2012). Although thin film solar (TPV) development primarily in the U.S. and China are showing improvement in efficiencies and comparable longevity to SVP, these are currently technologies still in development and not yet ready for commercialization (Jacoby 2016a).

Solar technology is developing rapidly. Dye-sensitized solar cells (DSSCs) are closely related to perovskite solar cells and also are the subject of major research attention (Jacoby 2015). Sunlight on a DSSC excites electrons in a layer of light-absorbing molecules referred to as dyes or sensitizers. These excited electrons are attracted into physically-attached semiconductor materials such as TiO_2, and then migrate to an electrode, producing an electric current. In addition, another new direction provides electrical current from heat. Photothermally activated pyroelectric polymer films for harvesting of solar heat with a hybrid energy cell structure further has the potential to increase solar energy conversion (Park et al. 2015). It is widely understood that solar panels can become heated by exposure to direct sunlight and that the higher temperatures lower efficiency of conversion of sunlight into electricity. Not only will additional electricity be directly derived from the heat by this new technology, but conversion of the heat also will have the effect of lowering temperature of the photovoltaics, which will increase their efficiency.

Because solar technology is available in modules, with 250 Wh panels being the standard panel for residential users, solar power is extremely scalable. As would be expected, solar is particularly good for distributed power production because it is easily transported and installed, with little visual or other impact on the environment. A breakdown of solar users indicates, however, that solar use in eleven major regions by 2020 will be dominated by a combination of utility-scale and commercial users. Africa goes against this trend with the largest percentage of residential users (Lux 2015).

Although solar is now a well understood technology that has a proven history, research into the use of less expensive materials and manufacturing, lifetimes as long as the extremely durable SPV technology, lower weight, increased flexibility, and most important, greater efficiency is underway in many countries. But PV performance is already good enough for its widespread adoption as a supplementary power source on individual dwellings and businesses and plans for large-scale solar power stations are being put into place. California may open up about half a million acres—about half the size of Rhode Island—for solar and wind farms (McDonnell 2015). It is envisaged that electricity from solar will rise from 7% (mid 2015) to 20% by 2030. As solar improves and first-generation PV is upgraded, it can be expected to become an even more important source of renewable energy.

Indirect production of electricity from sunlight also allows the energy from sunlight to be stored (Kearney et al. 2004). A prototype plant of the technology consists of an array of large mirrors arranged around a tower on which the mirrors direct the light. The focused light heats a non-toxic salt to over 566 °C. The salt is then pumped into holding tanks where it is used as a heat source to produce steam that drives electric generator turbines (Solarreserve 2016). These plants would appear to have to have a large footprint but probably a low environmental impact. Maintaining the stored energy in a liquid form while causing it to cool during energy production may be more difficult and costly than the promotional literature suggests as mirrors have to be kept clean and cloudy or rainy weather may also have an effect on energy production. We feel that this technology has promise but requires testing under natural conditions.

Wind Energy: Wind power is the least constant of the major renewable energy sources as it is only available when the wind is blowing within the range of wind speeds for which an apparatus is designed. Wind is notoriously difficult to predict on a long-term basis, although many locations have a certain periodicity to wind and its velocity and direction. For instance, in many coastal areas, especially adjacent to adjacent high-relief coastlines, wind will be essentially weak and without a firm direction near sunrise and sunset but during the day there may be strong onshore breezes while at night there may be strong offshore breezes.

Local coastline geographies and land morphology may have a strong impact upon wind, which may be funneled along certain paths, and increasingly it appears that offshore windfarms have many operational advantages, the most annoying of which is the noise generated by the large wind turbines. Also, there may be strong diurnal periodicity, which may be directly counter to peak demands. For instance, in the summer of 2011 the managers of the Texas electrical grid had to declare a

Level 1 Power Emergency as ten thousand megawatts of wind power became unavailable for a multiday period (Fowler 2011). There can be a dramatic difference between the installed wind capacity and actual wind output, which can vary much more than solar. The wind power MW output commonly cited is usually for a high wind speed, and more conservatively for a daily average. In fact, the output during peak demand for this period was higher than the level said to be legally "reliable", but it had dropped below the 8% level just 5 h earlier.

It is easy to become impressed by the emerging scale of large wind turbines (modern windmills). Not too long ago 5 MW wind generators were thought by many to be unbuildable, with low survivability in storms. At the writing of this book, they are up to 9 MW, with 15 MW wind Turbines on the horizon. Although it is easy, especially for people without personal experience of them or access to operating and maintenance data to discount wind energy, wind turbine performance is noteworthy and new turbines appear both impressive and stable.

Other Renewable Energy: Tidal, wave, geothermal, and other less abundant energy sources (Table 1.1) may be highly location specific, for instance along seacoasts and it might require transmission over long distances. Variable renewable energy (Table 1.1) can be treated as 'opportunistic energy' that can be fed into the grid when it is available, at which time it can displace any of the other power sources on either commercial or environmental grounds.

Non-intermittent and intermittent power. From an industrial point of view there are also two types of power supply, but these relate to the availability and dependability of the source. Industrially, sources are regarded as being either non-intermittent (produced on-demand) or intermittent (variable power and times of availability), also referred to as variable renewable energy (VRE) (Wiser et al. 2017). Fossil fuels (Table 1.1), nuclear and hydrokinetic (including pump storage) are non-intermittent power supply sources. In contrast, all renewable energy sources are intermittent, but with some degree of predictability for each type.

One of the issues that is usually forgotten in discussions of 'renewable' energy by non-specialists is that there are costs associated with intermittency that are very different from the non-intermittent power paradigm. Intermittent energy requires additional distribution and business costs, but most important, it has to be backed up so that demand can be met and excess power is not generated as this can also cause transmission problems. Dispatchability is an important industrial concept that is often ignored by non-specialists. Dispatchable generation can be uploaded to the transmission system at the request of power grid operators or of the plant owner. In a fully dispatchable system, generating plants that can be turned on or off, or can adjust their power output accordingly in order to keep the transmission system stable. Shepstone (2017a), for instance, describes how on 14 different days in March, the State of California produced so much solar power that it needed to pay a number of nearby states to take the power to avoid overloading its transmission system. Producing too much unplanned energy, however, is not as dangerous for the grid as other electricity sources can be scaled back as part of standard load balancing.

Non-intermittent energy sources may not necessarily be a 100% supply solution. Supply problems can occur with normally non-intermittent power supplies. Fuel availability can present problems to energy supplies, as was the case in Europe and the United States during the 1973 OPEC embargo or when the supply itself or transport is disrupted. Weather can affect hydrokinetic power supply during droughts, such as are being experienced now in the SW United States, particularly in California. Lower water head reduces the force through the turbines. Water flow and velocity is directly related to the amount of electricity produced. In the states of Oregon and Washington, for instance, peak hydrokinetic electricity generation has fallen 10% since 2011 (Hoza 2015), largely because of diminished precipitation. Although tidal energy production can be calculated years in advance, it varies steadily in accordance with tides.

Customers want full time, on-demand electrical power. Failure to provide what appears to customers to be an unlimited supply (whose breaks in service are usually weather or accident related) is commonly regarded in the United States and other countries used to a stable energy supply as something close to a catastrophic event, particularly in very hot or cold weather. Although private customers may become uncomfortable, industry and business must have dependable power or there will be a direct loss of income.

As with all intermittent power sources, storage of energy so that its use will not be tied to periods in which the energy is being generated is key to any long-term application. Energy storage technologies are improving the economics of accommodating more widespread use of renewable energy (ESA 2017). Hydrokinetic pump storage, pressurized systems (based for instance on Tata 2015 use in pressurized air powered vehicles) and mechanical systems (Putnam 2015) can store power, but all have significant disadvantages (cost–durability–convenience, etc.) compared to battery storage of electricity.

Battery Backup. Batteries are electrochemical means of storing electrical power. In addition there are energy storage devices based on kinetic energy storage and gravity, phase change materials, particularly in refrigeration, and managed buoyancey changes, amongst other suggested solutions; these are not discussed here as they do not have the mass-effect or potential wide-spread use of electro-chemical batteries.

There are many types of electro-chemical battery, from the old dependable lead-acid through the current lithium-based batteries of choice. There are many types of batteries that are used and recently batteries with enough energy storage to back up electrical grids are under development. Battery backup is presently the subject of considerable research in many countries, usually supported in the early stage of each new technology by government funding and programs. Most intellectual property rights can be captured in the early stages of a technology development when general concepts can be recognized. Battery backup is being applied now, principally for providing the additional energy required when electricity use is at its peak (Bade 2018). The Arizona Public Service will install a 50 MW, 135 MWh batter system that will deliver power between 1500 and 2000 each day after the power output from a nearby 65 MW solar farm is waning.

Batteries can usually be categorized for two main applications, (1) Light duty and (2) Heavy duty. Great strides have been made in improving light duty battery technology because these are immediately usable in mass market devices such as portable digital technology such as computers, mobile devices and phones, and in light vehicles. The secondary development of light duty battery technology provides a near term economic return for investment. Further improvements in increase energy density will are underway in each technology, for instance lithium-ion batteries, but these developments are usually more incremental. Step improvements in performance require new technologies. Using different combinations of metals and electrolytes as well as increasing the surface area of batteries by using a miniaturized 3D structure instead of plate-type batteries separated by increasingly thin membranes have the potential to improve light duty batteries (Parker et al. 2017).

There are two main applications for heavy duty battery use. The first is for heavier vehicles such as buses, trucks, and possibly trains. The second application is to back up electrical grids to provide power when renewable energy is not available. The power storage and delivery requirements of grid backup are so high that present technology is not sufficient for widespread heavy-duty use. Major backup of high-demand electric grids requires existing batteries to be assembled on a massive scale, which is problematic from a cost/benefit point of view. The use of fossil fuels, in particular natural gas combined cycle electricity generation to back up renewable energy, is still the only certain way to insure continuous energy on demand (Chap. 11).

The first step to smoothing the delivery of intermittent energy generation through mass storage has been put on the market. Patel (2017a) notes that limited grid backup battery storage deployments of all technologies grew to 336 MWh in 2016, doubling of megawatt-hours deployed in 2015 and deployment is a global project. Elon Musk, leveraging Tesla's expertise in battery technology and manufacture, is bringing a home or dwelling battery unit into the market (Bradford 2015). Larger applications for backing up transmission line supply are being funded by the U.S. Department of Energy. For instance, a new type of potentially inexpensive battery that can be constructed on very large scales is proving to be effective for high MW storage. These are 'redox flow' batteries, which store the electrochemical energy in the electrolyte, which can be kept in separate containers from the anode and cathode Jacoby 2016b). Cost-effective adequate electric storage for renewable energy will be disruptive to the current trends in wholesale, regional delivery of electricity to consumers. Battery backup could take the place of natural gas as a backstop on-demand energy resource for relatively short periods of time, but only if its cost could be brought down. The development of large scale commercially suitable battery backup, however, remains in the research and development phase.

Electricity storage in batteries or production of H_2 and O_2 for use in fuel cells by energy efficient electrolysis are both are important centers of current energy storage technology research. Development of suitable long-term energy storage will enable a widespread off the grid power paradigm that will have the potential to be highly disruptive to existing energy generation and distribution businesses. Major industrial players are already engineering electricity storage for scale (Doosan 2017). The effect of grid-scale energy backup for renewable energy is already having an effect in accelerating implementation of renewable energy. Fuel supply and infrastructure are becoming an increasingly important consideration in the new horizontally organized energy industry (Sect. 1.4) in which there is a rush to renewables backed by energy storage and gas-fired power stations (ISONE 2018).

The incentive for battery deployment appears to be largely the result of the Aliso Canyon natural gas storage reservoir. There appear to be irregular, short term changes in the amount of battery backup being deployed. Adding to the local power shortages are the shutdown of San Onofre nuclear plant and impending shutdown of the Diablo Canyon nuclear plant that both feed into the local grid base load. This is pushing California to take the lead on battery storage and to increase the amount of renewable energy, generally solar. Matter of how storage operators should be paid for different ancillary services is still being resolved.

From 2008 to 2016, nearly 35% of U.S. coal fired generating capacity has been shut down and Duke Energy, one of the larger energy producing companies, retired 5425 MW of coal fired generation since 2011 and has plans to close a further 2006 MW by 2024 (Trabish 2018a). Some plants are being closed and repurposed, others are converting to gas and or renewables. The energy industry appears to be on the verge of changing the power industry in favor of renewables worldwide.

In summary, the lowering cost of virtually all renewable energy types and support infrastructure such as increasingly capable batteries have passed a tipping point and the increased use of renewable energy is almost certainly unstoppable (Milford et al. 2018). Clean energy will shortly become a $50 trillion industry (Romm 2017). Enough intermittent energy has not been produced and installed so that direct comparisons on return on investment (ROI) as a function of either EROEI (a direct energy measure) or EROI (financial measure), can now be made (Chap. 11.3) and show that intermittent energy is competitive now. For instance, on 28 December 2017, Xcel Energy revealed that bids for constructing new solar and wind farms, including those with battery backup was as low as (a too good to be true scenario) $21/MWh or $02.1 per kWh, which is lower than most of the operating coal fired plants currently being operated in Colorado (EXGY 2017).

1.3 Electrical Load Characteristics

'Load' is a tem that describes the power being supplied to meet demand. For power grids, the loads are usually measured in thousands of megawatts (MW). In general, four types of load typically can describe the demand and the energy component or source that is used to meet load demand (DOE 2017). These are: (1) Base Load, which is the steady energy platform upon which demand is met. (2) Intermediate Load could be thought of as the upper part of base load that may vary according to both electricity availability. Intermediate Load follows load variations but is characteristically produced by base load power plants. (3) Peak Load must meet the daily period of highest energy demand. (4) Operating reserves are the power sources that are used to perfectly match power supply to demand (Fig. 1.4).

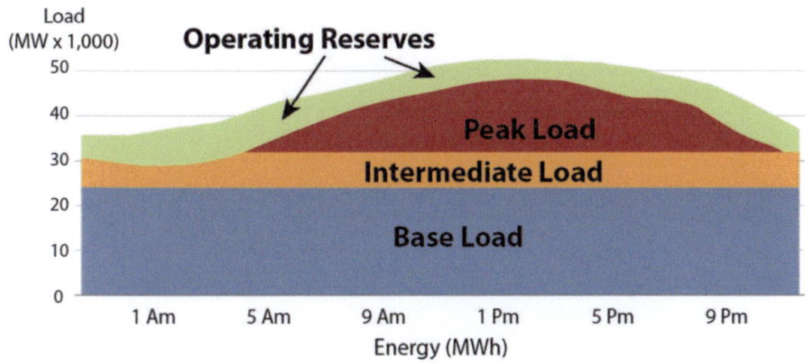

Fig. 1.4 Diagrammatic load component diagram of overall demand. Redrawn after Fig. 1.2 of DOE (2017)

Base Load (and much of Intermediate Load) generation relies on power plants that are run at sustained high output level, usually above 80% of capacity. Power output usually varies little, but can be cut back if intermittent energy is used because of cost or regulatory factors. This is the base of power supply. In the simplest cases where demand is steady, only base load power that can ramp up slightly to meet unexpected additional demand, is necessary. Base load power has been traditionally supplied by nuclear, and fossil fueled steam turbine generation. In particular, coal fired generation has been the long-term base load fuel, stretching back into the 19th century for electrical power and into the 18th century for steam power. A factor in the long-term dependability of nuclear, coal, and oil is that very large amounts of fuel can be stored on the power plant site, and as more gas pipelines are built and local storage improved, natural gas will also be able to assume more of a base load character.

Fossil fuel powered electricity generation confined to a restricted area provided the basic power supply system in most countries for the last 150 years. Because of the new availability of intermittent renewable energy, and the desirability of increasing the use of renewable or intermittent energy to reduce CO_2 and pollutant exhaust into the atmosphere, the energy supply business is becoming more complex. The general phase-out of coal-fired electricity generation and increased direct electrical storage (UD 2018) would appear to be at least a near-term trend.

1.4 Matching Power Supply to Demand

The power generation sector is the dominant world gas user (Fig. 1.5) and is expected to remain so for at least until 2020 (IEA 2015). Depending on competition with other energy sources that can have a base load character, such as nuclear energy and hydrokinetic power, natural gas will probably be the combustion fuel of choice for both economic and environmental reasons. Base load energy is so-called because it is very dependable and capable of producing predictable and controllable

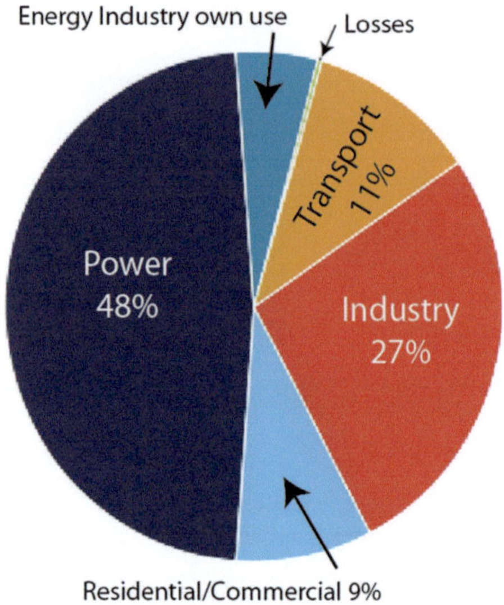

Fig. 1.5 Gas demand by sector, 2014–2020. Redrawn after IEA (2015). 'Industry' includes feedstock for the petrochemical industry. Transport is mainly where CNG is used as a replacement for liquid fuel in vehicles (CNG 2015), including LNG. Although LNG is planned for use in some ships, CNG is better utilized for vehicles that require relatively rapid filling that may have to stand unused for irregular periods, especially now that the weight of CNG tanks has dramatically decreased (Price 2015). CNG may also be a fuel of choice for home backup generator systems as propane becomes more expensive

amounts of power for extended periods of time, usually being out of service only for scheduled maintenance, and for being available to meet peak demands.

Gas use by the other sectors is also liable to increase, both through displacement of existing liquid fuels and as petrochemical industrial feedstock (and every chef knows that cooking on gas gives the best results). Natural gas is also the feedstock for the direct production of animal feed through microbial metabolic processes, which has already achieved government approval for safety (Le Page 2016). Use of natural gas to produce electric energy is also increasing and as spiker power becomes more important, it has the potential to increase more than increases in the other sectors of gas use.

The electrical power business is in a transition from a monopoly situation (Maize 2018) when generating, transmitting, and delivering on-demand power supplies within a distinct supply area. Small numbers of independent power suppliers existed here and there, but independent power generation has remained a very marginal business. In a monopoly, direct matching of supply to demand has been very profitable, even where power utilities have been regulated. Until recently, this vertical integration by area, with a single company responsible for power generation (operation of the power plant with secure fuel supplies), distribution (responsible for its own transmission lines) and delivery to customers (Fig. 1.6), was the norm.

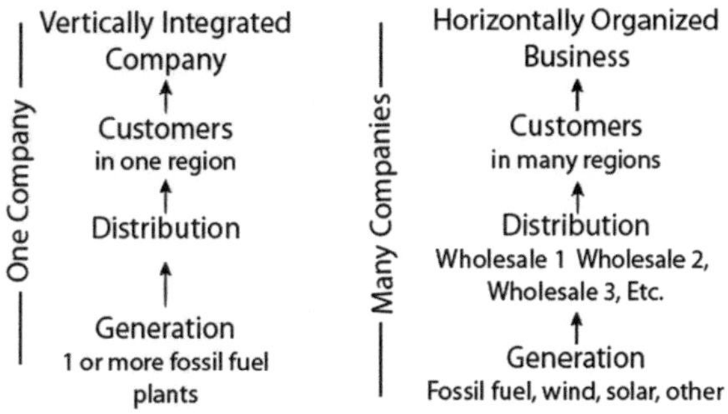

Fig. 1.6 Horizontally integrated business model for managing multiple energy sources including renewable energy

There are advantages to a monopoly situation, in that a single company is may have a very good continuing estimate of demand and a closer relationship to its customers.

The new energy paradigm is becoming a more regional, or 'horizontal', approach in which an increasing number of power inputs, including intermittent power supplies have to be accommodated as part of the energy mix. The monopoly model is breaking down, the number of independent power producers is increasing rapidly, particularly in renewable energy that has to be blended. This is happening both because of regulations, largely in response to climate change outside of the United states and partly within, and because the cost of renewable energy is becoming competitive with traditional on-demand power sources, especially coal. A regional rather than a traditional state licensing approach appears to lower cost by about 30% overall (Rybarik 2013), which suggests that while the monopoly model was internally efficient and profitable, competition amongst power suppliers had the anticipated effect of lowering cost to the consumer.

To an energy supplier in a small region with a small delivered electricity load, a single on-demand energy source would be preferable because matching supply and demand is straightforward. In addition, if a generation problem would develop, power can usually be transmitted from adjacent power supply areas because the amount of power to make up would be small. Larger areas, using many power supplies and delivery systems, vastly increase operational security.

Increasingly, more intermittent power is being used in the energy mix. Whereas energy delivery once involved relatively small areas supplied from single power source such as dams or combustion power generation, management now involves using power from different sources and 'smoothing' both input and availability in such a way that customers have continuous and dependable energy supplies. In the United States, for instance, there has been massive shift in the increased use of intermittent power from 1000 MW in 2000 to 18,000 MW in 2014 and acceleration in the use of intermittent power sources is anticipated (Rybarik 2013).

Renewable energy is inherently episodic or intermittent and is available at different levels depending on the time of day and not at all at certain times of the day, and is weather dependent. As a result, prediction of energy availability with any certainty more than a few days ahead is impossible for the energy source to demand relationship required. 'Big Data' which is a computer intensive technique for convolving predictions of electricity demand and production from all renewable sources with variations in existing supply from each is required to be able to input power on a cost-accessibility basis and keep ahead of demand so that backup power, most likely from gas-fired 'spiker' stations can be brought on and offline on a just in time basis (DW 2016).

In order to maintain smooth energy supplies when unconventional energy sources become dominant, larger and more robust and stable power grids (NAS 2016). The strength of a transmission system is important to reliability and regionalization allows for better power sharing. More transmission lines will be necessary in bigger regions. Articles in Wilson and Biette (2014), for instance, share a common view that the geographically related North American shared energy region including Canada, the United States, and Mexico could be successfully merged into a single power grid optimizing the resources of each Lusinyan et al. (2016). Regionalization and sharing between regions would smooth both uneven energy supply and demand (Fig. 1.7). In addition to smooth energy deliver, companies distributing electricity will use the most cost-effective supply mix.

The core of energy supply will remain non-intermittent power sources until intermittent power sources are implemented much more. Intermittent poser sources will continue their dramatic development. The development of regional wholesale energy markets and operators, for instance, the Midcontinent Independent Transmission System Operator (MISO) that presently is the power distributor in a central U.S. swath from Minnesota and South Dakota in the North to Louisiana in the south allows each power source operator to concentrate on their core business. Distribution that can regulate power inputs is becoming an important business area in itself. According to the independent system operators (ISO)s, organized wholesale competitive energy markets improve grid reliability, optimize transmission system use, introduce and lower wholesale prices, optimize power plants, while reducing a competitive market for clean energy resources and load demand providers (ISO 2015). Nonetheless, as energy is transmitted increasingly long

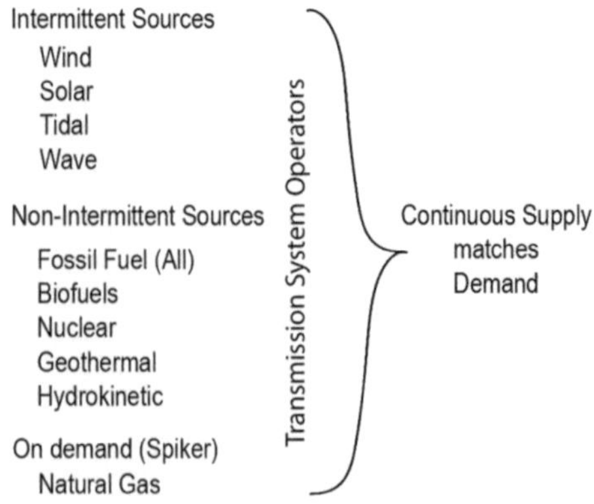

Fig. 1.7 Diagram of power sources and output

distances on electrical grids, increasing transmission losses occur. A large grid will use (or waste) considerable energy.

Power supply is composited by first using existing base level supplies, which are traditionally the most dependable, with intermittent sources usually incorporated opportunistically as they become available. This may allow some base level load to be reduced or to be used for charging energy storage (e.g., pump storage). The historic load model was that base load power should be capable of providing all power demand, although this model is a feature of a region with a dedicated non-intermittent power source such as a fossil fuel or nuclear plant supplying a particular area alone, except in emergencies. The energy model is changing to provide a mix of power sources, with renewable power becoming increasingly available, at the same time that some coal and oil-fired power stations are being retired. Increased intermittent energy and reduced non-renewable power might create situations where demand might not be able to be met if a region were too small to provide enough power at intermittent power lows.

Geothermal energy is CO_2-free renewable energy resource that has the character of base load electrical supply (Kagel et al. 2007; UCS 2014) and is also a large enough resource that with suitable new technology and investment, optimists project that it has the potential to provide all of today's U.S. electrical requirements. But unless high quality naturally occurring thermal sources are available, geothermal will probably remain a relatively minor source of electricity.

Environmentally, the combination of dependability and extremely low risk makes it a very attractive energy option. No fuel is necessary as the heat is produced by nature and the only waste product, other than unpolluted end-of-life generation plant, is hot water. There need be no exhaust other than water vapor, which can be captured and reheated, thus keeping water requirements low. Some generating technologies may directly exhaust water vapor.

Geothermal energy is accessed by drilling into hot rocks and using the heat to generate steam that is used to power turbine—electric generators. For countries lucky enough to have high quality geothermal energy sources available at shallow depths such as volcanic terranes such as Iceland and the 'ring of fire' around the Pacific Ocean, water can be boiled directly, without deep drilling, redrilling, or heat pumping. Where lower quality geothermal heat is available, drilling and other technical issues can be prohibitive.

Geothermal energy resources are very area specific. In the United States, for instance, geothermal energy is mainly concentrated in the rocky mountain belt and California, New Mexico and Colorado (Fig. 1.8). Except for California, where more than 40 geothermal plants provide nearly 7% of the State's electricity and are co-located with large population density, the sources are in sparsely settled areas. Elsewhere, this availability of CO_2—free renewable energy is not co-located with population and argues strongly for a new century electrical transmission network for the U.S. as a whole so that power from opportunistic energy resources can be delivered to markets.

Other countries also have geothermal plants. Mexico has almost 1000 MW of installed geothermal capacity, which is 3% of its total energy supply and includes

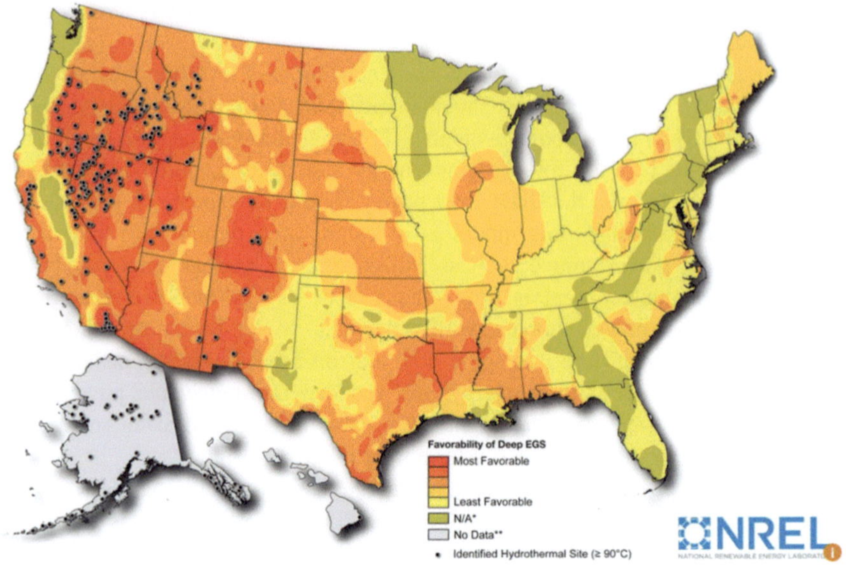

Fig. 1.8 Geothermal energy resource in the United States. From the U.S. National Renewable Energy Laboratory (NREL)

the world's largest geothermal plant at the Cerro Prieto Geothermal Power Station. Italy initiated the first geothermal project over a century ago and has over 800 MW of installed geothermal capacity, or 10% of its total energy production. Iceland has 575 MW of installed capacity, or 30% of its total energy production. Other countries such as the Philippines, Indonesia, Turkey, Chile, Nicaragua, New Zealand, Ethiopia, Ruanda, Kenya, and Panama, amongst others, have significant geothermal potential. Following the meltdown of one of the Fukushima Daiichi nuclear reactors, Japan is also seriously investigating geothermal energy.

A geothermal plant with a closed circulating water system has no systematic pollution and because a malfunctioning system can be shut down without releasing any long-term pollutants, as in the case of nuclear energy, environmental risk is low. Most geothermal facilities have closed-loop water systems, in which extracted water is pumped directly back into the geothermal reservoir after it has been used for heat or electricity production. In such systems, the water is contained within steel well casings cemented to the surrounding rock. There have been no reported cases of water contamination from geothermal sites in the United States, which are mandated as closed systems by regulation.

A byproduct of a geothermal system is a considerable amount of waste heat. The 'waste heat' of the water after energy generation can still have elevated temperatures. It can be recirculated and used for other purposes, such as heating buildings and dwellings including swimming pools, but possibly most important, greenhouses. Iceland, for instance, is currently co-locating enclosed agriculture enterprises with their geothermal plants and is growing substantial food that would otherwise have to be imported (and is suitable for export). Geothermal can be the base technology that could open up northern areas for continuous human habitation that would otherwise be too hostile.

Open system geothermal power plants can have impacts upon both water quality and downstream use. Hot water pumped from underground reservoirs often contains high levels of sulfur, salt, and other minerals, including hazardous metals. The distinction between open- and closed-loop systems is important with respect to air emissions. In contrast to closed loop systems that emit no emissions to the atmosphere or waste-water, open-loop systems emit hydrogen sulfide, carbon dioxide, ammonia, methane, and boron. Hydrogen sulfide, which has a distinctive "rotten egg" smell, is the most common noxious emission.

Tidal energy (Reddy and Affholder 2002) can also be used as base load power to provide energy backbone for the mixed energy future will feature dramatically increased reliance on renewable energy. This is because the tidal stream at any point can be calculated decades in advance and tidal generators situated so that virtually continuous power may be available within a grid region. In a country such as Ireland, for instance, which is surrounded by bodies of water that often have high tidal flows, a series of well-located tidal electrical generating installations and a national transmission system, can provide for predictable energy generation.

All non-tidal forms of intermittent energy are inherently unpredictable and can only be used to top up base level supply, to charge energy storage devices, or are simply wasted. Even solar energy, which may be calculated for a particular location depending on full latitude, season, diurnal potential, may be affected by cloudy or rainy weather that lowers available light energy.

Energy mixing: Increased use of intermittent energy sources has led to the further development of on-demand or 'spiker' power designed originally to meet short-term high demand periods (the term 'peaker' is also used). Base load power is meant to be steady and continuous and to only have its levels change slowly, if at all. Spiker power plants have traditionally been natural gas fired because they can be started and brought to full power quickly, usually within half an hour, and can operate at variable power output to meet demand. An attribute of spiker power is that the plants can be purchased and installed in an area of increased demand quickly and supplied with fuel from an existing nearby gas pipeline grid. Small gas electrical generating plants involve much smaller individual investment than other fossil fuels, which in any case are being retired, and they can be installed in a highly distributed fashion with fuel distributed by pipeline. Burning natural gas in large centralized power station to generate electricity to heat homes and provide for cooking and hot water is much less efficient than using distribution systems to transfer natural gas to local electricity generation systems or to buildings themselves (Gregg 2017). In a situation where all available intermittent energy sources in a region are low or go offline, 'spiker' power, that fills the supply—demand gap becomes critical to meeting demand.

Figure 1.9 shows a diagrammatic example of power matching using a variety of power sources that vary on a diurnal basis in the Electric Reliability Council of Texas region (ERCOT). This is a major power control company providing mixed

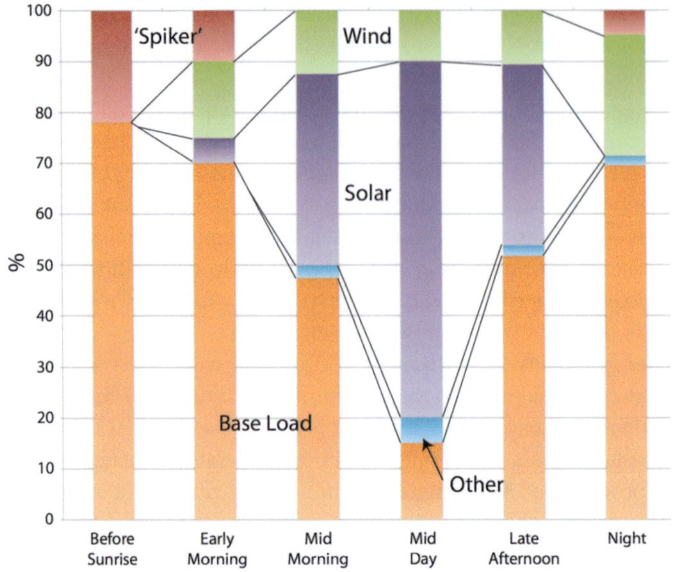

Fig. 1.9 Diagrammatic representation of diurnal power source inputs into regional electrical grid dominated by carryover base load capability

energy to the majority of state of Texas has natural gas, coal, nuclear, and wind as its major powers sources, and meets demand when more wind is available by reducing coal fired capacity (Bennett 2015). Mixed power supply is a much more complex issue for the power supplier than a smaller region tied to a single power plant. Increasingly into the future, as more renewable power sources become available and as their output increases, power swings from one source to another will become larger and more difficult to input to derive a steady power output to a region. In this environment, base load is adjusted down to accommodate as much intermittent energy as possible during the day when it is at a maximum.

Base load will be the most dependable power available, which can include any dependable power derived from any source from inside or outside the region. As the renewable energy becomes available, the spiker power is 'turned off'. Spiker power will most likely be gas-fired turbines and may be part of an installation providing base load power or a separate installation or installations responding to demand. Wind is shown as rising into the night or early evening while solar is at its maximum during the middle of the day, and could include both large installations and a myriad of smaller installations operated on or near private dwellings or businesses. 'Other' is shown as being available from mid-morning until the evening. Power storage is not shown as a separate source of power as this might depend on a sudden diminishment of power from any or all sources that would be used as needed, and if available. Only significant power sources are shown.

Another difficulty arises when multiple power sources and power matching are shared between regions. Available base load and spiker capacity may have to be brought up to maximum to meet own regional load and provide excess power to adjacent regions.

Two issues arise where substantial renewable energy has already been installed. In the first, public companies supplying power from fossil fuel plants can expect to generate and sell much less power. In the second, instead of a very predictable load based on experience of demand, the unpredictability of renewable energy supply now has to be taken into account. While the implementation of significant renewable energy is an overall environmental benefit, substantial demands are placed on the flexibility of base load substitution and spiker response.

The impact of increasing renewable energy supply on meeting daily demand, which in turn creates a need for energy on demand (Fig. 1.10) that can be rapidly brought into play places enormous strain on power grids unless the power can be produced locally. Although the requirements for non-renewable energy supply drops during the day when solar is at a maximum and wind is normally greater, when renewable energy drops with nightfall and changes in the diurnal wind pattern, non-renewable energy demands increase correspondingly, along with increased normal demand in the evening. If temperatures in the region continue to increase, which will increase air conditioning demands, while installed renewable

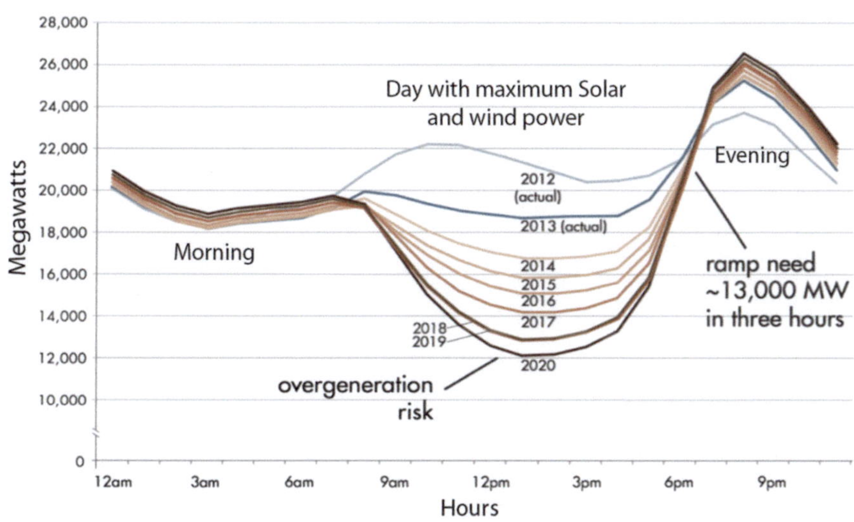

Fig. 1.10 The 'Duck Curve' (Modified from Fig. 2 of CISO 2015 [Licensed with permission from the California ISO. Any statements, conclusions, summaries or other commentaries expressed herein do not reflect the opinions or endorsement of the California ISO.]), in which the profile of diurnal energy demands is thought to resemble a duck. Net load, March 31, 2015 for the California CISO. Heavy brown line is non-renewable energy production. Years show impact of actual and estimated non-renewable energy demand because as a function of renewable energy substitution

energy also increases, the flexibility of ramping up baseload/spiker on demand power will also need to increase considerably. The greatest increase to the evening peak demand, however, may come from a dramatic increase in electric vehicles. Annual energy consumption from electric vehicles in the United States is expected to rise from "a few terawatt-hours" in 2017, to at least 118 TWh—and potentially as high as 733 TWh—by 2030, and the load will be greatest as people end their evening commute and begin to charge their vehicles (Walton 2018b).

A straight-forward energy trade situation that took place on 21 August 2017 occurred because the partial eclipse whose totality traversed the United States reduced solar energy production in California by 60–70%. This created an inverse dip in solar electricity production and a spike in thermal energy production, mainly from natural gas fired power plants to make up the energy shortage (Fig. 1.11). Although eclipses are rare, other naturally occurring phenomena, such ad dust clouds from volcanos, more cloudy days and thicker clouds reflecting a warming atmosphere that can hold more water vapor, or any atmospheric phenomena that has the potential to remove energy from solar radiation before it reaches a solar array on the ground, will have the same effect and require the same type of on-demand energy production as diurnal variation in energy source availability to make up or backstop renewables energy shortfall.

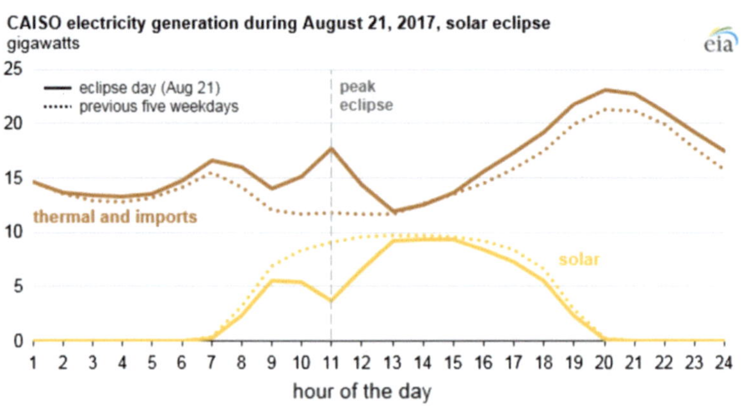

Fig. 1.11 Based on U.S. energy information administration and California independent systems operator (CAISO—Daily Renewables Watch) (EIA 17-8-24)

1.5 The 100% Renewable Energy Objective: Cost and Security Roadblocks

The prospect of 100% renewable energy that would power both electricity generation and vehicle would reduce the infusion of non-agricultural greenhouse gases to the atmosphere very significantly is a tantalizing objective. This would allow for other solutions for the reduction of CO_2 effusion to the atmosphere from other industries, such as cement fabrication, to be attacked as an individual issue, each in turn. Jacobson et al. (2015) illustrate energy roadmaps for the United States by which 100% renewable energy use can be obtained. These have been revised in a somewhat less optimistic manner, which affects the timeline but not the objective. Even in more highly planned economies, such as that of China, achievement of this objective may not be realizable. In more capitalist economies, interests are often in competition against each other and achievement of zero CO_2 emissions is only one of the competing objectives.

The problem of energy distribution remains a fundamental roadblock, unless local energy storage to bridge the energy deficiency gaps can be developed and brought into operation. Even if a full range of tidal, ocean energy, solar, hydrokinetic, wind, Ocean thermal energy conversion, OTEC (Kleute 2018), and other renewable energy resources were to be developed, the cost of the suitable electric grid and the maintenance of its dependability remains a very costly requirement, especially. The larger the grid the more energy is lost in transmission, which introduces a cost-minus formula for long distance transmission from an available energy source that may be the width of the continent distant. We do not wish to initiate a long argument about whether a 100% renewable energy objective is obtainable or any timing and cost pertaining to it. But there must be a reliable backup to insure energy to meet demand locally.

Renewable energy is characteristically intermittent and must be backstopped by on-demand energy of some sort. Presently, renewables, and energy storage of all sorts are either unpredictable or not sufficient to be a flexible a load shifter to meet peak demand. Natural gas is now available in volume increasingly on a worldwide basis (Chap. 11) and can reliably backstop variable renewable energy demand. It seems to us that there will be a need for natural gas for a considerable time into the renewable energy future. In addition to renewable energy being intermittent, it also has a low energy density in terms of mass necessary to produce power. Wind and solar power, which are developing into the backbone of the renewable energy industry, have very low energy densities in relation to natural gas (Wilson 2013). There appears to be no single renewable energy source that can backstop a completely renewable energy mix, unless long distance transmission could provide for energy delivery.

A local response to energy independence can be achieved by instillation of hydrid renewable/ storage projects. Although at present only a few utilities in the United States are pursuing hybrid projects that combine wind, solar and batter storage in a variety of ways, hybrid energy is already a fact (Trabish 2018b). Hibrid renewable energy, whether backed by natural gas or not, is a logical next step from a market-based transition from vertical to horizontal energy systems. But instead of an energy wholesaler mixing the energy to provide a dependable supply to consumers, local hydrid installations mix the energy at the source and require a much less robust electrical grid to supply spiker energy, if necessary.

There will be two consequences from a hybrid renewable paradigm. The first is that it will be highly disruptive to the existing energy system in which major energy producers focus on one of a number of renewable sources and the wholesalers distribute the energy. That is, if a distributed system of hybrid renewables is established, there will be less need for large energy producers. This will be a major disruptive business issue. The second consequence is that a distributed system will be much more secure both to the likelihood of continuing local energy supply and to national security.

Distributed energy and natural gas microgrids: Distributed energy based on natural gas requires widespread delivery and local microgrids, and these are already spreading (Willis 2017b; Baier et al. 2018). Weekly updates of working gas in underground storage is tracked and published on the U.S. Energy Information Administration website (http://ir.eia.gov/ngs/ngs.html). These not only provide for energy on demand, but it allows each local area within a micro-grid system to be somewhat self-contained and less at the mercy of giant power stations that may be at considerable distances and vulnerable transmission lines. A particularly efficient system, which allows the gas fired peaker station to operate at less than maximum load and often leads to wasted energy, has been implemented by Southern California Edison's Center Peaker and Grapeland Peaker plants. In this embodiment, a 10 MW/4.3-MWh battery system is operated concurrently with a 50 MW GE LM6000 aero-derivative gas turbine. Excess power can recharge the battery and battery backup provide instant power for demand spikes without stressing the gas turbines (Patel 2017b).

The energy industry is in the process of changing in many ways to accommodate cheaper and better utilization of renewable energy and energy storage that gives renewables increased electric grid resilience because the grid itself is fragmenting and each grid area is smaller. More efficient merging of electric power from different sources can be accomplished by better software and interconnectivity. Energy wholesalers are becoming more adept at meeting demand from a variety of variable electricity sources. Renewable energy is getting most of the attention but distributed power and micro-grids are also providing for more secure energy supply to customers. Without much media fanfare, for the first time, distributed electricity generation installations delivered more MW capacity than central power generation in 2013 and 2014 (LuxWB 2015).

Companies are pivoting towards renewable energy and distributed electricity generation with energy storage, primarily through the use of increasingly affordable batteries. Distributed power generation (DG) is virtually a requirement for providing energy during energy peaks, especially in where renewable energy is providing an increased share of local energy. The expanded demand for DG brings a new element to the energy supply value chain, but the business model is still under development.

An integrated DG model should give customers the means to break the 'use it or lose it' aspect of renewable energy, which can be transmitted within a region as necessary. Although the uninterruptable character of energy storage is convenient for homeowners, it may be vital for business where every minute of downtime can mean substantial lost income. For instance, server farms for internet nodes, investment and banking websites, and businesses that sell or book travel over the web, almost every commercial website today, must have full time power or their cash flow will suffer. Distributed energy generation is also likely to introduce new business opportunities. In 2015 the U.S. installed over 220 MWh of energy storage, which was up about 240% from 2014 at costs that were at least 25% less than 2014 (GTMR 2016).

The new level of computer interconnectivity has introduced both business and security issues that did not previously exist. Power companies are no longer vertically integrated and control and business communications are no longer isolated. Software firewalls cannot be made impenetrable. Each segment of the horizontally organized energy business has its own issues. For instance, power generation companies want to sell as much power as they can to increase their positive cash flow. Transmission companies minimize maintenance and upgrading to increase their income, and delivery companies are no longer responsible for the delivery of power as they were in the older vertically integrated paradigm, with the inevitable clashes of interest between customers and the energy industry.

In addition, distributed power generation in which many homes have solar or other renewable electric generating capability, runs counter to existing business models. Power companies will remain responsible for providing energy on demand, but their main generating periods will become shorter as the availability of renewable energy grows (Fig. 1.10). This will have the effect of reducing the income of power generating companies, while possibly increasing the costs of power distributing or wholesale electric companies (Fig. 1.6) in order to deal with more complex energy distribution requirements. Will increased costs fall on consumers implementing distributed renewable power generation? We believe that if the U.S. or any country is promoting increased use of renewables that a national plan for dealing with the disruption to the existing energy business model must be considered. Europe, particularly Britain and Germany, have developed national plans that could serve as models for U.S. national energy policy.

The new horizontally integrated, intercommunicative energy industry introduces a major national security issue as power transmission is becoming more important and resilience must be built into future grids (NAS 2017). Inherent in the interconnection of almost all power generation, transmission, and delivery companies in North America (Koppel 2015) is the danger that external intervention could disrupt power generation and distribution, potentially creating damage that would take weeks or months to repair. Except for Texas, which has its own power grid and a number of local 'micro-grids', the entire power system that connects over 3200 companies on the internet is extremely vulnerable to cyber-attack that could disable the system for weeks, if not longer.

At an extreme, cyber-attack could be used to damage equipment that would cause much more significant disruption of the system. A national system for energy security may run counter to some business interests, but is too important to ignore. For instance, in addition to transmission infrastructure suitable for sharing energy from multiple sources, local or regional energy supply at least on controlled reduced levels is a national security issue (Hutchinson 2016).

Brown (2016) suggests that small nuclear reactors backstopping regional base-load could be safer and more efficient than large nuclear power plants. Stanway (2017) draws attention to the Chinese initiative to develop small nuclear reactor (SNR) technology both for base load and as a supplemental power source for renewables. At the moment, there may be as many as 50 different SMR designs in most nuclear industrial countries. Unlike large reactors that take many years to build and cost in excess of $10 billion, SMRs can be fabricated at a factory, transported in sections by truck to a site, and brought into service quickly, similar to a combined cycle gas-fired turbine electrical generating facility but with a smaller footprint. SNRs have the potential to support distributed power systems, but do not require the local storage and distribution pipeline systems that gas-fired plants do. But—although the danger of a breach of a gas fired plant remains very low in virtually all circumstances, SMRs contain nuclear fuel and daughter products that are extremely dangerous if breached either accidentally or by terrorists (Chap. 10).

1.6 Energy Policy in a CO_2 Sensitive Power Future

The drive to reduce CO_2 emissions broadly is a response to projections that the increasingly large amounts of CO_2 in the atmosphere are causing atmospheric and oceanic warming. Even though CO_2 has the focus of environmentalists' interest today because of its likely effect on climate change (IPCC 2013), each of the renewable energy methods also has environmental impact that is relatively unimportant. Our challenge is to expand energy use, while diminishing the environmental negative consequences of the fossil fuels.

This Environmental Protection Agency (EPA) is a driving force for full power system evaluation, and unveiled its most recent rules for existing power plants as part of the Clean Power Plan in the autumn of 2015. This followed on existing guidance to industry with respect to allowable pollution (EPA 2014a, b). In anticipation of the formal plan, however, the power industry is acting on the principal factor of the reduction of CO_2 emissions. EPA clean power plant assessment is looking forward to introduction of carbon taxes and is hoping that the way these are applied in different States does not lead to market distortions. In addition, regulation of CO_2 emissions in many other countries and associations such as the European Union, are trying to balance reduction of carbon emissions with costs and impact to business and society. Although traditional energy companies are dramatically expanding their oil and gas activity and delivery system infrastructure in Asia and less developed countries, there is a push for renewable energy as the developing world 'goes green' (Klare 2015).

It is likely that new government regulations and fiscal policy in many countries will introduce new energy distribution issues. Market distortions may be induced by a carbon tax as a result of the law of unintended consequences. There are three approaches to carbon tax, (1) mass based, (2) rate-based approach, and (3) fee or tax based. The amount of CO_2, a variable valuation that may vary as the amount of CO_2 varies, and a tax on the gross revenue or profit can all be used as a basis for carbon taxes, but the simplest and most transparent system will undoubtedly be best. Individually, these introduce different parameters affecting costs and revenue decisions. All fiscal policy components (tax, subsidy, grants including government-backed loans) can introduce market distortions. A rate-based system interchanging energy with a mass-based system, for instance, will involve increased power transmission into the mass-based carbon region because it provides a higher income to the power wholesaler.

Carbon credits have been introduced as a first step to making it more expensive to emit CO_2 to the atmosphere. This has set industries and countries supporting the most emissions against each other and so many carbon credits have been issued that the price has begun to provide an incentive. The European Union is planning on refining its credit system as a market mechanism that will financialize the credits and allow them to be traded by private interests by 2019 to make the credits rarer and boost their price (EU 2015). Most countries do not operate a carbon credit or tax system. With carbon tax or credit system, hydrokinetic power becomes much more cost effective and tidal power becomes a more attractive investment. However, hydrokinetic power is limited geographically and with decreased precipitation projected for heavily populated regions of the United States, has limited possibilities of expansion, even with the full integration of Canadian hydrokinetic resources. Hydropower presently produces 62.5% of Canada's power; another 10% of Canadian hydropower is exported to the United States, with 5500 MW of capacity added between 2003 and 2013, and with another 7000 MW under development (Aarons and Vine 2015). Without a cap and trade system with a price on carbon, natural gas currently tends to win straight price competitions.

Thus, fiscal policy is commonly driven by the political response to scientific observations and predictions of the effect of increasingly large CO_2 emissions, although there is a substantial political undercurrent that rejects or ignores the science. Regardless of the political conflict on the issue, regulation of CO_2 emissions cannot be set aside and this results in a new energy paradigm from which important strategic issues emerge. As non-CO_2 emitting renewable energy production increases (DNV 2017), and the more coal and oil-fired power plants go offline, the greater will be the requirement for backup power to cover short-term insufficient renewable power generation.

1.7 Strategic Importance of Natural Gas in the New Energy Paradigm

The world's long-standing modern energy paradigm based on hydrocarbon fuels (that originally replaced burning of wood) is in transition to a new high technology renewable energy paradigm which will grow dramatically as part of the energy mix. It is entirely likely that research and development in CO_2 capture and sequestration will become viable in certain circumstances and will prolong the era of natural gas well into the future. The overlap on the timeline to 100% renewable energy is uncertain, not whether our humankind will make the energy transmission (so long as we are still around to do so). The simple reason for understanding this transition is that all fossil fuels are finite, and only a small part of the total resource is suitable to be commercially extracted; we will someday exhaust or price-out naturally occurring hydrocarbon fuels. But 'someday' could be a long way away, especially with respect to natural gas. The only combustible fuel that could be regarded at least in part as renewable is natural gas, which can be produced from sewage or captured from the natural biochemical cycle or by industrial processing of agricultural and waste.

Other liquid fuels such as ethanol, biodiesel, and moderately high energy density fuels derived from lignocellulosic biomass processes, waste, and waste oil processing, and other fuels will join the ranks of gasoline (petrol) and diesel. Blending of ethanol derived from corn was proposed in 2007 to mitigate the perceived volatility in the price of imported oil (Giles 2012), but the U.S., having developed shale oil may now be exporting oil. The logic of heavily subsidizing corn ethanol production therefore no longer exists. Nonetheless, the amount of corn-based ethanol in gasoline (petrol) was increased through an EPA regulation in late 2015 (Jalonick 2015). Many of these alternative liquid fuels, especially ethanol from corn, have the unintended consequence of affecting food prices, for instance where corn is used for animal feed, and probably could not survive without government regulation requiring its use as an automobile fuel.

The transition to a renewable energy future delivers considerable indirect cost savings. These include less pollution and consequent health benefits. Pollution from Chinese coal-fired power stations (some of which are burning American coal), for instance, has been detected in North America and radionuclide pollution has the potential to pollute whole ocean basins. For instance, the full impact of the leaks from the breached reactors at the Fukushima Daiichi power plant are not yet fully understood but are, as of the end of 2015, still flowing into the NW Pacific Ocean. Savings on political and military commitments aimed at safeguarding oil supplies may also be realized when countries or groups of countries can generate a substantial amount of their own renewable energy.

Renewable costs are decreasing and will continue to do so while fossil fuels are subject to substantial price fluctuations due to international politics, speculators, and the currently distorted markets. O&M costs for renewable energy are low and do not include fuel cost, as do non-renewables. Thus, renewable energy is largely inflation-proof. When dramatically increased renewable energy generation and use is combined with increased efficiency being delivered by improved and less costly technology, it seems likely that the transition to the renewable energy paradigm could easily be accelerated (Freeman and Parks 2016). As greater energy efficiency can be delivered, however, it can become disruptive to existing energy provider profit models. Also, financing efficiency on its own merits, especially where more efficient hardware is more expensive as a more expensive capital investment (CAPEX), may lack a convincing business model. Up-front spending to achieve long-term efficiency may appealing in a larger economic sense within a planned economy such as China rather than a more capitalist system such as in the United States. Measurement of an agreed baseline for energy efficiency and other downstream effects, such as the medical impact of lowering pollution, has achieved a consensus.

Rapid improvements in renewable technology are playing a big part in making a transition to a new energy paradigm not only possible, but attractive from both environmental and cost perspectives, although there may be a floor to renewable prices that may be higher than some of the rosier scenarios. Some sort of tipping point may have been reached in that even in a low-cost fossil fuel environment, investment in renewables is accelerating (Warrick 2016). It is important to focus on the transitional period we have entered in which coal and oil will be retired and non-CO_2-producing power sources and renewable energy will increasingly govern supply. The push toward low-carbon energy is intensifying and will increasingly shape investment, energy trade flows, and become an increasingly important geopolitical driver (Ladislaw 2015).

A further factor of making investment decisions is that within an energy company concerned with investment in exploration and new production, very often investment may be based on the particular opportunity, with little regard for the possibility of wider investment in energy or for the changing energy paradigm. For instance, Carbon Tracker (2014) considers the wider energy investment picture of carbon cost curves as a function of risk to capital expenditures. Because deepwater investments are necessarily long-term, and perhaps a decade could pass until cash flow begins to be realized, a long-term approach to energy investment now could be quite different from decisions made even ten years ago.

In concert with the transition to a low-carbon energy environment, natural gas could remain inexpensive for years and new technologies can be expected to open up previously inaccessible and new energy plays. This includes resources such as NGH, especially when they carry a very low environmental risk (Chaps. 8, 10). Because of its abundance and its superior low-carbon character of all fossil fuels, natural gas could serve as "a bridge fuel to a future" (Brown et al. 2009) in the "dash for gas" (Helm 2012) in order to reduce CO_2 emissions while retaining a capacity for uninterrupted electricity provision.

The new energy paradigm will consist of a CO_2 emission framework that is lower than today from a distributed base load mix of energy sources back stopped by energy storage. In situations where intermittent energy is unavailable, natural gas can act more widely as base load power supply, as it does in many locations today.

Integration of gas supply to plants as a response to local energy demand is being integrated now because it will be the on-demand backbone fuel for the new electrical power paradigm (Rybarik 2013).

Different countries and regions have different motivations or 'drivers' for increasingly using gas as a fuel, not just for new power installations but for replacing older plants using coal and oil. Europe is motivated by the need to diversify energy supplies and lower CO_2 emissions. India and China are both dramatically increasing their energy demand to increase their GDP and standard of living, but they have also given assurances that they will implement emissions reduction policies. China is also the world's largest investor in renewable energy and has given an undertaking to reduce CO_2 emissions wherever possible. China is also planning to upgrade and/or its coal-fired power plants to more than halve the amount of air pollutants they discharge by 2020 (Xinhua 1/15/16). China is planning on taking a lead in renewable energy and to have a low-carbon energy system by 2050 (Baeumler et al. 2012; Shell Global 2015). China has made a strategic choice for renewable energy and to manufacture the full range of renewable technology (Mathews 2016). Installation of renewables exceeds the European Union and perhaps the United States.

The United States in the final two years of the Obama Administration focused on lowering CO_2 emissions by gas replacing other fossil fuels, as well as continuing the drive against pollution from coal, particularly mercury (EDF 2011) and other heavy metals and radionuclides. Closing of coal fired power stations in the central New England region was simply a result of market forces causing outdated coal-fired power stations to be replaced by less expensive gas-fired stations. An additional benefit is the measurably lower mercury pollution in the downwind region as a direct result of reduced coal exhaust (Zhou et al. 2017). Despite the coal industry and political desire to 'bring back' coal, it can no longer compete on a purely fiscal basis. Independent analyses show that coal-fired electricity generation continues to decline (Negin 2017). Oil is also being replaced by natural gas.

As part of this renewed focus on the attributes of natural gas as a fuel, especially in Asia (Musarra 2015), gas that was previously regarded as stranded too far from markets to be economic is now being brought into the world gas supply (Guttulsrod 2015). It is likely that the developing world natural gas price based on much more widespread availability of LNG (Chap. 11) may prove to be the key to a broad backstopping of renewable energy production that will guarantee both base load and spiker power on demand. A dependable world supply of natural gas at predictable prices may also wean the world off other fossil fuels.

Natural gas currently plays a big part in the US energy mix (Table 1.1) and will become even more prominent as coal and oil plants are retired and more intermittent energy sources come online because the potential gap between available dependable base load and potential demand increases. In April 2015 natural gas overtook coal as the primary source of U.S. electrical generation, producing 31% of the national grid electricity to coal's 30% (Searles 2015). In 2009 there were 593 coal-fired power plants in the US. By 2013 the number had fallen to 518 with a total summer capacity of about 300 GW. EIA forecasts coal-fired power output will shrink by about 13 GW in 2015.

Use of natural gas is growing because of the speed and cost of building power plants and increased, low-cost fuel supply. Fourteen U.S. States used natural gas as their main fuel for power plants in 2015 (Muyskens et al. 2015). The switch to gas is already underway, driven mainly by market forces, but also by accelerated by regulations. For the first time since the U.S. Energy Information Administration (EIA) began collecting data in 1973, in 2015 natural gas has become the nation's number one fossil fuel for power generation. Despite economic growth and increased energy use in both China and the United States in 2014–2015 CO_2 emissions have declined in the power generation sector. Chinese emissions fell 1.5% as coal use dropped, in part due to greater plant efficiencies and to massive installation of renewable energy accounting for 90% of new electric production. On 19 March 2017 it was announced that the last coal-fired power station in Beijing, the Huaneng Beijing Thermal Power Plant, has suspended operations. The city's electricity is now being generated using natural gas (Phys 2015), which enacts a strategic decision by government (Chap. 10) to clean up Beijing's air (CER 2015). The increased use of natural gas is putting strain on the relatively small pipeline system in China, which requires considerable expansion (Clemente 2016).

In the U.S., an emissions decline of 2% was due to the transition from old coal plants to new gas-fired plants (Johnson 2016), which is leveraging the existing, large gas pipeline system. According to MIT (2013), natural gas electricity generation will continue to become much more important to electric power systems.

The repercussions of the accelerated switch from coal to natural gas have both positive and negative aspects. For traditional mining communities, the 'war on coal', which is regarded by the people and communities losing coal mining and related jobs as being driven by government regulation and to be harsh treatment of them. For consumers of electricity, it could mean lower electricity prices and/or higher profits for the power generation industry. For government concerned with promoting climate change agendas, it means the spectacular lowering of CO_2 emissions without any additional cost or tax implications. The impact of the switch to natural gas has meant that in 2015, coal retirements alone equaled as much coal capacity as the US retired in the 20-year period between 1990 and 2009 (SC 2015). The 2015 electric sector carbon emissions were projected in November 2015 to be 1983 million ton, which is 433 million ton (approximately 18%) below 2005 levels. This is within 5% of the level the Clean Power Plan of the U.S. EPA's target for 2022. In other words, market forces and regulation are dramatically reducing U.S. energy sector CO_2 emissions.

According to BP CEO Bob Dudley, speaking at a conference in May 2015, "we need to explain more clearly that gas is not so much part of the problem as part of the solution". Shell CEO Ben van Beurden referred to natural gas as being a "clean-burning ally to renewables such as solar and wind" (Ritter 2015). While energy companies support the UN climate change negotiations, they are not giving up on fossil fuels, but rather focusing on increasing the use of gas at the expense of other fossil fuels (Gallucci 2015). They note that they are preparing for possible limits on more CO_2-intensive fossil fuels and that use of gas will have the effect of reducing CO_2 emissions overall.

Technological advances continually redefine reality, and the natural gas and oil from the shale plays since 2008 have radically altered the oil and gas supply situation within the United States. Gas and oil from shale depends on fracking and other treatments to introduce permeability into the shale beds, which are actually thermally mature source beds in a petroleum system that has ceased evolving because the remaining hydrocarbons were trapped before they could migrate to a conventional trap or out of the system entirely.

In the new energy paradigm, natural gas recovered from conventional or unconventional resources will remain as the primary fossil fuel energy source. But it is not impossible that if the role of methane as an on-demand backup fuel in the energy mix in the future could prove to be valuable enough so that biogas (methane) production as a product of sewage treatment and other controlled processes could meet at least local requirements through being compressed and used when needed. In addition, this 'recovered' methane does not have to be combusted. Methane has the lowest carbon to hydrogen ratio (CH_4), which makes it an excellent feedstock for a future fuel for non-CO_2 producing hydrogen fuel cells using compressed hydrogen and either air or oxygen. A number of attributes characterize natural gas as a fuel:

- **Less CO_2 on a mole basis than any other combustible hydrocarbon fuel**
- The H:C ratio is higher in methane than any other hydrocarbon. For a given quantity of energy, which can be expressed as a function of the H_2O byproduct, less CO_2 is produced.
- **No particulates produced if properly burned**
- No insoluble residues or dissolved chemicals that will form soot are present in the methane. With proper conditions allowing complete combustion, zero particulate matter or biomedically harmful chemicals are produced from methane.
- **Easy and inexpensive fuel purification where necessary**
- Purifying gas is usually accomplished through gas diffusion and ionic separators. The purification process is more rapid in gas than liquids because of beneficial viscosity and diffusion rates.
- **No costly refinement process necessary**
- Crude oils must be refined before they can be used. This is often an expensive process, especially with heavy oil and tar sands. Methane gas does not require refinement or chemical processing involving similar energy inputs.
- **Best fuel consistency/character over a wide temperature range**
- Methane can be used in engines from the coldest Arctic to the hottest tropical or desert with no provision for special handling or storage. The character of methane fuel does not change significantly with temperature. Methane not solidify or clog in cold weather although propane may be more difficult to vaporize.

- **Usable in existing engine types with minor modifications**
- Most automobile engines can be retrofitted with methane to provide a dual fuel capability. This is a simpler process for carburetors than for fuel injection.
- **Very efficient in specially designed engines (methane gas turbines)**
- Engines that are specially designed for methane fuel.
- **Stable indefinitely in storage**
- One of the major problems any large user of fuel has is that stored fuel may separate into different fractions. Water may also form in the fuel. Liquid fuels inherently are often quite unstable. Methane will not separate into fractions. If water forms in the tank, it separates from the gas and is not carried as droplets within the fuel.
- **Stable at high temperatures**
- At temperatures up to around 1000 °C, in the absence of oxygen, methane remains chemically inert. It can provide a heat sink without deterioration.
- **Fuel for CH_4 fuel cells**
- Extremely efficient, high temperature fuel cells using low pressure methane and oxygen with a nitrogen buffer, have been successfully designed and tested.
- **Methane is the best point source for H_2**
- H_2 can be produced from methane for only 15% of the energy required to produce H_2 from water. The products of CH_4 reduction are inert carbon and H_2.
- **Methane can be transported as a gas, liquid and solid**
- Methane is conveniently and relatively safely transported to many homes, business places, and industrial sites as a gas. Methane, transport and use of more energy density-rich LNG is also common. Advanced experiments with artificially produced methane hydrate have been carried out up to full size industrial pilots.

Discussion: Questions such as how long will supplies of natural gas within the United States last or how long they will be available to the United States at or near the current low price become increasingly important. Strategically, indigenous natural gas will be the only dependable and allowable fuel available at short notice to supplement non-fossil non-renewable and renewable energy sources. The present model for energy security thus depends on the availability of indigenous natural gas from the shale play remaining high. Although the shale gas industry has popularized shale oil and gas as long-term resources on the scale of many tens of decades (Berman 2013), reserves may be much less than thought at the effective opening of the resource around 2008 (see Chap. 10). If it proves that the United States does not have a shale gas resource on the order of many tens of decades, then gas from other sources will become increasingly important in the nearer term.

Conventional gas continues to be discovered and brought into production along with the one-time unconventional gas resources, coalbed methane and shale but demand is also increasing. NGH (Chap. 2), which is the potentially greatest single sequester of natural gas on Earth (Johnson 2011; Beaudoin et al. 2014), is untested but very promising; exploration is only in the early stages and the resource remains (as of this writing) to be brought into production. NGH is just another gas source, and one that is regarded by the Department of energy as, "… a good candidate, sometime in the future, for another public-private success story …" (MIT 2011).

As recently as 2006 the U.S. chemical industry was in decline because conventional gas production was dropping and natural gas was becoming more expensive. No chemical manufacturing plants were under construction or planned for the U.S. Opening of the Shale resource changed this completely and by 2016 almost 100 chemical manufacturing plants were under construction or planned for the U.S. with jobs projected to grow by over 450,000 over the next decade (Alper 2016).

Renewable energy sources will play a more important role in energy supply, especially if there is more governmental emphasis on constraining atmospheric impact of burning fossil fuels. In response to the abnormal rise in CO_2 in that atmosphere, to which human anthropomorphic contributions appear to be an important part, the first successful international meeting to get agreement on national CO_2 reductions since the Doha Amendment to the Kyoto Protocol was held in Paris in 2015 (COP21 2015) officially tasked national targets. But because fossil fuels and nuclear energy are so dependable and well understood, with regard to non-catastrophic risk (Chap. 10), even mandates to implement renewables in their place cannot change the economics of low cost fossil fuels. Completely replacing fossil fuels with renewable energy may be very difficult to manage (Brooks 2015) and less important than focusing on the most environmentally friendly and lowest risk of the fossil fuels and transitioning to natural gas as part of the transition to a renewable energy future.

The life cycle costs of nuclear energy remain very high for the industry standard conventional light water reactors. With the downward trend of renewable energy costs and the dependability and relatively low pollution from natural gas, it is likely that conventional nuclear power will have no place in competitive electricity markets and will gradually be phased out (Bradford 2016). As older plants age out, they are unlikely to be replaced with the same nuclear technology. A number of new nuclear technologies, however, may change the equation of the cost-risk factors of nuclear energy, fossil fuels, and renewable (CATF 2017). Some of these systems have been installed and are operational so that accurate life cycle costs can be compared with pre-construction estimates.

Renewable energy depends on the sun shining, the wind blowing, the waves churning, and of important to baseload, tidal energy (which is still in development), most of which will be offshore in tidal streams rather than across the mouths of shallow bays. Hydrokinetic energy can fill in some energy gaps on a dependable basis, but many areas do not have plentiful hydrokinetic potential remaining, especially in drying climates. Backup or base load natural gas-fueled power generation will be necessary when renewables are not sufficient, as they can be supplied to meet demand. As part of an overall natural gas strategy, natural gas leakage should be reduced as part of a low-carbon strategy (Sanchez and Mays 2015).

Natural gas will be the fuel of choice, especially if there is widespread agreement that lowering greenhouse gas (CO_2) emissions is a global priority (NI 2015). Deyette et al. (2015), writing for the Union of Concerned Scientists, believe that becoming too reliant on natural gas poses numerous complex risks, the most important of which is that investing in gas rather than a completely renewable energy power mix will make it much harder to achieve long-term global warming emissions reduction goals. But it seems to us that they essentially ignore the difficulty there will be in transitioning to a completely carbon-free energy paradigm. Carbon Tracker (2015) forecasts shale gas, which it proposes will define the bottom of the gas market, at \$3–4 short term and \$4–5 long term. They also forecast LNG at \$10 for the foreseeable future.

Investment in energy in general is showing some new trends. Whereas investment in fossil fuels has for a long time appeared to be a safe bet, the post-2014 low price paradigm (Chap. 11) may be making investment in renewable energy more attractive. Divestment of fossil fuels by major companies, which is the flip side, is beginning to become significant while at the same time the cost basis of renewable energy is becoming very competitive (Watson 2016). In 2015 a solar project in Dubai went online at a rate of \$0.058/kWh. An offshore windfarm in Morocco priced electricity at \$0.30/kWh. Both are competitive with the EROEI (Chap. 11) of fossil fuels. Interestingly, Shell, which is one of the biggest international energy companies, has started to shed investments while biasing its future business toward renewables and natural gas (Kirka 2017). We suggest that whereas natural gas may remain as a profitable investment, both coal and oil investment may, because of direct price competition and higher CO_2 emissions and higher environmental risk, have reached a tipping point in their attractiveness for investment.

1.8 Natural Gas Backstop to Renewable Energy

From a strategic, practical, and increasingly a commercial point of view, natural gas is emerging as the premier fuel for electricity generation. Coal is in a slide that will continue to reduce its significance for electricity generation while natural gas use is

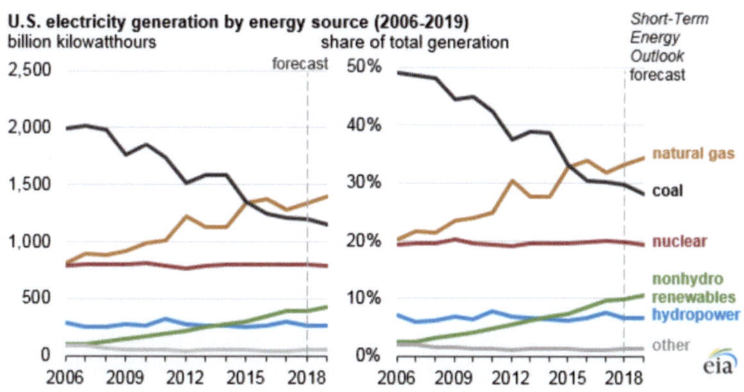

Fig. 1.12 U.S. energy production sources used for electricity generation. From EIA, 1/22/18 (2018)

trending rapidly upward (Fig. 1.12). As more gas pipelines are constructed, gas will become the dominant fuel for electricity generation and will persist in the future as a transitional backup for greater electricity production from renewables.

- Implementation of increasing renewable energy electrical generating capability requires a base load fuel that can be turned on and off quickly to make up for renewable deficiencies (wind stopping, night, etc.).
- Natural gas is the best fossil fuel for providing backstop energy for intermittent or renewable energy during the transition into a CO_2-free future energy system and possibly for the indefinite future.
- Natural gas produces the least contaminated exhaust as there is virtually no sulfur, dissolved metals, or other contaminants in the fuel to begin with in contrast with oil and especially coal.
- Natural gas is easy to store either in compressed, LNG, or NGH form.
- Natural gas will not deteriorate when stored or in long term transit whereas petroleum products deteriorate and coal in outdoor storage deteriorates and loses Btu value.
- Natural gas power plants can be started and turned off more rapidly and safely than either coal or oil fuels.

References

Aarons, K., & Vine, D. (2015). *Canadian hydropower and the clean power plan* (36 pp). Center for Climate and energy Solutions. http://www.c2es.org/docUploads/canadian-hydropower-04-2015.pdf. Accessed April 23, 2016.

ALA. (2011). *Toxic Air. The case for cleaning up coal-fired power plants*. American Lung Association, 12 pp.

Alam, M. S., Bala, B. K., Huo, A. M. Z., & Matin, M. A. (1991). A model for the quality of life as a function of electrical energy consumption. *Energy, 16*(4), 739–745. (Pergamon Press plc).

Alper, J. (Rapporteur). (2016). *The Changing Landscape of Hydrocarbon Feedstocks for Chemical Production: Implications For Catalysis: Proceedings of a Workshop*. National Academies Press, 136 pp. ISBN 978-0-309-44479-8. https://doi.org/10.17226/23555.

Arbib, J., & Seba, T. (2017). Rethinking transportation 2020–2030: The disruption of transportation and the collapse of the internal combustion vehicle and oil industries. A RethinkX sector disruption Report, May 2017. RethinkZ, 77 pp. http://www.ourenergypolicy.org/wp-content/uploads/2017/05/RethinkingTransportation_May_FINAL.pdf. Accessed May 14, 2017.

Bade, G. (2018). *APS to install 50 MW, 135 MWh solar-shifing battery D UtilityDive*. https://www.utilitydive.com/news/aps-to-install-50-mw-135-mwh-solar-shifting-battery/516850/ Accessed February 12, 2018.

Baeumler, A., Ijjasz-Vasquez, E., & Mehndiratta, S. (Eds.). (2012). *Sustainable low-carbon city development in China* (588 pp). The World Bank. ISBN: 978-0-8213-8987-4. https://doi.org/10.1596/978-0-8213-8987-4.

Baier, M., Bhaveraju, V., Murch, W., & Teleke, S. (2018). *Making microgrids work: Practical and technical considerations to advance power resiliency* (6 pp). Eaton Company. http://www.eaton.com/ecm/idcplg?IdcService=GET_FILE&allowInterrupt=1&RevisionSelectionMethod=LatestReleased&Rendition=Primary&dDocName=WP027009EN. Accessed March 25, 2018.

BBC. (2017). *Coal collapse drives down UK carbon emissions*. http://www.bbc.com/news/science-environment-39181753. Accessed March 7, 2017.

Beaudoin, Y., Dallimore, S., & Boswell, R. (2014). *A UNEP global outlook on methane gas hydrates V2* (96 pp). United Nations Environment Programme. ISBN: 978-92-807-3319-8.

Ben. (2008). The two most important graphs in the Energy Biz. In *The watt, energy Discussion for an efficient world* (3 pp). http://www.thewatt.com/node/168. Accessed 23 Apr 2016.

Bennett, M. (2015). *Surprise, clean power plan to lower natural gas demand*. http://oilpro.com/post/17329/surprise-clean-power-plan-to-lower-natural-gas-demand?utm_source=DailyNewsletter&utm_medium=email&utm_campaign=newsletter&utm_term=2015-08-07&utm_content=Article_13_txt. Accessed August 7, 2015.

Berman, A. E. (2013). *Let's be honest about shale gas* (2 pp). Houston Geological Society.

Biello, D. (2016). *Carbon capture may be too expensive to combat climate change* (pp. 59–65). Scientific American.

BNSF. (2013). *Guide to coal mines served by BNSF railway* (22 pp). BNSF Railway Coal Business Unit (June 12).

Bradford, A. (2015). *Tesla grid battery—The hype and impacts to natural gas* (6 pp). http://oilpro.com/post/12630/tesla-grid-battery-hype-and-impacts-to-natural-gas.

Bradford, P. (2016). *Compete or suckle: Should trobuled nuclear reactors be subsidized? The Conversation*. https://theconversation.com/compete-or-suckle-should-troubled-nuclear-reactors-be-subsidized-62069. Accessed March 7, 2017.

Brooks, A. (2015). *Paris conference message: We changed world's energy mix*. Oilpro 1/7/16. http://oilpro.com/post/21364/paris-conference-message-we-changed-worlds-energy-mix?utm_source=DailyNewsletter&utm_medium=email&utm_campaign=newsletter&utm_term=2016-01-07&utm_content=Feature_1_txt. Accessed January 7, 2016.

Brown, P. (2016). *A big push for small nuclear reactors*. Climate News Network. http://www.politico.com/sponsor-content/2016/04/energy-vital-role-in-economic-development?cid=20160app. Accessed April 27, 2016.

Brown, S.P.A., Krupnick, A.J., & Walls, M.A. (2009). *Natural gas: A bridge to a low-carbon future* (14 pp). National Energy Policy Institute Issue Brief 09-11.

Carbon Tracker. (2014). *Carbon supply cost curves: Evaluating financial risk to capital expenditures*. Carbon Tracker. http://www.carbontracker.org/wp-content/uploads/2014/09/CTI-Oil-Report-Oil-May-2014-13-05.pdf. Accessed January 16, 2015.

CAN. (2017). *Advanced energy and U.S. national security* (66 pp). Center for Naval Analyses, The CAN Military Advisory Board.

Cardarelli, R., & Lusinyan, L. (Eds.). (2016). *Power play, energy and manufacturing in North America* (176 pp). Washington, DC: International Monetary Fund. ISBN: 978-1-49836-497-9 9PAPER).

CATF. (2017). *Advanced nuclear energy: Need, characteristics, projected costs and opportunities*. Clean Air Task Force, 55 pp. http://catf.us/resources/publications/files/Advanced_Nuclear_Energy.pdf. Accessed April 23, 2017.

CER. (2015). *Beijing turns to natural gas as a cost-effective coal replacement* (3 pp). http://www.chinaeconomicreview.com/china-nurtures-natural-gas. Accessed March 19, 2017.

Chen, W., Wu, Y., Uye, Y., Liu, J., Zhang, W., Yang, X., et al. (2015). Efficient and stable large-area perovskite solar cells with inorganic charge extraction layers. *Science, 350*(6263), 944–948.

Clemente, J. (2016). China's Rising Natural Gas Demand, Pipelines, and LNG, 5 pp. Forbes https://www.forbes.com/sites/judeclemente/2016/04/24/chinas-rising-natural-gas-demand-pipelines-and-lng/#55d2762a5d00. Accessed 19 March 2017.

Cleveland, C. J., Kaufmann, R., & Lawrence, T. (2008). *Fundamental principles of energy*. http://www.eoearth.org/view/article/152893/. Accessed April 20, 2016.

CNG. (2015). Greenest CNG station opens in Tennessee. The Source. *American Gas Association, 8*(1), 18–19, 21–22.

COP21. (2015). *Conference of the Parties, Adoption of the Paris Agreement*. United Nations Framework Convention on Climate Change, Paris, 31 pp, 30 November–11 December 2015. FCCC/CP/2015/L.9.

DE. (2016). *Oil/gas-FIRED plants, steam and combined cycle*. Duke Energy. https://www.duke-energy.com/power-plants/oil-gas-fired.asp. Accessed April 23, 2016.

de Laleu, V. (2009). La Rance Tidal Power Plant. 40-year operation feedback—Lessons learnt. In *Presentation at BHA Annual Conference*, Liverpool, 14 & 15 October 2009. http://www.british-hydro.org/downloads/La%20Rance-BHA-Oct%202009.pdf. Accessed April 23, 2016.

de Quilettes, C. W., Vorpahl, S. M., Stranks, S. D., Nagaoka, H., Eperon, G. E., Ziffer, M. E., et al. (2015). Impact of microstructure on local carrier lifetime in perovskite solar cells. *Science 348*, 683–686. Accessed November 13, 2015.

Deyette, J., Clemmer, S., Cleetus, R., Sattler, S., Bailie, A., & Rising, M. (2015). *The natural gas gamble: A risky bet on America's clean energy future* (56 pp.). Union of Concerned Scientists.

DNV. (2017). *Renewables, power and energy use forecast to 2050* (68 pp). DNV-GL-Energy.

DOE. (1993). *U.S. coal reserves: an update by heat and sulfur content* (150 pp). Energy Information Administration, U.S. Department of Energy, DOE/EIA-0529(92) Distribution Category UC-98.

DOE. (2017). *Staff Report to the Secretary on Electricity Markets and Reliability* (187 pp). U.S. Department of Energy. https://www.energy.gov/sites/prod/files/2017/08/f36/Staff%20Report%20on%20Electricity%20Markets%20and%20Reliability_0.pdf. Accessed August 24, 2017.

Doosan. (2017). *Making utility-integrated energy storage a used, useful, and universible resource* (13 pp). Doosan Corporation Report. https://www.utilitydive.com/news/begin-with-the-end-in-mind-used-and-useful-utility-integrated-energy-stora/514551/. Accessed January 18, 2018.

DW. (2016). *Big data is about to transform renewable energy* (5 pp). DW.com (Deutsche Welle). http://www.dw.com/en/big-data-is-about-to-transform-renewable-energy/a-36189374. Accessed October 29, 2016.

EDF. (2011). *Mercury Alert: Cleaning up coal plants for healthier lives*. Environmental Defense Fund Presentation Report, 16 pp. http://www.edf.org/sites/default/files/EDF-mercury-report-27mar2011-revised.pdf. Accessed July 20, 2015.

EIA. (2016). Table 7. U.S. Coal Exports. U.S. Energy Information Administration. http://www.eia.gov/todayinenergy/detail.cfm?id=11791. Accessed May 16, 2016.

EIA. (2017). *Wind and solar in March accounted for 10% of U.S. electricity generation for first time*. U.S. Energy Information Administration. https://www.eia.gov/todayinenergy/detail.php?id=31632. Accessed June 17, 2017.

EIA 12/27/17. (2017). *Today in Energy. Monthly renewable electricity generation surpasses nuclear for the first time since 1984*. https://www.eia.gov/todayinenergy/detail.php?id=34292. Accessed December 28, 2017.

EIA 1/22/18. (2018). *EIA forecasts natural gas to remain primary energy source for electricity generation*. U.S. Energy Information Administration. https://www.eia.gov/todayinenergy/detail.php?id=34612. Accessed February 4, 2018.

EIA 17-8-24. (2017). *California increased electricity import as and natural gas generation during solar eclipse*. https://www.eia.gov/todayinenergy/detail.php?id=32632&src=email Accessed August 27, 2017.

El-Ashry, M. T. (2015). *The renewable energy era has already started* (5 pp). NewSecurityBeat. Wilson Center Environmental Change and Security Program. http://www.newsecuritybeat.org/2015/11/renewable-energy-era-started/?mkt_tok=3RkMMJWWfF9wsRonva3Pd%2B%2FhmjTEU5z16u0vWq%2B2hYkz2EFye%2BLIHETpodcMTcpgMbrYDBceEJhqyQJxPr3HLdkN18NoRhfmCw%3D%3D. Accessed April 23, 2016.

EPA. (2014a). *Carbon pollution emission guidelines for existing stationary sources: Electric utility generating units 79 FR 34829*. https://federalregister.gov/a/2014-13726. Accessed April 23, 2016.

EPA. (2014b). *National greenhouse gas emissions and sinks: 1990–2013*. Publication No. 430-R-14–003. http://go.nature.com/GSeOar. Accessed April 23, 2016.

EPA-1. (2015). *Clean energy*. U.S. Environmental Protection Agency. http://www.epa.gov/cleanenergy/energy-and-you/affect/air-emissions.html. Accessed August 3, 2015.

ESA. (2017). *35X25 a vision for energy storage*. Energy Storage Association. http://energystorage.org/system/files/attachments/esa_vision_2025_final.pdf. Accessed January 30, 2018.

EU. (2015). *EU market mechanism to boost carbon price*. https://euobserver.com/environment/128596.

EXGY. (2017). 2018 Electric resource plan, 2017 All source solicitation 30-day report (Public Version). Colorado Public Service Company Proceeding No. 16A-0396E, December 28, 2017, 4 pp. https://assets.documentcloud.org/documents/4340162/Xcel-Solicitation-Report.pdf. Accessed January 15, 2018.

Fowler, T. (2011). *Texas calls Level 1 Energy Emergency, new power record set*. FuelFix. http://fuelfix.com/blog/2011/08/02/texas-tops-power-record-another-may-fall-tuesday/. Accessed April 23, 2016.

Freeman, S. D., & Parks, L. Y. (2016). *All-electric America: A climate solution and the hopeful future* (236 pp). Solar Flare Press. ISBN-13:9780996174725.

Gallucci, M. (2015). *Oil and gas companies support UN climate change negotiations, but they're not ditching fossil fuels*. International Business Times. http://www.ibtimes.com/oil-gas-companies-support-un-climate-change-negotiations-theyre-not-ditching-fossil-2144637. Accessed October 18, 2015.

Giles, B. (2012). *Oil is far too interesting: How volatility threatens is supremacy and drives innovation*. http://quarterly.luxresearchinc.com/quarterly/?p=240. Accessed November 25, 2015.

Gregg, J. P. (2017). Electrify, Electrify. *American Public Gas Association. The Source, 9*(4), 20–21.

GTMR. (2016). 2015 Energy Storage Scorecard. GTM Research ESA U.S. Energy Storage Monitor: 2015 Year in Review. http://www.greentechmedia.com/research/subscription/u.s.-energy-storage-monitor. Presentation, 19 pp. Accessed April 23, 2016.

Guttulsrod, T.-I. (2015). FLNG design adapts to an evolving energy mix. *Offshore, 75*(5), 96–98. http://www.offshore-mag.com/articles/print/volume-75/issue-5/engineering-construction-installation/flng-design-adapts-to-an-evolving-energy-mix.html. Accessed May 25, 2015.

Haegel, N. M., et al. (2017). Terawatt-scale photovoltaics: Trajectories and challenges. *Science, 356* (6334), 141–143. Supplementary materials including all authors and affiliations at: sciencemag.org/content/356/6334/141/suppl/DC1.

Hall, C., Tharakan, P., Hallock, J., Cleveland, C., & Jefferson, M. (2003). Hydrocarbons and the evolution of human culture. *Nature, 426,* 318–322.

Harris, J., Goodwin, N., Nelson, J., & Institute, G. (2013). Economic growth. http://www.eoearth.org/view/article/151940. Accessed April 23, 2016.

Helm, D. (2012). *The Carbon Crunch* (304 pp, p. 206)). Yale Press.

Helm, D. (2017). *Burn out: The endgame for fossil fuels* (304 pp). Yale Universary Press. ISBN: 9780300225624.

Hoza, M. (2015). *No NW hydro? No problem. We got gas* (3 pp). Oilpro. http://oilpro.com/post/12785/no-nw-hydro-no-problem-we-got-gas?u…paign=newsletter&utm_term=2015-05-07&utm_content=Article_15_txt. Accessed April 23, 2016.

Hutchinson, L. J. (2016). *Energy's vital role in economic development reshaping cities*. Politico. http://www.politico.com/sponsor-content/2016/04/energy-vital-role-in-economic-development?cid=20160app. Accessed April 26, 2016.

IEA. (2014). *CO2 Emissions from Fuel Combustion Highlights 2014* (136 pp). Paris, France: International Energy Agency.

IEA. (2015). *Medium-term gas market report (MTGMR) 2015* (12 pp). International Energy Agency.

IPCC. (2013). Intergovernmental Panel on Climate Change, United Nations. Index page for downloads http://www.ipcc.ch/report/ar5/wg1/. Accessed March 10, 2016.

ISO. (2015). *California ISO Opening Access. What the duck curve tells us about managing a green grid*. http://www.caiso.com/Documents/FlexibleResourcesHelpRenewables_FastFacts.pdf. Accessed April 23, 16.

ISONE. (2018). *Operational fuel-security analysis* (56 pp). ISO New England. https://www.iso-ne.com/static-assets/documents/2018/01/20180117_operational_fuel-security_analysis.pdf. Accessed January 18, 2018.

Jacobson, M. Z., Delucchi, M. A., Bazouin, G., Bauer, Z. A. F., Heavy, C. C., Fisher, E., et al. (2015). 100% clean and renewable wind, water, and sunlight (WWS) all-sector energy roadmaps for the 50 United States. *Energy & Environmental Science, 8*, 2093–2117. (Includes a following paper: Jakobson, et al., 106 pp).

Jacoby, M. (2015). Iron complex harvests solar energy. *American Chemical Society C&EN, 93* (41), 5.

Jacoby, M. (2016a). Commercializing low-cost solar cells. *Chemical and Engineering News, 94* (18), 30–25.

Jacoby, M. (2016b). Electricity storage goes with the flow. *Chemical and Engineering News*, 26–28.

Jalonick, M. C. (2015). *EPA boosts amount of ethanol in gasoline supply*. PennEnergy November 2015. http://www.pennenergy.com/articles/pennenergy/2015/11/petroleum-news-epa-boosts-amount-of-ethanol-in-gasoline-supply.html?cmpid=EnlWeeklyPetroDecember42015&eid=288179069&bid=1248970. Accessed December 1, 2015.

Joglekar, M., Nguyen, V., Pylypenko, S., Ngo, C., Li, Q., O'Reilly, M. E., et al. (2015). Organometallic complexes anchored to conductive carbon for electrocatalytic oxidation of methane at low temperature. *Journal American Chemical Society, Article ASAP.*. https://doi.org/10.1021/jacs.5b06392.

Johnson, A. H. (2011). Global resource potential of gas hydrate—A new calculation. In *Proceedings of the 7th International Conference on Gas Hydrates (ICGH 2011)*, July 17–21, 2011, Edinburgh, Scotland, United Kingdom.

Johnson, J. (2016). Energy-related CO_2 emissions keep falling. *Chemical & Engineering News, 94*(13), 17.

Kagel, A., Bates, D., & Gawell, K. (2007). *A Guide to Geothermal Energy and the Environment* (p. 86). Washington, DC: Geothermal Energy Association.

Kaufman, A. C. (2015). This is the beginning of the end of the fossil fuel industry. *Huffpost Business* 12/2/15. http://www.huffingtonpost.com/entry/end-of-fossil-fuels_564f7457e4b0258edb316faf. Accessed April 23, 2016.

Kearney, D., Kelly, B., Herrmann, U., Cable, R., Pacheco, J., Mahoney, R., et al. (2004). Engineering aspects of a molten salt heat transfer fluid in a trough solar field. *Energy, 29*, 861–870.

Kirka, D. (2017). *Shell preparing for world economy that shifts away from oil*. PennEnergy. http://www.pennenergy.com/articles/pennenergy/2017/07/oil-and-gas-shell-preparing-for-world-economy-that-shifts-away-from-oil.html?cmpid=enl_pennenergy_pennenergydailypetroleum_2017-07-31&email_address=michaelmax1@mac.com&eid=288179069&bid=1825062. Accessed July 31, 2017.

Klare, M. T. (2015). Fossil fuels are all but finished: Renewable energies are the future, whether the GOP acknowledges it or not. Salon. http://www.salon.com/2015/12/25/fossil_fuels_are_all_but_finished_rene…ergies_are_the_future_whether_the_gop_acknowledges_it_or_not_partner/. Accessed December 27, 2015.

Kleute, B. J. (2018). OTEC market on the rise. *Sea Technology, 59*(4), 16, 18–19.

Koppel, T. (2015). *Lights out: A cyberattack, a nation unprepared, surviving the aftermath* (288 pp). Crown publishers. ISBN-13: 978-0553419962.

Ladislaw, S. (2015). Implications for a low-carbon future. In C. Cohen & M. G. Dalton (Eds.) *2016 Global forecast* (pp. 112–114). Center for Strategic & International Studies.

Laforge, J. (2018). *Nuclear reactors, bankrupting their owners, closing early*. Counter Punch. https://www.counterpunch.org/2018/02/06/nuclear-reactors-bankrupting-their-owners-closing-early/. Accessed February 6, 2018.

Lane, J.-E. (2014). GDP, Energy, and greenhouse gases—Some cross-sectional evidence. Redfame Publishing. *Applied Economics and Finance, 1*(2), 97–101. ISSN 2332-7294 E-ISSN 2332-7308. http://dx.doi.org/10.11114/aef.v1i2.566. Accessed May 3, 2015.

Larson, A. (2017). World's most–efficient combined cycle plant: EDF Bouchain. *Power (Magazine), 161*(9), 22–23.

Le Page, M. (2016). *Food made from natural gas will soon feed farm animals—And us*. London: Daily News. https://www.newscientist.com/article/2112298-food-made-from-natural-gas-will-soon-feed-farm-animals-and-us/. Accessed November 11, 2016.

Lee, M. M., Teuscher, J., Miyasaka, T., Murakami, T. N., & Snaith, H. J. (2012). Efficient hybrid solar cells based on meso-superstructured organometal halide perovskites. *Science, 338*(6107), 643–647.

Lusinyan, L., Mahedy, T., Muir, D., Papageorgiou, R., Pescatori, A., & Valencia, F. (2016). North American integration and the energy market (Chapter 8). In R. Cardarelli & L. Lusinyan (Eds,), *Power play, energy and manufacturing in North America* (pp. 131–147). International Monetary Fund.

Lux. (2015a). The solar market is positioned for disruptive, sustainable growth to 70 GWp in 2020, driven by better technology and business models. LuxPopuli Lux Research Analyst team newsletter of 8/16/15.

LuxWB. (2015b). The future of power is distributed: How to profit from the coming revolution in Energy generation webinar, Meeting number: 666-919-346. http://www.luxresearchinc.com. Accessed November 24, 2015.

Maize, K. (2018). What is the future of independent power? *Power Magazine, 162*(1), 35–38.

Maloney, P. (2018). EIA: Low gas prices set to drive decline in coal generation. D UtilityDive. https://www.utilitydive.com/news/eia-low-gas-prices-set-to-drive-decline-in-coal-generation/514556/. Accessed January 11, 2018.

Mathews, J. A. (2016). China's continuing renewable energy revolution—latest trends in electric power generation. *The Asia-Pacific Journal, Japan Focus, 14*(17–6).

Max, M. D., & Johnson, A. H. (2014). Hydrate petroleum system approach to hydrate natural gas exploration. *Petroleum Geoscience, 20*(2), 187–199. Geological Society of London (Published in Online First 21 March 2014). https://doi.org/10.1144/petgeo2012-049.

McDonnell, T. (2015). This map shows where the next clean energy told min is. *Mother Jones*, Wednesday, November 11. http://www.motherjones.com/print/289046. Accessed January 2, 2016.

Mellin, C. (2018). *U.S. carbon dioxide emissions down big-time since 2005*. Natural Gas Now. https://www.printfriendly.com/print?customCSSURL=&disableClickToDel=0&disableEmail=0&disablePDF=0&disablePrint=0&headerImageUrl=&headerTagline=&imageDisplayStyle=left&imagesSize=full-size&source=cs&url_s=uGGC_%7E_PdN_%7E_PcS_%7E_PcSAnGHEnytnFABJmBEt_%7E_PcSH-F-pnEoBA-qvBKvqr-rzvFFvBAF-ovt-Gvzr-FvApr-caaf_%7E_PcS. Accessed February 10, 2018.

Milford, L., Mullendore, S., Olinsky-Paul, T., & Sanders, R. (2018). *Jump-start. How activists and foundations can champion battery storage to recharge the clean energy transition*. The Clean Energy Group. https://www.cleanegroup.org/wp-content/uploads/Jump-Start-Energy-Storage.pdf. Accessed April 17, 2018.

Miller, S. (2017). *This is how big oil will die*. NewCoShift (Medium). https://medium.com/@sethmiller_59231/this-is-how-big-oil-will-die-38b843bd4fe0. Accessed June 4, 2017.

MISO. (2015). *Book 4 Regional Energy Information September 14, 2015, 2nd draft. Chapter 9 Regional Energy Information* (23 pp). https://www.misoenergy.org/Library/Pages/Results.aspx?q=regional%20energy%20distribution%20more%20efficient%20than%20local. Accessed July 12, 2016.

MIT. (2011). The future of natural gas. In *An interdisciplinary MIT study* (178 pp). Massachusetts Institute of Technology Report. <https://mitei.mit.edu/system/files/NaturalGas_Report.pdf. Accessed April 1, 2016.

MIT. (2013). Report from growing concerns, possible solutions: The interdependency of natural gas and electrical systems. Summary of An MIT Energy Initiative Symposium held on April 16, 2013. Massachusetts Institute of Technology, 240 pp. Accessed April 1, 2016.

Montgomery, J. (2013). Solar cell efficiency round-up: thin film closing the gap with silicon. Renewable Energy world.com (M17 May 2013). http://www.renewableenergyworld.com/articles/2013/01/solar-cell-efficiency-round-up-thin-film-is-closing-the-gap-with-silicon.html. Accessed May 17, 2015.

Musarra, S. P. (2015). Gas demand driving mega projects off Asia/Pacific. *Offshore* 3. 05/06/2015. http://www.offshore-mag.com/articles/print/volume-75/issue-5/international-field-development/gas-demand-driving-mega-projects-off-asia-pacific.html. Accessed May 25, 2015.

Muyskens, J., Keating, D., & Granados, S. (2015). *How the United States generates its electricity*. The Washington Post Sunday 2 August 2015, A6.

NAS. (2016). *Analytic research foundations for the next-generation electric grid* (162 pp). The National Academy Press. PDF at http://www.nap.edu/21919. Accessed April 19, 2016.

NAS. (2017). *Enhancing the resilience of the nation's electricity system* (282 pp). National Academies of Sciences, Engineering, and Medicine. Washington, DC: The National Academies Press. https://doi.org/10.17226/24836.

Negin, R. (2017). *A dwindling role for coal* (pp. 15–17). Catalyst (17, Fall 2017). Union of Concerned Scientists.

NI. (2015). *Market forces can't fix methane-gas emissions*. The National Interest. http://nationalinterest.org/print/feature/market-forces-cant-fix-methane-gas-emissions-13370. Accessed April 23, 2016.

Nicholls, T., de Guitaut, E., Brower, D., Gorst, I., Lewis, I., & Ruddy, M. (2014). *How the Energy Industry Works, The ultimate guide to energy 2014* (116 pp). www.energy-future.com. Accessed April 21, 2016.

OECD/EIA. (2013). *Redrawing the energy-climate map* (134 pp). World Energy Outlook Special Report. International Energy Agency.

Park, T., Na, J., Kim, B., Kim, Y., Shin, J., & Kim, E. (2015). Photothermally activated pyroelectric polymer films for harvesting of solar heat with a hybrid energy cell structure. *ACSNano*. https://doi.org/10.1021/acsnano.5b04042.

Parker, J. F., Chervin, C. N., Pala, I. R., Machler, M., Burz, M. F., Long, J. W., et al. (2017). Rechargeable nickel-3D zinc batteries: An energy-dense, safer alternative to lithium-ion. *Science, 356,* 415–418.

Patel, S. (2017a). Battery storage goes mainstream. *Power (Magazine), 161*(5), 25–29.

Patel, S. (2017b). Two SCE gas-battery hybrid projects revolutionize Peaker perforance. *Power (Magazine), 161*(9), 34–35.

PennEnergy 6-1. (2015). Oil and gas majors call for carbon pricing. 6/5/15. http://www.pennenergy.com/articles/pennenergy/2015/06/oil-and-gas-majors-call-for-carbon-pricing.html?cmpid=EnlWeeklyPetroJune52015&eid=288179069&bid=1089254.

Phys. (2015). *Beijing shuts last coal power plant in switch to natural gas* (3 pp). Phys.Org. https://phys.org/news/2017-03-beijing-coal-power-natural-gas.html. Accessed March 19, 2017.

Price, H. (2015). *CNG viability highlighted at NGVOK conference*. Oilpro. http://www.pennenergy.com/articles/pennenergy/2015/08/cng-viability-highlighted-at-ngvok-conference.html?cmpid=EnlDailyPetroAugust272015&eid=288179069&bid=1162381. Accessed April 23, 2016.

Putnam, C. S. (2015). *The mechanical battery*. Damn Interesting. http://www.damninteresting.com/the-mechanical-battery/. Accessed April 23, 2016.

Reddy, M. P. M., & Affholder, M. (2002). *Descriptive physical oceanography: State of the Art* (440 pp). Taylor and Francis. ISBN 90-5410-706-5.

Reed, J. (2015). *European major tone shift: From O&G to G&O*. Oilpro 6/2/15. http://oilpro.com/post/13886/og-go-you-need-to-know-europe-top-majors-discussing-week-paris?utm_source=DailyNewsletter&utm_medium=email&utm_campaign=newsletter&utm_term=2015-06-03&utm_content=Article_10_txt. Accessed June 4, 2015.

Ritter, K. (2015). *Oil and gas turn their backs on coal as climate deal nears*. PennEnergy 6/4/15. http://www.pennenergy.com/articles/pennenergy/2015/06/oil-and-gas-turn-their-backs-on-coal-as-climate-deal-nears.html?cmpid=EnlDailyPetroJune42015&eid=288179069&bid = 1087838. Accessed April 23, 2016.

Ritter, K. (2016). A new life for coal ash. *American Chemical Society C&EN* February 16, 10–14.

Romm, J. (2017). Clean energy will be a $50-trillion industry, but Trump policies mean the U.S. won't benefit. *Thinkprogress*. https://thinkprogress.org/clean-energy-charts-of-the-year-2016-93d781b55adb#.4ftblk5o9. Accessed January 4, 2017.

Rybarik, B. (2013). UP energy summit. MISO hot topics, 15pp. https://www.iaenvironment.org/webres/File/News%20%26%20Resources/Comments/NOI-2014-0001%20IEC%20ELPC%20IPL%20MEC%20Joint%20Comments%20Nov%202015.pdf. Accessed 12 September 2018.

Rüttinger, L., Smith, D., Stang, G., Tänzler, D., Janani Vivekananda, J., Brown, O., et al. (Eds.). (2015). *A new climate for peace. taking action on climate and fragility risks*. An independent report commissioned by the G7 members. adelphi, International Alert, Woodrow Wilson International Center for Scholars, European Union Institute for Security Studies, 2015, 150 pp.

Sanchez, N., II, & Mays, D. C. (2015). Effect of methane leakage on the greenhouse gas footprint of electricity generation. *Climatic Change, 133,* 169–178. https://doi.org/10.1007/s10584-015-1471-6.

Sandalow, D., Wu, J., Yang, Q., Hove, A., & Lin, J. (2014). *Meeting China's shale gas goals* (83 pp). September 2014 Working draft for public release (Attachment D). Columbia University SIPA, Center on Global Energy Policy.

Sanders, P. (2016). *Chile energy auction gives Bachelet a success to boast about*. Bloomberg News. https://www.bloomberg.com/news/articles/2016-08-18/chile-s-energy-industry-a-spot-of-light-in-a-darkening-economy. Accessed March 10, 2017.

SC. (2015). *Accelerating the US coal phase out: Leading by example in Paris and beyond* (4 pp). Sierra Club, November 3, 2015. https://content.sierraclub.org/creative-archive/sites/content.sierraclub.org.creative-archive/files/pdfs/1172-BeyondCoal-Paris15_Fact_05_low.pdf. Accessed November 9, 2015.

Scheiner, K. (2018). The future of nuclear power? Think small. *The Los Angeles times*. http://www.latimes.com/nation/la-na-oregon-nuclear-20180201-story.html?utm_source=Sailthru&utm_medium=email&utm_campaign=Issue:%202018-02-02%20Utility%20Dive%20Newsletter%20%5Bissue:13882%5D&utm_term=Utility%20Dive. Accessed February 2, 2018.

Searles, G. (2015). *Gas use in power stations*. Oilpro. http://oilpro.com/post/16474/natural-gas-overtakes-coal-first-time-ever-biggest-us-electricity?utm_source=DailyNewsletter&utm_medium=email&utm_campaign=newsletter&utm_term=2015-07-13&utm_content=Article_3_txt. Accessed July 13, 2015.

Seimens. (2016). Siemens gas turbine generator heads to SARB offshore Abu Dhabi. *Offshore* 5/11/16. http://www.offshore-mag.com/articles/2016/05/siemens-gas-turbine-generator-heads-to-sarb-offshore-abu-dhabi.html. Accessed May 12, 2016.

Sessolo, M., & Bolink, H. J. (2015). Perovskite solar cells join the major league. *Science, 350* (6262), 917.

Shell Global. (2015). *Is the Chinese Dragon green? The energy future, The colours of energy*. http://www.shell.com/energy-and-innovation/the-energy-future/colours/is-the-chinese-dragon-green.html. Accessed April 23, 2016.

Shepstone, T. (2017a). *If it's not dispatchable, it's not helpful, but hurtful*. Natural Gas Now. http://naturalgasnow.org/not-dispatchable-not-helpful-hurtful/. Accessed July 16, 2017.

Shepstone, T. (2017b). *Northeast air quality rises to new heighs with natural gas use*. http://naturalgasnow.org/northeast-air-quality-rises-new-highs-natural-gas-use/. Accessed August 13, 2017.

Shiao, M. J. (2012). *Thin film manufacturing in the sub-dollar-per-watt market: Part II*. Greentech Solar. http://www.greentechmedia.com/articles/read/thin-film-manufacturing-in-a-sub-dollar-the-watt-market-ii. Accessed April 23, 2016.

Silverstein, K. (2018). *Despite efforts to trip it up, new energy economy remains on sound footing*. https://www.forbes.com/sites/kensilverstein/2018/01/01/despite-efforts-to-trip-it-up-new-energy-economy-remains-on-sound-footing/#61766cf45a52 Accessed March 12, 2018.

Simonsen, B. (2017). *Nel and high water: the ultra-efficient electrolyser*. InnovOil July 29. http://www.innovoil.co.uk/single-post/2017/06/28/Nel-and-high-water-the-ultra-efficient-electrolyser. Accessed July 10, 2017.

Sivaram, V., Stranks, S. D., & Snaith, H. J. (2015). Outshining silicon. *Scientific American*, 55–59.

Slingsby, J. (2015). *Coal-to-gas switching in 2016*. Oilpro (Intecsea.com) 12/4/15. http://oilpro.com/post/20618/coal-to-gas-switching-2016?utm_source=DailyNewsletter&utm_medium=email&utm_campaign=newsletter&utm_term=2015-12-04&utm_content=Article_13_txt. Accessed April 23, 2016.

Smyth, J. (2018). Moody's report shows Tri-State's coal plants are more expensive than new renewable energy. *Clean Cooperative*, 2. http://www.cleancooperative.com/news/moodys-report-shows-tri-states-coal-plants-are-more-expensive-than-new-renewable-energy. Accessed April 27, 2018.

Solarreserve. (2016). *Technology*. http://www.solarreserve.com/en/technology.

Stanway, D. (2017). Enter the Nimble Dragon: China looks to small reactors for nuclear edge. *Reuters/Stringer*. https://www.reuters.com/article/us-china-nuclear-analysis-idUSKBN1950HT. Accessed July 2, 2017.

Tata. (2015). Tata airpod compressed-air car to launch in Hawaii this year: Report. http://www.greencarreports.com/news/1096772_tata-airpod-compressed-air-car-to-launch-in-hawaii-this-year-report. Accessed May 13, 2015.

Thakur, N. K., & Rajput, S. (2011). *Exploration of gas hydrates—Geophysical techniques* (281 pp). New York: Springer. ISBN 978-3-642-14233-8, https://doi.org/10.1007/978-3-642-14234-5.

Trabish, H. K. (2018a). *Shuttered coal plant fixer-uppers for sale all over the U.S.* D UtilityDive. https://www.utilitydive.com/news/shuttered-coal-plant-fixer-uppers-for-sale-all-over-the-us/514213/. Accessed January 16, 2018.

Trabish, H. K. (2018b). *Utilities take note: Hydrid renewables projects are coming*. https://www.utilitydive.com/news/utilities-take-note-hybrid-renewables-projects-are-coming/520319/. Accessed April 3, 2018.

TREIA. (2015). *Texas Renewable Energy Industries Alliance*. http://www.treia.org.

Turchin, P. (2015). Cultivating human culture; did food and fuel help shape our value systems (Book Review). *Science, 348,* 508.

UCS. (2014). *How geothermal energy works* (8 pp). Union of concerned scientists. http://www.ucsusa.org/clean_energy/our-energy-choices/renewable-energy/how-geothermal-energy-works.html#.ViALz9am3cZ.

UNDP. (2014a). Human Development Report 2014. *Sustaining human progress: Reducing vulnerabilities and building resilience* (239 pp). United Nations Development Programme. ISBN 978-92-1-126368-8, eISBN 978-92-1-056659-9, http://hdr.undp.org/sites/default/files/hdr14-report-en-1.pdf. Accessed April 23, 2016.

UNDP. (2014b). *Table 1: Human development index and its components* (7 pp). United Nations Development Program. http://hdr.undp.org/en/content/table-1-human-development-index-and-its-components.

Walton, R. (2018a). *New York clean energy proposal seeks 1.5 GW of storage, coal phaseout*. D Uutility Dive. https://www.utilitydive.com/news/new-york-clean-energy-proposal-seeks-15-gw-of-storage-coal-phaseout/513955/. Accessed January 4, 2018.

Walton, R. (2018b). *Time not on their side: Utilities ill-prepared for EV demand, SEPA finds* (9 pp). Utility Dive. https://www.utilitydive.com/news/time-not-on-their-side-utilities-ill-prepared-for-ev-demand-sepa-finds/519530/. Accessed March 22, 2018.

Warrick, J. (2015). U.S. exports emissions—As coal. *The Washington Post Friday*, October 16, 2015. A1, A10–A11.

Warrick, J. (2016). Wind, solar power soaring in spite of bargain prices for fossil fuels. *Washington Post 1* January 2016. https://www.washingtonpost.com/national/health-science/wind-solar-power-soar-in-spite-of-bargain-prices-for-fossil-fuels/2015/12/30/754758b8-af19-11e5-9ab0-884d1cc4b33e_story.html. Accessed January 2, 2016.

Wasserman, H. (2017). Goodbye nuclear power. Construction of two of four remaining planned U.S. *Plants Just Canceled.* http://progressive.org/dispatches/goodbye-nuclear-power-construction-of-two-of-four-remaining. Accessed August 1, 2017.

Watson, B. (2016). Have we reached the tipping point for investing in renewable energy? *The Guardian*, 3. http://www.theguardian.com/sustainable-business/2016/feb/13/renewable-energy-investment-fossil-fuel-divestment-investor-summit-climate-change. Accessed February 13, 2016.

WEF. (2016). *4th industrial revolution will be driven by renewable energy*. Davos, Switzerland: World Economic Forum (WEF). http://news.xinhuanet.com/english/photo/2016-01/22/c_135033673.htm. Accessed January 22, 2016.

Willis, J. (2017a). *World's first zero-emissions natural gas power plant* (5 pp). Natural Gas Now. http://naturalgasnow.org/worlds-first-zero-emissions-natural-gas-power-plant/. Accessed December 10, 2017.

Willis, J. (2017b). *Microgrid power plants powered by natural gas spreading*. http://naturalgasnow.org/microgrid-power-plants-powered-natural-gas-spreading/. Accessed August 13, 2017.

Wilson, R. (2013). The future of energy: Why power density matters. In *The Energy Collective*. http://theenergycollective.com/robertwilson190/257481/why-power-density-matters. Accessed April 23, 2016.

Wilson, C., & Biette, D. (2014). *Is geography destiny? A primer on North American relations* (68 pp). The Wilson Center. ISBN: 978-1-938027-35-2.

Wiser, R, Mills, A., Seel, J., Levin, T., & Botterud, A. (2017). *Impacts of variable renewable energy on bulk power stem assets, pricing, and costs* (105 pp). Argonn and Berkeley National Laboratories, U.S. Department of Energy.

Xinhua 1/15/16. (2016). *Energy chief promises less polluting power plants*. Xinhua English edition. http://news.xinhuanet.com/english/2016-01/15/c_135014080.htm. Accessed April 23, 2016.

Yang, W. S., Noh, J.-H., Jeon, N. J., Kim, Y. C., Ryu, S., Seo, J., et al. (2015). High-performance photovoltaic perovskite layers fabricated through intramolecular exchange. *Science, 348*, 1234–1237.

Ye, Q., & Lu, J. (2018). *China's coal consumption has peaked*. Brookings. https://www.brookings.edu/2018/01/22/chinas-coal-consumption-has-peaked/?utm_campaign=John%20L.%20Thornton%20China%20Center&utm_source=hs_email&utm_medium=email&utm_content=60346551. Accessed January 31, 2018.

Zhang, Y., Jaegle, L., & Thompson, J. A. (2014). Natural biogeochemical cycle of mercury in a global three-dimensional ocean tracer model. *Global Biogeochemical Cycles, 28*, 553–570. https://doi.org/10.1002/2014GB004814.

Zhou, H., Zhou, C., Lynam, M. M., Dvonch, T., Barres, J. A., Hopke, P. K., et al. (2017). Atmospheric mercury temporal trends in the Northeastern United states from 1992 to 2014: Are measured concentrations responding to decreasing regional emissions? *Environmental Science & Technology Letters*, 7. https://doi.org/10.1021/acs.estlett.6b00452. Accessed March 5, 2017.

Chapter 2
Economic Characteristics of Deepwater Natural Gas Hydrate

Abstract Natural gas hydrate (NGH) is unique among gas resources. In its natural state NGH is a pure, solid crystalline material formed by digenetic crystallization in permafrost regions and in marine sediments where temperature, pressure, and gas flux are suitable. NGH compresses methane and minor other gases by about 164 times (above 1 atm @ STP) within the crystal structure. Greater than 95% of the NGH resource is found sediments below seafloors that are deeper than 500 m in the open ocean. First order exploration objectives are concentrations of NGH occur generally in deepwater and ultra-deepwater host sandy turbidites within and near the base of the gas hydrate stability zone (GHSZ), the shallow sub-seafloor interval where the pressure and temperature conditions allow NGH concentrations to potentially form, and that have economic potential for production of its natural gas. Other types of NGH concentrations occur but are amenable to production using conventional methodology. NGH is stable within its sedimentary hosts and will only convert to its constituent gas and water phases relatively quickly if reservoir pressure-temperature conditions are artificially disturbed. Environmental change can induce conditions that may cause more NGH to form or dissociate, but relatively much more slowly. Induced dissociation has the potential to produce gas rapidly enough to support long-term natural gas production. NGH is potentially the largest of the three types of unconventional gas resources, with estimates of gas-in-place within producible reservoirs in excess of 40,000 Tcf.

2.1 Natural Gas Hydrate

We use the term "Natural Gas Hydrate" or 'NGH' as a general term to denote a clathrate or hydrate of methane-rich natural gas of any composition or crystal structure. The term NGH encompasses compound hydrate containing two or more natural gases and is synonymous with 'gas hydrate' and 'methane hydrate.' NGH is the last of the major naturally occurring fossil fuel resources from which natural gas may be produced, and it may prove to be the largest. Gases in NGH are in general the same as those found in natural gas used by consumers for cooking, heating, and

M. D. Max and A. H. Johnson, *Exploration and Production of Oceanic Natural Gas Hydrate*, https://doi.org/10.1007/978-3-030-00401-9_2

increasingly as a transport fuel; and industrially as a fuel for power plants and as an industrial feedstock for the petrochemical industry, although methane is by far the commonest natural gas present. NGH is a naturally-occurring storage medium that concentrates natural gas in host sediments in relatively near-seafloor marine sediments in deep and ultra-deep seas. We focus here on those aspects of the NGH system that we regard as key to exploration for NGH concentrations, valuation, and production of natural gas, in order to deliver NGH to a position as a natural gas resource. Industry tends to consider oil and gas deposits as "accumulations" rather than "concentrations". We use the term 'concentrations' to distinguish accumulations of a high enough grade to constitute a potential recoverable energy resource. A 'reservoir' is not defined until a full physical, economic, and financial analysis is carried out that fully defines its margins with cutoff values and gas-in place.

NGH is essentially a geologically 'modern' deposit in which the solid crystalline NGH concentrations are in a dynamic equilibrium with particular pressure and temperature conditions and with pore fluids in which the dissolved concentration of natural gas is high enough to drive crystallization or at least high enough to prevent dissolution of the NGH. In essence, there are no geologically ancient NGH deposits as there are in some older conventional gas deposits, no older geological history that bears on the current formation of NGH concentrations other than conditions that govern the recent and present influx of gas into a GHSZ. The thermodynamic conditions are naturally altering with climate and sea level changes, but only relatively slowly (Chap. 10).

Although natural gas is sequestered and can be produced from NGH deposits, they are not 'fossil fuels' in the true sense of conventional hydrocarbons that are the end product of a usually long series of chemical transformations usually involving significant heat. They are in many respects renewable in that most currently existing concentrations are still in the process of extracting some natural gas from the great biochemical life system of the earth. Prospects for the formation of large concentrations of NGH are good because the root of gas flux and migration that could naturally rise toward the seafloor can exist over the whole history of a thick, deepwater sediment pile.

The critical moment for conventional hydrocarbons is often a relatively short period of time in which fluid hydrocarbon generation, migration paths, and geological trap coincide in time and space to produce hydrocarbon concentrations. In contrast, gas generation and upward migration of natural gas from both thermogenic and biogenic sources can take place as part of all conventional hydrocarbon critical moments in a deepwater basin. In addition, gas produced by long-standing mechanisms such as repeating thermal and structural events in a sedimentary wedge formed by plate collisional processes can also migrate into the GHSZ.

Methane is a simple molecule with one carbon and four hydrogen atoms. It is stable chemically and does not have molecular substitution or polymorphs and is also stable in high temperatures approaching 1000 °C, in the absence of oxygen that would cause combustion (Max 2003). Methane is the dominant gas in NGH, especially on rifted, passive continental margins with shallow or only moderately deeply-rooted petroleum systems. Other gases may compose no more than traces, but where higher density hydrocarbons are present, other types of compound NGH can form. Because almost all of the NGH having commercial potential will be methane dominated, we do not discuss compound NGH in any detail, but note that the presence of higher density natural gases may substantially increase the commercial value of the gas.

NGH is a solid, crystalline compound composed formed by a chemical reaction between water molecules that form interlocking 'bucky-ball'-type cage structures surrounding voids that are predominantly occupied by hydrocarbon gas molecules in a body centered cubic crystal system (Sloan and Koh 2008). NGH of interest to us is dominantly methane or methane plus higher density hydrocarbon gases, but almost all gases can from hydrate. Non-hydrocarbon gas hydrates are not discussed in this book. Our discussion mainly concerns the commercial potential of producing natural gas from NGH. NGH sequesters hydrocarbon natural gases wherein hydrate formation extracts dissolved natural gas (mainly methane) from pore water. Methane hydrate is the most common clathrate found in nature. More details about all types of NGHs can be found in Sloan and Koh (2008), Max et al. (2006) and Collett et al. (2009), and other books and many scientific papers on the topic. Abstracts and bibliography of NGH papers published by Springer (Wood 2015) provide additional references.

One of the difficulties in understanding NGH is that unlike conventional hydrocarbons that are liquid or gas fluids, NGH is a crystalline solid. Its petrogenesis is actually closer to that of hydrothermal ore-forming processes, particularly those that involve circulating seawater and marine sediment pore water (Robb 2005, Part 2 Hydrothermal Processes). Seawater can be a hydrothermal fluid and the temperature of mineralizing solutions can be low, ranging to just above freezing. Hydrothermal or mineralizing fluids carry dissolved or free reactants to a location where spontaneous chemical reaction can take place in which two or more reactants for a third substance that is stable within a pressure-temperature field. Is NGH an ore of natural gas? In a strict sense, yes, as it forms in the same way as many other solid crystalline materials from which valuable material concentrated by the formation process can be recovered. In addition, its disposition in concentrations is variable just like most mineral deposits (Chap. 5), in contrast to conventional hydrocarbons that occupy porosity as fluid (gas or liquid) that tends to be disposed hydraulically and uniformly with respect to pore filling within the boundaries of its concentration.

When NGH forms, the natural gas, usually methane, is effectively compressed into a smaller volume. A single cubic foot of NGH comprised only of methane gas will yield up to approximately 164 cubic feet of gas (at STP) with a higher heating value (HHV)[1] energy density of about 180,000 Btu/ft^3. Even small amounts of higher density hydrocarbon gases can substantially raise the energy density of gas produced from NGH and render the NGH more valuable.

[1]The **HHV** of a fuel is defined as the amount of heat released at 25 °C and the products returned to a temperature of 25 °C, which takes into account the latent heat of water vaporization.

An important aspect of the NGH physical chemical system is that the reactions of growth and dissolution are highly reversible. Bonding of the water molecule cage structure and gas molecules in the cages in NGH is accomplished by the weak electrical van der Waals bonding. This renders the system highly responsive to changes in temperature and pressure so that it will either spontaneously grow and dissociate as slips in and out of stability (van der Walls 1956). Formation of NGH can be a very slow process, especially in low gas flux in which dissolved concentration results in only a weak driving force of crystallization. But artificially controlled dissociation offers the prospect of quickly producing the stored natural gas. No natural dissociation other than massive heat from direct igneous activity can dramatically drive rapid dissociation.

Biogenic Methane Natural production of biogenic methane is gigantic. NGH has probably been a feature of the earth's ocean biosphere since the development of methanogenic bacteria. In todays' oceans, the descendants of one of the oldest methanogenic prokaryotes, the 'archaea', remain as major contributors to global methane flux. Just these prokaryotes are estimated to produce one billion tons of methane per year, with an additional equal amount estimated to be oxidized by archaeal methanotrophs (Evans et al. 2015). The greater proportion of methane in NGH appears to be of biogenic origin (Paull et al. 1994; Kvenvolden 1995; Kvenvolden and Lorenson 2001), particularly on thickly sedimented passive rifted continental margins, which indicates the widespread presence of bacterial methanogens and prokaryotes. It is also possible that the heavier hydrocarbons are metabolized to produce methane by microbial activity as oil migrates upward, or even within a reservoir.

Thermogenic Methane In regions with strong and deeply rooted petroleum systems, such as on convergent or active plate margins where sediments are undergoing subduction tectonics, other hydrocarbon gases (ethane, propane, butane) produced thermogenically may rarely comprise from traces to up to 10–15% of the gas mixture in the NGH. H_2S and CO_2, amongst other gases, can also be constituents of NGH. Thermogenic gases are more common on active margins, particularly above subduction zones in continent margin sedimentary prisms (Max et al. 2006). Natural gas is low-density and usually includes only small amounts of dissolved substances material.

Conventional hydrocarbon source beds usually require deep burial and subjected to thermal maturation in order to generate oil and gas, which then migrate into sandy reservoirs. If there is a strong upward flux of natural gas from the source and migration pathways into and through a suitable sandy host bed within the GHSZ (Chap. 3), NGH concentrations are likely to form.

Abiotic methane produced by non-biological chemical processes is the rarest methane source that has been documented in NGH hosted in thin abyssal sediments (Johnson et al. 2015; Rajan et al. 2012). It is formed through serpentinization of basaltic oceanic crust, usually in hydrothermal systems associated with spreading centers, or subduction of oceanic crust beneath active margin sedimentary prisms. Hydrogen is produced by the interaction of basic igneous minerals with water, which then can chemically combine with carbon to produce methane. It is anticipated that more of these abyssal NGH deposits may be identified, but they are very unlikely to become NGH exploration targets. Methane produced by serpentinization at constructive plate margins in very deep water rarely has overlying sediments in which to migrate. An exception is along the Svalbard margin of the Fram Strait related to the northernmost North Atlantic slow-spreading constructive plate margin. Because this abiotic methane may migrate into host sediments and form concentrations, Johnson et al. (2016) invoke an abiotic methane window at serpentinization temperatures ranging from 200 to 350 °C. This new methane source can be accommodated within the NGH petroleum system of Max and Johnson (2014) because the source of gas migrating into a NGH concentration is not a critical factor.

Hachikubo et al. (2015) have shown that gas compositions associated with thermogenic methane in NGH can be more complex than measuring the gases from the NGH itself. They measured the molecular and stable isotope compositions of volatiles both bound in NGH and from headspace gases from sediments that had not formed NGH. Two sites were sampled; one consisted primarily of thermogenic methane, the other contained both thermogenic methane and microbial methane and was associated with a shallow sulfate–methane interface (SMI) and high microbial methane production beneath the SMI depth. At both sites, relatively high concentrations of propane and neopentane (2,2-dimethylpropane) were found in headspace gases but it is not known why compound hydrate (Max et al. 2006) had not formed, even though it is the more stable thermodynamic hydrate at the pressure–temperature conditions of the Class 1 NGH.

H_2S, which may introduce a hazard on dissociation, and CO_2 are the most common minor gases found in multi-gas NGH, commonly referred to as 'compound hydrate'. Because not all of the cage structures need to be filled for the mineralization to be stable, the amount of gas molecules in any particular volume of NGH can vary. But so long as the NGH is dominantly methane, the compression factor will only be affected trivially.

NGH is often described as being "ice-like", with the natural gas somehow absorbed in an ice-like structure, even though gas is almost insoluble in ice. This is a misleading analog since the formation and thawing of ice is a liquid-solid phase change. In contrast, the nucleation (Kashchiev and Firoozabadi 2002a) and growth of NGH involves a chemical reaction between two chemical components (gas and water) that combine to form a third compound. NGH also has a much higher heat of transformation (about 57 kJ/mol) than ice, which is about 6 kJ/mol (Pellenbarg and Max 2003, Table 2). When NGH or any similar NGH forms, a calculated 57 kJ/mol of heat has to be absorbed by the environment in which it forms (Sloan and Koh 2008). This is a controlling factor, along with elevated pressure, of why NGH forms in association with permafrost and in the deep, cold oceans. Dissociation, in contrast, is an endothermic reaction to which heat has to be supplied from its surroundings. When dissociation is forced by depressurization, heat must be applied to the system to prevent cooling that would cause temperature to fall and force dissociation to cease (Chap. 11).

Like many other chemical reactions, crystallization is governed by diffusional processes and the manner in which it will grow in the porosity of a sedimentary reservoir is significantly different from the disposition of a material undergoing phase change. A mischaracterization of NGH as a type of ice may lead to overlooking a number of important exploration (Chap. 3) and production (Chap. 11) characteristics that may provide opportunities for exploitation.

The term "methane hydrate" also can be misleading when discussing the broader field of naturally occurring hydrates, especially when compound hydrate composed of two or more hydrate-forming gases are present. The higher density hydrocarbon gases can add substantial commercial value as natural gas liquids and condensates, in the same way that they do in conventional gas deposits. Nonetheless, methane hydrate is used as the international standard for chemical reactions, heats of growth, dissociation, and dissolution (Max and Johnson 2011a) because it allows for unequivocal comparison.

2.1.1 Gas Transport Within a Marine Sediment Pile

One of the principle conditions for NGH formation according to the hydrate petroleum system analysis (Max and Johnson 2014), is a strong gas flux that extends back into the recent past and older. Two elements are required. First, considerable gas has to be created either by biogenic or thermogenic means. Second, the gas has to be able to migrate. The degree to which pervasive gas movement take place from depths up into the GHSZ is referred to as the gas flux. Gas is normally produced in relatively small amounts within very large volumes. When it begins to flow, it generally can be described as diffuse flow of small gas bubbles or as diffusion within pore water. Where the gas begins to concentrate and migrate in large volumes over relatively small courses, such as within faults, it is referred to as focused flow. Focused flow into a permeable horizon within the GHSZ has the potential to form large concentrations of NGH (Max et al. 2006). Faults capable of delivering high gas flux (Max and Dillon 1998 Fig. 2.7 this Chapter; Plaza-Faverola et al. 2015) may be a more important mechanism for bringing gas into or near the base of a GHSZ than long migration through permeable sedimentary strata. Gas can also be derived from subjacent gas concentrations in structural traps by leakage or at locations down flank from structural crests (Løseth et al. 2009).

Some models for NGH concentrations postulate that where biogenic methane is present in suitable host sands, the methane may be locally derived from adjacent muddy sediments. While we do not discount the possibility that some of the biogenic methane may be locally derived, this cannot be the case where thermogenic methane and higher density hydrocarbon gases are present in significant quantities in concentrations in host sediments that have never been heated to thermogenic temperatures (Max and Lowrie 1993). Also, when the amount of NGH in a concentration is compared to the amount of methane that may have escaped from adjacent source beds during compaction (and not all impermeable shales bounding suitable sand hosts have the quality of source beds), it seems to us that there is too much methane in the concentration for local transport mechanisms alone.

The general layout of prospective Class 1 & 2 deposits (Sect. 2.4) indicates gas sourced from below and trapped during ascent in a permeable sand bed. Locally derived methane that migrated laterally into a host bed could be expected to be concentrated in beds in the upper part of the GHSZ, where conditions of stability are greatest. The geometry of known NGH deposits, however, favor a model of upward transport of mineralizing solutions, which form NGH soon after rising into a GHSZ. Gas dissolved in pore water will increase in relative saturation as it rises from a deep source, so that even if gas bubbles are not present at depth, gas can form as the pore fluids naturally depressurize and supersaturation is achieved.

On the other hand, if free gas forms at depth, it can work its way up in permeability of dewatering sediments as they compact and expel water upward. Much more rapid upward transport of gas, however, can take place in chimneys that can form in the secondary porosity of joint systems and faults. Because the gas can be overpressured, it can assist with dilation of faults and enhance the rate of upward propagation of the chimney. Where gas/water venting is persistent and high volumes of gas and pore water are transported more rapidly than temperature can be depth equilibrated, the temperatures in the vicinity of the chimney/vent system will rise and the base of the GHSZ will be warped upward even to the point that the GHSZ in the immediate vicinity of the vent can thin to essentially nothing (Max et al. 2006, Figs. 4.16 and 4.17). Local upwarping of the base GHSZ around a chimney/vent is a strong proof of warm fluids/gas brought from subjacent sources to the seafloor.

2.1.2 NGH Stability

Temperature and pressure govern NGH stability growth and dissociation so long as the concentration of dissolved hydrate-forming gas in the enclosing pore water is high enough to drive NGH crystallization. The stability field of methane hydrate (Englezos and Bishnoi 1988; Sloan 1990) is usually taken as the gold standard or reference point for the stability of NGH (Fig. 2.1). When a pore-water solution containing sufficient dissolved gas is introduced to conditions of NGH stability, NGH will spontaneously nucleate and grow. Personal observation of laboratory experiments using natural seawater (a close analog for pore water in marine sediments) has shown consistently good nucleation and growth both in bulk and surface effect crystallization (MDM, personal observations).

NGH can form spontaneously under the right combinations of pressure and temperature conditions when there is sufficient natural gas flux. The temperature of the water column in the open ocean decreases with depth, which renders NGH stable in the sediments from water depths of about 500 m and greater. In the Arctic Ocean, water temperature at depth is at or below 0 °C, and the minimum depth of NGH stability is at about 250 m water depth. Figure 2.2 shows the methane hydrate

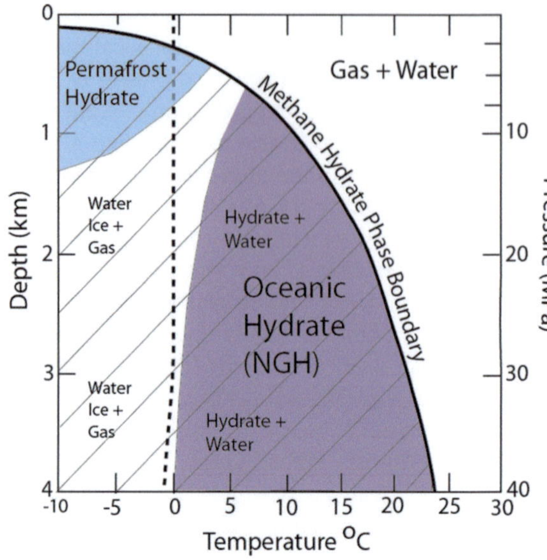

Fig. 2.1 Dashed near vertical line is approximate freezing point of ice in seawater. Experimental determination of methane hydrate stability (to left and below lines). Downward pressure increases for oceanographic-geological visualization, first used by Kvenvolden (1988). Redrawn from Englezos and Bishnoi (1988)

stability field as a combination of colder and higher-pressure conditions found in permafrost regions on land and in marine sediments in the oceans. We focus on oceanic NGH because Permafrost NGH occurs in geological traps rather than the thermodynamic traps common to oceanic NGH, oceanic NGH is estimated to contain about 95% of the entire resource, and it thus is the most economic target. In addition, that major gas importing countries such as South Korea, China, India, and many other energy resource-poor countries with continental margins within which NGH concentrations may occur, may have the NGH option available to them. But oceanic NGH is the primary target mainly because most of the new technology development that has the potential to dramatically lower the cost of exploration and production will not apply to permafrost hydrate.

The persistence of NGH, even substantially outside of its stability field, is well known. There are many examples of NGH cores being brought to the surface over a period of hours and dissociating relatively slowly to its constituent gas and water. Stern et al. (2003) measured decomposition rates of pure, porous methane hydrate at 1 atm, over the temperature range 195–290 K, which resulted in the discovery of a warm-temperature thermal regime (250–270 K) where methane hydrate can remain metastable preserved for extended periods of time. This inherent stability, or self-preservation attribute (Stern et al. 2003), acts against rapid dissociation.

NGH is stable in sediments as long as temperature and pressure conditions are maintained. If the temperature is increased (such as by drilling or by the flow of warm oil produced from deeper sediments) NGH may revert back to its component water and gas, a process termed "dissociation". In addition to temperature and pressure, NGH is sensitive to water chemistry. Dissociation may occur through increases in salinity as well as by the presence of certain chemical inhibitors,

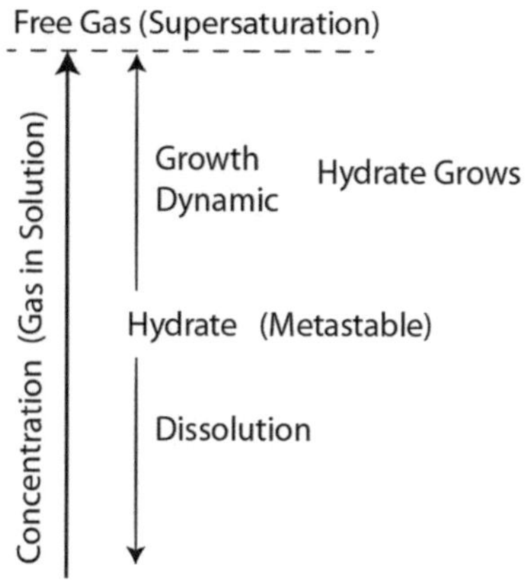

Fig. 2.2 Growth/dissociation tendencies controlled by relative dissolved concentration of hydrate-forming gas

including methanol and glycol. Because the hydrate cage structure concentrates natural gas, a cubic meter of NGH will yield 0.8 m of water and approximately 164 $m^3$of gas at standard temperature and pressure (STP). Thus, the dissociation of NGH has important implications for drilling safety (Chap. 10).

Because the NGH system is a highly reversible chemical reaction, altering stable pressure and temperature conditions to instability [i.e., from the left side of the phase boundary to the right of the phase boundary (Fig. 2.2)] by altering either temperature or pressure or both, will cause the destabilization required for conversion of the NGH to its constituent gas and water. This is a very important characteristic of the chemical system because warming or lowering pressure are the principle ways to initiate conversion of NGH into its constituent gas and water so that the gas may be produced.

This reversibility is also important to the part that NGH may play in the Earth's climate cycle by sequestering natural gas or releasing natural gas into the environment as climate cools or warms. In general, only concentrations of NGH are of interest for gas production while dispersed NGH may be of interest to climate concerns (Max et al. 2006; Archer 2007; Beaudoin et al. 2014a), but we are focused on NGH and new methods for exploration and production that could dramatically lower development costs of the low environmental-risk NGH resource. Brewer et al. (1998) have carried out experiments in the Pacific Ocean off southern California in which the predicted natural nucleation, growth and dissociation/dissolution of NGH have confirmed laboratory work for hydrate stability, while demonstrating that no chemical or thermal precursors commonly used in laboratories are necessary to rapidly nucleate and grow hydrate in deep sea conditions.

When a typical open ocean hydrothermal profile is superimposed on the stability diagram, the shallowest water depth at which the temperature passes into the stability field is the water depth of about 500 m at which NGH will begin to be resident on the seafloor or just beneath the surface. The seafloor temperature in most of the interconnected world ocean is around 2.5 °C or below. This is the feather edge of the Gas Hydrate Stability Zone (GHSZ), which is a zone of stability that

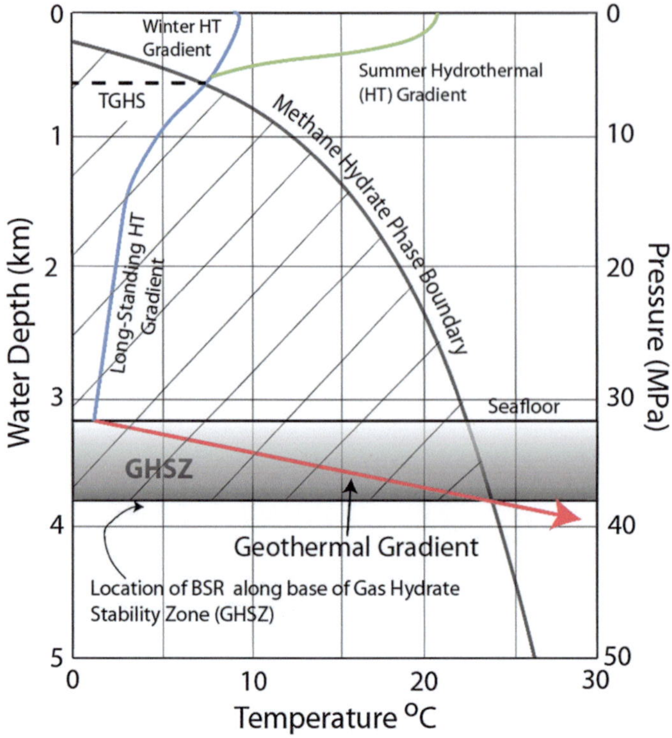

Fig. 2.3 Stability field of Methane hydrate simplified from Fig. 2.2, using 0 °C as a left boundary. Conditions approximately those for the Blake Ridge drilling site (Paull et al. 1998). GHSZ is thicker or thinner depending on location, water depth, seafloor temperature and geothermal gradient. TGHS is top of NGH stability in seawater. Linear geothermal gradient (red) used for illustration. Actual gradient detected in Blake Ridge was variable. Summer (green) and winter (blue) hydrothermal gradients. Stability field to the left of the phase boundary highlighted by diagonal lines

extends from the seafloor to some depth determined by temperature. The rise in temperature indicated graphically by the geothermal gradient (red line in Fig. 2.3) denotes temperature change with depth. Although NGH is known to form within seawater up to a depth marked by the phase boundary (Fig. 2.2), this hydrate will float upward and naturally pass out of its field of stability and dissociate. Thus, there is no possibility of a high concentration of NGH occurring within seawater. NGH that has been observed to form mounds on the seafloor where it has bound with the seafloor sediments that prevent them from rising is not considered by us to have potential for natural gas production on a long-term commercial scale for reasons discussed elsewhere (Max et al. 2006).

2.1.3 Solution Concentration Controls Growth

Similar to many other chemical reactions, crystallization is governed by diffusional processes, wherein the supply of reactant to NGH nuclei control the conditions for growth, dissociation, and dissolution (Rehder et al. 2004; Max et al. 2003, 2006). The relative concentration of dissolved hydrate-forming gas in pore water, in contact with NGH determines a tendency to crystallize, dissolve, or remain relatively the same. Where the vapor pressure of the hydrate-forming gas in pore water in contact is sufficiently greater than its vapor pressure in the NGH, a growth dynamic will drive crystallization (Kashchiev and Firoozabadi 2002b). When the vapor pressure in the surrounding water is sufficiently less, dissociation or dissolution (if dissolved concentration remains low) will take place. In other words, hydrate will form when the natural gas concentration in pore water solution exceeds the local solubility of the natural gas in the hydrate phase (Davie and Buffet 2001). Diffusion processes across the surface contact of the NGH with pore water controls both growth and dissociation/dissolution. When the concentration of the hydrate-former in the hydrate and surrounding medium is at or near vapor pressure equilibrium, the system is stable within its environment (Fig. 2.2). The degree of over- or under-saturation, with respect to the hydrate, will affect the growth and dissolution dynamics.

Therefore, without a strong flux of methane and other hydrocarbon gases, no hydrate will be present in sediments. Even though there may be a zone of hydrate stability in the sense that pressure and temperature are suitable for hydrate formation and perseverance, without sufficient hydrate-forming gas in solution, the concentration of dissolved natural gas will be too low to allow nucleation and growth.

The hydrate reaction for hydrocarbon and other common hydrate-forming gases is very reversible. That is, very little change in pressure and temperature across the phase boundary can alter the stability of NGH so that it can change from formation to dissociation and vice versa, very quickly. Although compound hydrate composed of more than one hydrate-forming gas can react more complexly, an essentially single-gas NGH such as methane hydrate is very sensitive to its environment. In addition, NGH that is in equilibrium with its pore water media will either grow or dissociate/dissolve with relatively small changes in dissolved concentration of the hydrate-forming gas.

Where good, permeable sandy strata that would be an ideal NGH host occurs in a GHSZ but is not mineralized, it indicates one of two things: that the ratio of gas production to water transport locally was not and is not high enough and the dissolved concentration of the hydrate-forming gas is too low to allow for significant NGH nucleation and growth, or the subjacent plumbing system is not bringing mineralizing solutions to the GHSZ. Natural gas in high dissolved concentration can nucleate and grow spontaneously and rapidly when introduced into conditions of NGH stability, whether there are geological traps or not. The concentration of

NGH primarily depends on the existence of a GHSZ, which is a 'thermodynamic trap' throughout which causes NGH to crystallize.

If NGH occurs in geological traps such as sealed anticlinal sand, it could indicate that natural gas arrived as free gas and concentrated and was converted into NGH in the same manner that permafrost NGH appears to show (Max et al. 2006). In the case of concentration of gas prior to NGH forming, some of the gas might not convert if all water was taken up and only a solid diffusion mechanism was available for allowing water molecules to reach the gas and form NGH. Gas bubbles flowing into the GHSZ would acquire thin NGH shells, which would dramatically retard dissolution, a precondition of growth. If primary or secondary permeability was high enough to allow the equivalent of permeable flow rather than slower percolation, it is possible for the gas to be swept away from the base of the GHSZ where concentrations might form in a structurally higher low porosity zone as well as being vented.

From the position of NGH concentrations near the base of the GHSZ, it would appear that these form when upward migrating methane-charged pore water passes upward into a suitable geological host strata within the gas NGH stability zone (GHSZ). Although NGH can form anywhere within the GHSZ, its concentration near the base of the GHSZ suggests an overwhelmingly important subjacent supply of NGH reactant. The subjacent groundwater plumbing system is extremely important to exploration for NGH because it is the means by which natural gas-rich pore waters are delivered into the GHSZ. In passive margin sediments, water movement is driven mostly by compaction dewatering while in active margins, much of the dewatering is driven by tectonic compaction that is a result of formation of an imbricated sediment prism. It is not necessary to fully understand water movement within the prism as a whole, but for exploration purposes, regions of the lower GHSZ that have both abundant sand strata and relatively high pore-water input and for perhaps 1 km below would be first order exploration zones. A key to NGH exploration is to identify the most promising strata passing through the base of the GHSZ in which groundwater flow is focused.

2.1.4 The Gas Hydrate Stability Zone

The term GHSZ was defined as an economic geological term that referred to the zone of NGH stability within seafloor sediments in which large concentrations of NGH could form (Max et al. 2006) and from which the natural gas might be recovered. The top of the zone is at the seafloor, in the absence of biogeochemical action (Sect. 2.1.5). The base of the GHSZ is defined by temperature. The depth at which the geothermal gradient intersects the phase boundary defines the lower limit of hydrate stability. This is the depth below seafloor at which rising temperature defines a lower limit to NGH stability and is also a phase boundary. The two phase boundaries together fully encompass the field of pressure–temperature NGH stability. Although NGH is stable within seawater to the left of the 'general' phase boundary (Fig. 2.3), NGH is positively buoyant, and if it forms in seawater above the seafloor (Barnard et al. 2015), or otherwise is unbound to the seafloor, particles will rise naturally and begin to convert to water and gas when it becomes unstable. As particles may warm as warmer water is encountered as they rise through the seawater, according to the temperature variation of the hydrothermal gradient, the actual depth of incipient conversion will be to the right of the stability intersection and slightly deeper.

The GHSZ is the only region within the sediment (Fig. 2.3) in which NGH concentrations will form. Methane is almost the only natural gas that occurs within the GHSZ. The GHSZ is presently the focus of exploration but it is entirely possible that higher density hydrocarbon hydrates occur at greater depth because they are stable at higher temperatures and pressures.

Heat flow is important as it defines the thickness of the GHSZ. The angle of a geothermal gradient vector (Fig. 2.3) is shallower in higher heat flow regimes, for instance in marine sediments overlying young oceanic crust, and is steeper in lower heat flow regimes such as thickly-sedimented passive margins over cold basement. Heat flow vectors can also change with depth depending on the thermal conductivity of the sediments, and to some extent on the amount of low conductivity NGH present. Below the base of the GHSZ, which is the approximately seafloor parallel 'geological' phase boundary for hydrate stability, the phase boundary curve shown is only a projection and no longer represents the actual position of a phase boundary. The GHSZ is thus encompassed by the seafloor and the intersection of the general and geological phase boundaries.

Although the base of the GHSZ is usually shown to be about parallel at some relatively constant depth below the seafloor, local anomalies in heat flow, such as over diapirs and in the vicinity of seeps that bring warm water and gas through the GHSZ cause the base to rise. Occasionally, with sufficient heat flow, the GHSZ may actually thin to zero thickness (Wood et al. 2004; Max et al. 2006, Fig. 4.17).

In general, higher density hydrocarbon gases have 'larger' hydrate stability fields that allow them to be stable at higher temperatures and lower pressures, but have yet to be found in nature without an abundance of methane present. Nonetheless, their presence has the potential to alter the position of a methane reference phase boundary, which may be one of the reasons that multiple GHSZs or relict GHSZs may be identified on seismic records (Max et al. 2006).

Oceans in Polar Regions are colder than in more temperate regions and have a smaller temperature range than in some temperate regions or in restricted bodies of water such as the Mediterranean and Black Seas (Chap. 4). In particular, the Arctic Ocean only has a deep-water connection to the open ocean through the Fram Strait between NE Greenland and Svalbard on the NW corner of the Barents Sea continental shelf. This constriction and the shallow water shelf across the narrow Bering Strait between Alaska and easternmost Siberia that blocks the flow of deep cold water is responsible for the colder water in the ocean and a lower temperature at the seafloor, except on the shallow continental shelf where summer water temperatures can warm appreciably (Max et al. 2013).

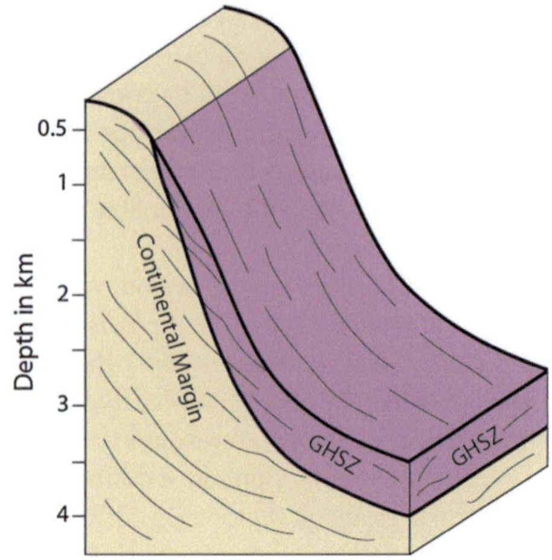

Fig. 2.4 Diagrammatic representation of thickening of GHSZ with increasing water depth and pressure (constant geothermal gradient assumed)

In the Arctic Ocean, NGH can be found in sediments as shallow as 250 mbsf year-round. Around Antarctica the continental slopes are also somewhat colder than in more temperate regions but there are no major constrictions and mixing of cold, saline water produced during winter descends and mixes with deeper water of the open ocean around the entire circumference of the polar continent. In general, the feather edge of NGH stability in the world ocean is about 500 m water depth. The GHSZ thickness increases with water depth to seafloor. This is because increasing pressure moves the general phase boundary (Fig. 2.4) to the right (Fig. 2.2) and the vector of any geothermal gradient will intersect the general phase boundary at greater depths yielding a thicker GHSZ. Low geothermal gradients will promote thickening of the GHSZ and a high geothermal gradient will thin it. The shallow feather edge of the GHSZ (Fig. 2.4) is susceptible to migration up and down slope with seafloor water warming and cooling. Thus, buildup of NGH will not take place from the feather edge of stability down to some water depth at which the hydrothermal gradient is relatively stable. From the depth at which the hydrothermal gradient is stable, concentrations of NGH can form, depending on the availability of hydrate-forming gas. A thicker GHSZ may encompass more suitable host sediment, which increases the likelihood for an increased number of NGH concentrations.

The GHSZ is a worldwide feature of deep continental shelves, slopes and rises to some abyssal regions that vary in thickness depending on sediment thickness, seafloor temperature, and pressure related to water depth (Fig. 2.5). The GHSZ is the focus for exploration because NGH only occurs within it (Chap. 3). The seaward thinning of the GHSZ usually indicates thinner sediments and a lower likelihood of NGH concentrations because the sediments tend to be muddy and have poor reservoir character. In general, only those areas in which the seafloor is within depth ranges in the ultra-deepwater that are currently being explored or will be explored, and where some production is taking place or is likely to take place, are of

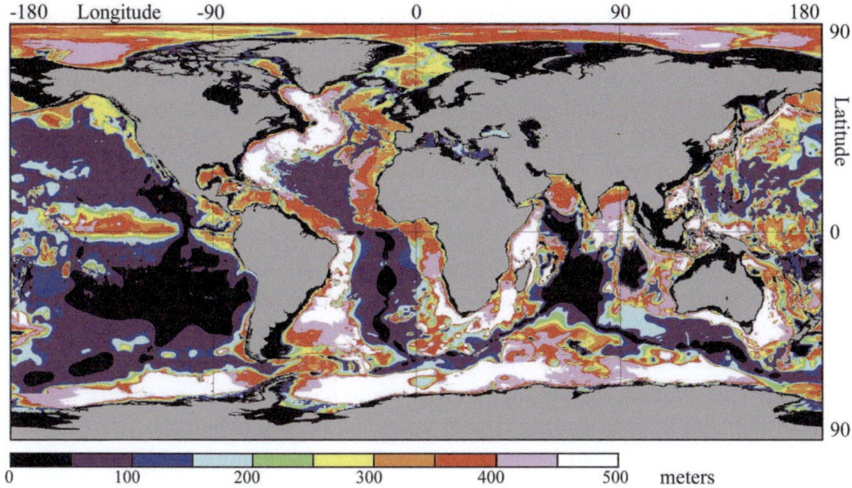

Fig. 2.5 Methane hydrate Stability zone thickness. Update from Wood and Jung (2008)

first order interest for NGH gas exploration and production (Crager 2015; Total 2016), at least until dedicated NGH technology can be introduced that may cut the cost of exploration and production (Chaps. 7 and 8).

As the geothermal gradient varies considerably, the thickness of the GHSZ varies on a global scale. Rapid lateral changes in the thickness of the GHSZ are rare except near salt diapirs over which, and in the vicinity of, heat flows may be abnormally high even over short distances laterally. Locally, high heat flow associated with venting may warp the base of the GHSZ to the surface upward (Max et al. 2006), but NGH concentrations associated with venting are dependent on transitory mineralizing solution delivery and are less likely to be associated with large NGH concentrations that may have commercial potential.

The Polar oceans, the Atlantic margin of the Americas (Kliewer 2015), including the internal basins and the heavily sedimented ring of islands marking the eastern margin of the Caribbean Sea (Henke et al. 2008) and Africa, the continental margins in the Indian Ocean, and the southeast margin of Australia, have the thickest and most prospective GHSZs (Fig. 2.5.). Applying NGH petroleum system analysis (Chap. 3) allows more favorable regions within the GHSZ to be identified.

Because a certain amount of sediment overburden is required to support a depressurization conversion of the NGH in a production scenario, we would regard the ~200 m thick GHSZ of the Japanese Nankai deposit (Matsumoto et al. 2004; Takahashi and Tsuji 2005; Fujii et al. 2015), identified during the Japanese MH-21 first phase (RCMHRJ 2008), as a working minimum safe thickness. NGH concentrations can occur anywhere in the GHSZ, and are, in fact, more stable toward the top of the zone owing to colder temperatures nearer the seafloor (Fig. 2.5). The gas flux appears to be mainly from below, so NGH will tend to be concentrated in the lower part of the GHSZ, particularly in porous strata near the base of the GHSZ in which the gas first encounters NGH stability conditions. If vents, chimneys, and faults can release the natural gas higher in the GHSZ, concentrations could be found well away from the base of the GHSZ (Max et al. 2006). NGH has been found at all depths within a GHSZ, but large concentrations have been found to date only at or near its base.

NGH is responsive to forming or dissociating as environmental conditions may be altered, for instance by warming and cooling and rise and lowering of sea level during glacial-interglacial episodes and longer scale tectonic restructure of continental plates, during sea level fall. Therefore, the thickness of the GHSZ has almost certainly changed throughout geological time. Deepwater and ultra-deepwater ocean seafloor warming would also thin GHSZs. This would produce natural gas that in shallower GHSZs would probably dissolve in the ocean if it escaped the sediment. This cyclical gas production also could have resulted in gas produced by NGH conversion at an ascending GHSZ base where it would buoyantly rise into the overlying, thinner zone and recrystallize, further concentrating NGH. In any case, with substantial new gas arriving at the base of a GHSZ, concentrations of NGH can form relatively quickly.

2.1.5 The Seafloor May not Be the Top of the GHSZ

The top of the GHSZ theoretically occurs at the seafloor, and where methane flux is very high, NGH may occur on the seafloor and in the topmost sediments (Fig. 2.3). In regions with very active petroleum systems and prominent venting through the seafloor, solid masses of NGH may occur on the seafloor and are useful because samples can be taken for analysis that can throw light on the nature of the gas generation (Sassen and MacDonald 1997; Sassen et al. 2001). NGH on the seafloor or just below it has elicited calls for 'seafloor mining' that are considered by us to be non-starters for exploration and production. The deposits are very small in contrast to the size of concentrations that could be considered to have commercial potential, and are associated with specific biota (Fisher et al. 2000) that would likely be decimated during mining or in situ conversion operations.

But where methane flux is low, a zone at the top of the GHSZ may not contain NGH. Marine sediments are deposited with sulfate-bearing seawater at concentrations of 2.6–3.2 mg/L. Biogenic activity feeds on the natural gases seeping up from below. This results in a sulfate/methane transition (SMT) zone that may extend downward from the seafloor into the subsurface, with sulfate-bearing sediment above and methane-bearing sediment below. Methane and sulfate will react to form carbonate and sulfide:

$$CH_4 + SO_4^{2-} \rightarrow HCO_3^- + HS^- + H_2O$$

Because of this reaction, sulfate and methane will not exist together in subsurface sediments. If the methane has been consumed by biota or chemical reaction, the hydrate-forming gas concentration will be too low for NGH to form. Thus, in locations with a relatively thin GHSZ and a very low natural gas flux, the SMT may be deeper than the base of the GHSZ, and no hydrate will be present even though pressures and temperatures would be sufficient for hydrate stability. Foucher et al (2009) and Bhatnagar et al. (2008) discuss the SMT in more detail, along with the relationship of venting to chemosynthetic biota along with the relationship of methane to sulfate, which affects the overall solubility of methane in the GHSZ.

Where gas flux is sufficient to promote NGH growth but where there are few vents that may affect the base of the GHSZ or to deposit NGH directly at the seafloor, a relatively thin SMT on the order of meters or less indicates prospectivity for NGH within the GHSZ. In general, however, the upper part of GHSZs are not first order prospects for NGH concentrations. The depth of an SMT may be used as an exploration indicator of the strength of the local petroleum system that is important for forming NGH concentrations.

2.2 NGH Stability Within the GHSZ: Implications for Gas Production Cost

Controlled conversion of NGH to its constituent gas and water is critical to production of natural gas from the resource and is probably the most important variable factor governing production cost. Because the energy cost of conversion depends on the level at which the NGH occurs in the GHSZ, the location of the NGH deposit is critical to production cost. The closer to the base of the GHSZ a concentration occurs, the lower the energy cost of conversion across the range of conversation technologies (Max and Johnson 2011a). This is because the further away from the phase boundary, the more stable the NGH. Although pressure decreases slightly, in an ultra-deepwater GHSZ the pressure drop is less important than the temperature drop toward the seafloor, at which temperatures are at their coldest. Higher in the GHSZ, more energy is required either by heating, depressurization, or any other means, to cause dissociation. For the known deposits that have been proven—Walker Ridge in the northern Gulf of Mexico and the Nankai deposit off Japan—as well as all other drilling verification of NGH occurrence, the highest concentrations of NGH are found at and immediately above the base of the GHSZ. Although it may be too early to be certain, as a working model the base of the GHSZ is the present focus of exploration for first-order NGH deposits. If this exploration paradigm is established, it will mean that the minimum energy for NGH conversion will be the case for most oceanic NGH deposits (Chap. 11).

The base of the GHSZ is often marked by a bottom simulating reflector (BSR), which is a reflection seismic feature caused by the acoustic impedance between sediments within the GHSZ and underlying sediments containing some free gas. This reflector is often at a constant depth beneath the seafloor, cutting across bedding planes and is of diagenetic or younger origin. As this reflector is often at a constant depth beneath the seafloor, often passing across bedding planes, it was termed a "Bottom-Simulating Reflector" or BSR. A BSR is primarily a relatively sharp transition zone to subjacent strata that have free gas in their pore space below the base of the GHSZ (Max 1990; Paull et al. 1998; Paull and Ussler 2001). Gas lowers the overall wave velocity (Vp) of the gas-enriched sediment. The impedance contrast with the underlying gas-rich zone or the 'negative reflection coefficient' may be accentuated by the presence of high Vp NGH at or immediately above the base of the GHSZ and is most pronounced where NGH directly overlies gas in a porous/permeable horizon.

BSRs have been mistakenly used to infer the presence of NGH, whereas they are in fact caused by the presence of the top of free gas below the GHSZ (Max 1990; Max and Johnson 2014; Hillman et al. 2017). The Nankai NGH deposit off SE Japan was drilled on the basis of a BSR, but they were lucky as there was gas beneath the NGH developed in turbidite sands. Numerous drilling programs since, however, have shown that while a BSR denotes the base of the GHSZ, it provides little information about nature of the sediments within the GHSZ or the presence of or degree of NGH saturation. BSRs are often rare or absent in areas in which NGH has been shown to be present. In many basins the strongest, most continuous BSRs occur in fine-grained sediments containing only 3–5% NGH by volume (see Chap. 3).

2.3 Geology Controls NGH Paragenesis

Paragenesis is an economic geological term that pertains to the origin of minerals or mineral deposits and the order in which the minerals have crystallized. NGH is a diagenetic crystalline material whose crystallization is similar to low temperature strata-bound mineral deposits, which distinguishes it from other hydrocarbon fossil fuel deposits. Geology is the key to understanding the principal sources of methane, the strata in which it forms and its migration in marine sediments to the GHSZ. NGH paragenesis can be described as having its own 'petroleum system' that is useful for exploration (Chap. 3), but it can also be described as a mineral deposit that concentrates natural gas from mineralizing solutions through a chemical reaction.

In general, methane is common in marine sediments on both active and passive continental margins (Claypool and Kaplan 1974). The natural gas tends to migrate upward through the sediments along both primary and secondary porosity and will tend to be vented into the ocean as part of a long-standing biogeochemical process. Methane produced by microbial agents is termed 'biogenic' while methane and higher density hydrocarbon gases produced at greater depths under higher temperature conditions are 'thermogenic'. Commonly, both biogenic and thermogenic natural gas will be found in areas such as the northern Gulf of Mexico (Richardson et al. 2004, 2008), the Cascadia margin of the central west coast of North America, and the western Pacific Japanese continental margin, amongst other locations.

Natural gas can be captured and sequestered over a long period of time in the GHSZ. The gas can be produced locally or be transported in aqueous systems from considerable depths and laterally, given the right set of conditions. Geology governs the degree to which NGH might reach sufficient concentrations to be a natural gas resource. From the point of view of the likelihood of first order, high grade NGH concentration formation, the nature of the marine sediment is the primary governing factor. NGH is found in two basic types of deposits, dispersed NGH in a generally fine-grained muddy sediment host, and concentrated NGH that occurs in permeable sediment

Dispersed or isolated NGH Muddy sediments have very low permeability. Water movements are often only local in extent and related to the formation of temporary low-pressure zones that require sediment movement. Migration of dissolved components is controlled by short range diffusion processes. Local concentration of NGH in shales ranges from small dispersed nodules to larger, mostly irregular shaped discontinuous bodies whose form may have some vague stratigraphic control. Migration of water into secondary porosity pore filling veins usually is due to local structural control that has caused local pore waters with dissolved or free natural gas to migrate to lower pressure locations. Water flow from a remote source is usually directly related to faulting or vent formation.

According to Davie and Buffett (2001), local hydrate accumulation is unlikely to exceed an average pore volume fraction of 7% in the absence of pervasive fluid migration. Their modeling results confirm that the dispersed NGH found in the Blake Ridge on ODP Leg 164 could be explained by local generation, which suggests that high concentrations of NGH that have potential for economic exploitation must have be formed with additional fluid/gas migration into a host bed in the GHSZ. Muddy sediment is not likely to have NGH natural gas reservoir characteristics from the point of view of existing production scenarios.

Thick sequences of muddy sediments such as in the Blake Ridge seem to have mostly authigenic NGH, with usually no more than about 7% NGH, which seems to be the limit. That represents dispersed NGH formed in a diffuse flow environment. Reworking of NGH near and at the base of a muddy sediment sequence, such as at the Blake Ridge, will normally take place during climate swings that thicken and thin the GHSZ. For instance, when the seafloor warms during interglacial conditions, the GHSZ thins and NGH in the abandoned zone naturally dissociates. This produces free gas that can migrate upward into the GHSZ utilizing the many small faults that appear to cluster near the base of the GHSZ (Fig. 2.7).

Depending on the volume of free and dissolved gas arriving from below, free gas produced at the base of the GHSZ, and the remaining permeability in any beds having good primary porosity prior to NGH mineralization, free gas may build up in porosity. Further, depending on the ability of collected gas to equilibrate with otherwise normal formation pressures, pooled gas may develop pressure in excess of normal formation pressures. Although it is more likely that gas build up may take place beneath a large area of muddy sediments with poor permeability where gas-enriched mineralizing solutions cannot migrate easily into the overlying GHSZ, pressure in gas pools can be expected to be higher than in permeable sediments with focused flow. Overpressured gas may be responsible for faulting near the base of the GHSZ, which in extreme cases may extend to the seafloor (Hornbach et al. 2004).

Concentrated NGH Concentrated NGH appears to only occur in sandy beds through which water can percolate. The cleanest NGH-free turbidite sands in the Walker Ridge NGH drilling program in the northern Gulf of Mexico that are associated with NGH concentrations (Boswell 2012a, b) have permeability contrasts of 5–6 orders of magnitude with bounding shale beds. This is sufficient to create focused flow and NGH crystallization from either nearby shales or deep sources (Daigle et al. 2015). Although even a small fraction of fines in the sands can have a strong effect on permeability, sands are permeable pathways for fluid and gas migration and shales are not. Turbidite sands in marine sediments are presently the primary stratigraphic host for concentrated NGH in marine sediments (Max and Johnson 2014).

Significant volumes of natural gas must migrate into a host sand in order for a high grade NGH concentration to form, and it must migrate in sufficient concentration in the pore water to promote crystallization. The migration paths for conventional hydrocarbons are also faults and permeable strata that allow the formation of conventional gas deposits. Normally, gas charged pore water and freed gas moves more steeply upward along secondary permeability induced by faulting and laterally up dip within permeable strata, which in large structures and usually are not steeply dipping. Both conventional and NGH sand reservoir hosts are confined within the strata by shales or other less permeable strata that confine the migrating hydrocarbon to the discrete, permeable horizons. This confinement is not a trap in the sense that conventional gas is trapped by being stopped physically from further upward migration by less permeable strata in both structural (closure) and stratigraphic (sand gives way laterally to shale) traps. Although one could consider that when NGH concentrations become sufficiently impermeable they will carry out the same function of blocking fluid migration, when the blockage is due to crystallization of the migrating mineralizing solution, it is a very different thing. When the migration path reaches the base of the GHSZ, NGH crystallization initiates up-dip.

The gas and gas-enriched pore fluids can migrate over long distances using the groundwater plumbing system (Talukder 2012) which, on active margins, is driven largely by tectonic dewatering and on passive continental margins by natural compaction of the sediment pile. There must be a groundwater feeder system that allows initially dispersed hydrocarbon to coalesce or to form very high concentrations in pore water that then migrates. As in conventional gas deposits, sands generally provide the permeability that allows fluid and gas to percolate through them. Sands are the primary hosts of concentrated NGH deposits. In particular, sands near the base of a GHSZ will be primary exploration targets (Max and Johnson 2014).

In a stratified sequence, gas locally produced in muddy sediments, where most of the organic material resides, can migrate short distances across strata boundaries into permeable beds that effectively 'drain' the sediment pile. Micro-faulting and sediment deformation may accompany this dewatering although sandy strata usually have a higher sediment strength, at least with respect to volume change. Diffuse flow migration into the permeable beds creates a focused flow from which NGH concentrations can form. All this local gas is biogenic. From further away, focused flow from below the GHSZ, which is draining subjacent biogenic gas production in shales, flows through the permeable horizon. When it crosses into the GHSZ, it introduces further potential to add to the concentrated NGH. Thermogenic gas must come from higher temperatures sources that are usually related to deep-seated very active petroleum systems. These sources can occur in both passive margin (e.g., northern Gulf of Mexico) and active margin situations (e.g., Cascadia).

Because the bulk of concentrated NGH occurs at the base of GHSZs, it follows that the source of the NGH concentrations has been introduced from below. If the bulk of the natural gas would be produced locally in the muddy sediments of a stratigraphic sequence bounding a sand, the presence of NGH in permeable beds would be expected to be about the same regardless of structural level within the GHSZ, or even increase upward because of greater stability toward the top of the GHSZ (Fig. 2.6). The presence of higher density hydrocarbon gases almost certainly indicates a very deep source below the biogenic zone, but high level igneous intrusions can also cause thermogenic fraction to form in organic rich beds at higher levels.

Both conventional NGH sand marine sand bed reservoir hosts are confined within the strata by shales or other less permeable strata that form ducts for the migrating hydrocarbons. This confinement is not a trap but part of the migration path system, unless there is structural closure or stratigraphic pinch out.

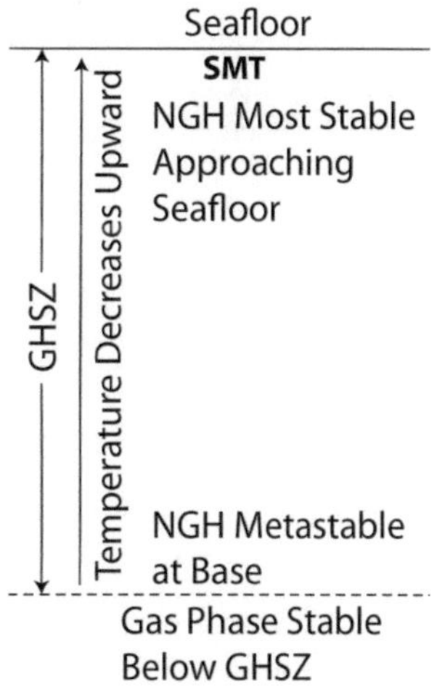

Fig. 2.6 Stability of NGH within the GHSZ. Compare with GHSZ in Fig. 2. SMT = sulfate − methane transition zone

2.4 NGH May Be the Largest Natural Gas Resource on Earth

Early estimates of the amount of oceanic NGH were unreasonably large because they included all sediments, even those with no viable means of producing gas from the NGH to several hundred thousand trillion cubic feet (Tcf). Most NGH occurs in low concentrations (1–5%, rarely to 8–10%), disseminated in shale and muddy sediments. These sediments do not constitute gas resources and are not included in our NGH resource base, although they may be important to global climate. NGH can both sequester and release natural gas produced in sediment and crustal rocks in response to changing climate and sea level. Early estimates concerned only potential gas-in-place and took little notice of either recoverability or commerciality of the resource. There was little known about pore space occupation, gas flux, the ease with which NGH could form, its mechanisms of nucleation and growth, and other aspects governing NGH paragenesis, volumes, and concentration.

As with any resource that becomes better understood with time, volume estimates have become more precise and, hopefully, accurate. The estimates for total gas in place have a very wide range from 0.1 to 1.1 million Tcf (Beaudoin et al. 2014a, b). Most of the NGH, however, occurs in marine sediments that have limited permeability or in seafloor and near-seafloor nodules and veins (Fig. 2.7). NGH in the Blake Ridge off the U.S. east coast (Dickens et al. 1997; Paull et al. 1998) occurred as dispersed NGH that occupied no more than 8–10% of the porosity but occurred in mechanically weak sediments. Although it is now known how these NGH occurrences can be drilled into safely, they cannot be produced without disrupting the sediment in which because of their mechanical failure during conversion of the NGH to its constituent water and gas. These have no immediate viability as natural gas resources because current operating technology does not provide a means for these NGH deposits to be concentrated or recovered in volume, although Max et al. (2006) have suggested some techniques related to mining that might be used in muddy, low permeability sediments.

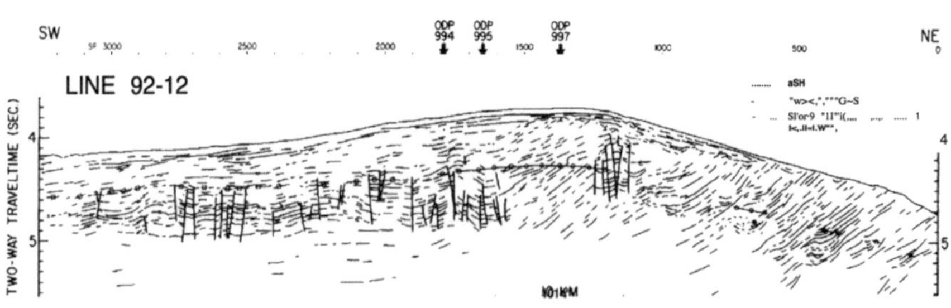

Fig. 2.7 U.S. Geological survey seismic interpretation of faulting in the vicinity of the base of the GHSZ. Note the glacial episode GHSZ would be thicker and the base more closely associated with faulting to the SW. Supplied by W. Dillon, USGS

Petroleum system analysis (Max and Johnson 2014) has allowed the focus for the NGH resource to turn to sandy turbidite marine sediments (Boswell and Collett 2006). Using these systematics, Frye et al. (2013), for instance, estimate 6000+ Tcf natural gas in NGH along the northern Gulf of Mexico and Max et al. (2013) estimate 6000+ Tcf in the continental slopes of the Arctic Ocean. An evaluation conducted for the Global Energy Assessment reported a conservative estimate of 43,000+ Tcf for NGH in place in sands within the GHSZ (Johnson 2012). We need to better understand the local relationships between erosion, sediment transport to the shelf edge and the depositional environment to better calibrate volume estimates. Because geological processes act similarly along all continental slopes, all of the world's continental margins constitute potential areas for exploration. National NGH programs of the United States, Japan, South Korea, India, and some commercial institutions in China envisage that natural gas may be commercially extracted from high grade NGH deposits in the first instance and possibly from more dispersed lower grade deposits as expertise and new technology allows.

The application of a first order NGH petroleum system analysis has restricted the volume of NGH to only those permeable strata that could host large concentrations of NGH. The primary exploration target for high-grade NHG deposits is in turbidite sands within the GHSZ. Coincidentally, it is exactly the type of sand sediment that also hosts conventional oil and gas deposits at deeper depths below the GHSZ. Because of this, the energy industry also possesses the bulk of the tools for NGH exploration and production. While the producible marine sediment target is only a small part of the total NGH, its projected volume compares well with conventional natural gas (Fig. 2.8).

Johnson (2012) applied a NGH petroleum system approach (Max and Johnson 2014) to estimating the potential amount of the NGH resource base assumptions similar to those commonly used in the energy industry for conventional natural gas

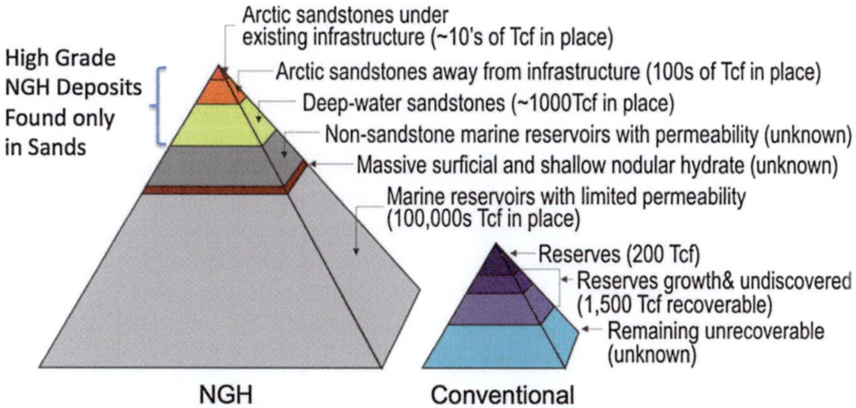

Fig. 2.8 NGH resource 'pyramid' and host strata types (left) after Boswell and Collett (2006) and Max and Johnson (2014). To-scale pyramid of conventional reserves pyramid to right. 'Arctic sandstones' (topmost smallest sub-pyramid) are onshore permafrost NGH

and oil. Estimates are only for gas-in-place in NGH, with no estimates made for trapped gas related to the NGH that might be produced along with the NGH gas. Further calculations of theoretically and economically recoverable gas have been made, again with reference to standard energy industry practice (assuming complete NGH conversion to constituent water and natural gas. Estimates have only been made for continental slopes (the continental margin zone of Max and Johnson 2014), but other zones, such as more structurally isolated bathymetric highs and deep troughs and passes within continental shelves may also host NGH deposits.

The criteria governing estimation of NGH gas-in-place that Johnson (2012) used and which we use here are relatively straight forward and are based on industry experience:

Prospect Areas Over Continental Slope Estimates were computed for a zone along the continental slope for about a width of 100 km basinward from the 800 m isobath. For trough mouths, an area of no more than about 50 km from a line across the trough mouth connecting slope breaks, and no other depositional environments other than turbidite sands were considered. Note that the area of the slope on which NGH might occur is greater than the areal projection of the steeply dipping prospect zone. Also, the areal extent of the prospect zone may be wider or narrower in different places; we have used the 100 km wide zone as a workable estimate. More detailed workups, especially using GIS, will refine our estimates.

GHSZ Thickness Early estimation of the GHSZ thickness was based upon heat flow measurements and seismic surveys. There are large areas of continental slope and deep continental shelf where data is insufficient and estimations have wider ranges. The world estimate was based on the GHSZ thickness as determined by Wood and Jung (2008), as applied to the defined aerial prospect areas. More detailed maps of calculated GHSZ thickness exist in the literature, and these can be used to improve accuracy, but by using the best available (in our opinion) map of world GHSZ, we at least insure that errors will relate to a common datum.

Host Sediments The percentages of sands within the GHSZ that are available for ascending fluid transport have been estimated for a number of areas from examination of reflection seismic data. Extrapolation of suitable sand bed estimates to regions that have little seismic data, such as the Arctic Ocean continental slopes, was based on a general projection of erosional history on land combined with likely eroded sediment composition, winnowing, combined shelf–slope depositional history. Clearly, more detailed assessment with fuller consideration of the impact of sea level and climate changes as indicated from sea level variation will tighten the estimates for available NGH host sediments. Earlier estimates noted by Johnson (2012) were modified using more recent estimates of sand on other continental shelves where they became available (Frye et al. 2013).

Industrial Workable Water Depths We consider that the exploration and development that will continue to be carried out in ultra-deep-water depths within the next 5–10 years will define the base of the NGH prospect zone for the foreseeable future. If any type of resource cannot be drilled and produced, it has no economic interest. Conventional reservoirs will probably to continue to be drilled from the sea surface because of the need to control the reservoir conditions. Using seafloor drill sites offers the opportunity to drill very quickly to the NGH resource without the high costs and operation issues of working so far above the resource total depth.

Information Base Enough is now known from three locations (Nankai. Gulf of Mexico and drilled permafrost NGH deposits) and from the success of the geotechnical approaches to seismic data processing (Chap. 3) to identify NGH concentrations and to provide good estimates of NGH saturation percentages (Max and Johnson 2014). We see no reason why active exploration, along with development of new technologies and methodologies for low-cost NGH exploitation should not begin more widely in earnest. Additional NGH occurs as seafloor mounds where vents bring hydrate-forming gases from depth, and as veins in impermeable sediments. These are also not included in our resource estimates. Several estimates have been published in recent years for NGH filling sand bodies from which commercial expertise exists to recover natural gas from the NGH. Deepwater NGH-bearing marine sands may have as much as 43,300 Tcf gas-in-place, of which 50% may be technically recoverable (Johnson 2012). SBC (2015) considers this estimate, based on NGH petroleum system analysis (Max and Johnson 2014) is consistent with other studies and corresponds to realistic and substantial resource potential. This contrasts an approximately similar amount of conventional resource (Fig. 2.7) and with gas-in-place resource estimates for coalbed methane of 9000 Tcf, tight gas of 7400 Tcf (NPC 2007), and technically recoverable shale gas of 665 Tcf, which is about 30% of U.S. natural gas production and 32% of world shale gas resources, exclusive of NGH's unproven potential (EIA 2015).

Interestingly, the United States is estimated to potentially have the greatest store of natural gas in the NGH resource (Fig. 2.9), although this is in part due to its being drilled for NGH, lowering the risk of its occurrence. However, the Gulf of Mexico has substantial sand deposition into the GHSZ by the Mississippi River along with a proven petroleum system. The Arctic Ocean, which was treated by us separately (Max and Johnson 2014) is also regarded by Equinor as having high NGH potential (Reichel and Gallagher 2015) and is not included in the totals of countries bordering the Arctic Ocean.

When environmental factors for production of each of the unconventional gas resources are taken into account (Chap. 10), production from NGH alone could provide for superior environmental performance, having a lower potential for pollution, blow-out risk, and water quality damage, among other issues. Potentially, NGH could support global energy security as a sole resource. It is not yet known how much NGH exists in richer deposits in which commercialization of the resource will begin. Yet, the resource appears to be so large that even if a very small

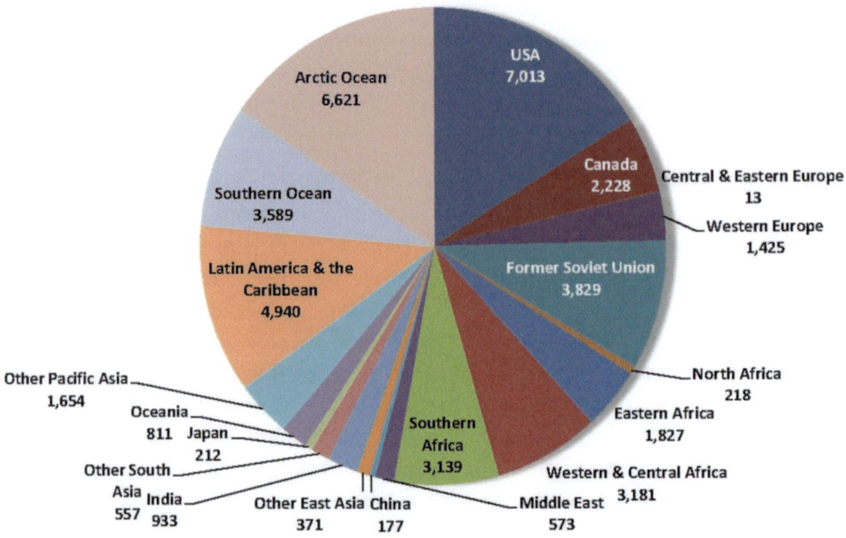

Fig. 2.9 Mean estimates for NGH in marine sands by country and region. After Johnson (2011). Copyright-free DOE publication

percentage of it could be commercialized (i.e., produced at cost and profit margins that would allow it to compete with other gas resources), the resulting production would be significant on a world scale.

As the cleanest and most environmentally friendly of the fossil fuel options, natural gas could be an important source of energy for any future energy scenario (Chap. 1). NGH could occur in such abundance on continental slopes and in areas of thick deposition on continental shelves around the world. If this potential is confirmed, it will become highly valued as local energy resources, particularly for nations with limited conventional domestic energy options. Although some would argue that much more scientific understanding of the implications and environmental consequences of NGH production needs to be developed before commercialization (Beaudoin 2014b), and before there is strong pressure for wide-scale production, just a little of the entrepreneurialism shown in the development of coalbed methane and shale gas has the potential to dramatically accelerate NGH commercialization.

2.5 Other NGH Concentrations that May Be Producible

While concentrations of NGH in permeable strata comprise the first order exploration and production objective, other types of concentrations have been identified that may one day comprise producible NGH deposits. The first type of concentrated

NGH are plug-shaped bodies associated with venting (Matsumoto et al. 2017). The second type consists of strata-bound NGH vein networks (Boswell et al. 2012b). The third of these would be similar to the Blake Ridge NGH that were drilled on ODP Leg 164 (Collett 2001; Dickens et al. 1997; Dillon et al. 1996; Paull et al. 1998). While current offshore energy industry methodology and technology are sufficient to undertake NGH production from permeable turbidite sands for which seismic interpretation techniques exist for identification and preliminary valuation, no production methodology currently exists for other types of NGH concentrations.

2.5.1 NGH Vent Plugs

Concentrations of NGH in the upper part of fluid/gas vents was recognized by Wood et al. (2004) in the northern Gulf of Mexico, along with pull-up of the base GHSZ, which was regarded by Max et al. (2006) as a thermal effect of the upwardly migrating warmer fluids and gases from below. More recently, Matsumoto et al. (2017) have reported from drilling on similar-appearing structures from the shallower sea areas to the west of the main island of Japan and north, west, and east of the north island, Hokkaido. They have found that NGH concentrations in nodular and lenticular aggregates in a mousse-like muddy matrix that seem to develop along bedding planes extend much deep into the vent chimney and often to the chimney apex bases of 100–120 m below sea floor. This suggests that drilling on selected vents in the northern Gulf of Mexico, in which it was concluded that only restricted NGH caps occurred in the vent chimneys could also identify larger NGH concentrations of this type.

In addition, Matsumoto et al. (2017) regard the pull-up of the apparent base GHSZ within and adjacent to the chimney structure to be at least in part a seismic effect of the higher Vp in the NGH-rich upper chimney structure. At least one of their wells penetrated the base GHSZ in the approximate core of a seismic pull-up, which provided ground truth. Thus, the real shape of the base GHSZ in the chimney structure and outside of it may not be so exaggerated as it seems as the real depth to the base outside of the chimney will be closer to the seafloor. The relationship between the apparent and real position of the base GHSZ associated with chimney structures is important as it affects the valuation of a concentration (Chap. 6).

Well logs show that NGH concentration can persist to the base GHSZ, although it is likely from the selective concentrations of NGH in some of the wells that there may be a bedding or secondary influence to NGH concentration. Concentration within the apex of the chimney structures ranges from 35–85% of combined thermogenic and biogenic methane, which suggests at least some deep sourcing to the chimney and vent. Although the participants of the Japanese NGH program may have evidence of the nature of the lateral transition away from the chimney culminations that would allow estimation of NGH volumes in each plug structure.

These deposits may contain enough NGH to warrant consideration as economic resources. Although the current assumption of first-order producible NGH is permeable strata (Fig. 2.8), this is based not only on existing technology and methodology for production but also on the nature of proven deposits. For instance, permeable sandy turbidites have been identified as the NGH concentration host in the Nankai deposit off SE Japan and in the Walker Ridge area of the northern Gulf of Mexico. Where concentrations of NGH in another setting such as in the NGH chimneys are large enough to warrant production, then these can be considered as second-order exploration targets, for which production scenarios have yet to be proven.

In economic geology, the shape of the deposit is often used to define a genotype. Although the concentrations occur in a vent chimney, which is the path for focused flow of water, gas, and potentially some fine sediment particulates, the general shape of the NGH concentration resembles that of a number of igneous mineral deposits in which the chemical constituents of the ore have been transported from below. Although the general form is similar to many salt diapirs, the NGH deposits do not result from mass movement of existing material but rather of crystallization in place. Therefore, we refer to these deposits as NGH plugs, as the best descriptor of their content and shape.

2.5.2 Stratabound Secondary Porosity NGH Concentrations

NGH that is common in primarily muddy sediments as nodules and veins may reach moderate concentrations of total sediment. Where it these concentrations occur in a bed-differentiated stratigraphic sequence, such as in the uppermost unit containing NGH drilled at Walker Ridge in the northern Gulf of Mexico (Boswell et al. 2012b), a possibility exists that natural gas produced from NGH conversion could be contained within bedding and produced. At present, however, the technically recoverable volume of NGH within the unit has not been estimated. In any case, there is no existing industrial methodology for conversion/production as conversion would promote mass flow redistribution of sediments. As the deposit at Walker Ridge is close to the seafloor in the region where it has been drilled, containment of mobile sediments with higher than formation gas pressures would be difficult. Where large enough NGH concentrations of this type have greater overburden and possibly mechanically stronger rocks or are more tightly bounded by different sediments from gas leakage, these might be considered to be between a second and third order type of deposit.

2.5.3 Blake Ridge Type Deposits

Almost entirely biogenic NGH is concentrated in nodules and veins along the lower part of the GHSZ in the bathymetric ridge and in the subjacent base of the GHSZ culmination. The sediments within the GHSZ at the Blake Ridge are essentially undifferentiated with no significant bedding and are fine grained or 'muddy'. They have high porosity but only limited permeability and diffusion is almost certainly the primary means of natural gas migration. It is possible that most, if not all of the NGH in the dispersed NGH region of the middle and upper parts of the GHSZ could be locally produced methane. In general, the GHSZ contains dispersed NGH that increases in percentage downward. In the lower part of the GHSZ where percentages of NGH run as high as 8–10%, the amount of natural gas to form this much NGH probably did not have a local source but suggests some mechanism of concentration operated. As is the case with permeable sand horizons that appear to act as migration paths from below (Max and Johnson 2014), it is possible that methane-rich gas or pore fluids largely migrated upward into this pseudo-geological trap from below.

The more concentrated NGH in the lower part of the NGH may also be due to a long-term reworking of NGH at the base of the GHSZ as it varied up and down in response to climate change that raised and lowered sea level and alternatively may have warmed the seafloor water based on the alternation from glacial to interglacial events (Max and Dillon 1998). In their model, when the base of the GHSZ moved upward, NGH conversion would naturally take place and the gas would tend to rise buoyantly into the suprajacent GHSZ. When the base would again move downward, free gas would be converted to NGH in a more concentrated form. Repeated raising and lowering of the base GHSZ provides a method for long-term concentration of natural gas from any sources.

Although these deposits appear to contain huge amounts of natural gas, they do so over very large volumes of sediment that have no inherent permeability. Because the GHSZ at the Blake ridge is nearly 600 m thick, there is scope for promoting controlled sediment movement through carefully planned focused conversion whose aim would be to allow separation of natural gas and water and production of the gas. Max et al. (2006) have suggested a mining-style analog for safely converting the NGH and recovering the natural gas from concentrations of this type, but they must be considered to have only third-order economic.

2.6 NGH in the Spectrum of Conventional and Unconventional Oil and Gas Resources

In general, the use of the terms 'conventional' and 'unconventional' simply differentiates between what the workers in the very conservative energy industry were used to dealing with from what they were not. Over time, gas and oil that flowed from a wellbore became the familiar hydrocarbon resource that became characterized as 'conventional'. Even though as deeper drilling and secondary recovery techniques became necessary, and the character of the hydrocarbon altered from low temperature and moderate pressures at shallow depths to high temperature and pressures, it was still regarded as 'conventional' if it was a gas or liquid. All other types of resource were lumped as 'unconventional'. Geologically, the term 'unconventional' refers to any hydrocarbon resource that must be stimulated in some way or converted from a solid or entrapped form, is commercialized, and becomes a significant contributor to national gas production, it ceases to be 'unconventional' in any practical sense. Commercially, coalbed methane and shale gas are being produced at scale and are no longer commercially unconventional. Only NGH can still be regarded as both geologically and commercially unconventional, and that may soon change.

According to Holditch (2013), a conventional reservoir is a high quality, high permeability reservoir in which all that is necessary to produce hydrocarbons is drill a vertical hole, perforate the productive interval, then let the well will flow under its own pressure flow under its own pressure in order to recover commercial volumes of hydrocarbons. According to Schlumberger (2015), however, fewer than 5% of wells flow naturally at desired commercial production rates and nearly all of the world's oil and gas production relies on artificial lift to maintain production rates, especially on older wells. Artificial lift is a process that is used mainly on oil wells to increase pressure within the reservoir and force oil to flow faster to the surface (Rigzone 2015). We regard early artificial lift as inherently different from secondary production methods that are designed to affect the physical properties of the reservoir or the resource. Any artificial stimulation, for instance hydraulic fracturing, acid treatment, among other methods, is regarded by the authors as a secondary recovery technique. In conventional wells, secondary production methods are designed to prolong production of a conventional play. In unconventional resources, application of secondary recovery methods is critical to freeing the gas for production at the initiation of production (see Sect. 10.2.3).

In deep and ultra-deepwater 'conventional' fields, however, pressure in the reservoir to be too low to drive the hydrocarbons to the surface, regardless of the quality of the reservoir. Self-flowing may have been common in wells on land and in shallow water, but in deepwater the pressure of the seawater and low formation pressures may strongly affect the tendency to self-flow. For instance, in the deepwater oil and gas at Parque das Conchas fields off Brazil, Shell has had to install oil and gas separators and high-pressure lift pumps on the seabed (Shell 2015). Elsewhere, seafloor lift pumps are increasingly being used to accelerate production rates. This further marks a tendency to place production and other equipment on the seafloor (Chap. 8).

Coalbed Methane The first of the unconventional hydrocarbons to be understood and commercially produced in significant volumes was 'coalbed methane', whose production depends on pumping water from the reservoir to decrease pressure in buried coal measures (much of which may be in thin partings in shales, too deep to mine, and not economic as a minable coal resource). This allows natural gas absorbed in the beds to exsolve, concentrate, and be extracted. This differentiates the pumping from that used for conventional deposits in which the gas is already present in concentrated gaseous form in the reservoir and in which the pumping is mainly to provide accelerated lift to the surface.

Shale and Tight Gas The second of the unconventional resources to be commercially produced are 'shale gas' and 'tight gas', which differ in paragenesis but are grouped together here because they require essentially the same production techniques. These resources are locked in shales and sandstones with having low permeability in the same range as road concrete (generally less than 1/10th of a millidarcy). Fracking (hydraulic fracturing) is used to fracture the rocks and establish the permeability required to allow the hydrocarbons to flow. Production of this low-permeability resource has changed the face of energy resource availability within North America because of its abundance and has done much to introduce the real possibility of energy independence for the continent.

NGH is the last of the recognized unconventional resources of natural gas and volumetrically perhaps the greatest. The principle aspect of NGH that has driven most of the research into it is that NGH deposits may be the largest repositories of natural gas on Earth (Kvenvolden and Lorenson 2001). NGH is unique among gas resources because it is the only one that is a stable solid in which natural gas is an essential part of the crystalline structure. Its physical nature and occurrence, the potential scale of the resource, and other key parameters allow for exploration and production opportunities that do not apply to the other unconventional gas deposits or to conventional resources. This allows for new technology and practices to be applied that will have a strong bearing on the commerciality of the resource. From the point of view of exploration and production of NGH, however, NGH may not be so unconventional (Smith 2009). All existing exploration and production technologies, most of which were developed for conventional deepwater hydrocarbons can be used for NGH without significant modification SBC (2015). If our contention that new technology can lead to a dramatic reduction of NGH exploration and production costs (Chaps. 7 and 8), the impact could be similar to the rapid ramp up of shale oil and gas that was also technology dependent. Permafrost NGH may be only slightly more expensive to produce than associated conventional gas and new technologies and practices in deepwater NGH (Barton 2015) may allow it to be fully competitive (Chaps. 7 and 8).

Reservoir Quality Holditch's definition of an unconventional reservoir also includes its being, "a low-quality reservoir… containing large volumes that are difficult to develop" that must be stimulated in some way to produce at commercial flow rates and volumes. Clearly, this observation is based on early drilling results, for instance on the Blake Ridge off the US East Coast (Paull and Ussler 2001), and predates the outlining of a NGH petroleum system in which specific higher quality marine strata are the exploration targets for NGH concentrations (Max and Johnson 2014). The sedimentary host of a marine NGH deposit is similar to sediment hosts of conventional deepwater deposits.

Although NGH is an unconventional resource, its reservoirs within continental margin sediments will be the same turbidite sandy beds as comprise many of the more deeply buried 'conventional' reservoirs in deep- and ultra-deep water (Chap. 3). While the term 'unconventional energy' is useful for describing the group of hydrocarbon resources that do not have a tendency to flow under their own pressure when penetrated from the surface by drilling the meaning is no longer relevant to the quality of the reservoir. We regard the term ‘unconventional‘ with respect to hydrocarbon deposits as outdated.

Also, the nature of the reservoir in the case of coalbed methane and shale hydrocarbons was not a ‘difficulty’ in the sense of presenting unsurmountable problems. Once the technological approach to mobilizing the hydrocarbons was understood and applied, it showed that the industry again had risen to a formidable challenge. Previously unconventional resources are already a significant part of the commercial gas base and projected to dominate North American gas supply in the future (EIA 2015). Their commercialization now renders them as a critical component of the hydrocarbon resource with proven reserves.

Because two of the three one-time unconventional energy plays (coalbed methane and shale gas) are now contributing strongly to the U.S. gas supply and the methodology for producing NGH is on the cusp of being initiated, the main difference between conventional and unconventional lies primarily in the requirement for inducing a flow of gas and liquid hydrocarbons. In any case, the oil and gas produced from these ‘unconventional‘ resources is the same or within the range produced from conventional energy sources.

Thus, we regard a ‘conventional‘ resource as one that consists of concentrated accumulations of gas and/or liquid in a reservoir that will flow, even though lift of the hydrocarbons to the surface from the seabed or water/gas injection may have to provide assistance. An unconventional resource is one in which the hydrocarbons are essentially stable in their reservoirs, whatever the nature of the host, and which must be stimulated in some way by artificial means. Without unlocking the reservoir prison with some technique (Chap. 10), simply drilling into the reservoir will have no productive effect.

2.7 Low Environmental Risk Character of the NGH Resource

It cannot be emphasized enough that NGH is stable in its natural habitat and will not dissociate unless it is disturbed in one of a number of ways, such as depressurization, heating, inhibitor stimulation, dissolution, and molecular exchange (Max and Johnson 2014, Sect. 11). Stability will persist even during drilling. For instance, even if NGH is destabilized, the rate of destabilization is limited because diffusion exchange of hydrate-forming gas molecules with pore water across the NGH grain boundaries controls the process. Even when it is placed under conditions that are extremely outside of stability, NGH will not crumble or explosively decompress. This is demonstrated when core samples are quickly retrieved from a deep well bore to the drilling deck of a drill ship or a production platform. When conditions of instability are slowly introduced, dissolution will be favored over dissociation and the formation of free gas can be prevented if pore water can be maintained in an undersaturated state by a combination of diffusion and water flow that will keep the level of dissolved natural gas from reaching supersaturation.

There is little danger of unanticipated gas release from NGH within a reservoir during drilling if cold seafloor water is used to maintain low temperatures in the wellbore and absorb heat produced by drilling activities. There is some risk of hitting relatively small free gas zones but these are unlikely to be very highly pressurized within the GHSZ. Gas concentrations below NGH in permeable horizons at the base of GHSZ appear to be no more than about 5% above reservoir pressure (Max and Johnson 2014). An early model for production was to depressurize the gas below NGH, which would cause depressurization of the immediately suprajacent NGH. This would result in a secondary recovery method wherein the reservoir would have a positive recharge rate.

2.8 Could Low-Salinity Water Be a Valuable Byproduct?

Although the main economic product of NGH production is natural gas, a very large amount of fresh water on earth is tied up in NGH. This water is produced along with the natural gas when the NGH is converted. It has been recognized that submarine strata contain aquifers holding very large amounts of fresh and brackish water. There is speculation that this water may be important as water sources in the future (Post et al. 2013).

We do not suggest that the fresh water produced along with the natural gas could now be regarded as an economic product, but if it could be produced along with the natural gas along coasts that were severely water stressed, it could have considerable value. This would be especially true if the gas was involved in lifting the water so that energy costs could be minimized. The water produced from NGH is relatively pure. Even mixed with some normally saline pore water it might be suitable for agricultural purposes and could require only superficial polishing to be graded as potable.

It has been estimated that in excess of 45,000 Tcf of gas in place in NGH exists in producible geological situations. This is equivalent to about 220 Tcf of fresh water in the NGH reservoir (0.8 ft^3 H_2O for each 164 ft^3 of methane). This is a lot of fresh water.

References

Archer, D. (2007). Methane hydrate stability and anthropogenic climate change. *Biogeosciences Discussions, European Geosciences Union, 4*(2), 993–1057. hal-00297882.

Barnard, A., Sager, W. W., Snow, J. E., and Max, M. D. (2015). Subsea Gas Emissions from the Barbados Accretionary Complex. *Marine and Petroleum Geology, 64*, 31–42. https://doi.org/10.1016/j.marpetgeo.2015.02.008.

Barton, C. M. (2015). *Industry continues to provide solutions for deepwater production challenges* (8 pp). Oilpro 25 May 2015.

Beaudoin, Y. C., Dallimore, S. R., & Boswell, R. (Eds.). (2014b). *Frozen heat: A UNEP global outlook on methane gas hydrates* (Vol. 2, 96 pp). United Nations Environment Programme, GRID-Arendal. United Nations Environment Programme, 2014. ISBN: 978-92-807-3319-8 Job No: DEW/1633/NO.

Beaudoin, Y. C., Waite, W., Boswell, R., & Dallimore, S. R. (Eds). (2014a). *Frozen heat: A UNEP global outlook on methane gas hydrates* (Vol. 1, 80 pp). United Nations Environment Programme, GRID-Arendal. United Nations Environment Programme, 2014. ISBN: 978-92-807-3429-4 Job No: DEW/1866/NO.

Bhatnagar, G., Chapman, W. G., Dickens, G. R., Dugan, B., & Hirasaki, G. J. (2008). Sulfate-methane transition as a proxy for average methane hydrate saturation in marine sediments. *Geophysical Research Letters, 35,* L03611. https://doi.org/10.1029/2007GL032500.

Boswell, R, Collett, T. S. (2006). The gas hydrates resource pyramid: Fire in the ice, Methane hydrate newsletter, US Department of Energy, Office of Fossil Energy, National Energy Technology Laboratory, (Fall Issue) *6*(3), 5–7.

Boswell, R., Collett, T. S., Frye, M., McConnell, D., Shedd, W., Dufrene, R., et al. (2012a). *Gulf of Mexico gas hydrate joint industry project leg II: Technical summary*. NETL, Morgantown WV: US Department of Energy.

Boswell, R., Collett, T. S., Frye, M., Shedd, W., McConnell, D. R., & Shelander, D. (2012b). Subsurface gas hydrates in the northern Gulf of Mexico. *Marine and Petroleum Geology*, 21. https://doi.org/10.1016/j.marpetgeo.2011.10.003.

Brewer, P. G., Orr, F. M., Friederich, G., Kvenvolden, K. A., & Orange, D. L. (1998). Gas hydrate formation in the deep sea: In-situ experiments with controlled release of methane, natural gas, and carbon dioxide. *Energy & Fuels, 12,* 183–188.

Claypool, G. W., & Kaplan, I. R. (1974). Methane in marine sediments. In I. R. Kaplan (Ed.), *Gas in marine sediments* (pp.73–99). Plenum.

Collett, T. S. (2001). Quantitative well-log analysis of in-situ gas-hydrate-bearing sediments, In C. K. Paull, & W. P. Dillon (Eds.), *Geophysical monograph 124* (pp. 189–210). American Geophysical Union.

Collett, T. S., Johnson, A. H., Knapp, C. C., & Boswell, R. (2009). In T.S. Collett, A. H. Johnson, C. C. Knapp, & R. Boswell (Eds.). Natural gas hydrates: A review. *American Association of Petroleum Geologists Memoir 89*, 146–219. ISBN13: 918-0-89181-370-5.

Crager, B. (2015). Slide 9 subsea production systems—Market update. *Endeavor Management* (2014). http://oilpro.com/gallery/249/3665.

Daigle, H., Cook, A., & Malinverno, A. (2015). Permeability and porosity of hydrate-bearing sediments in the northern Gulf of Mexico. *Marine and Petroleum Geology, 68,* 551–564. https://doi.org/10.1016/j.marpetgeo.2015.10.004.

Davie, M. K., & Buffett, B. A. (2001). A numerical model for the formation of gas hydrate below the seafloor. *Journal of Geophysical Research, 106*(B1), 497–514.

Dickens, G. R., Paull, C. K., Wallace, P., & ODP Leg 164 Scientific Party. (1997). Direct measurement of in situ methane quantities in a large gas-hydrate reservoir. *Nature*, *385*, 426–428.

Dillon, W. P., Hutchinson, D. R., & Drury, M. (1996). 4. Seismic reflection profiles on the Blake Ridge near sites 994, 995, and 997. In C. K. Paull, R. Matsumoto, & P. J. Wallace, et al. (Eds.), *Proceedings of the ocean Drilling Program* (pp. 47–56), Initial reports 164.

EIA. (2015). The growth of U.S. natural gas: An uncertain outlook for U.S and world supply. In *Presentation at EIA Energy Conference June 15, 2015* (44 pp), Washington D.C. http://www.eia.gov/conference/2015/pdf/presentations/staub.pdf. Accessed January 6, 2016.

Englezos, P., & Bishnoi, P. R. (1988). Prediction of gas hydrate formation conditions in aqueous electrolyte solutions. *American Institute of Chemical Engineers, 34*(10), 1718–1721.

Evans, P. N., Parks, D. H., Chadwick, G. L., Robbins, S. J., Orphan, V. J., Golding, S. D., et al. (2015). Methane metabolism in the archaeal phylum Bathyarchaeota revealed by genome-centric metagenomics. *Science, 350,* 434–438.

Fisher, C. R., MacDonald, I. R., Sassen, R., Young, C. M., Macko, S. A., Hourdez, S., et al. (2000). Methane ice worms. *Hesiocaeca methanicola colonizing fossil fuel reserves: Naturwissenschaften, 87,* 184–187.

Foucher, J., Westbrook, G., Boetius, A., Ceramicola, S., Dupré, S., Mascle, J., et al. (2009). Structure and drivers of cold seep ecosystems. *Oceanography, 22,* 92–109.

Frye, M., Shedd, W., & Schuenemeyer, J. (2013). *Gas hydrate resource assessment atlantic outer continental shelf* (57 pp). Board of Ocean Energy Management (BOEM) Report RED 2013-01.

Fujii, T., Suzuki, K., Takayama, T., Tamaki, M., Komatsu, Y., Konno, Y., et al. (2015). Geological setting and characterization of a methane hydrate reservoir distributed at the first offshore production test site on the Daini-Atsumi Knoll in the eastern Nankai Trough, Japan. *Marine and Petroleum Geology, 66*(1), 310–322. https://doi.org/10.1016/j.marpetgeo.2015.02.037.

Hachikubo, A., Yanagawa, K., Tomaru, H., Lu, H., & Matsumoto, R. (2015). Molecular and isotopic composition of volatiles in gas hydrates and in sediment from the Joetsu Basin, eastern margin of the Japan Sea. *Energies*, 4647–4666. https://doi.org/10.3390/en8064647.

Henke, T., Muller, C., Marquardt, M., Hensen, C., et al. (2008). Integrated gas hydrate quantification of Nicoya Peninsula—Costa Rica. In *Proceedings of the 6th International Conference on Gas Hydrates (IGCH 2008)*, Vancouver, British Columbia, Canada, 6–10 July 2008.

Hillman, J. I. T., Cook, A. E., Sawyer, D. E., Jüçük, H. M., & Goldberg, D. S. (2017). The character and amplitude of a 'discontinuous' bottom-simiulating reflections in marine seismic data. *Earth and Planetary Science Letters, 459,* 157–169.

Holditch, S. A. (2013). Unconventional oil and gas resource development—Let's do it right. *Journal of Unconventional Oil and Gas Resources, 1–2,* 2–8. https://doi.org/10.1016/j.juogr.2013.05.001.

Hornbach, M. J., Saffer, D. M., & Holbrook, W. S. (2004). Critically pressured free-gas reservoirs below gas-hydrate provinces. *Nature, 427,* 142–144.

Johnson, A. H. (2011). Global resource potential of gas hydrate—a new calculation. U.S. Department of Energy. *Fire in the Ice, 11*(2), 1–4.

Johnson, A.H. (2012). Gas hydrate. In *GEA, 2011: The Global Energy Assessment, IIASA* (pp. 35–43), Laxenburg, Austria and Cambridge University Press, Cambridge, United Kingdom and New York, NY, USA.

Johnson, J. E., Mienert, J., Plaza-Faverola, A., Vadakkepuliyambatta, S., Knies, J., Bünz, S., et al. (2015). Abiotic methane from ultraslow-spreading ridges can charge Arctic gas hydrates. *Geology, 43*(5), 371–374. https://doi.org/10.1130/G36440.

Johnson, J. E., Waghorn, K. A., Mienert, J., & Bünz, S. (2016). The Potential for abiotic methane in arctic gas hydrates. *U.S. Department of Energy Fire in the Ice, 16*(1), 9–12.

Kashchiev, D., & Firoozabadi, A. (2002a). Nucleation of gas hydrates. *Journal of Crystal Growth, 243,* 476–489.

Kashchiev, D., & Firoozabadi, A. (2002b). Driving force for crystallization of gas hydrates. *Journal of Crystal Growth, 241*(1), 220–230.

Kliewer, G. (2015). Ultra-deepwater to capture most of Brazil's offshore capex. *Offshore, 75*(5), 46.

Kvenvolden, K. A. (1988). Methane hydrate-A major reservoir of carbon in the shallow geosphere? *Chemical Geology, 71,* 41–51.

Kvenvolden, K. A. (1995). A review of the geochemistry of methane in natural gas hydrate. *Organic Geochemistry, 23,* 997–1008. https://doi.org/10.1016/0146-6380(96)00002-2.

Kvenvolden, K. A., & Lorenson, T. D. (2001). The global occurrence of natural gas hydrate. In C. K. Paull, & W. P. Dillon (Eds.), *Natural gas hydrates: Occurrence, distribution, and dynamics* (Vol. 124, pp. 3–18). American Geophysical Union, Geophysical Monograph.

Løseth, H., Gading, M., & Wensaas, L. (2009). Hydrocarbon leakage interpreted on seismic data. *Marine and Petroleum Geology, 26,* 1304–1319.

Matsumoto, R., Tanahashi, M., Kakuwa, Y, Snyder, G., Ohkawa, S., Tomaru, H., & Morita, S. (2017). Recovery of thick deposits of massive gas hydrates from gas chimne structures, eastern margin of Japan Sea: Japan Se Shallow Gas Hydrate Project. *U.S. Department of Energy Fire in the Ice, 17*(1), 1–6.

Matsumoto, R., Tomaru, H., & Lu, H., (2004). Detection and evaluation of gas hydrate formation in the Eastern Nankai Trough by geochemical and geophysical methods, *Resource Geology, 54*, 53–68.

Max, M.D. (1990). Gas hydrate and acoustically laminated sediments: probable environmental cause of anomalously low acoustic-interaction bottom loss in deep ocean sediments (p. 68). Naval Research Laboratory Report 9235.

Max, M. D. (Ed.). (2003). Natural gas hydrate: In oceanic and permafrost environments (2nd ed., 422 pp). London, Boston, Dordrecht: Kluwer Academic Publishers (now Springer).

Max, M. D., & Dillon, W. P. (1998). Oceanic methane hydrate: Character of the Blake ridge hydrate stability zone and potential for methane extraction. *Journal of Petroleum Geology, 21,* 343–358.

Max, M. D., & Johnson, A. H. (2011a). Methane Hydrate/Clathrate Conversion. In M. R. Khan (Ed.), *Clean hydrocarbon fuel conversion technology* (pp. 413–434). Woodhead Publishing Series in Energy No. 19. Woodhead Publishing Ltd. Cambridge, U.K. ISBN 1 84569 727 8, ISBN-13: 978 1 84569 727 3.

Max, M. D., & Johnson, A. H. (2014). Hydrate petroleum system approach to hydrate natural gas exploration. *Petroleum Geoscience, 20*(2), 187–199. Geological Society of London (Published in Online First 21 March 2014), https://doi.org/10.1144/petgeo2012-049.

Max, M. D., Johnson, A. H., & Dillon, W. P. (2006). *Economic geology of natural gas hydrate* (p. 341). Berlin, Dordrecht: Springer.

Max, M. D., Johnson, A. H., & Dillon, W. P. (2013). *Natural gas hydrate arctic ocean deepwater resource potential* (113 pp). SpringerBriefs in Energy.

Max, M. D., & Lowrie, A. (1993). Natural gas hydrates: Arctic and Nordic Sea potential. In T. O. Vorren, E. Bergsager, Ø. A. Dahl-Stamnes, E. Holter, B. Johansen, E. Lie, & T.B. Lund (Eds.), *Arctic Geology and Petroleum Potential, Proceedings of the Norwegian Petroleum Society Conference*, pp. 27–53. 15–17 August 1990, Tromsø, Norway: Norwegian Petroleum Society (NPF), Special Publication 2 Elsevier, Amsterdam.

NPC. (2007). Topic Paper #39 Unconventional Gas. Working document of the NPC Global Oil & Gas Study (July 18) National Petroleum Council, U.S. Department of Energy, p. 54.

Paull, C. K., Borowski, W. S., & Rodriguez, N. M. & ODP Leg 164 Shipboard Scientific Party. (1998). Marine gas hydrate inventory: preliminary results of ODP Leg 164 and implications for gas venting and slumping associated with the Blake Ridge gas hydrate field. In J.-P. Henriet, & J. Mienert (Eds.), *Gas hydrates: Relevance to world margin stability and climate change* (Vol. 137, pp. 153–160). Geological Society London Special Publication.

Paull, C. K., & Ussler, W. (2001). History and significance of gas sampling during DSDP and ODP drilling associated with gas hydrates. In Paull, C. K., & Dillon, W. P. (Eds.), *Natural Gas Hydrates. Occurrence, Distribution, and Dynamics. Geophysical Monograph* (Vol. 124, AGU, pp. 53–65).

Paull, C. K., Ussler, W., & Borowski, W. S. (1994). Sources of biogenic methane to form as hydrates, in situ production or upward migration. In E. D. Sloan, J. Happel, & M. A. Hnatow (Eds.), *International Conference on Natural Gas Hydrates, Annals NY Academy of Sciences* (Vol. 715, pp. 392–409).

Pellenbarg, R. E., & Max, M. D. (2003). Chapter 1 introduction, physical properties, and natural occurrences of hydrate. In M. D. Max (Ed.), *Natural gas hydrate: In Oceanic and permafrost environments* (2nd ed., pp. 1–8). London, Boston, Dordrecht: Kluwer Academic Publishers (now Springer).

Plaza-Faverola, A., Bunz, S., Johnson, J. E., Chand, S., Knies, J., Mienert, J & Franek, P. (2015). Role of tectonic stress in seepage evolution along the gas hydrate-charged Vestnesa Ridge, Fram Strait. *Geophysical Research Letters* 42, 733–742, https://doi.org/10.1002/2014GL062474.

Post, V. E. A., Groen, J., Kooi, H., Person, M., Ge, S., & Edmonds, W. M. (2013). Offshore fresh groundwater reserves as a global phenomenon. *Nature Insight Review, 504,* 71–78.

Rajan, A., Mienert, J., Bünz, S., & Chand, S. (2012). Potential serpentinization, degassing, and gas hydrate formation at a young (< 20 Ma) sedimented ocean crust of the Arctic Ocean ridge system. *Journal Geophysical Research, 117,* B03102. https://doi.org/10.1029/2011JB008537.
RCMHRJ. (2008). *PHAE 1 Comprehensive Report of Research Results August 2008 Edition* (72 pp). Research Consortium for Methane Hydrate Resources in Japan.
Rehder, G., Kirby, S. H., Durham, W. B., Stern, L. A., Peltzer, E. T., Pinkston, J., et al. (2004). Dissolution rates of pure methane hydrate and carbon-dioxide hydrate in undersaturated seawater at 1000-m depth. *Geochimica et Cosmochemica Acta, 68*(2), 285–292.
Reichel, T., & Gallagher, J. W. (2015). Global screening of gas hydrates—How does the Arctic score? In *Abstract in program of 3P Arctic The Polar Petroleum Potential Conference and Exhibition* (p. 92), Stavanger, Norway 29 September–2 October 2015.
Richardson, G. E., Leanne, S., French, L. S., Baud, R. D., Peterson, R. H., Roark, C. D., et al. (2004). *Deepwater Gulf of Mexico 2004: America's expanding frontier* (151 pp) (Offshore Continental Shelf Report MMS 2004-021). U.S. Department of the Interior, Minerals Management Service New Orleans.
Richardson, G. E., Nixon, L. D., Bohannon, C. M., Kazanis, E. G., Montgomery, T. M., & Gravois, M. P. (2008). *Deepwater Gulf of Mexico 2008: America's offshore energy future* (103 pp)(OCS Report MMS 2008-013). U.S. Department of the Interior, Minerals Management Service New Orleans.
Rigzone. (2015). *How does artificial lift work?* Rigzone http://www.rigzone.com/training/insight.asp?insight_id=315&c_id=4. Accessed April 23, 2016.
Robb, L. (2005). *Introduction to ore-forming processes* (373 pp). Blackwell Publishing Company, U.K. & Australia. ISBN 0-632-06378-5.
Sassen, R., & MacDonald, I. R. (1997). Thermogenic gas hydrates, Gulf of Mexico Continental Slope: Preprints, 213th ACS National Mtg., San Francisco, CA, 42(2), 472–474.
Sassen, R., Sweet, S. T., Milkov, A. V., DeFreitas, D. A., & Kennicutt, M. C. II. (2001). Stability of thermogenic gas hydrate in the Gulf of Mexico: Constraints on models of climate change. In C. K. Paull, & W.P. Dillon (Eds.), *Natural gas hydrates occurrence, distribution, and detection* (Vol. 124, pp. 131–143). American Geophysical Union Geophysical Monograph.
SBC. (2015). *Natural Gas Series FactBook* (99 pp). Gas Hydrates SBC Energy Institute (SBC-EI) Schlumberger. https://www.sbc.slb.com/~/media/Files/SBC%20Energy%20Institute/SBC%20Energy%20Institute_Gas%20Hydrates_FactBook.pdf. Accessed January 4, 2016.
Schlumberger. (2015). *Integrated lifting solutions for enhanced well production*. http://www.slb.com/services/production/artificial_lift.aspx. Accessed April 23, 2016.
Shell. (2015). *Tapping into deep-water reservoirs*. http://www.shell.com/global/aboutshell/major-projects-2/bc-10/technology.html. Accessed April 23, 2015.
Sloan, E. D. (1990). *Clathrate hydrates of natural gases* (p. 641). New York: Marcel Dekker.
Sloan, E. D., & Koh, C. A. (2008). *Clathrate hydrates of natural gases* (3rd ed., 752 pp). Chemical Industries Series 119. CRC Press, Taylor & Francis Group. ISBN 9780849390784 - CAT# 9078.
Smith, T. (2009). Gas hydrates—Not so unconventional. *GEOExPro, 6*(2), 19. http://www.geoexpro.com/articles/2009/02/gas-hydrates-not-so-unconventional.
Stern, L. A., Kirby, S. H., William, B., Durham, W. B., Circone, S., William F., et al. (2003). Laboratory synthesis of pure methane hydrate suitable for measurement of physical properties and decomposition behavior. In M. D. Max (Ed.), *Natural gas hydrate: In Oceanic and permafrost environments* (2nd edn., pp. 323–248). Springer, London, Boston, Berlin.
Takahashi, H., & Tsuji, Y. (2005). Multi-Well Exploration Program in 2004 for Natural Hydrate in the Nankai-Trough Offshore Japan. In *OTC 17162, 2005 Offshore Technology Conference* held in Houston, TX, U.S.A., 10 pp, 2–5 May 2005.
Talukder, A. R. (2012). Review of submarine cold seep plumbing systems: leakage to seepage and venting. *Terra Nova, 24,* 255–272. https://doi.org/10.1111/j.1365-3121.2012.01066.x.
Total. (2016). *Deep offshore reservoirs: tapping growth*. Total. http://www.total.com/en/energies-expertise/oil-gas/exploration-production/strategic-sectors/deep-offshore/challenges/overcoming-harsh-conditions-ocean-deeps?%FFbw=kludge1%FF.

van der Walls, J. H. (1956). Statistical mechanics of Clathrate compounds. *Transactions of the Faraday Society, 52,* 184–193. https://doi.org/10.1039/TF9565200184.

Wood, D. A. (2015). Gas hydrate research advances steadily on multiple fronts: A collection of published research (2009e2015). *Journal of Natural Gas Science and Engineering, 24*, A1–A8. http://dx.doi.org/10.1016/j.jngse.2015.04.019. Accessed March 2, 2016.

Wood, W. T., Gardner, J., Hagen, R. A., Coffin, R. B., Pohlman, J. W., Hart, P. E. & Hutchinson, D. R. (2004). Modeling Heat and Fluid Flux of Seafloor Mounds in the Gulf of Mexico (Abs.). In *Proceedings 85(47) AGU Fall Meeting*, San Francisco, Ocean Sciences General Contributions: Gas Hydrates I OS34B-08.

Wood, W. T., & Jung, W.-Y. (2008). Modeling the extent of Earth's marine methane hydrate cryosphere. In *Proceedings of the 6th International Conference on Gas Hydrates (ICGH 2008)*, Vancouver, British Columbia, Canada, 8 pp, July 6–10, 2008. Image is a more detailed model supplied by Warren Wood to M.D. Max, with thanks.

Chapter 3
Exploration for Deepwater Natural Gas Hydrate

Abstract The potentially large volume of oceanic natural gas hydrate (NGH) resources, combined with new understanding of the NGH petroleum system and demonstrated exploration techniques, suggests that near-term extraction of natural gas from the resource is feasible. Sandy marine turbidite sediments in the Nankai (Japan) and Walker Ridge (US—Northern Gulf of Mexico) localities have been proven to host high concentrations of NGH. NGH is a stratigraphic play insofar as the present primary exploration targets are marine turbidite sands. Sediments of this type, deposited on deep continental shelves and slopes, are a primary target for NGH extraction potential. Similar, more deeply-buried turbidites are prominent first-order hosts for conventional gas and oil deposits. Concentrations of NGH cluster along the base of the GHSZ although NGH is more stable toward its top. These are proven hydrocarbon resources and amenable to extraction of gas and oil. We infer that conversion of NGH to its constituent gas and water will allow the natural gas to be recovered. Exploration tools are sufficient to identify and value NGH concentrations.

3.1 NGH Exploration

Natural gas hydrate (NGH) occurs in marine sediments in regions that have active, leaky petroleum systems. Often significant conventional gas or oil accumulations may occur at greater depths. Therefore, NGH should be viewed as just a unique unconventional natural gas deposit that extends the geological range of continental flank hydrocarbon deposits to very near the seafloor.

NGH was originally considered to be a physical hazard where it occurred on or near the seafloor as it could induce instability impacting drilling safety, mooring, and placement of facilities on the seafloor. In addition, the unwanted formation of NGH in pipelines and other apparatus could impact production and income. Only since the 1970s was NGH envisaged as a potential resource, and only since the 1990s has enough understanding of the NGH resource made possible real exploration activities, which it is hoped will lead to commercialization of the resource.

M. D. Max and A. H. Johnson, *Exploration and Production of Oceanic Natural Gas Hydrate*, https://doi.org/10.1007/978-3-030-00401-9_3

NGH stability zone (GHSZ) conditions of temperature and pressure are found below about 500 m water depths that occur on continental margins worldwide. The general area of the exploration region can be easily located on a bathymetric map of the continental margins. Although NGH may occur throughout the GHSZ because local gas production is significant enough to form low percentages of dispersed NGH in primary or secondary porosity (Chap. 2), the paragenesis of large NGH concentrations with commercial potential is different from that of conventional hydrocarbon deposits. Exploration methods for NGH can borrow from conventional industry exploration methods, although they must be reconfigured and optimized for NGH. In the case of seismic survey and analysis, major steps have already been taken.

Only large concentrations of NGH are of interest as a potential gas resource; however, the dispersed NGH that can occur at relatively low percentages of pore volume over large areas in muddy sediments may retain some interest for their impact on climate due to their response to environmental change (Beaudoin et al. 2014). This chapter considers the factors that can cause large concentrations to form, along with the exploration and evaluation and the production of those deposits. The approach is, and will be, similar to that outlined by Boswell (2014).

In order for a continental margin to be prospective, it must have (Max and Johnson 2014): (1) A NGH stability zone (GHSZ) of suitable thickness and overburden to provide for adequate structural competency to allow for drilling and production. (2) Sufficient existing and historical gas flux and a 'plumbing system' suitable to allow gas and pore water to migrate upward into the GHSZ where it might form large amounts of NGH if other conditions are met. SBC (2015) refers to this coincidence of GHSZ and sufficient gas flux as the NGH occurrence zone (GHOZ), while we simply refer to first order prospective regions as 'NGH prospect zones' within the GHSZ to avoid another acronym. (3) Suitable host sediments in which large concentrations of NGH may crystallize from migrating mineralizing solutions. Once these conditions are recognized, there are certain very specific seismic attributes that may indicate the presence of NGH concentrations. It is important to note that there is not necessarily a geological trap as there has to be for conventional hydrocarbons to pond. Although the suitable host bed may be bounded by essentially impermeable strata, the formation of the NGH is due to a chemical reaction when the ascending mineralizing solutions migrate into a GHSZ.

3.1.1 Deepwater and Ultra-Deepwater

Conventional energy exploration was historically confined to land or the shallow water areas that could be treated essentially as flooded land. Exploration and production costs were low and the resources generally accessible. Exploration and production of gas and oil has extended into deeper water following WWII, a natural progression as conventional exploration targets were exhausted in the onshore and shallower water areas. Examples include the Louisiana and Texas continental shelves, the deeper water in the North Sea and off NW Australia continental shelves. Exploration for giant fields has now been forced to pursue much higher cost areas beyond the continental shelf and onto the continental slope and possibly the lower slope. Although giant gas fields are still being found and production opening in shallow water (Reed 2016), the majority of the new production is liable to come from fields in very deep water. For instance, the deepest US production is from the Jack/St. Malo Complex in the Walker Ridge Area of the northern Gulf of Mexico, in 7000 ft. of water, and the Lucius Field in the Keathley Canyon Area is in 7100 ft. of water. Chevron's "Toledo" well was drilled in 10,000 ft. of water.

Thus, exploration for conventional gas and oil, in which the internal pressures of the already concentrated gas and oil is high enough to force it to the surface, has pushed into water depths and distances from land in which very high costs must be incurred. These costs are still acceptable because of the high value of the products and the requisite size of the deepwater conventional resources. The increased costs for conventional hydrocarbon development opened the way to development of unconventional resources. Unconventional energy resources are those that will not automatically flow but must first be stimulated or converted in some way so that the gas and/or oil will flow, concentrate, and be producible in large enough quantities and at high enough rates to be commercially produced.

Increasing demand for energy, combined with national desires to be as energy independent as possible, has led to defining, exploring, and producing unconventional energy resources, especially in North America. Oil shales of the Green River Formation of SW Wyoming and adjacent states were the initial targets. These consist of immature hydrocarbons in which the organic material has not yet been converted to gas and oil. Huge tonnages of this resource occur at or near the surface and are amenable to open pit mining. It has remained uneconomic because of environmental issues and the relatively high cost of converting the immature hydrocarbons into usable gas and oil, particularly as other unconventional hydrocarbons have entered production.

When hydrocarbon exploration, drilling, and production first went offshore, simple analogs for drilling on land could be practiced from 'jackup' platforms. In fact, in many respects drilling from a shallow water sandy seafloor could be simpler than drilling from barges in channels in coastal wetlands. However, as exploration and production moved into increasingly deeper water farther offshore, entirely new technology had to be developed, and with each successive move to ever-deeper water, increasingly sophisticated engineering, materials and practices had to be mastered.

Industry has proved itself adept at providing solutions for deepwater and ultra-deepwater operations. During the evolution of offshore technology, each step devised increasingly robust equipment and practices. Offshore exploration and production technology is now capable to reach to any water depth, although considerable technology development and improvement to existing frontier solutions remains (Barton 2015).

Offshore drilling and production have to be far more resilient to industry cycles and price fluctuations than on land because of the greater investment and longer time necessary from discovery to production. But offshore fields and resident facilities are commonly larger than the majority of most onshore plays, with longer production runs. Exploration, including the acquisition of offshore acreage, seismic survey and interpretation, and drilling that confirms field discovery may take years. The elapsed time from discovery to project approval is commonly more than five years for spar facilities and in excess of eight years for tension leg platforms (TLPs). These are currently the production units most commonly used in the deepest water; with floating production, storage, and offloading facilities (FPSOs) installed in some locations for production support. The time from project approval to first production commonly exceeds three years. Thus, it is common for investment to support development for a decade before revenue income is achieved. Deep and ultra-deepwater costs commonly exceed $1 billion. Despite the high cost, ultra-deepwater production is set for steady growth (Offshore 1/19/15). In some deepwater regions, such as offshore Brazil, most of the developmental capital expense (CAPEX) will be spent in ultra-deepwater

Continental Margin Economic Depth Zones For the purposes of discussing offshore exploration and development of hydrocarbon resources, it is important to constrain the transitional depths from shallow to deepwater, and to ultra-deepwater as NGH can occur in marine sediments in all three zones.

There is little agreement on the water depth at which shallow water gives way to deep water. According to the U.S. Minerals Management Service (MMS, now Bureau of Ocean Energy Management, BOEM), the threshold separating shallow from deep water can range from 200- to 457 m (656–1500 ft.) in water depth, although they used a transitional depth of (MMS 2008) of 305 m (1000 ft.), close to that used by Crager (2015). Schlumberger (2015) uses the depth of 200 m ($\sim$600 ft.) as this has long been regarded as the approximate geomorphic depth at which most continental shelves give way to continental slope. Infield (2015) uses a transitional depth to deepwater of 500 m ($\sim$1650 ft.).

Troughs within continental shelves that host GHSZs, such as those in the Arctic Ocean (Max et al. 2013), such as the St. Anna Trough, Bear Island Trough, and a number of other troughs in the Barents Sea, have GHSZs up to $\sim$200+ m in thickness. The Canadian Arctic shelf is even more broadly incised with deep troughs that pass hundreds of km to the south, passing between large Arctic islands. These also have GHSZ $\sim$ 200+ m in thickness.

The troughs in the Arctic Ocean continental shelves and incised into the SE Canadian shelf (such as the Laurentian Trough and the Flemish Pass deep between Flemish Cap and the main continental shelf off Newfoundland) also have GHSZ up to 600 m thick (Majorowicz and Osadetz 2003) and a prolific petroleum system (Beckman 2015). NGH deposits in troughs will likely be in a more geotechnically stable situation than the Nankai deposits, which are on a steep slope very near the shelf edge in turbidite sands that ponded in a near shelf-edge fault basin (Egawa et al. 2013). Thus sediment-filled deep troughs with GHSZ of $\sim$150 m or thicker constitute a relatively unexplored NGH play that may prove to be sedimentologically and structurally a better host for NGH concentrations than continental slope marine turbidites.

Complex continental margins can also have large areas deeper than the common 200 m shelf break edge. For instance, bathymetric deeps in the Grand Banks off Newfoundland) are as thick as the shelf-margin Nankai NGH play about 70 km SW of Tokyo Bay, which argues for their having comparable NGH potential. The Nankai deposit is a special case. Geomorphologically it is part of the shelf edge, although it is on a deep, relatively level plateau step-down from the shallow shelf closer to the Japanese shore line (GM 2016). The Nankai NGH deposit occurs just to the west of the junction of the Japan Trench, and its geographic prolongation to the SW, the Sagami and Nankai Trenches, at about the point on the bathymetrically complex margin of the Eurasian Continental Plate (Fujii et al. 2015, Fig. 1) where the Izu-Bonin Arc and Trench diverge to the SE (Taira 2001). Not only is the entire Japanese SE continental margin an active subduction zone, it includes the confluence of younger and older very active accretionary margins.

Major energy companies generally use an operationally oriented deeper transitional depth for deepwater. For instance, ENI (2016) uses 450 m, while Fang and Duan (2014) quote water depth boundaries for deepwater of 400, 600, 915, 1070 m but note that at the 2002 World Petroleum Congress in Brazil, water deeper than 400 m is regarded as deepwater with ultra-deepwater deeper than 1500 m. From a perusal of current usage, the defined depth of deepwater appears to be following Infield's (2015) designation of 500 m. This is fortuitous for dealing with NGH, which only becomes stable at depths of about 500 m in the open oceans, and all

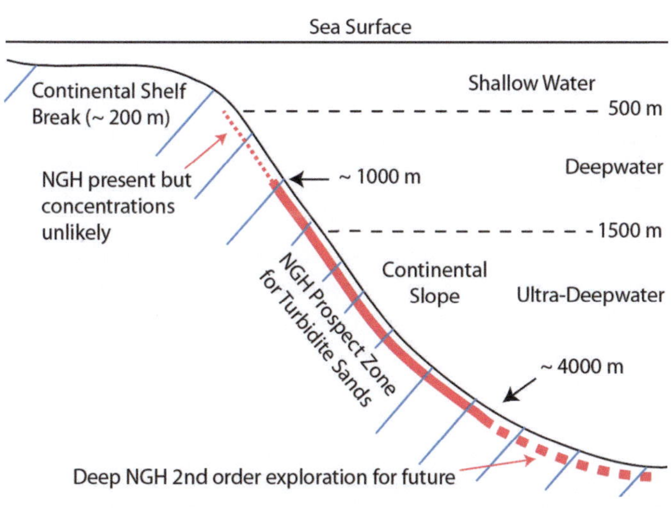

Fig. 3.1 Oceanic exploration target and general ocean depth zones. Geomorphic slope breaks are usually taken at about 200 m water depth but deep continental shelves having very cold seafloor water can have NGH

potential commercial oceanic NGH deposits will be found below this depth, and probably no shallower than the Nankai deposit off SE Japan, whose seafloor is a little less than 1 km deep. Relic permafrost NGH may be found in shallow water, but only in Polar Regions (Max and Johnson 2014). There is better agreement on the transition from deep to ultra-deepwater. Infield (2015) generally follows MMS (2004), which placed the transitional depth from deepwater to ultra-deepwater at about 5000 ft. (1524 m).

To further complicate depth zone definitions, the U.S. Government Deep Water Royalty Relief (DWRR) zones (200, 400, 800, and 1600 m) depth zones (MMS 2004) do not relate in any way to the commonly used terms for exploration and production water depth zones.

Thus, there is no absolute definition for the transition depth from shallow to deepwater to ultra-deepwater regions. In some classifications, a ‘midwater’ zone is included within a shallow–midwater–deepwater–ultra-deepwater classification system for conventional hydrocarbon exploration and production. We follow Infield (2015) and use the easy to visualize ≥ 500 and ≥ 1500 m for the transition from shallow to deepwater to ultra-deepwater (Fig. 3.1) because it fits well with NGH systematics in which NGH commonly occurs at a depth of about 500 m and exploration potential for NGH deposits will be common from 1500 m even where moderately high geothermal gradients exist (Chap. 2). The history of deepwater oil and gas E&P began in 1947 in the northern Gulf of Mexico and is well summarized by Nixon et al. (2016).

3.1.2 Basin Modeling

As with conventional exploration, early characterization of sedimentary basins, the first exploration step is to carry out identification of sediment-filled basins in which there is not only a GHSZ but sediment that can generate sufficient natural gas. There may be areas in which basement blocks that cannot generate hydrocarbons occupy much of the near-seafloor or are found pervasively downward from shallow burial depths. There may also be areas where igneous or volcanic rocks are profuse. But on passive continental margins the sediment pile is commonly thick and contains source, migration, and trap components of petroleum systems. The tectonic history of the basin need not be known in the same detail and timing as conventional basins, and the deeper parts of the basin may not be that important to exploration of NGH concentrations if the relationship between the uppermost plumbing system and the most important mineralization migration paths are known. Of the many faults and permeable strata that may exist in an NGH prospective area, the migration path for mineralizing fluids migrating into the GHSZ is the key to localizing the potential locations for NGH concentrations.

Sediment will tend to be thicker and less disrupted on passive continental margins and sediment continuity within the GHSZ should be relatively continuous and related to only the uppermost sediments which will tend to lie laterally with a more pronounced continuity than in a disrupted sediment pile. On active continental margins, even though total sediment may be thicker in an imbricated sediment pile, the nature of the sediment in each tectonic slice within and just below the GHSZ should be understood with respect to the depositional environment in which it formed before tectonic disruption and imbrication. Basin analysis from the point of view of establishing the likelihood of strata that could host high-grade NGH concentrations may be more complicated in active continental margins.

3.1.3 NGH Prospect Zone

Identification of a NGH prospect zone is based on the location of a GHSZ that is thick enough, and has enough overburden to host a large concentration of NGH, a sufficient focused natural gas flow, and suitable host sediments in which large concentrations of NGH could occur. Although NGH is stable at depths from about 500 m downward, large concentrations of NGH can be expected in marine sediments downward from about 1 km water depth. Our general model for the NGH prospect zone is on the continental slope because the slope break between shelf and slope is commonly about 200 m (Fig. 3.1), but it is important to note that there are deep continental shelf breaks and internal troughs that could also host NGH deposits. For instance, the Nankai NGH deposit occurs slightly shallower than 200 m below the seafloor in slightly in less than 1 km water depth near the edge of the continental shelf. In any case, exploration should probably be focused in water depths below 1 km, except where seawater is very cold, in which case 800 m water depth is probably a reasonable upper limit to the NGH prospect zone.

The NGH economic zone (Max and Johnson 2014) is essentially the full thickness of marine sediment below the seafloor in which the factors relevant to NGH exploration pertain. This zone will rarely be deeper than 1.2 km beneath the seafloor. The economic zone includes all of the GHSZ and enough of the subjacent sediment to support fluid movements and the general geology relevant to replacing the heat deficit caused by depressurization conversion of NGH to its water and natural gas components, with depressurization being the most likely of 5 NGH conversion methods to be used commercially. For NGH evaluation it is not usually necessary to conduct deeper seismic surveys, so less powerful sources and possibly shorter receiving arrays may be used. In addition, it is not necessary to understand the thermal history of an entire basin, and it is not necessary to do exploratory drilling deeper than this zone. All of these elements of exploration are significantly different from exploration for conventional hydrocarbons. When exploration tools are used that provide all that is necessary for NGH exploration, considerable cost savings over the heavyweight systems required for conventional hydrocarbons should be possible.

NGH can be expected to occur within sediments on most continental margins of the world. This is because GHSZ conditions are a simple function of pressure, temperature, and heat flow that naturally occur wherever the conditions for NGH stability occur. The two main variables bearing on the likelihood of large NGH concentrations in GHSZs are the presence of suitable host strata and adequate gas flux. Because only large concentrations of NGH are of interest as a potential gas resource, the dispersed NGH that can occur at relatively low percentages of pore volume over large areas in muddy sediments may retain some interest for their impact on climate due to their response to global warming (Beaudoin et al. 2014), but our focus is on the factors that can cause large concentrations of NGH to form, exploration, evaluation of a deposit for production and production itself.

Jackup rigs can only operate in shallow water along with submersibles, ROVs guided from the rig, barges, and occasionally semi-submersibles in the deeper part of shallow water (Fig. 3.1). New construction of jackups is focused on drilling in 100–125 m water depths, which have no application to NGH exploration and production. Early generation semi-submersibles and drillships can operate effectively in the deeper part of shallow water and into deeper water, but these limited water depth rigs are no longer being constructed, as the new frontier is deepwater. Ultra-deepwater prospects are the future of ocean hydrocarbon energy but pose the greatest challenges. Thus, these rigs are the most technically capable and expensive. Drilling by drillships and a variety of semi-submersible platforms is now possible in 3000 m water depths. Thus, drilling platforms and accompanying technology necessary for NGH exploration and production, including moving more processing and drilling to the seafloor, have already been developed for conventional hydrocarbons. These provide a starting point for early NGH exploration, although the development of new technology operating on the seafloor promises to be much less expensive in the long run.

Potentially economic deposits of NGH will be found in wherever the combination of water depth, seafloor temperature, and geothermal gradient combine to produce a GHSZ thick enough to provide for reservoir stability during production. Normally these deposits will be found in deepwater or ultra-deepwater (Fig. 3.1). However, the Nankai deposit lies on an abnormally deep continental shelf at about 1000 m near the deep shelf margin off SE Japan. Because of the relatively shallow water depth, and high geothermal gradient, the Nankai GHSZ is only about 230 m thick. Other NGH concentrations have been found in deeper water nearby downslope. With only a thin sediment cover, the confining sediments are relatively weak and more liable to disruption caused by NGH conversion and gas production activity. This will require production to be very closely controlled and geotechnically monitored.

The lower water-depth limit of the NGH prospect zone will be at some depth at which exploration and production activities are limited by industrial capability, although we suggest that the maximum seafloor depth that will be accessible may be greater than that for conventional hydrocarbons (see Chaps. 7 and 8).

We take the base of the NGH prospect zone somewhat arbitrarily at about 4 km water depth, although for practical purposes it may be closer to 3 km, especially along constructive plate margins (Johnson et al. 2015). This depth is liable to be the ultra-deep depth limit for extractive operations simply because of the cost and logistical factors that arise because thinner, more distal sediments offer a lower possibility of suitable host strata, and because there appears to be a higher natural concentration of NGH in the continental slope sediments above this depth, as observed for at least at one locality off Puerto Rico (Henke et al. 2008). The cause of the depth-related relative abundance of NGH is not known with any certainty and the relative abundance of NGH may be different from place to place. Nonetheless, the most likely causes for the variation would most likely be in a differential gas flux, with more gas reaching the GHSZ above thicker sediments, or some factor of physical chemistry, such as higher partial pressures of natural gas with increasing depth that affect the concentration of dissolved hydrate-forming gas with relation to NGH, such that the NGH growth dynamic (Max et al. 2006) drops so that only very slow growth is possible.

3.2 NGH Petroleum System Analysis

Petroleum system analysis is a process of incorporating diverse natural system information to assist petroleum and natural gas exploration (Ligtenberg and Neves 2008), especially in the early-stage evaluation of continental slope deposits (Okui et al. 2008). Conventional petroleum system analysis comprises those factors that drive the generation, migration, and accumulation of hydrocarbons. The NGH petroleum system consists of relatively few basic elements, all of which have to be active or interactive at present and in the immediate geological past. The NGH petroleum system describes the paragenesis of NGH formation, with emphasis on identification of the set of conditions that are necessary for large concentrations to form in host sediments and from which the natural gas may be extracted. The NGH petroleum system is similar to the conventional system in that the natural gas migrates into a location in which it can be concentrated, but there are important distinctions from the conventional petroleum system (Max and Johnson 2014). The NGH petroleum system consists of gas supply, migration path, suitable host, and thermodynamic trap.

The NGH petroleum system is similar to the conventional hydrocarbon petroleum system but differs in three ways. The first is that NGH only occurs in GHSZs that extend from seafloor down to some depth determined by rising temperature, whereas conventional deposits can occur to great depths in the sediment. The second is that NGH is a solid in its reservoir, rather than as a compressible gas or a flowable liquid. Formation pressure is essentially hydrostatic with a light sediment load, whereas deeper conventional hydrocarbon deposits can be highly pressurized (geopressured) with respect to seafloor pressure. The third is that there must be a current or a recent gas flux in the vicinity of a NGH deposit or trapped pore water in which the vapor pressure of the dissolved natural gas will be equilibrated with the NGH. In other words, in an open hydraulic system, if gas flux that has caused NGH to form becomes insignificant over time and the dissolved concentration of natural gas falls enough, the NGH will dissolve and disappear. This is in contrast to conventional deposits, some of which may have remained in geological traps for hundreds of millions of years.

Concentrations of liquid, solid, and gaseous hydrocarbons form fossil fuels, upon which our energy-driven economies have depended and will continue to depend even following implementation of significant renewable energy resources (Chap. 1). The concept of a petroleum system was first used to describe the genesis of gas and oil deposits, in which source beds produce mobile hydrocarbons that migrate by their buoyancy along a migration routes to a trap consisting of a geological seal that contain the concentration. All hydrocarbon resources such as coal, coalbed methane, and shale gas and oil, which do not reflect the entire process, also are likely to have their individually differing hydrocarbon systems.

The classical conventional petroleum system elements consist of, (1) A source bed rich in hydrocarbon-prone organic material that is subjected to heat and pressure (thermal maturation) sufficient to produce and liberate hydrocarbon liquids and/or natural gas, (2) A permeable strata or fault system through which the liberated gas and liquids can migrate from the source bed, (3) A reservoir rock in which the gas and liquids can accumulate, and (4) A structural or stratigraphic trap which is sealed by impermeable strata. Understanding the physical and temporal elements of a petroleum system is the key to exploration.

The NGH petroleum system is similar to the conventional system in that the natural gas migrates into a location in which it can be concentrated, but there are important distinctions from the conventional petroleum system, which is discussed in more detail by Max and Johnson (2014). The NGH petroleum system consists of gas supply, migration path, suitable host, and trap.

Gas Supply In conventional hydrocarbon evaluation source beds often can be directly linked with oil and gas concentrations. In addition, different source beds within a basin may have different critical moments of source bed production and hydrocarbon release. In NGH concentrations, gas flux simply must be recent and/or current so that the host beds are essentially flooded with dissolved gas at high enough concentrations to drive NGH crystallization (Chap. 2). The source of the gas, and whether it is thermogenic or biogenic, and when and where it formed or how it migrated, may not provide any information for further exploration as this type of information would for conventional hydrocarbons. Exploration to determine a region with gas a strong petroleum system and gas flux can map bubble plumes (Barnard et al. 2015) or the seafloor, where gas affects both the morphology and biology, and can be discerned using seafloor reflectivity (Roberts et al. 2006).

However, in NGH, a biogenic source could indicate that there is a high probability that it is derived from local subjacent shales and perhaps sourcing sands over a broad area of the GHSZ from nearby source beds. If the gas is strongly thermogenic, the seismic data needs to be used to identify the system fault(s) that brought the gas up from depth (Trèhu et al. 2004.) There is a strong likelihood that sands in the GHSZ closer to main migration concourse fault would have higher NGH concentrations.

Migration Path Similar to the migration paths in conventional gas deposits, the gas moves upward along secondary permeability induced by faulting and laterally up-dip within permeable strata. There must be a groundwater feeder system that allows initially dispersed hydrocarbon to coalesce or to form very high concentrations in pore water that then migrates. As in conventional gas deposits, sands generally provide the permeability that allows fluid and gas to percolate through them. Muddy sediments have very low permeability, although NGH may commonly occur within shales in secondary porosity, filling vugs and veins, and forming nodules. The dispersed NGH in muddy sediments is more likely to be largely locally derived rather than having a remote, deeper source. Both conventional and NGH sand reservoir hosts are confined within the strata by shales or other less permeable strata that confine the migrating hydrocarbon to the discrete, permeable horizon. This confinement is not a trap but part of the migration path system.

Trap Conventional gas accumulates when it meets a structural or stratigraphic trap that forms a barrier to further up-dip migration. The structural traps include anticlines and fault closures, or a combination of the two. Stratigraphic traps include unconformities, pinch-outs, lithofacies traps in which a sand transitions to an impermeable shale. The process of hydrocarbon accumulation depresses the boundary of gas and/or oil with subjacent water as more hydrocarbon reaches the trap.

In contrast, NGH does not require a geological trap into which hydrocarbons migrate and are concentrated. The NGH 'trap' can be described as thermodynamic in nature, rather than a geological trap that acts as a container. However, the presence of a trap may retard the throughput of free or dissolved gas, prolonging the time over which NGH crystallization can work. High NGH saturations have been observed through drilling programs to be higher in sands where there is a geologic trap. Thus far these have been documented in permafrost NGH (Max et al. 2006; Sloan and Koh 2008), but should be expected with marine NGH deposits.

In order for NGH to form, there must be a GHSZ (Chap. 2) of sufficient thickness so that overburden over the deposit can provide a long-term seal to the deposit during production. As the hydrate-forming pore fluid migrates upward into the GHSZ, it reaches temperature and pressure conditions suitable for the nucleation and growth of NGH. As long as the concentration of dissolved hydrate-forming gas is high enough (Chap. 2), NGH will crystalize directly from solution. The physical chemistry of the system that will control the degree of concentration is not discussed in detail here, but the concentration zone or zones can be modeled as giant, naturally occurring crystallizers.

In the absence of any of the required components of the NGH petroleum system, potentially producible concentrations of NGH will not be found, or are unlikely to be found in a geological situation in which first order production techniques that largely exist now, can be applied. Second order production systems for low percentage NGH occurrences have been suggested by Max et al. (2006), but only first order high grade NGH deposits will be of interest until considerable production from them has taken place.

3.2.1 NGH and Conventional Hydrocarbon System Analysis

The plumbing and migration system is the most important facet of the conventional petroleum system for NGH petroleum system analysis. In conventional analysis, the plumbing system is important because without a way for hydrocarbons to migrate from the source beds into reservoirs, there will be no recoverable deposits in gas or liquid form. In NGH system analysis, source beds can be assumed but their strength in producing gas will vary, and no physical trap is required for NGH (although higher NGH saturations may be expected if a structural trap is present). Fujii et al. (2016) have modeled the petroleum system for the eastern Nankai Trough deposit SE central Japan and constrained their predictions with 3D seismic and well data. Crutchley et al. (2016) demonstrate how Vp analysis can be used as a tool for exploring for NGH in the New Zealand sea area.

One of the biggest issues that all energy companies focus on in basin evaluation is figuring out the thermal history of a basin and when the source rocks went into the "oil window" and when the temperatures exceeded it. The question then became when the trap formed. In many cases the traps formed after source beds where done expelling oil and the traps would be empty of hydrocarbons. Conventional hydrocarbons have critical moments. In NGH, the critical moment is 'now' and recently (Max and Johnson 2014).

The hydrocarbon energy industry has been debating the importance of leakage for a long time. Exxon and Shell took the approach that the shales act as seals and the tightness of their seal was a measure of the security of a trap. They worked to identify conduits for fluid movement. BP took the approach that all systems leak to some extent, and that pay zones were places where the hydrocarbons were going in faster than they were leaking out. If the hydrocarbons are going into the trap reservoir faster than they are leaking out, the concentration will enlarge. They noted that most giant fields were related to some kind of surface seep. Even in the offshore they tended to fly over large areas with lasers looking for reflections off the water surface that pointed to underlying oil and gas reservoirs. While in some areas this mainly pointed to vents, they found that even the trap shales had significant amounts of hydrocarbons above commercial reservoirs.

Another factor for conventional hydrocarbon analysis is leakage character. For an NGH deposit, a very leaky system is liable to have the largest concentrations of NGH. This is because the size of the deposit is dependent on the scale of the fluid movement into and through the lower part of a GHSZ. In strong petroleum system areas where dissolved hydrate-forming gas concentration in the ascending pore fluids will be high (Løseth et al. 2009), the more dissolved gas that can be brought into the zone, the higher the likelihood that the scale of the mineralization will be greater. In a NGH concentration, leakiness provides a short term or transitory trap for part of the gas (Chap. 2) being transported by having it immobilized in NGH. The Class 2 NGH deposit consists of mobile water and is essentially an open fluid and gas conduit, the best example of which we know today is the Nankai deposit with sands that are actually venting to the seafloor updip the NGH enriched horizons. In this case the groundwater system is active through the hydrate concentration and physical mixing and transport as well as diffusion are at work. The key to both a leaky conventional reservoir and a NGH concentration is that a significant part of a large concentration be recovered.

3.3 Marine Sediment Host for NGH Deposits

Turbidite sand-silt systems within the NGH stability zone (GHSZ) on deep continental shelves and slopes are widely regarded to be the most likely marine sediment hosts for large concentrations of NGH. These constitute the primary exploration target for NGH (Boswell et al. 2012a, b, 2014; Max and Johnson 2014). To date, not only have the deepwater NGH deposits identified to date been hosted by these deposits, but they are identical to more deeply buried turbidite systems hosting conventional hydrocarbon deposits. This is important because industry already knows how to explore and produce from these sands. Turbidite sands and silts (Shepard 1954) have been proven to be good deepwater reservoirs, having acceptable production rates for conventional gas and oil deposits and are projected to also have acceptable NGH production characteristics following conversion of NGH to its constituent natural gas and water. Performance potential has been substantiated, at least in part, by a technical production test of the Nankai NGH deposit outer continental shelf of Southeastern Japan (JOGMEC 2013).

The first step in exploration is to identify the likelihood for marine turbidite systems that could host NGH concentrations similar to those described by Frye (2008) for the U.S. Gulf of Mexico and Frye et al. (2013a) for the U.S. Atlantic margin. Because the sand systems are much more common in the Gulf of Mexico and the northern part of the U.S. Atlantic offshore continental slope, exploration there is liable to be a more productive use of resources than in the southern part. Understanding the geology of turbidite systems will be important. For instance, the depocenters in the Texas-Louisiana continental margin of the northern Gulf of Mexico migrated to the east from the Late Cretaceous to Recent (Piggott and Pulham 1993). Many of the NGH targets may occur in channel complexes and levees, along with delta-like basin floor fans, although sandier beds may be found where they have ponded in bathymetric lows that up-slope. Sedimentation may have filled in these one-time lows and thus a depositional model of sand host beds must be undertaken (Waltham et al. 2008). As with assessment for conventional hydrocarbons that occur in more deeply buried old turbidite systems, for instance Labourdette et al. (2006), an understanding of the depositional setting and the time slice bathymetry is an important aspect of exploration. Higher quality reservoir sands may host NGH (Chap. 4).

3.4 NGH Hydrocarbon Component Expectations

The NGH petroleum system is a process by which hydrate forming material, mainly dissolved or free methane gas is brought into the GHSZ along permeable strata (e.g., turbidite sands) or in secondary permeability (faults and vent courses). Turbidite sands can be produced although the long-term production volume/rates have yet to be proven to be competitively commercial. Open and closed NGH turbidite sand plays have been identified. These are purely geological classifications that are not related to thermodynamic models for NGH production (Chap. 11). The manner in which the NGH petroleum system acts in type may result in a difference in the composition of hydrocarbons in the reservoir and in the produced gas and water. Further exploration will undoubtedly discover a wide range of geological expressions of closed and open NGH concentrations, as well as transitional types, but even at this early stage of NGH exploration it may be important to characterize them because content of the material in the reservoirs may be different enough to warrant different processing.

3.4.1 Closed NGH Concentrations

These are geological structural (anticlinal closure and fault traps) and stratigraphic traps in the GHSZ into which hydrate-forming materials can be delivered. These can be similar in every way to conventional traps, except they will occur only in the GHSZ, which is much higher in the stratigraphic succession than conventional deepwater traps. These closed traps with reservoir potential have only recently been recognized (Kumar et al. 2016) off the eastern coast of India. Because they are closed, that is there is no means for material that may rise buoyantly into them to migrate out, large concentrations of NGH will bear a superficial resemblance to permafrost hydrate deposits in Alaska (Max et al. 2006). The NGH may cluster largely upward in some host strata from the base of the GHSZ, or either up or downward from the point at which the hydrate-forming material is introduced into a permeable horizon (only currently recoverable concentrations are considered here), hydrate concentrations may not fill the potential reservoir from the top down as is the case for conventional reservoirs. However, all of the material that is brought into the reservoir will remain there. Virtually all of the natural gas that can reach the potential reservoir in the GHSZ will be trapped there as NGH or dissolved or free gas, along with any other material that has also been transported into the geological trap.

3.4.2 Open NGH Concentrations

Open NGH concentrations characteristically form in the lower part of a GHSZ and entrap or sequester the natural gas entirely as part of a thermodynamic process. In an open reservoir, pore water may continue migration past the location of crystalizing NGH concentrations and may actually vent from the seafloor. Although controlling flow of produced gas during production may be more difficult than in a closed reservoir from which the gas can only relocate up-buoyancy, the manner of NGH formation may be important because the NGH may be purer. Thus, the produced gas and water from an open NGH concentration may be purer. Oil droplets and dissolved and mico-dots of oils, waxes, and naphthenic impurities migrating with the pore water that carries the gaseous hydrate-forming materials can potentially be removed in the pore water that bypasses NGH concentrations as it migrates along the migrating pore water. This type of separation that may control the relative volume of extraneous material that may be expected in closed reservoirs and that may not be concentrated in open NGH reservoirs.

3.5 NGH Exploration Methods

The same combination of exploration methods used for conventional offshore conventional gas and oil can be used for NGH exploration with little modification of the equipment to reflect optimization for the relatively shallow NGH targets. The principal exploration method is seismic data acquisition and analysis, carried out both from the sea surface and deep towed. Electromagnetic methods are presently not as capable of resolving details within NGH deposits, although they can provide substantiation for the likelihood of gas and NGH prior to drilling. Gravity and magnetics studies are useful only for defining sedimentary basins of interest, and will not be further discussed. A wide range of seafloor morphology and subsurface exploration tools are available (Max and Miles 1999) and have proved successful in NGH exploration (Boswell 2014), but we only discuss here the most relevant NGH exploration methods.

3.5.1 *Seismic Survey and Analysis*

The primary geophysical survey methods for NGH exploration are the same as that for conventional hydrocarbons. Data processing techniques, however, are different in detail. This is because the pressure and shear wave velocities and attenuation of NGH, amongst other geotechnical properties of sediments with variable amounts of NGH, must be used to identify and quantify NGH. Sediment properties vary considerably with depth and overburden. Conventional seismic processing can be used to estimate gas volumes subjacent to NGH in the permeable bed. Technical descriptions of the seismic exploration apparatus and survey can be found in Thakur and Rajput (2011), as well as many other specialist geophysical publications.

NGH has Direct Hydrocarbon Indicators (DHI) or seismic indicators or anomalies that typically are caused by the presence of gas and oil. In fact, they are inferential rather than being direct in the geological sense, but they are regarded by industry as virtually direct seismic indicator after decades of experience. NGH also has unique seismic indicators that are formed by the acoustic response of high Vp, bulk and shear modulus NGH and the often subjacent associated low Vp, high attenuation-scattering low bulk modulus gas. DHI may come to be used for seismic identification while NGH will be used when direct data from drilling confirms NGH in place. Their form on a seismic section will often more resemble that of a conventional hydrocarbon stratigraphic trap than a structural closure, especially where a permeable horizon that is a migration path for gas and dissolved gas crosses into a GHSZ (Chap. 2). Because NGH recognition on seismics is still a relatively new field, misidentification of authigenic carbonates, silicates, and other diagenetic materials is an issue that requires interpretive care until more drilling ground truthing can be achieved.

Most reflection seismic surveys are carried out using a vessel towing an acoustic (sound) source and one or more hydrophone receiver arrays. There were three basic sound sources: (1) Electric 'sparkers' that gave a narrow wavelength band, generally higher frequency sharp noise that was most useful for higher resolution at shallower depths down to several hundred meters, (2) airguns, in which a pulse of air was exploded into the water, and, (3) For the greatest depths where the most energy had to be used for the shot, many tens of kilos of explosive. Receivers were originally waterproofed microphones which were replaced early by hydrophone receivers in oil baths on a hydrophone array cable. Some solid-state hydrophones are now being used.

A great deal of the seismic data available around the world is comprised of old single and multi-channel 2D surveys. In many places where there are no 'modern' seismic data available, these records are often surprisingly useful for identifying seismic features associated with natural gas and NGH in sediments, such as bottom simulating reflectors and blanking in the lower part of the GHSZ (Chap. 2). Very commonly, none of this older seismic data would have been inspected for indications of NGH because its possible presence was virtually unknown to the geophysical community as a whole. Any evaluation of an area for NGH exploration should begin with review of all the seismic data available, including any older seismic records that can be found.

In 'modern' seismic data, the acoustic data is stored on digital systems where it can be processed in a number of ways using purpose-written software. Like other branches of science, computers now capture the data for subsequent processing on computer workstations. In addition to being able to image reflection seismic images similar to older paper record seismic data, much more information can be gained. For example, reflection processing can be made using different resolutions and frequency bands that can reveal different aspects of the geology and acoustic structure. Interval and individual stratigraphic velocity analyses are particularly useful for NGH exploration and valuation of deposits. These are used to determine acoustic velocity on a small scale in detail while measuring other acoustic parameters such as attenuation and reverberation (Chap. 5). Individual strata can be mapped using time and depth slices about parallel with the general plane of bedding. Sequence stratigraphic units can be better defined, often with respect to depositional environment and its control on NGH formation and concentration (Komatsu et al. 2016). Detailed seafloor morphology and vented gas in the water column, which may be important to estimating gas flux to assist NGH exploration, can be mapped in detail as a data subset along with ship-borne multi-beam bathymetric instruments.

The relatively shallow NGH target area in the GHSZ allows for superior resolution even at higher frequencies that are usually not available for analysis of deeper targets (Gabrielsen et al. 2013) can both be processed at higher levels of resolution, not only for structure, as revealed by reflection events, but for the detailed velocity and more complex geotechnical evaluation of NGH. Older surveys can also be reprocessed for attributes such as interval amplitude data because when those digital datasets were processed, the shallower data were usually clipped or ignored. High-resolution seismic data over the shallow GHSZ have successfully been extracted at relatively low cost from existing two- and three-dimensional (2D and 3D) conventional seismic surveys acquired for the exploration of conventional hydrocarbons. 3D seismic may substantially reduce risk when detailed valuation is carried out (Chap. 5) as a drilling plan for production can be implemented to take account of NGH continuity, grade, and nature of the host sediment. New seismic methods to produce higher resolution within NGH reservoirs will be necessary for reservoir engineering, but because of the shallower nature of the NGH targets within the GHSZ, surveys using seafloor transducers or OBS (3.4.2) should deliver the necessary resolution.

Geotechnical Processing Conventional data processing to determine gas or oil volume estimates is fundamentally different from that which is required for NGH. In a conventional reservoir, both gas and liquid will separate from water and will occupy different an interval in a trap. Within their interval, they will fill porosity, minus bound water and some other factors, so that if an average porosity were to be estimated, a volumetric estimate can be made by calculating the interval, its available space in porosity, and 3D lateral extent (plus formation pressure). Processing for NGH must not only identify the limits of a concentration, but it also the pore fill variation within the NGH interval. NGH will not fill an interval or a volume to the limit of porosity as will oil and gas.

We use the term ‘geotechnical processing‘ for NGH because the geotechnical effect of variable amounts of solid NGH displacing fluid when it forms incrementally changes the geotechnical properties of the reservoir host sediments. There is a subtle but direct relationship between NGH % and distribution, and the way in which the geotechnical properties slowly change, particularly data in the 50–250 Hz range as outlined by Haines et al. (2013a). This acoustic inversion processing techniques is based on the geotechnical character of the solid, high-pressure and shear-wave acoustic velocity of NGH (McConnell and Zhang 2005).

In general, the seismic response of NGH-bearing sediments relates to higher bulk and shear modulus introduced by the presence of NGH. Because the acoustic parameters of NGH are quite different from oil or gas deposits, the range of variation in properties and their combinations yield fine detail of the NGH % in the reservoir. Geotechnical parameters related to analysis input have been determined from downhole logging while drilling (LWD) (McConnell et al. 2009), with interpolation between data points. Industry analytical techniques all remain usable but with a higher level of processing detail required for NGH evaluation. These include attribute analysis, elastic inversion and spectral decomposition that were developed for identifying sandy units in a continental slope environment, and for evaluating the content of the gas or liquids filling porosity.

Workstation processing of digital data using standard seismic analysis software and hardware has been successfully used in the northern Gulf of Mexico (Frye 2008; Boswell et al. 2011, 2012a, b; Frye et al. 2012), in the Nankai deposits off SE Japan (Noguchi et al. 2011), and in 2015 drilling off SE India, where sands within the GHSZ have now become the primary exploration target rather than the more random earlier drilling there. Relative and absolute pressure wave velocity and amplitude anomaly full waveform analysis (Bachrach et al. 2004), based on both measured and estimated physical properties of NGH (Worthington 2010; Shelander et al. 2012), was used for detection, classification, and identification of drilling targets for the Gulf of Mexico Joint Industry Program. The objective was to identify variability of NGH saturation estimates in the northern Gulf of Mexico (Frye 2008; Boswell and Collett 2011) that could be ground-truthed by drilling.

Although improvement to exploration seismic acquisition and processing continues (Haines et al. 2013a, b), drilling verification of processing, based on NGH and reservoir host response, indicated that seismic analysis techniques are have reached the point for NGH characterization and valuation because the variance from seismic processing estimates are small enough to allow for reliable estimates of gas-in-place to be made for zones thicker than 15 m and with at least 40% NGH saturation. The predictive capability of seismic analysis has also been confirmed by the extensive 2015 drilling program in the marine area east of India where predictions for NGH occurrence and porosity saturation (Chap. 5) were confirmed by drilling using logging and core sampling (Kumar et al. 2016). Work continues to quantify thinner zones (below the current resolution of the seismic data) and zones with lower saturation.

Riedel et al. (2010a) and Waite et al. (2012) also have summarized NGH physical properties for incremental full waveform processing of NGH, and have also compared processed data estimations of NGH saturation to cores and artificial NGH. In particular, Spence et al. (2010) have modeled the pressure and shear-wave seismic effects of NGH induced to crystalize in sediment pore space in their laboratory apparatus GHASTLI, which was decommissioned in 2014. These actual concentrations were used to calibrate seismic processing to provide volumetric estimation models. Drilling has validated the predictive accuracy of geophysical processing techniques by confirming not only the location of NGH concentrations but also estimates for saturation that can be made on a cellular scale (Boswell et al. 2011, 2012b). Thus, NGH seismic processing techniques are ready to support NGH exploration.

Data acquisition, analysis, environmental impact studies, leasing, plans of exploration, and then drilling is essentially no different in principle for NGH than it is for conventional offshore oil and gas exploration. Care has to be taken in evaluating older seismic data, however, because most of the existing data shot in the 70s and early 80s, and that even with more recent reprocessing may not be suitable for quantitative NGH exploration. In particular, this data was optimized for deeper conventional hydrocarbons, with gain controls often used to smooth the data in the GHSZ shallow region to reduce return noise. Only a fully digital seismic dataset will allow for proper NGH processing.

In summary, potential reservoir turbidite sands can be identified in marine sediment successions. Seismic mapping of the immediate sub-GHSZ region can shed light on the groundwater system that could point toward focused groundwater flow in which NGH deposits could form in a permeable sand at or near the base of the GHSZ. Both, NGH and free gas can be detected and volumetrically estimated using acoustic processing effects. Initially, the sands can be distinguished by higher pressure wave velocity (Vp), but on a smaller scale by their upper positive and negative lower reflection coefficient, and by their amplitude, usually assuming the physical properties of more shale-rich bounding strata that contain the focused flow in the permeable sands.

In a sand below the base of the GHSZ, *Vp* will be low and attenuation high, reflecting the presence of free gas; while, in the NGH-enriched section in the GHSZ, *Vp* will tend to be higher while attenuation will generally be lower. Relative velocity amplitude analysis has been used successfully to estimate NGH saturation, and further improvements in estimating NGH percentages are likely using more geotechnical parameters such as relative *Vp* and attenuation alteration. *Vp* will decrease upwards in both the gas and the NGH-enriched zone, and with a relative negative reflection coefficient at the junction between the gas and NGH-enriched zones reflecting relative gas and NGH saturation.

Special deep-tow seismic surveys can yield much higher-resolution (Gettrust et al. 2005) and can also yield higher-resolution data but these surveys are slow, 2D and are only useful if targeted upon NGH concentrations in wide area exploration. These surveys were made during the early research period of defining GHSZ character, but are probably no longer necessary. In any case, the level of NGH identification and resolution now achieved by the geotechnical approach to NGH identification and classification has proved adequate for economic purposes. High resolution seismic (HRS) data has advanced considerably in recent years and it has become fairly standard to process pre-stack and depth migration, SRME DE multiple and AVO studies using near, mid, and far traces (Stennett et al. 2014). HRS and ultra-high resolution seismic (UHRS) surveys have been used principally to investigate top-hole conditions for conventional wells. Although the principle role has been identification of geohazards, the methods could be applied to the whole of the GHSZ, which could be especially useful for valuation of NGH deposits (Chap. 5). HRS in the frequency range 18–372 Hz was used for fine resolution of NGH drilling targets in the northern Gulf of Mexico (Haines et al. 2014a, b).

3.5.1.1 BSR (Bottom Simulating Reflector)

Initial efforts to identify NGH deposits were focused on a single, broad reflector that may be found on some reflection seismic data. This reflector was used to infer the presence of NGH above, and the reflector's absence inferring a lack of NGH. The seismic event occurs at the hydrate phase boundary and is caused by the acoustic impedance contrast between sediments in the GHSZ above and free-gas-bearing sediments below. In regions where the geothermal gradient does not vary appreciably, this reflector is often at a constant depth beneath the seafloor, cutting across bedding planes; hence it was termed a "Bottom-Simulating Reflector" or BSR. It is now understood that the occurrence of an obvious BSR on a reflection seismic record is an indication of the top of the free gas zone.

Acoustic propagation modeling has shown that the strong negative impedance introduced at the top of the free gas zone (coincident with the base of the GHSZ) was primarily responsible for the BSR as a reflection seismic feature (Max 1990a, b). This was demonstrated using three cases: (1) typical marine sediment with no hydrate or gas, essentially the Hamilton geoacoustic profile (Hamilton 1980), (2) typical marine sediment with a range of NGH but no gas, and, (3) typical marine sediments with a range of NGH and a subjacent zone containing some free gas in porosity. Many publications, including Thakur and Rajput (2011), illustrate further details of acoustic interaction and response with detailed modeling of seismic cases involving different thicknesses of NGH and subjacent gas. From an exploration and quantification point of view, geotechnically-based seismic analysis can now not only identify NGH, but also quantify it well enough for valuation (Bachrach et al. 2004; Boswell et al. 2014).

Drilling has shown that while a BSR may be coincident with the base of the GHSZ, it provides little information about the presence or percentage of NGH in sediments within the GHSZ. In fact, it is very difficult to even estimate what percentage of gas is in a particular sediment. We now know that over large areas marine sediments, the strongest, most continuous BSRs occur in fine-grained sediments containing characteristically only 3–5% dispersed NGH by volume. Not only are small percentages of NGH present, but they also occur in sediments from which no gas can be recovered because of low mechanical strength and permeability. Therefore, in general, BSRs over large areas are almost certainly not highly prospective areas for NGH concentrations. Although BSRs alone, especially when expressed over a large area are not regarded as a primary factor for identification of a potential large NGH deposit (Max and Johnson 2014), the Japanese NGH program drilled on BSR near the margin of the deep continental margin to the SW of Tokyo, SE Japan and discovered the Nankai NGH deposit (Max 2003a).

Drilling on the Blake Ridge off the U.S. SE coast during the course of the Ocean Drilling Program Leg 164 (Fig. 3.2) has confirmed the results of acoustic modeling with respect to the major cause of a strong BSR. It has shown that NGH may occur where there is no BSR (Site 994) while sites 995 and 997 suggest that the BSR is a

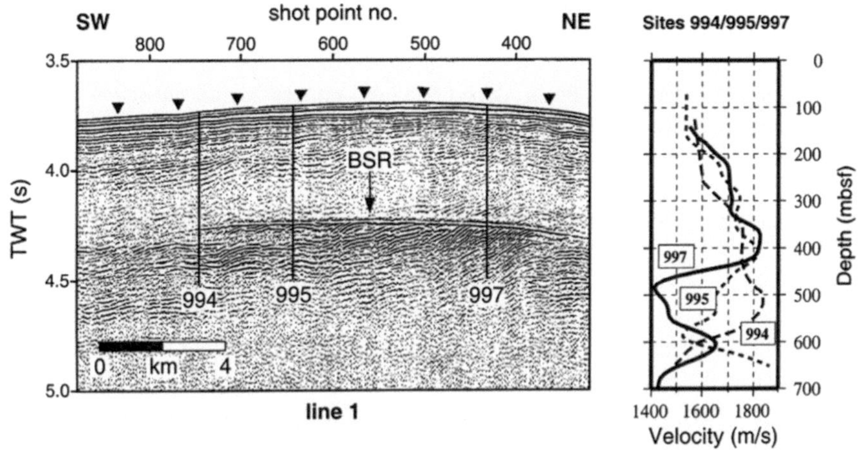

Fig. 3.2 Reflection seismic section (left) and Vp from VSPs on the Blake Ridge. The BSR at 995 and 997 is at ~440 mbsf. Note the zone of high reflectivity beneath the BSR. This is probably caused by slight variations of free gas concentration in sediments having different potential for producing or holding free gas that might have a subjacent source. After Holbrook et al. (1996)

factor of free gas in pore space below the base of the GHSZ (Holbrook et al. 1996). Nearby seismic sections suggest Pleistocene venting of large amounts of gas below an apparent collapse structure within the GHSZ (Dillon et al. 2001). Boswell et al. (2012a, b, 2014) demonstrate that small BSRs confined to individual porous and permeable sandy turbidite beds are good indicators of concentrations of NGH. The most prospective areas for NGH are associated with reflection seismic data on which the base of the GHSZ is often difficult to pick because it is often discontinuous (Hillman et al. 2017). In contrast to early exploration guidance in which prominent BSRs were the object of searches for NGH, the 'string-of-pearls' BSR (Max and Johnson 2014) on which a discontinuous BSR, caused by the presence of free gas below concentrated NGH in permeable horizons marks the base of the GHSZ in sandy beds, which constitute the primary exploration targets in marine sediments.

For purposes of exploration, a BSR indicates the presence of a gas province, but a long continuous BSR indicates muddy sediment host with poor stratigraphic differentiation. Very deepwater sediments with no likelihood of having a suitable host bed, for instance, will not constitute a likely target for NGH development even though NGH is inferred to be present (Rajan et al. 2012). 'String-of-pearls' BSRs, which are much less obvious, indicate the presence of permeable strata in which focused groundwater flow has brought NGH mineralizing solutions to the level of the base of the GHSZ (Max and Johnson 2014). Upon closer examination of these seismic 'pearls', the extent of NGH concentrations and subjacent free gas can usually be interpreted. With 3D seismics, the volume of NGH can be estimated with some confidence.

3.5.2 Ocean Bottom Seismometers

A type of 3D dataset can be produced for high-quality seismic analysis through deploying a network of OBS in order to capture shear-wave information that can be convolved with the reflection seismic data to produce much a higher resolution of the NGH concentrations (Haines et al. 2013b). Ocean bottom seismometers (OBS) are excellent tools for determining detailed seismo-acoustic response and were used for many years for detailed acoustic analysis of shallow marine sediments. They may be effectively superseded by 3D seismic surveys not only because the level of detail can be about the same but because most OBS processing is still a highly expert manual process whereas software packages and dedicated workstations are available for 3D processing. OBS measure three acoustic responses on three axes, and because they rest on the seafloor, they directly record not only Vp, but also shear wave returns (Vs). Other, slower, seismic waves can also be generated and detected, but these are not used for analysis for hydrocarbon reservoirs. Also, where seafloor sources are used, direct pressure and seismic wave energy can be generated that can be used to also directly measure both pressure and shear wave attenuation. OBS is only used over small areas in which interest has been determined by standard seismic analysis techniques. OBS is not commonly used for delineation of hydrocarbon resources because when OBS receivers are placed in a small area the angle of separation from source to more than one receiver is usually too small to provide for adequate reflection resolution. In addition, carrying out a survey, especially in ultra-deepwater, consumes considerable time with respect to any additional information that might be gained.

OBS cannot be used for regional survey because it is much more expensive than conventional towed seismic. Where OBS is gaining use is in 4-D analysis for producing fields, with a seismic survey periodically reshot while the seismometers remain in place. This allows an operating company to monitor the depletion of individual reservoirs as production proceeds. By placing OBS receivers on the seafloor, the resulting data is characterized by improved suppression of multiples which results in improved imaging. In addition, the shallow NGH target allows for a higher degree of angular separation of acoustic raypaths to each OBS, which will result in a very low confidence level for positioning (Fig. 3.3). In a NGH production scenario, monitoring changes in the reservoir as NGH is converted and natural gas is extracted can be done using a similar approach and leveraging conventional practices and software.

3.5.3 Electromagnetic (EM) Survey

In addition to seismic surveys, electromagnetic surveys can also be carried out, although without seismic data it is of much less value as an exploration tool. Drilling safety can be directly affected by near-seafloor NGH dissociation related to

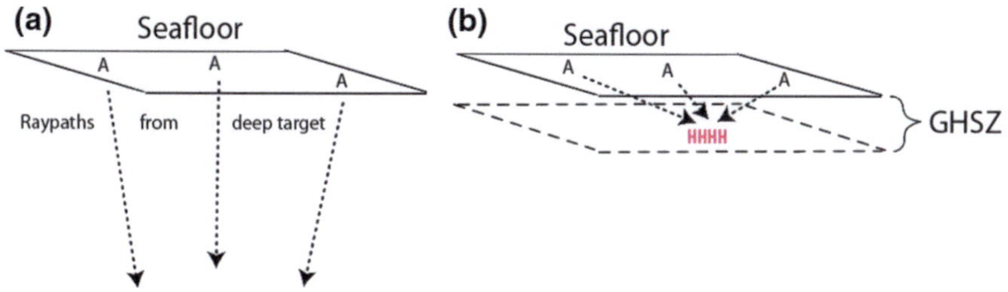

Fig. 3.3 Diagram of difference in angular separation of raypaths to OBS for deep conventional gas deposit and potential NGH deposits in the relatively shallow GHSZ. 3D processing uses much larger datasets, which is another way to approach the issue of resolution. Angular separation to a target in a GHSZ from a surface towed array is much less

drilling activity. Industry always carries out pre-drill seafloor and shallow sediment survey and analysis to de-risk well-understood potential drilling hazards of shallow water and gas flow and sediment mass movement (Aird 2016). Industry now uses a combination of shallow seismic and EM surveys. Induced electrical technology (controlled-source electromagnetic or CSEM) and magnetotelluric (MT) methods measure relative resistivity and allows relative identification porosity, and resistivity in pore fluids in shallow marine sediments. Of the two, MT has been found to be the most useful (Durham 2017). A fuller description of the electromagnetic methods, history and application is described by Thakur and Rajput (2011).

Both 2 and 3D mapping of high resistivity zones that appear to be associated with deposits of natural gas is being used increasingly to provide an external perspective to seismic survey in shallow gas surveys. Equipment used by EMGS Company, for instance, identifies somewhat large and indistinct frequency anomalies that are interpreted as indicating natural gas in the sediments. Electromagnetic Geoservices ASA is the licensee for the TMGS data library. The technique has been used in many places such as the Mediterranean and the Barents Seas (Yakymchuk et al. 2015; Feather 2007), where anomalies have been distinguished, but the anomalies are very general in appearance with almost gradational transitions and much less detail than obtained through seismic survey and analysis. In addition, the data does not appear to have the capacity to distinguish mineralization grade and thus has no value for valuation of a deposit, much less describing the actual shape of the anomaly.

There are two modes of survey (May 2015). In the seafloor node method, receivers are deployed on the seafloor and an EM dipole source is towed by a vessel above the seafloor receiver array. Obviously, this is not a wide area survey method as detectors have to be distributed, located (for instance on a GIS base), and then recovered following survey. In contrast, the towed EM streamer system uses a towed source and a receiver streamer and areas can be surveyed at about the same rate as a seismic survey. Results, shown as 2D resistivity sections appear to have excellent detail of the unconformable top of a faulted gas-charged shallow sediment, but similar to other anomalies we have viewed, internal detail and the detail of the base (or possibly even the base of the gas zone) is too indistinct to be useful for detailed analysis. Levashov et al. (2014, 2015) suggest that experiments with frequency–resonance surveys over known gas deposits in Barents and Mediterranean Sea sediments yielded improved resolution of resistivity anomalies, compared with a more total field approach, but again, the method shows only gross anomalies with little detail.

The main application of electromagnetic methods is for detection of shallow gas that might be a drilling hazard, although these are more likely to be detected in water that is too shallow and/or warm to have a stable GHSZ (Chap. 2) in which shallow NGH concentrations could be found. Danielsen and Bekker (2011), however consider that improvements in data quality and basic processing to resolve smaller targets and maintain accuracy to greater depths, especially where the resistivity of the pore fluid in a particular horizon can be determined.

3.5.4 NGH Ground-Truthing: Drilling

Drilling is the final exploration step and includes well logging and coring. It provides physical data that leads to a decision on whether to produce, and the plan for production. Detailed assessment of cores from the Japanese Nankai deposit (Fujii et al. 2015), for instance, allowed wider area recalibration of seismic interpretation and the outline for the technical experiments of the 2013 short production test (JOGMEC 2013).

Drilling provides data from petrophysical logs, side wall cores, and cuttings samples for reserve calculations, extraction modeling, among other direct sampling and measurements. However, the economics differ from valuing a conventional deposit. For instance, conventional reserves in a high-quality reservoir are normally characterized by a single hydrostatic pressure regime with the hydrocarbon usually equally spread through the saturated pore spaces without significant changes in physical character. In contrast, NGH concentrations can be expected to be anisotropic, with variation within porosity throughout a reservoir. Whether NGH forms in bulk, that is, unsupported and entirely within pore water, or affixed to a surface, it is a solid whose formation increases sediment strength and the bulk modulus.

In all of the high-grade NGH deposits drilled throughout the world to date, the first sand above the phase boundary appears to become filled first, and if there are stacked sands at different elevations within the NGH, the lower sand has the greater accumulation of NGH and successively higher ones have proportionally less. In a weak petroleum system with a thick GHSZ containing multiple sands, only the deepest sands may have NGH concentrations suitable for production. If the GHSZ is thin, and only contains a single producible sand, that sand may constitute an excellent NGH target as ascending fluid will tend to be concentrated in order to form a passage from deep sources of fluid to the seafloor. Where thick sands lie on the base of the GHSZ, and especially where gas flux or water percolation is rapid, the NGH may be distributed over a larger area but in lower local saturations because of less focused flow.

Logging while drilling (LWD) in our opinion, has the potential to resolve the best rock physics data from NGH-enriched strata to use as input for seismic analysis. Measurement in-place is particularly important for NGH, which has the potential to recrystallize, even in a pressure core and especially when subjected to temperatures well below those it naturally forms into compensate for lower pressure.

No matter how carefully drilling is controlled, some alteration of borehole temperature and salinity in the hole is inevitable during drilling. NGH disruption can be minimized by keeping temperatures in the hole at or below ambient; pressure will be ambient with depth in the hole. Drilling muds, if used, will only serve to stabilize NGH near the hole. It is important to minimize alteration caused by drilling and sampling because there might be enough recrystallization, formation of ice phase, or most important, alteration of the fine fabric of microscopic water-courses that may remain in natural high-grade NGH concentrations (Chap. 5).

LWD takes place so soon after the ambient conditions of a NGH accumulation are encountered that alterations affecting logging responses within the NGH reservoir can be expected to be minimal. The in situ permeability is especially important as a key factor for initiation and maintenance of wide-spread, controlled dissociation (Max and Johnson 2011). The highly reversible reaction of the NGH system, and its known responsiveness to changes in pressure–temperature and other environmental conditions, argues strongly for downhole measurements rather than even well-taken pressure cores.

3.5.4.1 Core Sampling: Bringing Unchanged NGH to the Drillship

Solid, crystalline NGH is stable in its host sediment (Chap. 2), but once it is removed from that environment by either lowering the pressure or increasing the temperature, it becomes unstable and dissociates. In order to study it properly, samples of the NGH and sediment host must be recovered to the drill ship as little changed as possible. That is, the pressure and temperature of the samples will ideally remain as they were when they were sampled. By preserving the NTH at it reservoir pressure and temperature can any alterations induced by its recovery be prevented.

Only this preservation will allow the NGH and its habit and all the fine details of its physical state to be understood as if the samples were still in their natural environment, where they had come to thermodynamic equilibrium. If the temperature were raised but pressure increased and the NGH remained stable, or if the temperature were dramatically lowered, for instance by exposure to liquid nitrogen (a common refrigerant for both organic and inorganic samples, subtle changed in grain boundary interaction, NGH-sediment grain interaction, and other alterations responding to the new thermodynamic regime, could leave sample unrepresentative

Fig. 3.4 Borehole drill cores containing massive gas hydrate taken from 'chimney structures' using non-pressure wireline coring tools. *Photo courtesy of Geotek Coring Ltd.* Cores recovered by wireline about 20 min from start of assent and out of the GHSZ zone in the water column ($\sim$500 m water depth, see Fig. 2.1) to core lay out on ship. See cover image of pressure core laboratory. White areas of cores are concentrated NGH. Note that the dissociation is a relatively slow process as the cores are very far out of their stability field

of the reservoir. In order to recover ambient condition NGH and maintain the physical conditions of the NGH reservoir, a portable integrated borehole-to-laboratory pressure core solution for gas hydrate investigations has been developed and has been used for sampling and for transport of the reservoir ambient condition cores to laboratories (Schultheiss et al. 2014, 2017).

The pressure core analysis and transfer system (PCATS) is the link between the NGH concentrations in the borehole and the laboratory, where vastly more information can be determined than from downhole logging methods. In addition, non-pressurized cores (Fig. 3.4) can also provide substantial information even though the more detailed investigations possible with pressure core for instance, 3D permeability and sediment grain distribution, details of NGH grain boundary contacts, very small residual gas bubbles not normally visible to the naked eye, and other reference points that may be of value in determining production potential.

Concentrated NGH holds its shape even after considerable dissociation while disseminated NGH in muds wetting clay matrices result in liquefaction of muds and loss of the core shape. Undistorted muddy matrices presumably have little or no disseminated NGH. More concentrated NGH core regions are holding the external form of the core.

3.5.4.2 Picking Drilling Targets

Normally, our advice would be to target apparent NGH concentrations at or near the base of the GHSZ because the energy required, especially in easily controlled depressurization, will be the lowest for NGH conversion. The further away a NGH concentration is in pressure-temperature space on the phase diagram (Chap. 2), the greater the energy that is required to maintain conversion for natural gas production. For instance, in IODP Leg 311 off Vancouver (Reidel et al. 2006), the section drilled was mainly shale but the lowest sand well above the base GHSZ had high pore saturations of NGH. Although this was an isolated and non-commercial deposit, it indicates that NGH concentrations need not occur immediately at the base GHSZ and each concentration must be evaluated for its production potential as a function of energy input for conversion.

Where mineralizing solutions with a high dissolved concentration of methane having a strong growth dynamic have migrated along permeable horizons into the GHSZ, NGH concentrations will form when conditions of NGH stability and favorable migration pathway/host sediment are first encountered. Where mineralizing solutions have a lower growth dynamic, significant crystallization may not occur until the solutions have risen further into the GHSZ, in which NGH stability becomes more profound upward toward the seafloor (Chap. 2). NGH concentrations at the base of the GHSZ in dipping permeable strata may free gas below the NGH. In this case the BSR will be highly discontinuous with distinctive but often difficult to recognize mini-BSRs that have been termed a 'string-of-pearls' BSR (Max and Johnson 2014), NGH concentrations of NGH higher in the GHSZ may have crystallized without the presence of free gas and may not have even small associated BSRs. Picking drilling targets from geophysical data is going to be easier and surer near the base of the GHSZ.

If NGH occurs in geological traps such as sealed anticlinal sand, it could indicate that natural gas arrived and was first concentrated as free gas. Conversion into NGH could then be accomplished in the same manner as permafrost NGH (Max et al. 2006). Forming gas concentrations that then crystalized to NGH in place could have taken place during an interglacial when the GHSZ was thinner or insubstantial or if the gas flood was extremely rapid. Trapped gas converted to NGH has been observed in at least one marine deposit in the Gulf of Mexico. While drilling the Alaminos Canyon Block prospect in water depths of 2744 m for Paleocene and Eocene (Wilcox) objectives, the Chevron #1 well encountered a Frio-age (Oligocene) NGH-bearing sand. Late Oligocene compressional folding created a structural closure permitting the sand to fill with natural gas. Continued folding uplifted the crest of the closure to within 460 m of the seafloor, within the GHSZ. The gas-bearing sands at the crest of the structure subsequently were converted to NGH (Latham et al. 2008).

The first offshore production test from a NGH deposit was conducted in March 2013 in order to evaluate production of gas from marine NGH using depressurization dissociation. The test was performed by the Research Consortium for Methane Hydrate Resource Development in Japan (MH21) at the AT1 site located on the northwestern slope of the Daini-Atsumi Knoll in the eastern Nankai Trough, Japan. During the pre-drilling campaign conducted in 2012, extensive geophysical logging and pressure coring using a hybrid system were conducted to control well positions and determine expectations. Two wells (AT1-MC and AT1-C were located about 25 m from the test well that was depressurized in order to help identify the dissociation envelope.

3.5.5 *State of NGH Exploration*

While the strict meaning of petroleum is "rock oil" with ancient Greek roots, 'petroleum' is often used synonymously for "oil". It is also often used as a catch-all term for oil, gas, and condensate. We use "Petroleum System" to describe the overall process of oil and gas generation, migration, and trapping, with the proviso that the petroleum system for NGH is very different than for conventional hydrocarbons (Chap. 2).

The basic principles of the NGH petroleum system are understood to be able to analyze 'basins' for their NGH potential. Exploration methods are also available that can be used directly for large area exploration with focusing on targets for further evaluation now possible because the parameters for advanced NGH geotechnical processing have been proven by ground truthing. These methods will be improved with experience, but evaluation of gas-in-place is not currently accurate enough to allow for valuation (Chap. 6).

The U.S. Bureau of Ocean Energy Management (BOEM) has undertaken petroleum system evaluations of NGH potential for the US Pacific, Atlantic, and the Gulf of Mexico coasts that were summarized by Frye et al. (2013b). The methodology is essentially the same as for conventional hydrocarbon exploration in that a basin analysis is carried out in which the relative proportion of favorable host beds is established. For NGH exploration only the identification of sands in the GHSZ are required, whereas deeper sands have to be identified for conventional hydrocarbon exploration. Seismic survey and analysis is the principle exploration tool. The presence of sands establishes that suitable migration pathways and NGH sediment hosts exist (Max and Johnson 2014). The second exploration step is to confirm the presence of a sufficient gas flux, and its biogenic or thermogenic character. Identification of NGH concentrations, exploration drilling including NGH pressure coring, technical and production tests, preparation of the deposit for production if justified, and production essentially will follow familiar industry procedures for risk reduction (Chap. 10).

Although we now have the tools to be able to identify the in-place technical and commercial value of an NGH deposit, the costs of NGH exploration and production need to be significantly decreased to make NGH more commercially attractive (Chaps. 7, 8, and 11).

3.6 National NGH Programs and Company Interest

The energy potential of NGH is now the primary motivating agent for hydrate research at the national level of the United States and most other countries that are making significant investments in hydrate research (SDN 2015). This thrust follows a long period (from the 1930s) during which the primary research interest in hydrate was driven by the energy industry's concerns in the field of flow assurance, or mitigation and remediation of unwanted hydrate that formed in pipes carrying wet hydrocarbons. Drilling safety and flow assurance appear to be the main concerns of most energy companies, many of which are involved with the government-driven NGH-related energy research, while the carbon cycle and global climate modeling appears to be the research driver in other countries, particularly those which do not have a likelihood of hydrate energy resources in their oceanic (or permafrost) areas.

The U.S. Department of Energy (DOE) initiated the first national NGH program in 1982. The DOE made extensive use of contractors and was based at the Morgantown, WV DOE laboratory the precursor to the present National Energy Technology Laboratory (NETL). The program was active until 1992, after $8 million had been well invested, but was terminated owing to the low price of conventional energy sources and DOE policies. The program was invaluable for transforming the field of hydrate science to a potential energy program and for establishing the framework for further development worldwide. The Japanese and the Indian governments built on the results of the U.S. program and initiated national hydrate programs in the mid-1990s. The United States established a formal national hydrate research program in 2000 with the passing of the Methane Hydrate Research and Development Act. Since then, a number of countries having energy or foreign currency issues have initiated hydrate research programs, or at least have raised their level of awareness about the potential of a NGH energy resource.

3.6.1 Exploration Activity in Regions and Countries

The results of NGH programs and progress are being closely monitored by both governments and private companies. An excellent diagram of historic events in NGH research can be found in Fig. 1.2 in Beaudoin et al. (2014, p. 7).

Because of the current (2015+) over-supply of natural gas, and the increased LNG transport capability that is rapidly pulling down delivered gas prices due to increased competition, NGH programs are largely confined to major importing countries such as Japan, South Korea, Taiwan, and India, which seek indigenous natural gas supplies. Energy supply is not simply a commercial matter. In the case of highly industrialized, relatively wealthy countries such as South Korea and Japan, an indigenous supply would mean that more of their foreign capital could be used for more productive purposes. In the case of India, among many other less prosperous countries, resources available to purchase energy are much more limited. Other countries having few natural gas resources and a lower technology base will be dependent on successful NGH exploitation being developed by other countries' research programs.

Russian Federation Little work appears to have done on oceanic NGH, probably because the country holds the largest conventionalnatural gas reserves in the world (Kable 2015). Exploration on the Siberian continental shelf promises to substantially enlarge their gas resource base, as does their potential deepwater conventional hydrocarbons. In addition, Russia has considerable potential for shale gas and oil that has not yet been well evaluated.

Seismic indications of NGH have been seen in the Arctic Ocean, the Sea of Okhotsk, and possibly elsewhere along the Russian Pacific margin. NGH may occur in the Russian sector of the Black Sea (Chap. 4). Identification of NGH in Russia (Soviet Union of the time) was based on a chance blowout of a well in thick permafrost of the Messoyakha region of the western Siberian gas fields, and in difficult to explain reservoir performance in which it appeared that the reservoir had a mysterious natural recharge capability. Although this may have been the first identification of NGH, it was not made public for many years, by which time NGH had been identified in the Alaskan North Slope permafrost and Russian Arctic offshore (Max et al. 2006). NGH is known to exist in the seafloor of Lake Baikal, which is the world's deepest lake at about 1635 m depth. Most other deep lakes around the world have not been explored for NGH. As NGH does not appear to be prominent in Russian natural resources research, we have not updated the situation here.

Canada: was one of the world leaders in NGH research from about 1996 to 2010 when world attention was focused on the Mallik gas hydrate research site in Canada's Mackenzie Delta. The Geological Survey of Canada (GSC) and the Japan National Oil Corporation (JNOC) have led this work, with the U.S. DOE an active participant. Among the participants were the GSC, JNOC, USGS, GeoForschungsZentrum Potsdam (GFZ), India Ministry of Petroleum and Natural Gas (MOPNG)/Gas Authority of India Limited (GAIL), and the ChevronTexaco-BP-Burlington joint venture group. The advances made during this period were substantial and included NGH identification in the Canadian continental margins. Canada has since abandoned its national program of NGH research, with the Canadian Geological Survey focused onto what are apparently considered to be nearer term energy resources. Some NGH research, mainly of a physical chemical rather than an exploration nature, is taking place in some Canadian universities.

Norway: has no formal National NGH Program. In 2012, Equinor placed their small NGH team under 'exploration' rather than 'basic research', but with the post mid-2014 down turn in exploration activity, Equinor has de-emphasized its internal NGH program. Molecular modeling of the NGH processes is being carried out at the University of Bergen and the newly established Centre for Arctic Gas Hydrate, Environment and Climate (CAGE) at UiT. The Arctic University of Norway (at Tromsø) has grown out of earlier, groundbreaking NGH research carried out by the University of Tromsø. The Arctic University has a dedicated 'Centre of Excellence' specializing in NGH. Additional research is being carried out at other Norwegian universities, particularly by Bjorn Kvamme, at the University of Bergen (Chap. 5). Equinor has carried out a global screening of continental margins to classify regions for their NGH hydrate potential (Reichel and Gallagher 2014, 2015). They conclude that there may be NGH concentrations that could be commercially attractive, and that the greatest NGH resources may lie in the Arctic Ocean.

United States: The Methane Hydrate Research and Development Act of 2000 has been operated by the U.S. DOE using a program plan developed earlier (DOE 1998, 1999). The program restarted the initial hydrate research program that had been initiated at the DOE laboratory at Morgantown, WV in the late 1980s with basic research (Max 2003b). The program is currently focused on exploration and production issues (NAS 2010), with the aim of carrying out a production test within three years (for which funding must be found). NGH research efforts have been concentrated on the northern Gulf of Mexico, the North Slope of Alaska, and along the Arctic Ocean continental slope of Alaska.

The United States Geological Survey (USGS) has maintained a continuous, broad research program in NGH studies since 1990 and, along with DOE, holds the largest repository of information on NGH in the U.S. This work has included extensive field seismic studies on the Atlantic, Gulf of Mexico, Pacific, and Alaskan continental margins of the U.S. The United States NGH program promotes publishing of results, in contrast to some other national programs that are largely closed. In the late 1990s, the Naval Research Laboratory (NRL) developed a collaborative NGH program with the University of Hawaii (Hawaii Natural Energy Institute) and NRL formed an international consortium for NGH research. This early international collaboration grew to include five nations (U.S., Canada, Chile, Germany, and Japan), and along with the joint industrial program of DOE, was investigating the continental slopes of the US, Canada, Japan and Chile, with particular input recently into joint work with India in their offshore area (Waite 2016).

The NGH development program of the U.S. operated by DOE has been very successful in developing and proving exploration and valuation technology for NGH. The DOE NGH program has been reactivated following the successful conclusion of the JIP in the northern Gulf of Mexico, which proved a large NGH concentration in a stratigraphic host suitable for gas production (Boswell 2014). New funding amounting to $15 million was put into NGH research in 2014. The object of the developing program is to operate a production test well within 3–5 years in the northern Gulf of Mexico, although further funding will be required. Preparatory work, such as pressure core drilling and core analysis along with calibration and improvement of seismic analysis techniques for identifying and valuing deposits (Chap. 5) are the immediate focus of the program.

A particularly innovative in situ experiment to test molecular substitution of CO_2 for CH_4 in NGH, in order to produce natural gas and sequester CO_2, was carried out by ConocoPhillips under the auspices of U.S. DOE, with the participation of JOGMEC (Japan) from 2008 through 2013 (Hester et al. 2011). A suitable test site was identified and access permissions were obtained for a field trial from 2008 to 2011. The Iġnik Sikumi #1 test well targeted a well-known permafrost hydrate deposit in the Eileen formation and was drilled from a temporary ice pad in early 2011. An injection/production test was performed in early 2012. Production operations took place from January 2012 to May 2012 (Schoderbek et al. 2013). The well was then plugged and abandoned.

CO_2 and co-injection of N_2 and CO_2 were used in separate experiments in the Ignik Sikumi tests. All experiments demonstrated how CH_4–CO_2 exchange is a favorable approach where CH_4 gas production takes place at the same time as sequestration of CO_2 (Birkedal et al. 2015). There was no apparent dissociation during this solid-liquid-solid transition, which suggests that geomechanical stability was maintained while the replacement process was accomplished essentially by diffusional transport with apparently no significant destruction of the water cage structure in the NGH (Chap. 2). Replacement of CH_4 by CO_2 is therefore demonstrated to be a valid approach to NGH gas production (Max and Johnson 2011).

The present U.S. NGH program has a two-pronged approach, focusing on both the North Slope of Alaska and the Gulf of Mexico. The next U.S. field test is being developed under the leadership of the University of Texas at Austin, in partnership with The Ohio State University, University of New Hampshire, Columbia University-Lamont-Doherty Earth Observatory, the Consortium for Ocean Leadership and the U.S. Geological Survey. Their four-year exploratory program will characterize prospective drilling locations in the Gulf of Mexico. In 2018 a field excursion to drill and collect pressure cores and well logs as well as conducting short-duration pressure drawdown tests is scheduled to be undertaken.

On the North Slope, the program is pursuing two options with Japan for a long-term test. The program is evaluating unleased acreage the State of Alaska has set aside temporarily for this test and is also exploring the possibility of testing within one of the producing units. The program continues to fund various researchers to evaluate NGH dynamics in response to changing environments and implications for ocean-ecology and climate change feedback.

The U.S. program had been severely scaled back in recent years; however, it has been revitalized by the announcement in October 2014 of a four-year program that includes $41,270,609 of DOE funding and a cost share of $17,030,884. The DOE development plan to support NGH activities (DOE 2015) envisages supporting activities to establish metrics for a production test and 2015–2020 out-years planning for the program is underway. The current Gulf of Mexico program completed the evaluation of prospective drill sites in October 2015, and has officially transitioned into Phase 2 with a full-scale, land-based test of the project's existing pressure coring system anticipated for December 2015. During Phase 2, the project will initiate efforts to develop needed core storage, transport, handling and analysis capabilities, and will conduct a marine field test of the system. That test ideally would take place in 2018 will likely take cores from some of the locations where NGH was logged with LWD tools in the 2009 program led by Chevron. The Phase 2 results will guide a large-scale field expedition slated for Phase 3. The project is planned to conclude at the end of September 2020, with a total cost of $81,022,246, of which DOE is contributing in excess of $50 million (DOE 2015). On 27 March 2018 the DOE budget for 2019 markup was for about $20 million for the Gas Hydrate Program.

Japan: Japan has been at the forefront of NGH research since the mid-1990s and has maintained a focused, well-funded program. The initial NGH work was encompassed by a research consortium for methane hydrate resources in Japan (also known as the MH21 Research Consortium). About 250 people in thirty mainly Japanese organizations were working on the MH21 program by 2005, but there has been almost a complete lack of programmatic and staffing information on the program outside of Japan. The program has been extended past its original 2011 termination date as the program has transitioned from being research oriented to more production oriented (Tanahashi 2011).

Japan's NGH program continues to evolve from basic through applied science to engineering. In a move to become a commercial rather than governmental project, a new industrial joint venture corporation was formed. The "Japan Methane Hydrate Operating Company" (JMH) was formed with the agreement and capital participation of 11 companies engaging in oil and natural gas development and in plant engineering. Commercial production of natural gas from hydrate is follows the MH21 plan and is scheduled to initiate from Nankai in Japan's Pacific margin in 2018.

The successful short five-day technical test carried out at the Nankai site about 60 km to the SW of Tokyo Bay in March 2013 by the Japan Oil, Gas and Metals National Corporation (JOGMEC) was the first of its kind. This test proved the feasibility of low-energy depressurization using technology that was readily available from the offshore industry as the choice of conversion method (Max and Johnson 2011) for controlled NGH dissociation to initiate gas production. Innovative methods for gas–water separation were also proved. Depressurization was achieved in this pioneering test flow of methane from hydrates offshore Japan earlier using a specially designed electric submersible pump system able to separate the natural gas from water and move them to the drillship through separate production strings.

A production rate of 706,000 cft^3/d (or about 0.7 MMcf/d or 20,000 m^3/d) was obtained from a drill stem test using depressurization. While this production rate was about a fourth of what a good conventional commercial well would produce, this was just a vertical drill stem and technical test through the producing horizon. Sustained commercial production could be expected to be significantly higher. The test was stopped because of sand flow issues, but it is unclear whether these were purposefully stimulated or not as part of the production rate testing. In any case, this was technical test of depressurization conversion, and not a long-term production test. In a full production test, a longer intersection with the NGH deposit would be used, as would wellbore liners, a more elaborate drilling plan, and sand screens. The Nankai test was accomplished with the top of the well cased but the bottom of the well in the NGH deposit open hole, although with a gravel pack in place (Kawamoto 2013).

To date, no NGH wells have been brought into production or even followed by a long-term production text after a drill stem test (DST) . This makes prediction of NGH production rates uncertain and is a strong argument for a production test. The key aspects of a DST are the pressure measurements recorded during the test that will give data on 'flowability' over time and other production factors, in addition the actual volume of gas or oil produced. For radius of investigation (equivalent to

NGH 'dissociation envelope') and productive capacity to be determined, it is necessary to know the drawdown pressure history and other parameters. Interpretation of production from a DST (i.e., the only DST at Nankai in March 2013) will be very different for a NGH well than a conventional gas well in that artificial depressurization is likely to be the primary method for NGH conversion used for NGH wells. Results of the 2013 Nankai DST (JOGMEC 2013) included flow rates but not a test pressure profile. Even so, given the available Nankai results, it would not be unreasonable to expect a long-term flow rate of 2–5 million cubic feet per day, and possibly higher.

We anticipate that the original MH21 intended 2018 target date for production from the Nankai NGH deposit may not be met because of unexplained apparent delays in new testing that would ramp up to production. We expect the testing to follow a fairly normal test sequence with resolution of individual issues, such as sand control and equipment-specific technical issues, but also to include issues that can be expected to resolve differently from conventional natural gas, such as stopping and restarting NGH conversion, which may be necessary for commercial or technical reasons, such as routine maintenance or other issues requiring a shutdown. Tests exploring the energy input for conversion versus gas production and other issues will also be assessed. We anticipate that the well-planned and thorough Japanese NGH program will be well-executed, whatever the timeline.

The lack of public information (Spring 2016) about their NGH program may be a consequence of the steep fall in LNG delivered prices from about under $20 per million cubic feet of gas. Imported gas is purchased by the Japanese government and is then retailed by the government importing agency to private industry at about double the price. And the political impact of the remaining nuclear power stations and their potential restart is still not known. Since there is no global shortage of gas with more becoming available, Japan may be content to keep importing, at least so long as the current low gas prices and the developing world price for natural gas remains low. Nonetheless, we regard it as highly likely that the program is moving forward to further production testing with the usual closely-held information about their program progress.

Japan has expanded its NGH program beyond the Nankai area and is funding programs to explore the Sea of Japan and conduct an assessment. The areas being explored in the Sea of Japan are off the Joetsu, Akita, and Yamagata regions (Iwata 2014) and good evidence for deeper water play related to Nankai has been identified (Taladay and Moore 2015). Based on the initial results of a June, 2014 drilling program, the assessment is focused on "shallow" hydrates, the term used by Japan's Agency for Natural Resources and Energy (ANRE) for chimney/fracture-fill occurrences. This represents a new development since throughout the world all commercial considerations for production of gas from NGH have involved hydrate-bearing sands.

India: India's national NGH research program (NGHP) has moved rapidly from an early phase of preliminary identification of NGH resources in the eastern side of the Bay of Bengal sector of the northern Indian Ocean (Riedel et al. 2010b; Sahay and Johnson 2014). It would appear that after a relatively slow start (1998) due to contested ownership of the NGH program among Indian interests, India's program may be currently more aggressive than that of Japan. Probably because of its lack of foreign currency reserves, India has been very serious about developing its indigenous natural gas resources. In addition, a major use for natural gas in India is the production of urea fertilizer, a necessary component of India's focus on being self-sufficient in food production. Although large quantities of NGH were identified in the Kerala-Konkan offshore region as having significant hydrate shows, these were mostly in secondary porosity in fine-grained muddy deposits (Sain and Gupta 2012). More recent drilling has been for sandy strata that will provide host for reservoirs, following the NGH petroleum system model for productive strata hosts (Max and Johnson 2014).

The yearly budgeted funding of the NGH program of India has more than doubled since Narendra Modi, Prime Minister of India, took office in May 2014 (MES 2015). This follows closely on a decision made by the previous government of Prime Minister Singh who in 2013 set a goal of slashing India's energy import bill by 50% in seven years and reducing it to zero by 2030 (Vickery 2014). The Modi government's support for NGH exploration accelerates the strategy of seeking energy independence for India. Increased NGH funding aimed at exploration and production issues is leveraging the usefully large knowledge base of more than 25 years of research by other nations, in particular, the U.S. NGH program and international support in their drilling and prospect evaluation program. It is likely that if the current (2014–2016) drilling and analysis program for NGH concentrations are successful India's NGH program will be further accelerated.

Results of India's first NGH drilling program (NGHP01) were published in December 2014 in Marine and Petroleum Geology. That expedition included a 113.5-day voyage from April 28 to August 19, 2006 during which the expedition cored or drilled 39 holes at 21 sites (one site in the Kerala-Konkan Basin, 15 sites in the Krishna-Godavari Basin, four sites in the Mahanadi Basin, and one site in the Andaman deep offshore areas). The drilled holes penetrated a total of more than 9250 m of sedimentary section and recovered nearly 2850 m of core. Twelve holes were logged with logging-while-drilling (LWD) tools and an additional 13 holes were wireline logged.

The second leg (NGHP-02) of India's NGH drilling program utilized the drillship *D/V Chikyu* from 3-March-2015 to 28-July-2015 drilling in water depths of from 1519 to 2815 m. This cruise has proven some interesting prospects. It was conducted off the eastern coast of India (Kumar et al. 2016). The program drilled seismic targets interpreted as NGH-bearing sands with a focus on reservoir delineation and resource assessment. The first two months of the cruise involved LWD operations of 25 wells. The following three months involved coring operations at the 10 most promising sites. As was the case in the Walker Ridge region of the northern Gulf of Mexico (Boswell et al. 2012a), predictions made from seismic interpretation for NGH occurrence and pore fill (Chap. 5) closely matched pre-drill predictions. Two areas in the Krishna-Godavari Basin were determined to host significant 'world-class' NGH accumulations and represent ideal sites for NGH production testing that may lead to full production. India appears to be very serious about cutting CO_2 emissions and is also carrying out accelerated energy research in addition to their NGH program. The current state of their related ocean thermal energy conversion (OTEC) program for energy and seawater desalination (Max 2002, 2008) in a summary of international NGH activity Max et al. (2006) is understood to be still in progress but new achievements have yet to be made public. In an effort to leverage the largest thorium reserves of any country, India is carrying out research on the use of thorium, which involve less radioactive waste hazard, as a primary nuclear fuel and is planning a next-generation thorium-based 300 MW reactor at Terapur to be built sometime in the 2020s (Bagla 2015). Norway and the United States are also reinvestigating thorium reactors.

South Korea: The Korean NGH program is using its universities and government research agencies to carry out research after a beginning with U.S. scientists. NGH potential has been identified in the Ulleung Basin. The targets are the NGH-bearing sand reservoirs that were found during the Second Ulleung Basin Gas Hydrate Drilling Expedition (UBGH2) in 2010 (Lee et al. 2011; Ryu et al. 2014), although it is likely that their estimates of 30 + Tcf of gas in sands may be high. This lies in the constricted sea area between Korea and Japan, but the source of clastic sediment was mainly from the north and more promising sediments may lie in the offshore area of North Korea and Russia. There is no likelihood of NGH to the west of Korea in the 'Yellow Sea' because it is too shallow for a GHSZ to form. Because Korea is one of the world's largest LNG importers, displacing these with indigenous natural gas is possible if exploitable NGH deposits can be identified. The Korean government continues to invest in NGH research. Research projects using NGH separation properties in a chemical engineering approach to seawater desalination (Park et al. 2011) are similar to U.S. Government supported activities of the same type (Max and Osegovic 2007; Max 2003b).

Taiwan: McDonnell et al. (2000) and Liu et al. (2004) have recognized BSRs and blanking attributed to the presence of NGH in the northern sector of the South China Sea near Taiwan. Although the Taiwan NGH program is not very active, it is likely that NGH and subjacent gas is present. The Gas Hydrate Master Program (GHMP) has been incorporated in the Taiwan National Energy Program—Phase II (GHMP), which is the current funding vehicle for NGH research.

Africa: Because most of the countries in Africa have only recently emerged from colonial status, they are suffering from a combination of traditional lack of investment in science and indigenous natural resource investigations since these countries received their independence, and a lack of finance. Traditionally resource rich countries in Africa, such as Nigeria in which hydrocarbon discoveries are relatively recent and civil associations and government experience weak, government attention is mainly focused on exploiting and dividing the new wealth rather than implementing investment in science and technology. African countries have traditionally depended on foreign countries and companies for technology. Widespread BSRs have been identified along the southern periphery of the Orange River delta off South Africa. The seafloor in the region appears to have many pockmarks and mud volcanoes indicating a strong, gas-rich petroleum system.

Latin America: Latin America is a late starter in NGH research. There is a desire for NGH research in some Latin American countries, most of which have substantial continental margins.

Chile: Southern Chile produces a small amount of gas, but most of the long Chilean margin has not been well explored for either conventional gas or NGH. The entire western continental margin of South America is very narrow. There are no large tectonically undisturbed sedimentary basins along the active plate margin that marks the base for the steep continental slope, which has curtailed conventional exploration and the amount of incidental information that would bear on the presence of NGH deposits. NGH investigations have been limited to a few expeditions by an international consortium (Max et al. 2006) using piston coring, heat flow measurements, and collection of both normal and deep-tow seismic data. NGH has been recovered from some shallow cores. This indicates a thin sulfate reduction zone and a high gas flux, and the Chilean government approved an expanded program to investigate the national NGH resource potential. But despite further investigations by a subsequent international consortium, which included some high-resolution seismic survey, no significant discoveries have been made. In May (2015), the Mining Committee of the Chilean Senate reconsidered NGH with the intention to ask the country's president to initiate a Chilean national NGH program.

Brazil: The passive margin deepwater basins off SE Brazil and the continental slope sediments have a very active petroleum system with both oil and natural gas deposits in water depths between 600 and 2800 m. Brazil is currently supporting considerable exploration and development of its abundant deep-water hydrocarbon resources, but the emphasis is on development of conventional hydrocarbon resources, especially its sub-salt and pre-salt oil. There is currently no country-based national NGH research program, although several geologists have been sent to countries with NGH expertise, such as Japan, to complete advanced degrees specializing in NGH, and Petrobras (the Brazilian national energy company) has NGH investigators on staff and works with some Brazilian universities.

Mexico: Mexico is currently opening its offshore and onshore areas to energy companies in order to expand exploration and production. This is expanding the role of the national energy company, PEMEX, to joint ventures and supervision of activities. Some of the structural energy issues may now be assessed. For instance, Mexico imports LNG to their Pacific coast while they have an excess of natural gas on their Gulf Coast—and no pipeline across the country. At present Mexico is a net importer of natural gas. We understand that Pemex has been transferring explorationists from their deepwater exploration and production teams to the onshore and shelf assignments (2015) mainly because they have been finding primarily gas rather than oil in deeper water.

On the basis of thickest GHSZ, the most prospective areas for NGH in the Gulf of Mexico and the Caribbean lie north of a line from SE Cuba to about Central America just north of the coastline of Honduras where the GHSZ ranges from 300 to 500 m in thickness over large continental slope and deeper areas (Wood and Jung 2008). Turbidite sands are also common within the GHSZ (Boswell et al. 2012b). The geology of the Mexican deepwater east coast has many similarities to the U.S. Gulf of Mexico, including diapiric and allochthonous salt, although there is no sediment supply on the order of the Mississippi River. Natural oil seeps are present throughout the deepwater area. In addition, the whole Gulf region benefits from a very active petroleum system, has considerable existing infrastructure that is now extending into deepwater and the ultradeep, and the regulatory system for U.S. waters is well established so that it is legally possible to explore, drill, and produce.

Other: Amongst the other South and Central American countries, Uruguay's National Oil Company (ANCAP) along the South Atlantic coast and the Instituto Colombiano del Petroleo (ICP) along its more thickly sedimented Caribbean coast (Ochoa 2013) have initiated NGH energy research and exploration activities. Several joint international cruises have taken place off the Pacific coast of South America but did not involve more than 2D seismic acquisition and seafloor sampling. A number of researchers in South American universities have initiated NGH research.

Australia and New Zealand: New Zealand also has an active research program investigating the resource potential of New Zealand's NGH deposits. The program is led by GNS Science (GNS 2015), in collaboration with NIWA, the University of Otago, and the University of Auckland; with funding from the Ministry of Business, Innovation, and Employment. The current program builds on a 2010–2012 pilot program funded by the Foundation for Research, Science, and Technology.

As part of a larger project focused on understanding the dynamic interaction of NGHs and slow moving active sediment mass flows, a joint New Zealand-German research team mapped a large area of hydrate-bearing sediment off New Zealand's eastern coast in April and May, 2014. The project utilized 3D and 2D seismic data and found evidence of NGH along with 99 gas plumes venting from the seafloor. The plumes formed columns extending up to 250 m into the water column. The venting and the presence of NGH have significant implications for slope failure along New Zealand's coastal margin as they do elsewhere (Talukder 2012).

The key objectives for the resource assessment program are to study the regional distribution of NGH and to characterize individual NGH reservoirs. The initial area of investigation is a zone outside of the Hikurangi Margin. This characterization effort is utilizing analysis of seismic data to improve the understanding of gas-hydrate reservoir rocks and investigation of gas-hydrate-formation mechanisms. Initial production modeling has been completed as well as a first assessment of seafloor communities that may be affected by NGH production. The overarching goal within the current program is to identify targets for scientific exploration drilling. In its 2015 lease offering for the Offshore Pegasus and East Coast Basins, the Petroleum and Minerals Department of New Zealand's Ministry of Business, Innovation and Employment, noted that the basins are prospective for oil, gas, and for NGH.

Australia's ocean territory is about 16 million km^2, about twice as large as its land area. There are extensive continental slope and marginal basin sediments in which NGH can be expected to be found. Australia has emerged as a major supplier of natural gas, which is exported as LNG. New Zealand is about halfway through a six-year national NGH program. If results are promising, a successor program will probably be established. A number of potential NGH provinces have been identified in the huge New Zealand offshore area. Australia presently has no national NGH program. There is considerable activity among university marine scientists.

European Union (EU): In the past, the European Union funded NGH research mainly as part of environmental issues, with the notable exception of Ireland, which was the first country in Europe to carry out a preliminary evaluation of NGH of its continental slopes. A preliminary assessment of Irish oceanic NGH potential was carried out using existing seismic, heat flow, and other marine geological data during the early part of 2005 with the participation of Hydrate Energy International, working with Providence Resources under the auspices of the Geological Survey of Ireland.

In 2014 the EU established the "Marine gas hydrate—an indigenous resource of natural gas for Europe" (MIGRATE) program that is implemented as ESSEM COST Action ES1405 (https://www.migrate-cost.eu/home). This program follows the main NGH programs that have energy as their focus. The EU is now investing in hydrate R&D to explore and identify resource potential and develop technologies for gas production from NGHs. Stefan Bünz, associate professor at Centre for Arctic Gas Hydrate, Environment and Climate (CAGE) at The Arctic University of Norway, was elected the Vice Chair of the action.

The MIGRATE program is based on the proposition that Natural gas from indigenous NGH deposits should play an important role in the future European energy system. It could (1) enhance the security of energy supply, (2) contribute to the reduction of CO_2 emissions by replacing coal and oil as well as providing the fossil fuel back up for renewable energy production by providing both base load and spiker capability (see Chap. 1). Ideally, NGH could be brought on stream to provide adequate natural gas Europe's as conventional gas resources deplete in the coming decades. This would provide for a longer bridge period to implement an integrated renewable energy matrix while reducing the growing dependence of Europe on natural gas imports. Study areas currently are focused on the European open ocean continental margins, the Black Sea, the Mediterranean Sea (Chaps. 3 and 4), and as a participant in international consortia.

The COST action MIGRATE is designed to integrate the expertise of a large number of European research groups and industrial players to promote the development of multidisciplinary knowledge on the potential of NGH as an energy resource. In particular, MIGRATE aims to (1) estimate the European inventory of exploitable NGH, (2) evaluate current NGH technologies for exploration, production and monitoring, (3) assess environmental risks, and (4) prepare a field production test in European waters. National efforts will be coordinated through Working Groups (WG) and will focus on: (1) Resource assessment, (2) Exploration, production, and monitoring technologies, (3) Environmental challenges, and, (4) Communication, information dissemination to enhance public perceptions of program objectives and achievements.

Turkey: Turkish scientists are attending NGH research meetings and are reported as having initiated at least preliminary hydrate assessment programs.

In March 2015, Hydrate Energy International led a meeting of the Turkish National Oil company (TPAO) and marine scientists from Dokuz Eylül University in Izmir. We understand that NGH evaluation is now part of the Turkish energy matrix, particularly in the Black Sea.

After many years of planning, Turkey has begun an extensive evaluation of the nation's NGH potential in the Black Sea. The program is being led by Dokuz Eylül University in conjunction with TPAO. This comprehensive program includes depositional modeling that integrates onshore and offshore studies, hydro-acoustic and geophysical surveys (multibeam, sonar, chirp, high-resolution seismic acquisition, and bathymetry surveys), water column sampling, sediment sampling, laboratory studies, and computer modeling. The initial NGH cruise was scheduled to begin in March 2015 and data collection will continue for more than one year. After three years, a second phase is planned that will include 3D seismic and electromagnetic data acquisition, along with evaluation and development of production technology.

China: In 2000, three Chinese institutions took an interest in different aspects of NGH. This research built on earlier surveys to identify NGH that were undertaken in 1999 by the Guangzhou Marine Geological Bureau. In 2001, a NGH project was established (GMGS2 Program), and the Geological Survey of China initiated a number of marine research projects focusing on the identification of hydrate (Wang et al. 2006). The first Chinese scientific program meeting of this project was held in Beijing in November 2003, with mainly Chinese and Japanese scientists attending. In May 2004, the Center for Hydrate and Natural Gas Research was established in the Guangzhou Institute of Energy Conversion (Chinese Academy of Sciences), and this still appears to be the center of NGH activity, with the active participation of GEOMAR (of Germany). After seismic analysis, including drilling (Yang et al. 2015). In 2002 a national NGH project was initiated with the equivalent of 100 million dollars allocated as start-up funding. The Second Institute of Oceanography of the State Oceanic Administration was also originally involved with NGH research.

The Chinese research in NGH began by funding considerable research through universities, mainly of a rock mechanics and physical chemistry character. There appears to have been no oversight body and programs were not well coordinated although much of the work was of very high quality. More recently, the resource potential aspects of NGH have become more important and have attracted research funding. In May and June of 2017, a number of NGH projects were carried out in the Chinese EEZ to the SE of the Pearl River estuary off southern China (Ziguo et al. 2017) under the auspices of the Geological Survey of China. In particular, over an 8-day period in May, an industrial pilot production from NGH took place in 1266 m of water in the Shenhu area of the Pearl River region continental slope. The silt-clay reservoir was between 201 and 277 m below the seabed. Natural gas production reaches as high as 3.5×10^4 m^3 per day and averaged 1.0×10^4 m^3 per day with steady gas flow for 7 days and 19 h. Biogenic methane content was up to 99.5% of gas by volume. These results are very promising, especially because the grain size of the sediment was less than the sand size considered optimal for NGH production of natural gas.

South China Sea (SCS) Other: Away from the Chinese EEZ, other countries bordering the SCS such as Vietnam and the Philippines, amongst others to the south, will also be interested in hydrocarbon energy. Although China has claimed a large area of the SCS, its economic division is a political issue that may be settled through the U.N. Law of the Sea, but more likely by direct negotiation. As this is not an energy issue, we do not discuss it further.

Rising energy demand in Asia, particularly in China, has focused attention of States bounding the South China Sea (SCS), particularly along the coasts of China, Vietnam and the southern SCS margin, but not the eastern margin of the SCS including the Philippine Sea area (Laursen 2013). China is one of the world's major importers of hydrocarbons and its demand is growing. The SCS is the subject of political jockeying for EEZ rights and is probably one of the key issues for countries bordering on the SCS. Vietnam, the Philippines, Malaysia, and China have all occupied islands and reefs in the SCS with a view to enhancing their EEZ claims, although most of the emphasis in the western press has concerned the dramatic conversion by China of a number of reefs into islands with sheltered ports and airfields. Conventional oil and gas discoveries have been made in the SCS, and hydrocarbon exploration, which is promising, is still in its early stages. It is likely that the SCS has a moderate to strong petroleum system and should be significant regional source of hydrocarbons in the near to medium term. NGH has been recognized along the Chinese and Taiwanese margin of the SCS (McDonnell et al. 2000; Wang et al. 2006; Trung 2012). NGH resources are presently a strong exploration issue.

3.7 Frontier Regions

Resources nearest to markets tend to be utilized first, with increasing costs and transport distances added as more distant resources are required to meet demands. This is usually associated with increasing value to support the increased costs. Frontier regions are those that, at this stage of development of the hydrocarbon resource base, have yet to be explored or even considered as practical because of harsh conditions (ultra-deepwater, polar regions), the degree of stranding with no nearby land that can be used to support logistics, and outright distance from markets.

North Polar Region: The Arctic is a true frontier region in which the most long-term R&D is being carried out. Although territorial claims have yet to be completely resolved under the Law of the Sea treaty, this process is underway (ECS 2016), and all Arctic countries have now submitted their geological evaluations supporting their claims to expanded EEZs under the terms of the Law of the Sea (which the U.S. still remains the only outstanding signatory). The Arctic is becoming an area of active exploration because of climate change that is responsible for dramatically shrinking summer ice cover. Deeper water areas that were inaccessible year-round because of weather and ice conditions are now seeing new exploration activity even before the migration of exploration across some of the shallow water shelves is complete. Ice-resistant floating platforms from which to carry out exploration and production activities are already entering service in the Arctic (Phillips 2014). The Arctic has some major economic factors to overcome, including remoteness, access, and logistics (Chap. 7).

Seismic acquisition and exploration in general is difficult in the Arctic because of sea ice, the tendency to icing of equipment which affects stability, and often inhospitable weather conditions. Exploration in the Arctic is already taking advantage of autonomous survey capabilities beneath the sea ice (Millar and Mackay 2015). Although full scale seismic surveys have not yet been attempted, even limited 3D surveys are possible using swarms of AUVs acting together. The advantage of autonomous technology is that the work can proceed beneath the ice and weather, with the two main issues being powering autonomous vehicles and communicating of data and new operating instructions.

The Arctic Ocean has immense potential for oceanic NGH (Max et al. 2013) but apart from identification of BSRs on virtually all Arctic continental margins (which are not discussed in detail here), there has been little focused exploration such as has taken place around Japan (Noguchi et al. 2011) and the northern Gulf of Mexico (Boswell et al. 2012a, b).

Antarctic Polar Region: There has been little NGH related research in the Antarctic other than BSR identification on seismic sections. The Antarctic is not regarded as a potential natural gas resource area in the same way that is the Arctic (Chap. 9), except it is much more remote from the major industrial countries in the Earth's northern temperate region. Technically, it is less of a frontier region than it is a region in which no NGH exploration is likely to be carried out until virtually everywhere else on Earth has been explored.

References

Aird, P. (2016). *Assessing geo-hazards*. (Drilling consultant, BP Norge, date of download). http://www.offshore-mag.com/articles/2016/01/det-norske-studying-impact-of-ice-on-subsurface-geology.html. Accessed January 29, 2016.

Akçar, N., Yavuz, V., Ivy-Ochs, S., Kubik, P. W., Vardar, M., & Schlüchter, C. (2008). A case for a downwasting mountain glacier during Termination 1, Verçenic valley, Northeaster Turkey. *Journal of Quaternary Science, 23*(3), 273–285.

Aksu, A., Hiscott, R. N., Mudie, P., Rochon, A., Kaminski, M. A., Abrajano, T., et al. (2002). Persistent Holocene outflow from the Black Sea to the Eastern Mediterranean contradicts Noah's Flood hypothesis. *GSA Today, 12,* 4–10.

Algan, O., Gokasan, E., Gazioglu, C., Ycel, Z. Y., Aopar, B., Guseysu, C., et al. (2002). A high-resolution seismic study in Sakarya Delta and submarine canyon, southern Black Sea shelf. *Continental Shelf Research, 22,* 1511–1527. https://doi.org/10.1016/S0278-4343(02)00012-2.

Aloisi, G., Pierrea, C., Rouchyb, J.-M., Foucherc, J.-P., Woodside, J., & the MEDINAUT Scientific Party. (2000). Methane-related authigenic carbonates of eastern Mediterranean Sea mud volcanoes and their possible relation to gas hydrate destabilisation. *Earth and Planetary Science Letters, 184*(1), 321–338. https://doi.org/10.1016/S0012-821X(00)00322-8.

Argnani, A. (1993). Neogene tectonics of the Strait of Sicily. In M. D. Max & P. Colantoni (Eds.), *The geological development of the Sicilian-Tunisian Platform. Conference Proceedings of International Meeting* held in Urbino, Italy, 4–6 November, 1992. UNESCO Technical Report in Marine Sciences, 55–60.

Bachrach, R., Beller, M., Liu, C. C., Perdomo, J., Shelander, D., Dutta, N., et al. (2004, April). Combining rock physics analysis, full waveform prestack inversion and high-resolution seismic interpretation to map lithology units in deep water: A Gulf of Mexico case study. *The Leading Edge*, 373–383.

Bagla, P. (2015). Thorium seen as nuclear's new frontier. *Science, 350*(6269), 726–727.

Bard, E., Hamelin, B., Fairbanks, R. G., & Zindler, A. (1990). Calibration of the 14C timescale over the past 30,000 years using mass spectrometric U-Th ages from Barbados corals. *Nature, 345,* 405–410.

Bard, E., Jouannic, C., Hamelin, B., Pirazzoli, P., Arnold, M., Faure, G., et al. (1996). Pleistocene sea levels and tectonic uplift based on dating of corals from Sumba Island, Indonesia. *Geophysical Research Letters, 23,* 1473–1476.

Barka, A., & Reilinger, R. (1997). Active tectonics of Eastern Mediterranean region: Deduced from GPS, neotectonic and seismicity data. *Annali Di Geofisica, X2*(3), 587–610.

Barnard, A., Sager, W. W., Snow, J. E., & Max, M. D. (2015). Subsea gas emissions from the Barbados Accretionary Complex. *Marine and Petroleum Geology, 64*, 31–42. https://doi.org/10.1016/j.marpetgeo.2015.02.008.

Barton, C. M. (2015). Industry continues to provide solutions for deepwater production challenges. *Offshore Magazine. 75*(5). https://www.offshore-mag.com/articles/print/volume-75/issue-5/engineering-construction-installation/industry-continues-to-provide-solutions-for-deepwater-production-challenges.html. Accessed 24 September 2018.

Barton, C., Hambling, H., Albaugh, E. K., Mahlstedt, B., & Davis, D. (2015). Deepwater solutions & records for concept selection. Going deeper with production technology. *Poster May, Offshore Magazine inset, 1 sheet.*

Beaudoin, Y. C., Waite, W., Boswell, R., & Dallimore, S. R. (Eds.). (2014). *Frozen heat: A UNEP global outlook on methane gas hydrates* (Vol. 1). United Nations Environment Programme, GRID-Arendal. United Nations Environment Programme, 2014, 80 pp. ISBN: 978-92-807-3429-4. Job No: DEW/1866/NO.

Beckman, J. (2015). Newfoundland and Labrador opens frontier Flemish Pass blocks for licensing. *Offshore Magazine November*, 34–35.

Birkedal, K. A., Hauge, L. P., Graue, A., & Ersland, G. (2015). Transport mechanisms for CO_2-CH_4 exchange and safe CO_2 storage in hydrate-bearing sandstone. *Energies, 8,* 4073–4095. https://doi.org/10.3390/en8054073.

Boswell, R. (2014). Developments in marine gas hydrate exploration. Paper OTC-25192-MS. In *Proceedings of the Offshore Technology Conference*, held in Houston, Texas, USA, 5–8 May 2014, 8 pp.

Boswell, R., & Collett, T. S. (2011). Current perspectives on gas hydrate resources. *Energy & Environmental Sciences, The Royal Society of Chemistry, 4,* 1206–1215.

Boswell, R., Collett, T. S., & Frye, M. (2011). *Gulf of Mexico Gas Hydrate Joint Industry Project Leg II: Technical summary* (26 pp). Morgantown, WV: US Department of Energy, NETL.

Boswell, R., Collett, T. S., Frye, M., Shedd, W., McConnell, D. R., & Shelander, D. (2012a). Subsurface gas hydrates in the northern Gulf of Mexico. *Marine and Petroleum Geology, 34,* 4–30. https://doi.org/10.1016/j.marpetgeo.2011.10.003.

Boswell, R., Collett, T. S., Frye, M., Shedd, W., McConnell, D. R., & Shelander, D. (2012b). Architecture of gas-hydrate-bearing sands from Walker Ridge 313, Green Canyon 955, and Alaminos Canyon 21: Northern deepwater Gulf of Mexico. *Marine and Petroleum Geology, 34,* 134–149.

Boswell, R., Saeki, T., Shipp, C., Shedd, B., Collett, T., Shellander, D., et al. (2014). Prospecting for gas hydrate resources. *Fire in the Ice, 14*(2), 9–13.

Boswell, R., Shipp, C., Reichel, T., Shelander, D., Saeki, T., Frye, M., et al. (2015). Prospecting for marine gas hydrate resources. In: Special section: Exploration and characterization of gas hydrates. *Interpretation, 4*(1), SA13–SA24.

Bourry, C., Chazallon, B., Charlou, J.-L., Donval, J. P., Ruffine, L., Henry, P., et al. (2009). Free gas and gas hydrates from the Sea of Marmara, Turkey Chemical and structural characterization. *Chemical Geology, 264,* 197–206.

Bruneton, B., Konofagos, E., & Foscolos, A. E. (2012). *The importance of Eastern Mediterranean gas fields for Greece and the EU* (28 pp). Pytheas Investors Service.

Caiti, A., Ingenito, F., Kristensen, Å., & Max, M. D. (1996). Geoacoustic models for selected shallow-water areas. *SACLANT Memorandum, 305,* 20.

Çifçi, G., Dondurur, D., & Ergün, M. (2002). Sonar and high resolution seismic studies in the Eastern Black Sea Basin. *Turkish Journal of Earth Sciences, 11,* 61–81.

Colantoni, P. (1988). *Bathymetric-Morphological-Lithological Map of the Adventure Bank (Strait of Sicily). Scale 1:2000,000.* Bologna, Italy: Consiglio Nazionale Delle Ricerche (CNR).

Crager, B. (2015). *Benchmarking of subsea decommissioning, Part 1.* http://oilpro.com/post/4421/green-field-subsea-boosting-life-of-field-boosting-unconstrained. Accessed April 23, 2016.

Crutchley, G. J., Maslen, G., Pecher, I. A., & Mountjoy, J. J. (2016). High-resolution seismic velocity analysis as a tool for exploring gas hydrate systems: An example from New Zealand's southern Hikurangi margin. In: Special section: Exploration and characterization of gas hydrates. *Interpretation, 4*(1), SA1–SA12.

Dahlmann, A., & the Anaximander Team. (2005). Gas hydrates in the Eastern Mediterranean: Occurrence and biogeochemical environment compiled from detailed sampling of the Anaximander Mountains. *Geophysical Research Abstracts 7*, 05176. SRef-ID: 1607-7962/gra/EGU05-A-05176.

Danielsen, J., & Bekker, R. (2011). The future of marine CSEM. *First Break, 29,* 77–81.

Dillon, W. P., Nealon, J. W., Taylor, M. H., Lee, M. W., Drury, R. M., & Anton, C. H. (2001). Seafloor collapse and methane venting associated with gas hydrate on the Blake Ridge—Causes and implications to seafloor stability and methane release. In C. K. Paull & W. P. Dillon (Eds.), Natural gas hydrates occurrence, distribution, and detection (pp. 211–233). Geophysical monograph 124, American Geophysical Union.

DOE. (1998). *A strategy for methane hydrates research & development* (58 pp). U.S. Department of Energy DOE/FE-0378.

DOE. (1999). *National methane hydrate multi-year R&D program plan* (34 pp). U.S. Department of Energy.

DOE. (2015). *National gas technologies program* (59 pp). Gas Hydrates Research and Development. Science and Technology Assessment (2016–2020) draft 3/16/2015.

Dondurur, D., Küdük, H. M., & Çifçi, G. (2013). Quaternary mass wasting on the western Black Sea margin, offshore of Amasra. *Global and Planetary Change, 103,* 248–260.

Durham, L. S. (2017). *EM Tech as a cost-saving complement to seismic.* AAPG Explorer. http://www.aapg.org/publications/news/explorer/emphasis/Articleid/41392/em-tech-as-a-cost-saving-complement-to-seismic: Accessed August 2, 2017.

ECS. (2016). *Continental shelf, the last maritime zone* (36 pp). UNEP/GRID-Arendal (originally published 2009). ISBN: 978-82-7701-059-5.

Egawa, K., Furukawa, T., Saeki, T., Suzuki, K., & Narita, H. (2013). Three-dimensional paleomorphologic reconstruction and turbidite distribution prediction revealing a Pleistocene confined basin system in the northeast Nankai Trough area. *AAPG Bulletin, 97*(5), 781–798. https://doi.org/10.1306/10161212014.

ENI. (2016). *Deep water. Innovation & technology.* http://www.eni.com/en_IT/innovation-technology/eni-projects/deep-water/deep-water-project.shtml. Accessed March 19, 2016.

Fairbanks, R. G. (1989). A 17,000-year glacio-eustatic sea-level record: Influence of glacial melting rates on the Younger Dryas event and deep-ocean circulation. *Nature, 342,* 637–642.

Fang, H., & Duan, M. (2014). *Offshore operation facilities: Equipment and procedures* (1196 pp). Gulf Professional Publishing. ISBN: 0123972841, 9780123972842.

Feather, K. (2007). The rapid adoption of seabed logging. *Scandinavian Oil and Gas Magazine, 5* (6), 37–38.

Finetti, I., Bricchi, G., Del Ben, A., Pipan, M., & Xuan, Z. (1988). Geophysical study of the Black Sea. *Bolletino di Geofisica Teorica ed Applicata, 30,* 197–324.

Frye, M. (2008). *Preliminary evaluation of in-place gas hydrate resources: Gulf of Mexico outer continental shelf* (136 pp). US Department of the Interior Minerals Management Service Resource Evaluation Division, OCS Report MMS 2008-0004.

Frye, M., Shedd, W., & Schuenemeyer, J. (2013a). *Gas hydrate resource assessment, atlantic outer continental shelf* (57 pp). Bureau of Ocean Energy Management (BOEM) Report RED 2013-01.

Frye, M., Schuenemeyer, J., Shedd, W., Piper, K., & Herman, B. (2013b). Gas hydrate resource assessment on the United States Outer Continental Shelf: A mass balance model. In *Proceedings of the 7th International Conference on Gas Hydrates (ICGH 2011)*, Edinburgh, Scotland, United Kingdom, July 17–21 (13 pp).

Frye, M., Shedd, W., & Boswell, (2012). Gas hydrate resource potential in the Terrebonne Basin, Northern Gulf of Mexico. *Marine and Petroleum Geology, 34,* 150–168.

Fujii, T., Aung, T. T., Wada, N., Komatsu, Y., Suzuki, K., Ukita, T., et al. (2016). Modeling gas hydrate petroleum systems of the Pleistocene turbiditic sedimentary sequences of the Daini-Atsumi area, eastern Nankai Trough, Japan. In: Special section: Exploration and characterization of gas hydrates. *Interpretation, 4*(1), SA95–SA111.

Fujii, T., Suzuki, K., Takayama, T., Tamaki, M., Komatsu, Y., Konno, Y., et al. (2015). Geological setting and characterization of a methane hydrate reservoir distributed at the first offshore production test site on the Daini-Atsumi Knoll in the eastern Nankai Trough, Japan. *Marine and Petroleum Geology, 66*(1), 310–322. https://doi.org/10.1016/j.marpetgeo.2015.02.037.

Gabrielsen, P. T., Abrahamson, P., Panzner, M., Fanavoll, S., & Ellingsrud, S. (2013). Exploring frontier areas using 2D seismic and 3D CSEM data, as exemplified by multi-client data over the Skrugard and Havis discoveries in the Barents Sea. *First Break, 31*(1), 63–71.

Garcia-Castellanos, D., Estrada, F., Jinénez-Munt, I., Gorini, C., Fernàndez, M., Vergés, J., et al. (2009). Catestrophic flood of the Mediterranean after the Messinian salinity crisis. *Nature Letters, 462*(10), 778–781.

Garziglia, S., Ioulalen, M., Migeon, S., Ducassou, E., Mascle, J., Sardou, O., et al. (2007). Triggering factors and tsunamicenic potential of a large submarine mass failure on the western Nile margin (Rosetta Area, Egypt). In V. Lykousis, D. Sakellariou & J. Locat (Eds.), Submarine mass movements and their consequences (pp. 347–355). Springer.

Gautier, F., Clauzon, G., Suc, J. P., Cravatte, J., & Violanti, D. (1994). Age and duration of the Messinian salinity crisis. C.R. Academy of Science. *Paris (IIA), 318,* 1103–1109.

Gettrust, J. F., Wood, W. T., Coffin, R., Pohlman, J., & Pandey, R. (2005). Coordinated, high-resolution studies of hydrated sediments. OTC Paper 17586. In *Proceedings of the Offshore Technology Conference*, Houston, TX, U.S.A., 2–5 May 2005 (9 pp).

Gillet, H., Lericolais, G., Rehault, J.-P. (2007). Messinian event in the black sea: Evidence of a Messinian erosional surface. *Marine Geology, 244*(1–4), 142–165. http://dx.doi.org/10.1016/j.margeo.2007.06.004, http://archimer.ifremer.fr/doc/00000/3324/.

Giosan, L., Filip, F., & Constantinescu, S. (2008). Was the Black Sea catastrophically flooded in the early Holocene? *Quaternary Science Reviews, 28,* 1–6.

GM. (2016). Google earth maps, SE Japan projection. (No PDF). https://www.google.com/maps/place/Japan/@33.0115944,134.3508365,1224364m/data=!3m1!1e3!4m2!3m1!1s0x34674e0fd77f192f:0xf54275d47c665244!6m1!1e1.

GNS. (2015). *Gas hydrates as an energy resource*. http://www.gns.cri.nz/Home/Our-Science/Energy-Resources/Gas-Hydrates/Current-Research/Gas-hydrates-as-an-energy-resource. Accessed October 13, 2015.

Haeckel, M., Bialas, J., Klaucke, I., Wallmann, K., Bohrmann, G., Schwalenberg, K., et al. (2015). Gas hydrate occurrences in the Black Sea—New observations from the German SUGAR project. *Fire in the Ice, 15*(2), 6–9.

Haines, S. S., Hart, P., Shedd, W., Frye, M., Agena, W., Miller, J., et al. (2013a). New seismic data acquired over known gas hydrate occurrences in the deepwater Gulf of Mexico. *Fire in the Ice, 13*(2), 3–6.

Haines, S. S., Hart, P., Shedd, W., Frye, M., Agena, W., Miller, J., et al. (2013b). Seismic investigation of gas hydrates in the Gulf of Mexico: Results from 2013 OBS and high-resolution multicomponent seismic surveys. Poster OS21A-1618, presented at *the American Geophysical Union Fall Meeting*, San Francisco, 9–13 December 2013.

Haines, S. S., Hart, P., Shedd, W. W., & Frye, M. (2014a). Seismic investigation of gas hydrates in the Gulf of Mexico: 2013 Multi-component and high-resolution 2D acquisition at GC955 and WR313, Paper OTC-25318-MS. In *Proceedings of the Offshore Technology Conference* held in Houston, Texas, USA, 5–8 May 2014 (18 pp).

Haines, S. S., Hart, P., Ruppel, C., et al. (2014b). Cruise report for P1-13-LA, USGS Gas Hydrates Research Cruise, R/V. *Pelican*, April 18 to May 3, 2013, Deepwater Gulf of Mexico. United States, Geological Survey Open-File Report 2014-1080, (38 pp).

Hamilton, E. L. (1980). Geoacoustic modeling of the sea floor. *Journal of the Acoustical Society of America, 68,* 1313–1340.

Henke, T., Muller, C., Marquardt, M., Hensen, C., et al. (2008). Integrated gas hydrate quantification of Nicoya Peninsula—Costa Rica. In *Proceedings of the 6th International Conference on Gas Hydrates (IGCH 2008)*, Vancouver, British Columbia, Canada, 6–10 July 2008.

Hester, K. C., Stevens, J. C., & Howard, J. J. (2011). Composition studies to determine rate and extent of CO_2 exchange in a hydrate-bearing core. In *Proceedings of the 7th International Conference on Gas Hydrates*, Edinburgh, UK, 17–21 July (7 pp).

Hillman, J. I. T., Cook, A. E., Sawyer, D. E., Kúçúk, H. M., & Goldberg, D. S. (2017). The character and amplitude of 'discontinuous' bottom-simulating reflections in marine seismic data. *Earth and Planetary Science Letters, 459,* 157–169.

Holbrook, W. S., Hoskins, H., Wood, W. T., Stephen, R. A., Lizarralde, D., & Leg 164 Science Party. (1996). Methane hydrate and free gas on the Blake Ridge from vertical seismic profiling. *Science, 273,* 1840–1843.

Infield. (2015). *Deep & Ultra-deepwater technologies & projects 2008/12*. http://www.infield.com/Downloads/Deepwater_Technologies_And_Projects_Report_2008.pdf.

Iwata, M. (2014). New type of methane hydrate confirmed in Japan Sea. *The Wall Street Journal*. http://blogs.wsj.com/japanrealtime/2014/12/26/new-type-of-methane-hydrate-confirmed-in-japan-sea/.

JOGMEC. (2013). *News release: Gas production from methane hydrate layers confirmed*. Japan Oil, Gas and Metals National Corporation, Tokyo. http://www.jogmec.go.jp/english/news/release/release0110.html.

Johnson, H. P., Miller, U. K., Salmi, M. S., & Solomon, E. A. (2015). Analysis of bubble plume distributions to evaluate methane hydrate decomposition on the continental slope. *Geochemistry, Geophysics, Geosystems, 16*, 15 pp (Online article). https://doi.org/10.1002/2015gc005955.

Jongsma, D., van Hinte, J. E., & Woodside, J. M. (1985). Geologic structure and neotectonics of the North African Continental Margin south of Sicily. *Marine and Petroleum Geology, 2,* 156–179.

Kable. (2015). *The world's biggest natural gas reserves*. Kable Intelligence Limited. http://www.hydrocarbons-technology.com/features/feature-the-worlds-biggest-natural-gas-reserves/.

Kawamoto, T. (2013). *The first offshore MH production test. PDF presentation.* http://energy.gov/sites/prod/files/2013/06/f1/Takami%20Kawamoto%20-%20The%20First%20Offshore%20Production%20Test.pdf. Accessed December 22, 2015.

Komatsu, Y., Suzuki, K., & Fujii, T. (2016). Sequence stratigraphy and controls on gas hydrate occurrence in the eastern Nankai Trough, Japan. In: Special section: Exploration and characterization of gas hydrates. *Interpretation, 4*(1), SA73–SA81.

Kumar, P., Collett, T. S., Vishwanath, K., Shukla, K. M., Nagalingam, J., Lall, M. V., et al. (2016). Gas hydrate-bearing sand reservoir systems in the offshore of India: Results of the India National Gas Hydrate Program expedition 02. U.S. Department of Energy. In *Fire in the Ice, 16*(1), 1–8.

Labourdette, R., Poncer, J., Seguin, J., Temple, F., Hegre, J., & Irving, (2006). Three-dimensional modelling of stacked turbidite channels in West Africa: Impact on dynamic reservoir simulations. *Petroleum Geoscience, 12,* 335–345.

Lamb, M. P., & Mohrig, D. (2009). Do hyperpycnal-flow deposits record river-flood dynamics? *Geology, 37*(12), 1067–1070. https://doi.org/10.1130/G30286A.1.

Lambeck, K. (1993). Glacial rebound of the British Isles—I. Preliminary model results, II. A high-resolution, high precision model. *Geophysics Journal International, 115,* 941–990.

Lambeck, K. (1996). Glaciation and sea-level change for Ireland and the Irish Sea since Late Devensian/Midlandian time. *Journal of the Geological Society, London, 153,* 853–872.

Lambeck, K., Rouby, H., Purcell, A., Sun, Y., & Sambridge, M. (2014). Sea level and global ice volumes from the last Glacial Maximum to the Holocene. *Proceedings of the National Academy of Sciences, 111*(43), 15296–15303. www.pnas.org/cgi/doi/10.1073/pnas.1411762111.

Latham, T., Shelander, D., Boswell, R. Collett, T., & Lee, M. (2008). Subsurface characterization of the hydrate bearing sediments near Alaminos Canyon 818. In *Proceedings of the 6th International Conference on Gas Hydrates (ICGH 2008),* Vancouver, British Columbia, Canada, July 6–10, 2008.

Laursen, W. (2013). South China Sea offers opportunities, challenges. *Offshore Magazine* 09/18/13. http://www.offshore-mag.com/articles/print/volume-73/issue-9/asia-pacific/south-china-sea-offers-opportunities-challenges.html. Accessed April 23, 2016.

Lee, S. R., Kim, D. S., Ryu, B. J., Bahk, J. J., Yoo, D. G., Kim, G. Y., et al. (2011). Recent developments of gas hydrate program in Korea: Ulleung Basin gas hydrate drilling expedition 2. In *Proceedings of the 7th International Conference on Gas Hydrates (ICGH 2011),* Edinburgh, Scotland, United Kingdom, July 17–21 (7 pp).

Lericolais, G., Bourget, J., Popescu, I, Jermannaud, P., Mulder, T., Jorry, S., et al. (2013). Late quaternary deep-sea sedimentation in the western Black Sea: New insights from recent coring and seismic data in the deep basin. *Global and Planetary Change, 103,* 232–247. http://dx.doi.org/10.1016/j.gloplacha.2012.05.002.

Lericolais, G., Popescu, I., Guichard, F., & Popescu, S. M. (2007). A Black Sea lowstand at 8500 yr B.P. indicated by a relict coastal dune system at a depth of 90 m below sea level. *Geological Society of America Special Papers, 426,* 171–188.

Levashov, S. P., Yakymchuk, N. A., Korchagin, I. N., Bozhezha, D. N. (2014). Frequency-resonance method of remote sensing data processing: Approbation on hydrocarbon field of Barents Sea offshore. In *76th EAGE Conference & Exhibition 2014 Amsterdam RAI,* The Netherlands, 16–19 June 2014, ResearchGate (6 pp). https://doi.org/10.3997/2214-4609.20141265. http://www.researchgate.net/publication/266635578. Accessed October 24, 2015.

Ligtenberg, H., & Neves, F. (2008). Integrating disciplines for petroleum system analysis—From seismic interpretation to stratigraphic and basin modelling. *Petroleum Geoscience, 14*(3), 219–221. https://doi.org/10.1144/1354-079308-809.

Lister, C. J., Hiscott, R. N., & Mudie, P. J. (2015). Compositional trends through the Holocene mud succession of the southwestern Black Sea shelf: Implications for sedimentary provenance and water-level history. *Sedimentary Geology, 316,* 13–25.

Liu, C.-S., Schnurle, P., Hsiuan, T.-H., Yang, T., Lin, S., Wang, Y., et al. (2004). Gas hydrate distribution offshore southwestern Taiwan and its geophysical and geochemical characteristics

(Abs.). In Collett, T. & Johnson, A. (Eds.), *Natural gas hydrates: Energy resource potential and associated geologic hazards, AAPG Hedberg Research Conference*, September 12–16, Vancouver BC, Canada (2 pp).
Løseth, H., Gading, M., & Wensaas, L. (2009). Hydrocarbon leakage interpreted on seismic data. *Marine and Petroleum Geology, 26,* 1304–1319.
Lüdmann, T., Wong, H. K., Konerding, P., Zillmer, M., Petersen, J., & Flüh, E. (2004). Heat flow and quantity of methane deduced from a gas hydrate field in the vicinity of the Dnieper Canyon, northwestern Black Sea. *Geo-Marine Letters, 24,* 182–193.
Lykousisa, V., Alexandria, S., Woodsideb, J., de Langec, G., Dählmannc, A., Perissoratisd, C., et al. (2009). Mud volcanoes and gas hydrates in the Anaximander mountains (Eastern Mediterranean Sea). *Marine and Petroleum Geology, 26*(6), 854–872. https://doi.org/10.1016/j.marpetgeo.2008.05.002.
Majorowicz, J. A., & Osadetz, K. G. (2003). Natural gas hydrate stability in the east coast offshore-Canada. *Natural Resources Research, 12*(2), 93–104.
Maslin, M. A., & Thomas, E. (2003). Balancing the deglacial global carbon budget: the hydrate factor. *Quaternary Science Reviews, 22*(2003), 1729–1736.
Max, M. D. (1990a). *Gas hydrate and acoustically laminated sediments: Probable environmental cause of anomalously low acoustic-interaction bottom loss in deep ocean sediments* (68 pp). Naval Research Laboratory Report 9235.
Max, M. D. (1990b). Natural gas clathrates: Geoacoustic inducement of low bottom loss. Science and technology for a new oceans decade, Vol. III. *Proceedings of MTS '90,* Washington, D.C., September 26–28, 1990 (pp. 578–583). Marine Technological Society.
Max, M. D. (2002). Desalination using positively buoyant or negatively buoyant/assisted buoyancy hydrate. U.S. Patent (24 December) No 6,497,794 B1, 35 pp.
Max, M. D. (2003a). Chapter 18: Hydrate as a future energy resource for Japan. In Max, M. D. (Ed.), *Natural gas hydrate: In oceanic and permafrost environments* (2nd ed., pp. 225–238). London, Boston, Dordrecht: Kluwer Academic Publishers (now Springer).
Max, M. D. (Ed.). (2003b). *Natural gas hydrate: In oceanic and permafrost environments* (2nd ed., 422 pp). London, Boston, Dordrecht: Kluwer Academic Publishers (now Springer).
Max, M. D. (2008). Hydrate-based desalination using compound permeable restraint panels and vaporization-based cooling. US Patent 008/0028781. (Filed 8 May 2007, Issued 7 February 2008), 52 pp.
Max, M. D., & Johnson, A. H. (2011). Methane hydrate/clathrate conversion. In Khan, M. R. (Ed.), *Clean hydrocarbon fuel conversion technology. Woodhead Publishing Series in Energy No. 19.* Cambridge, U.K.: Woodhead Publishing Ltd. ISBN 1 84569 727 8, ISBN-13: 978 1 84569 727 3, 413–434.
Max, M. D., & Johnson, A. H. (2014). Hydrate petroleum system approach to hydrate natural gas exploration. *Petroleum Geoscience 20*(2), 187–199. Geological Society of London. https://doi.org/10.1144/petgeo2012-049. Published in Online First March 21, 2014.
Max, M. D., & Johnson, A. H. (2015). A fresh look at the Mediterranean and Black Sea basins: Potential for high-quality hydrate reservoirs. *Fire in the Ice, 15*(1), 15–18.
Max, M. D., Johnson, A. H., & Dillon, W. P. (2006). *Economic geology of natural gas hydrate* (341 pp). Berlin, Dordrecht: Springer.
Max, M. D., Johnson, A. H., & Dillon, W. P. (2013). *Natural gas hydrate arctic ocean deepwater resource potential* (113 pp). SpringerBriefs in Energy.
Max, M. D., Kristensen, Å., & Michelozzi, E. (1993). Small-scale Plio-Quaternary sequence stratigraphy and shallow water geology of the west central Malta Plateau. In M. D. Max & P. Colantoni (Eds.), *The geological development of the Sicilian-Tunisian platform, Conference Proceedings* (pp. 117–122). UNESCO Technical Report in Marine Sciences.
Max, M. D., & Miles, P. R. (1999). Marine survey for gas hydrate. In *Offshore Technology Conference Paper OTC 10769* (12 pp).
Max, M. D., & Osegovic, J. P. (2007). Hydrate-based reduction of fluid inventories and concentration of aqueous and other water-containing products. U.S. Patent U.S. 7,255,794, August 14, 2007) (15 pp).

May, J. (2015). Towed streamer EM—Resistivity in context. Tech Talk, Petroleum Exploration Society of Australia (PESA) News, August–September, 2–4.

McConnell, D., Boswell, R., Collett, T. et al. (2009). *Gulf of Mexico gas hydrate joint industry project leg II: Walker ridge 313 site summary*. US Department of Energy, Washington, DC. www.netl.doe.gov/technologies/oil-gas/./WR313SiteSum.pdf.

McConnell, D., & Zhang, Z. (2005). Using acoustic inversion to image buried gas hydrate distribution. *Fire in the Ice, 5*(4), 3–5.

McDonnell, S. L., Max, M. D., Cherkis, N. Z., & Czarnecki, M. F. (2000). Tectono-sedimentary controls on the likelihood of gas hydrate occurrence near Taiwan. *Marine and Petroleum Geology, 17*(8), 929–936.

MES. (2015). *Gas hydrates (with budget table)*. Indian Ministry of Earth Sciences. http://moes.gov.in/programmes/gas-hydrates. Accessed April 23, 2016.

Millar, G., & Mackay, L. (2015, April). Maneuvering under the ice. *Sea Technology*, 35–38.

MMS. (2004). *Deepwater Gulf of Mexico 2004: America's expanding frontier*. OCS Report MMS 2004-021, U.S. Department of the Interior Minerals Management Service, 166 pp.

MMS. 2008. *Preliminary evaluation of in-place gas hydrate resources: Gulf of Mexico outer continental shelf.* OCS Report MMS 2008-004, 136 pp.

Niranjan C., & Nanda, N. C. (2016). Chapter 5. Direct hydrocarbon indicators (DHI). In *Seismic data interpretation and evaluation for hydrocarbon exploration and production. A Practitioner's guide* (pp. 103–113). Berlin: Springer.

NAS. (2010). Realizing the energy potential of methane hydrate for the United States. Committee on Assessment of the Department of Energy's MHR&DP. National Research Council of the National Academy of Sciences, 205 pp.

Naudts, L., Greinert, J., Artemov, Yu.G & De Batist, M. 2005. Subsurface and morphologic setting of 2778 methane seeps in the Dnieper paleo-delta in the northwestrn Black Sea. (Poster). In *International Gas Hydrate Workshop*, Victoria, BC, Canada, 8–11 May, 2005.

Nicholas, W. A., & Chivas, A. R. (2014). Chapter 14: Late Quaternary sealevel change on the Black Sea Shelves. *Geological Society London, Memoir, 41*, 199–212. https://doi.org/10.1144/m41.14.

Nixon, L., Kazanis, E., & Alonsos, S. (2016). *Deepwater Gulf of Mexico December 31, 2014*. Bureau of Ocean Energy Management OCS Report BOEM 2016-057, 84 pp.

Noguchi, S., Furukawa, T., Aung, T. T., & Oikawa, N. (2011). Reservoir architecture of methane hydrate bearing turbidite channels in the eastern Nankai Trough, Japan. In *Proceedings of the 7th International Conference on Gas Hydrates (ICGH 2011)*, Edinburgh, Scotland, United Kingdom, July 17–21, 2011, 9 pp.

Levashov, S. P., Yakymchuk, N. A., Korchagin, I. N., Bozhezha, D. N. (2015). Operative assessment of hydrocarbon potential of Zohr Prospect in the Mediterranean Se (Shorouk block, Egypt offshore) by the frequency-Resonance method of the remote sensing data processing and interpretation. *ResearchGate*, 36 pp. http://www.researchgate.net/publication/282184034. Accessed October 24, 2015.

Ochoa, A. E. C. (2013). Characterization of gas hydrate reservoirs in the offshore Caribbean region of Columbia. *Fire in the Ice, 13*(2), 7–11.

Offshore 10/15/15. (2015). *Lukoil discovers potential giant gas field offshore Romania*. http://www.offshore-mag.com/articles/2015/10/lukoil-discovers-potential-giant-gas-field-offshore-romania.html. Accessed April 23, 2016.

Okui, A., Kaneko, M., Nakanishi, S., Monzawa, N., & Yamamoto, H. (2008). An integrated approach to understanding the petroleum system of a frontier deep-water area, offshore Japan. *Petroleum Geoscience, 14,* 223–233.

Park, K-n, Hong, S. Y., Lee, J. W., Kang, K. C., Lee, Y. C., Ha, M.-G., et al. (2011). A new apparatus for seawater desalination by gas hydrate process and removal characteristics of dissolved minerals (Na+, Mg2+, Ca2+, K+, B3+). *Desalination, 274,* 91–98. https://doi.org/10.1016/j.desal.2011.01.084.

Perissoratis, C., Ioakim, Chr., Alexandri, S., Woodside, J., Nomikou, P., Dählmann, A., et al. (2011). Thessaloniki mud volcano, the shallowest gas hydrate-bearing mud volcano in the

Anaximander mountains, Eastern Mediterranean. *Journal of Geological Research 2011*, Article ID 247983, 11 pages. https://doi.org/10.1155/2011/247983.

Phillips, A. (2014). In Russia, world's first ice-resistant oil platform starts production. Climate Progress. http://thinkprogress.org/climate/2014/04/21/3429130/russia-arctic-offshore-oil-platform/. Accessed April 23, 2016.

Piggott, N., & Pulham, A. (1993). Sedimentation rate as the control on hydrocarbon sourcing, generation, and migration in the deepwater Gulf of Mexico. In *Proceedings, Gulf Coast Section SEPM 14th Annual Research Conference*, 179–191.

Pirazzoli, P. A. (1993). Global sea-level changes and their measurement. *Global and Planetary Change, 8,* 135–148.

Plag, H.-P., Austin, W. E. N., Belknap, D. F., Devoy, R. J. N., England, J., Josenhans, H., et al. (1996). Late Quaternary relative sea-level changes and the role of glaciation upon continental shelves. *Terra Nova, 8,* 213–222.

Poort, J., Vassilev, A., & Dimitrov, L. (2005). Did postglacial catastrophic flooding trigger massive changes in the Black Sea gas hydrate reservoir? *Terra Nova, 17,* 135–140. https://doi.org/10.1111/j.1365-3121.2005.00599.x%3e.

Popescu, I., Lericolais, G., Panin, N., de Batist, M., & Gillet, H. (2007). Seismic expression of gas and gas hydrates across the western Black Sea. *Geo-Marine Letters 27*, 173–183. http://dx.doi.org/10.1007/s00367-007-0068-0. Accessed August 13, 2015.

Praeg, D., Geletti, R., Wardell, N., Unnithan, V., Mascle, J., & Camerlenghi, A. (2011). The Mediterranean Sea: A natural laboratory to study gas hydrate dynamics? In *Proceedings of the 7th International Conference on Gas Hydrates (ICGH 2011)*, Edinburgh, Scotland, United Kingdom, July 17–21, 2011, 8 pp.

Rajan, A., Mienert, J., Bunz, S., & Chand, S. (2012). Potential serpentinization, degassing, and gas hydrate formation at a young (G20 Ma) sedimented ocean crust of the Arctic Ocean ridge system. *Journal of Geophysical Research, 117,* B03102. https://doi.org/10.1029/2011JB008537. 14 pp.

Reed, J. (2016). *Here's one supermajor that's NOT planning job cuts*. 1/20/16. http://oilpro.com/post/21661/here-one-supermajor-not-planning-job-cuts?utm_source=DailyNewsletter&utm_medium=email&utm_campaign=newsletter&utm_term=2016-01-20&utm_content=Article_6_txt. Accessed April 23, 2016.

Reichel, T., & Gallagher, J. W. (2014). A global review of gas hydrate resource potential. *Fire in the Ice, 14*(2), 5–8.

Reichel, T., & Gallagher, J. W. (2015). Global screening of gas hydrates—How does the Arctic score? In *Abstract in Program of #P Arctic the Polar Petroleum Potential Conference and Exhibition*, Stavanger, Norway 29 September–2 October 2015, 92.

Riedel, M., Collett, T. S., Malone, M. J., & Expedition 311 Scientists. (2006). Site U1327. In *Proceedings of the Integrated Ocean Drilling Program*, Vol. 311. https://doi.org/10.2204/iodp.proc.311.105.2006.

Riedel, M., Willoughby, E. C., & Chopra, S. (Eds.). (2010a). In *Geophysical Development Series: Vol. 14. Geophysical characterization of gas hydrates* (pp. 321–328). Society of Exploration Geophysicists.

Riedel, M., Collett, T. S., Kumar, P., Sathe, A. V., & Cook, A. (2010b). Seismic imaging of a fractured gas hydrate system in the Krishna Godavari Basin offshore India. *Marine and Petroleum Geology, 27*, 1476–1493.

Roberts, H. H., Hardage, B. A., Shedd, W. W., & Hunt, J., Jr. (2006). Seafloor reflectivity—An important seismic property for interpreting fluid/gas expulsion geology and the presence of gas hydrate. *The Leading Edge, 25*(5), 620–628.

Ryu, B.-J., Yoo, D. G., Kang, N.-K., Kim, G. Y., Bahnk, J. J., & Lee, J. Y. (2014). In-place gas hydrate resources in the Ulleung basin, East Sea of Korea. *Fire in the Ice, 14*(1), 9–10.

SBC. (2015). *Natural Gas Series FactBook*. Gas Hydrates SBC Energy Institute (SBC-EI) Schlumberger, 99 pp. https://www.sbc.slb.com/~/media/Files/SBC%20Energy%20Institute/SBC%20Energy%20Institute_Gas%20Hydrates_FactBook.pdf.

Sahay, V. K., & Johnson, A. H. (2014). Gas hydrate deposits of Krishna Godavari Basin, India: Issues and potentiality in exploration and commercial production. Paper OTC-25137-MS. In *Proceedings of Offshore Technology Conference* held in Houston, Texas, USA, 5–8 May 2014, 6 pp.

Sain, K., & Gupta, H. (2012). Gas hydrates in India: Potential and development. *Gondwana Research, 22,* 645–657. https://doi.org/10.1016/j.gr.2012.01.007.

SDN. (2015). *Developing nations join R&D race for gas hydrates*. SciDevNet. http://www.scidev.net/global/energy/feature/developing-nations-join-r-d-race-for-gas-hydrates.html#.

Shepard, F. P. (1954). Nomenclature based on sand-silt-clay ratios. *Journal of Sedimentary Petrology, 24,* 151–158.

Shelander, D., Dai, J., Bunge, G., Singh, S., Eissa, M., & Fisher, K. (2012). Estimating saturation of gas hydrates using conventional 3D seismic data, Gulf of Mexico Joint Industry Project Leg II. *Marine and Petroleum Geology, 34,* 96–110.

Schlumberger. (2015). Continental shelf. Oilfield Glossary. http://www.glossary.oilfield.slb.com/en/Terms/c/continental_shelf.aspx.

Schoderbek, D., Farrell, H., Hester, K., Howard, J., Raterman, K., Silpngarmlert, S., et al. (2013). *Oil & natural gas technology ConocoPhillips gas hydrate production test final technical report*, October 1, 2008–June 30, 2013. 208 pp.

Schultheiss, P., Roberts, J., Druce, M., Priest, J., Holland, M., Yamamoto, K., & Yang, S. (2014). PCATS and PCATS triaxial: Further development and recent field experience making core measurements under pressure. In *Proceedings of the 8th International Conference on Gas Hydrates (ICGH8-2014)* (pp. 10). Beijing, China, July 28–August 1, 2014.

Schultheiss, P., Holland, M. Roberts, J., Bigalke, N. & Mimitz, M. 2017. Advances in wireline pressure coring, core handline, and core analysis related to gas hydrate drilling investigations. In *Proceedings of the 9th International Conference on Gas Hydrates (ICGH 2017)*, Denver, Co. USA, June 25–30 2017, 14 pp.

Sloan, E. D., & Koh, C. A. (2008). *Chemical Industries Series 119: Clathrate hydrates of natural gases* (3rd ed., 752 pp). CRC Press, Taylor & Francis Group. ISBN 9780849390784 - CAT# 9078.

Sorgente, R., Drago, A. F., & Ribotti, A. (2003). Seasonal variability in the central Mediterranean Sea Circulation. European Geosciences Union. *Annales Geophysicae, 21,* 299–322.

Soulet, G., Ménot, G., Garreta, V., Rostek, F., Zaragosi, S., Lericolais, G., et al. (2011). Black Sea "Lake" reservoir age evolution since the Last Glacial—Hydrologic and climatic implications. *Earth and Planetary Science Letters, 308*(1–2), 245–258. http://dx.doi.org/10.1016/j.epsl.2011.06.002.

Spence, G. D., Haacke, R. R., & Hyndman, R. D. (2010). Seismic indicators of natural gas hydrate and underlying free gas. In M. Riedel, E. C. Willoughby & S. Chopra (Eds.), *Geophysical Developments Series: Vol. 14. Geophysical characterization of gas hydrates* (pp. 39–71). Society of Exploration Geophysicists.

Stanley, D. J. (1995). A global sea-level curve for the late Quaternary: The impossible dream? *Marine Geology, 125,* 1–6.

Stennett, I. A., Wakefield, N. D., & Games, K. P. (2014). Applications of detailed velocity field studies from high resolution seismic (HRS) data. Paper OTC-25096-MS. In *Proceedings of the Offshore Technology Conference* held in Houston, Texas, USA, 5–8 May 2014 (15 pp).

Sunamura, T. (1992). *Geomorphology of rocky coasts* (p. 302). Wiley: Chichester, England.

Taira, A. (2001). Tectonic evolution of the Japanese Island arc system. *Annual Reviews Earth and Planetary Science, 29,* 109–134.

Taladay, K. B., & Moore, G. F. (2015). Concentrated gas hydrate deposits in the Kumano Forearc Basin, Nankai Trough, Japan. *Fire in the Ice, 15*(1), 1–4.

Talukder, A. R. (2012). Review of submarine cold seep plumbing systems: Leakage to seepage and venting. *Terra Nova, 24,* 255–272. https://doi.org/10.1111/j.1365-3121.2012.01066.x.

Tanahashi, M. (2011). Present status of Japanese methane gas hydrate research and development program. Institute for Geo-Resources and Environment, AIST/Geological Survey of Japan, presentation at *Hydrate in Japan CCOP Hydrate Workshop.* 2 March 2011, 32 pages.

Thakur, N. K., & Rajput, S. (2011). *Exploration of gas hydrates, geophysical techniques* (281 pp). Berlin, Heidelberg: Springer. ISBN 978-3-642-14234-5.

Tonarelli, B., Turgutcan, F., Max, M. D., Akai, T. (1993). Shallow sediments at four localities on the Sicilian-Tunisian Platform. In M. D. Max & P. Colantoni, (Eds.), *The geological development of the Sicilian-Tunisian platform, Conference Proceedings* (pp. 123–128). UNESCO Technical Report in Marine Sciences.

Trenhaile, A. S. (1972). The shore platforms of the Vale of Glamorgan, Wales. *Transactions of the Institute of British Geographers, 56,* 127–144.

Trenhaile, A. S. (1974). The geometry of shore platforms in England and Wales. *Transactions of the Institute of British Geographers, 62,* 129–142.

Trenhaile, A. S. (1987). *The geomorphology of rocky coasts* (384 pp). Oxford: Oxford University Press.

Trèhu, A. M., Glemings, P. B., Bangs, N. L., Chevallier, J., Gràcia, E. Johnson, J. E., et al. (2004). Feeding methane vents and gas hydrate deposits at south Hydrate Ridge. *Geophysical Research Letters 31*, 4 pp. https://doi.org/10.1029/2004gl02128.

Trung, N. N. (2012). The gas hydrate potential in the South China Sea. *Journal of Petroleum Science and Engineering, 88–89,* 41–47.

Tsimplis, M. N., Proctor, R., & Flather, R. A. (1995). A two-dimensional tidal model for the Mediterranean Sea. *Journal Geophysical Research, 100*(C8), 16223–16239.

UNEP. (2015). Sicily Channel/Tunisian Plateau: Topography, circulation and their effects on biological component. United Nations Environment Programme Mediterranean Action Plan. UNEP(DEPI)/MED WG.408/Inf.23, 29 pp.

Vassilev, A. & Dimitrov, L. (2003). Model evaluation of the Black Sea gas hydrates.Comptes rendus de l'Academie bulgare des Sciences (Bulgarian Academy of Sciences) Tome 56(3) Geophysique (Oceanologie), 15–21.

Vickery. (2014). *India energy the struggle for power* (125 pp). Washington, D.C.: Asia Programme of the Wilson Institute.

Waite, W. (2016). USGS scientists participate in Indian ocean gas hydrates drilling. USGS sound waves. https://soundwaves.usgs.gov/2016/12/fieldwork.html. Accessed January 4, 2017.

Waite, W. F., Santamarina, J. C., Rydzy, M., et al. (2012). Overview of the interlaboratory comparison of wave velocity measurements in sand with gas hydrate and other pore-filling material. *Fire in the Ice, 12*(1), 16–21.

Waltham, D., Jaffery, N., MacLean, S., & Zampetti, V. (2008). Stratigraphic modelling of turbidite prospects to reduce exploration risks. *Petroleum Geoscience, 14,* 273–280.

Wang, S., Yan, W., & Song, H. (2006). Mapping the thickness of the gas hydrate stability zone in the South China Sea. *Terrestrial, Atmospheric, and Ocean Sciences, 17*(4), 815–828.

Wingfield, R. T. R. (1995). A model of sea-levels in the Irish and Celtic seas during the end-Pleistocene to Holocene transition. In R. C. Preece (Ed.), *Island Britain: A quaternary perspective* (pp. 209–242). Geological Society Special Publication No. 96.

Wingfield, R. T. R., & Evans, C. D. R. (1998). The significance of the shoreface ramp for coastal development: Holderness, eastern England, U.K. In J. M. de Prat (Ed.), *Sustainable waterfront and coastal developments in Europe: Socioeconomics, technical and environmental aspects* (pp. 662–678)., Proceedings 4th Eurocoast (European Coastal Association of Science and Technology) at LITTORAL '98 Barcelona, Spain: Suport Servais.

Wood, W. T., & Jung, W. Y. (2008). Modeling the extent of earth's marine methane hydrate cryosphere. In *Proceedings of the 6th International Conference on Gas Hydrates (ICGH 2008)*, July 6–10, 2008, Vancouver, British Columbia, Canada, 8 pp.

Worthington, P. F. (2010). Petrophysical evaluation of gas-hydrate formations. *Petroleum Geoscience, 16,* 53–66. https://doi.org/10.1144/1354-079309-018.

Yakymchuk, N. A. Levashov, S. P., Korchagin, I. N., Bozhezha, D. N., & Pydlysna, I. S. (2015). New experience of direct geophysical method used for hydrocarbon potential assessment of large prospecting blocks in Arctic. Section in Korchagin et al. 2015. Research Gate. https://doi.org/10.13140/rg.2.1.2917.9360, https://www.researchgate.net/publication/282184034.

Yang, S., Ming, Z., Liang, J., Lu, J., Zijiang, Z., Holland, M., et al. (2015). Preliminary results of China's third gas hydrate drilling expedition: A critical step from discovery to development in the South China Sea. *Fire in the Ice, 15*(2), 1–5.

Zahno, C., Akçar, N., Yavuz, V., Kubik, P. Wl, & Schlüchter, C. (2010). Chronology of Late Pleistocene glacier variations at the Uludağ Mountain, NW Turkey. *Quaternary Science Reviews, 29,* 1173–1187.

Ziguo, H., Hongcai, F., Qingqing, H., & Lian, L. (2017). China has successfully conducted its first pilot production of natural gas hydrates. *Acta Geologica Sinica (English Edition), 91*(3), 1133–1134.

Chapter 4
Potential High-Quality Reservoir Sediments in the Gas Hydrate Stability Zone

Abstract In addition to sandy turbidite exploration targets that are common along many continental margins, other marine sediments having NGH reservoir potential may also exist. These terrigenous sediments would have been deposited in continental rifts that evolved into oceanic zones. Terrigenous continental margin sediments, known for their high-quality reservoir characteristics, may have subsided along continental flanks where they are now overlain by thick marine sediment sequences. On the contrary, some of these high-quality reservoir sediments might have been lifted to the gas hydrate stability zone (GHSZ) due to tectonism. During re-flooding of continental basins related to the development of passive margins, shorelines progressed up slope. During periodic halts in sea level rise, paralic deposition can take place along temporary shorelines, often interacting with fluviatile sediment supply or reworking older terrigenous sediments depositing upon the flooding terrigenous sediments. Mass flow deposits related to re-flooding may also occur. The majority of the subsided terrigenous and paralic sediments are likely to be deeply buried and be below the GHSZ. Here they comprise exploration targets for conventional hydrocarbons. Younger high-quality reservoir sediments may occur in the GHSZ. Continental margins starved of modern course grained marine sediment may contain these high-quality terrigenous and paralic sediments. Reworking of sediments after deposition has the potential to increase permeability and porosity of dominantly fine-grained sediments by deflation caused by current activity. High-quality reservoirs not formed in marine turbidite systems are liable to be rare, they could provide high-quality reservoir hosts for large and producible NGH concentrations.

4.1 High-Quality Sand Reservoirs on Continental Margins

High-quality sedimentary reservoirs that are not part of marine turbidite systems, commonly occur beneath deepwater and ultra-deepwater marine sediments deposited on passive continental margins. These are composed of high porosity sediments than occur in marine turbidites that are currently considered to be the principle

M. D. Max and A. H. Johnson, *Exploration and Production of Oceanic Natural Gas Hydrate*, https://doi.org/10.1007/978-3-030-00401-9_4

exploration hosts for NGH deposits (Chap. 3). Nonetheless, there is potential for coarse grained turbidites, where sediments are close to a suitable siliciclastic sediment source. For example, in the Walvis Basin offshore Namibia, NW African margin, where course and well sorted sandy sediments have been deposited in marine environments since the Early Cretaceous (Intawong and Nodgson 2017). Exploration in the relatively shallow GHSZ has not been carried out but along continental margins that have suitable siliciclastic sediment sources and relatively short transport distances to deepwater deposition, there is always the possibility that high grade turbidite reservoirs may exist.

Porous and permeable carbonates were also deposited in similar shallow-water depositional environments that are now subsided (Bryant et al. 2012; PG 2014). These sediments may also be elevated into near-seafloor positions along complex passive margins affected by tectonic activity. They reflect depositional environments that preceded and marked the opening of a new plate margin and contain 'rift' sediments deposited before the basins were flooded and amalgamated in the interconnected world ocean. Most potential high-quality reservoirs now found in marine environments represent subaerial and shallow water depositional environments. These sediments were deposited as the grabens or sags overlying continental crust transitioned through further depression and subjacent extension.

There are three general depositional environments that can form high-quality reservoirs in early passive margin basins prior to complete ocean flooding. First, before and during the depression, clastic sediments infilling the subsiding basin can form thick off-lapping successions that drape the incipient continental shelves. Second, sediments deposed under aeolian conditions by reworking of sediment exposed on the depressing flank of older continental sedimentation. Third, paralic sediments deposited prior to or associated with fluviatile activity and prograding sea levels that reflect temporary sea level stands during flooding of the basins as the region transitioned to open ocean seafloor.

Active plate margins also may have hydrocarbon potential. Current hydrocarbon frontier areas include active continental margins. There are many tectonically active continental margins where thick continental slope marine sedimentary successions have excellent potential (Chap. 3). Coarse grained sediments may commonly be found in constructive sedimentary prisms where narrow continental shelves are near the 'sand machine', coastal regions where there is for rapid erosion from coastal mountains with deposition in the continental margin nearby. For instance, much of SE Asia, in which rifting of continental masses was not part of their early history of tectonics and sedimentation, may have considerable good-quality reservoirs (Crager 2014, p. 5; Offshore 2/12/16).

Sands deposited in shallow water and subaerial depositional environments commonly have high porosity and permeability. This is particularly true for well-sorted aeolian sands that may contain very little of the fine-grained material that usually degrades reservoir performance. These sands will have the characteristics of high-quality reservoirs, although they will be found beneath and marginally interbedded with deep and ultra-deepwater sediments. These high-quality reservoirs constitute major exploration targets.

In addition to exploration techniques used to locate marine turbidite systems that could host NGH concentrations (Frye et al. 2013), the geological history of passive crustal margin deepwater prospect areas, in which subsided clastic-dominated sedimentary basins now flank oceanic crust, are important if they occur in deep-water GHSZs. The age and history of rifting and thermal events in the basins are less important for NGH than for conventional hydrocarbons (Chaps. 2 and 3). Early, pre-marine sediments may contain considerable high-quality reservoir sands and porous and permeable carbonate sediments. Pre-and early-rift sediment packages are most likely to be deeply buried, where they can host conventional hydrocarbon deposits. But where faulting and thrusting can bring the high-quality reservoir sands into current gas hydrate stability zones (GHSZ), they can also host NGH concentrations.

4.2 Subsided Rift-Related Sediments and Basin Inversion

Passive margins developed when tectonic processes caused continental masses to rift and separate (Scrutton 1982; Bott 1995; Gibson et al. 2015). Rift tectonics led to the establishment of new accreting constructive plate margins in which oceanic crust is formed in an axial corridor that is the nexis of spreading. Rifting can be relatively simple (Fig. 4.1), with the onset of rifting driven by a single tectonic framework or with more than one phase of rifting and reorientation of the major

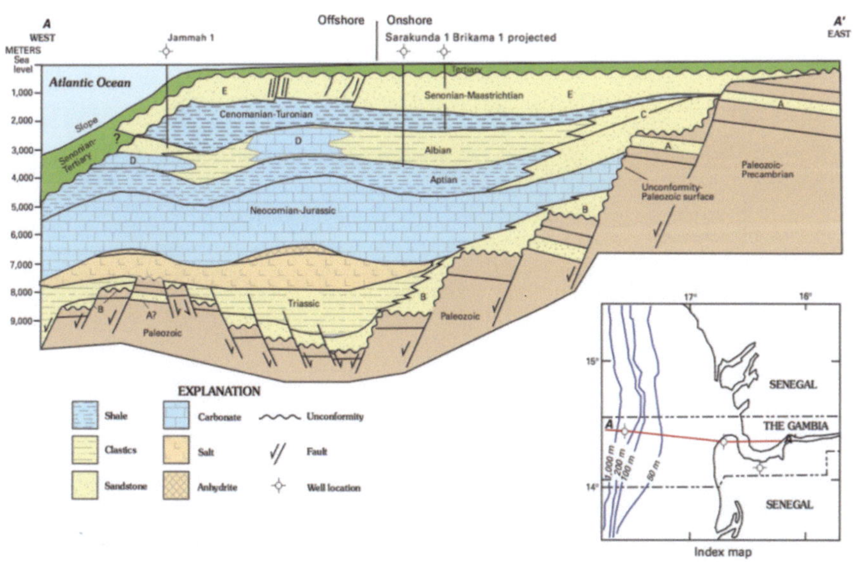

Fig. 4.1 Rifted margin off Gambia, NW Africa, modified from Brownfield and Charpentier (2003) (USGS report)

stress fields over time. Although Fig. 4.1 shows rifting under sediments that have prograded oceanward, some subsided margins have narrow sediment infill zones with rifting and rift sediments commonly out to the base of continental slopes, where they may only underlie thin marine sediment successions. The length of time over which rifting takes place is important, as is the climate during rifting. Generation of clastic sediments will be enhanced if summers are characterized by very hot days and cold nights and wet winters with substantial freeze-thaw conditions to provide siliciclastic sediments. The relative altitude between rift floors and upthrown blocks is also important.

Potential high quality reservoirs of the rift phase are generally the most deeply buried where a continental margin is characterized entirely with subsidence. However, a significant number of early continental margin basins show inversion. When compressional or transpressional zones develop in the subsided early rift sediments, zones can invert or be structurally raised in anticlines or fault blocks. For instance, in the Juba Basin off southern Somalia, Davison and Steel (2018) have identified a number of inversions. None of these appear to have lifted sandstones into the local gas hydrate stability zone (GHSZ) but even the formation of weak anticlines that may persist all the way to the seafloor have the potential to focus gas flow, which is a necessary component of the NGH petroleum system (Chap. 2).

The margins of most of the continents that were once part of crustal fragments of Pangea that broke into the more familiar continental cores we see today usually have characteristically good reservoir rocks of clastic character (Condie 1997). This breakup began in the Triassic period of the Mesozoic era and the oldest oceanic crust in the pull-apart zones appears to be Jurassic in age. North and South America separated from Europe and West Africa while East Antarctica pulled away from the other continents and Australia (Fig. 4.2). Rifted continental margins are the sites of

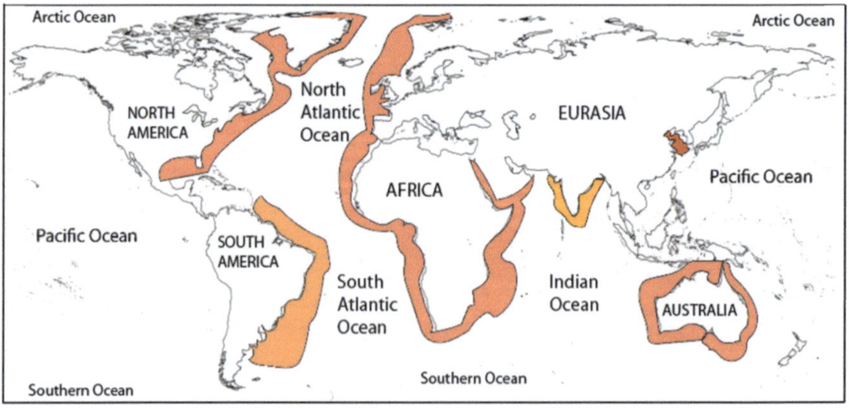

Fig. 4.2 Generalized passive margins (Orange shaded areas) on which high-quality reservoirs could exist subjacent to marine sediments. Based on NAS (1989). Volcanic and non-volcanic margins not distinguished. Structurally complex margins in high latitudes not shown owing to projection exaggeration and lack of detail

many clastic and carbonate sediments that have high-quality reservoir character that were deposited in fault basins straddling the incipient continental margins (Gibson et al. 2015). The subsided clastic rift sediments are now overlain by marine sediments, which were deposited after the rifts were depressed below sea level and were flooded. Breakup is continuing with the separation of North Africa and Arabia in early Tertiary times with oceanic crust spreading in the Red Sea. Activity in the region has continued into modern times with activity along the East Africa Rift, which is likely to define a line of future breakup (Mohriak and Leroy 2012).

The coast of NW Africa (Ghana, Senegal, Republic of Congo, etc.) has well developed high-quality reservoirs in down-thrown fault blocks, overlain by high-stand high-quality sands and deeper water turbidite sands (Brownfield and Charpentier 2003). Virtually the whole of the Atlantic passive margin zone is highly prospective for rift margin reservoirs, including the U.S. East Coast, on which there is currently a Federal ban on all hydrocarbon exploration. Offshore Senegal (W. Africa) has good plays in Jurassic and Cretaceous platform margin plays that are now inland through younger midwater plays to Cretaceous deepwater clastic sediments (Offshore 1/12/16). The North Atlantic opened somewhat later (Ravnås et al. 2000), with the northernmost part of the Mid-Atlantic Ridge plate margin extending into and forming the axial core of the Eurasian Basin that terminates against the Laptev Sea continental margin (Max et al. 2013).

In addition, many other marine margins are floored by shallow water and continental sediments along the continental margin and which now lie beneath thick marine sediment cover. In many cases these terrigenous sediments contain beds having substantial organic material, including coal, in which bacteria are actively producing biogenic methane (Inagaki et al. 2015). Thus, high-quality terrigenous clastic sediments and high-quality source beds can both occur in close juxtaposition beneath a marine sedimentary succession, in which a different set of potential reservoir sediments can coexist (Max and Johnson 2014).

From late Carboniferous to late Triassic-early Jurassic, the Mozambique Channel off East Africa was deformed by recurrent rifting, which resulted in Karoo continental deposition. During a post-Karoo early Jurassic transgression, clastic, carbonates and evaporites were deposited along the pre-breakup plate margins. A subsequent marine transgression took place in mid-Jurassic with widespread deposition of shelf carbonates and basin sediments. The onset of drift probably occurred during the Bajocian.

The Madagascar–India–Antarctica plate separated from Africa in the early Aptian. During the drift phase, predominantly marine deposition took place throughout the region. The post-drift followed an Aptian hiatus in sedimentation and resulted in the deposition of thick marine sediments. Offshore Mozambique and Tanzania very large deepwater natural gas deposits are found in thick clastic reservoirs (Kelly 2012), but because of a small local market, adding value depends on exporting LNG, which is scheduled to begin in 2018.

The Petroleum Agency of South Africa has authorized a multi-client airborne gravity and magnetic survey program that will extend across two offshore blocks, totaling about 78,000-line km (48,500 mile), offshore SW Cape Agulhas

(West Bredasdorp) and Durban (Eastern Margin) (Offshore 2/9/16). This survey will allow the picking of smaller regions in which to carry out seismic surveys to identify potential hydrocarbon potential. Almost all exploration activity on the African margin is for conventional oil and gas.

4.3 Paralic Reservoirs

Paralic depositional environments produce high-quality clastic sediments at or close to sea-level. Paralic successions consist of deltas, estuary, sand bars and lagoon systems, and may be related to aeolian deposition to landward. Associated sub-environments offer definite evidence of the orientation of one-time shoreline systems. These include a wide range of high-quality reservoir sand bodies including braided and infilled incised distributary channels, splays and shoreline sands, and barrier islands and point bars.

Transgressive-phase shoreline sandstones are shoreline-parallel and in low-energy marine receiving basins where their lateral extent may be controlled by inherited, pre-transgression delta morphology. Their width and thickness are a function of relative sea level history and wave energy. Sediment reworking is continuous during regressive and progressive phases. The rate of sea level change will be a major control on the degree of reworking of clastic sediments. The end-members of the paralic depositional system to seaward are freshwater coals and marine limestones.

When interpreting seismic data, it is important to use a geological—geographical overview as the shape and orientation of the different paralic features, which define the distinctive nature of paralic oil and gas fields. There can also be considerable short-term but important storm and tide-generated heterogeneity in paralic systems. Because sea level change is a fundamental aspect of sedimentation, dividing shallow water marine from subaerial sedimentation, paralic successions are very sensitive to sea-level change. On the geological time scale, sea level change drives transgressive and regressive shifts in facies belts. This depositional environment results in stratigraphically interleaved successions that commonly have very good discontinuous bed definition and are strongly layered.

Paralic reservoirs have been important in petroleum system analysis as they have provided reservoirs from which oil and gas production has taken place in many regions throughout the world, currently contributing around 30% of global conventional hydrocarbon production (Reynolds 2015). For this reason, paralic reservoirs are a primary high-quality exploration target. However, because their formation involves the interaction of complex processes and highly variable sediment input and reworking, the morphology and size of paralic reservoirs is difficult to predict.

This group of sediments includes deltas and estuarine systems that directly reflect sediment input from rivers and shoreline systems including longshore bars, beaches, and sand islands. Sediments within the reservoirs are dominated by laminated sandstones with identifiable sedimentary structures from which their depositional environment can be deduced. Fluid and gas migration in these clastic sediments is also well understood and relatively easy to control during production, especially if the sands are sealed by bounding low permeability shales.

Paralic reservoirs often consist of stacked sand bodies, with prograding sequences marking a landward migration of shorelines associated with rising sea levels. In ideal cases, individual sand lenses, for instance related to migrating longshore bar sands, have permeability paths in the form of faults or sand 'bridges' between major bodies so that fluid and gas flow will develop in the stacked complex as a whole. Regressive sequences may include relics of prograding systems where sea level fall is rapid and near-coastal erosion is weak, but more commonly the sands in the stacked paralic reservoir systems may be consolidated into a larger paralic system of a recessional shore line (Reynolds 1999). Paralic systems have been recognized in many prospective sedimentary basins, such as the Rovuma Basin of Mozambique (Smelror et al. 2006), where they overlie subsided rift sediments. The pre-late drift continental deposits (pre-Lower Cretaceous) found onshore in the northwest Rovuma Basin are generally expected to show mostly low net-to-gross ratios due to diagenesis, but the contemporaneous paralic and shallow marine clastics towards the east and southeast are expected to have fair to good reservoir potential, with moderate to minor reduction of porosity. The Aptian–Albian succession comprises reservoir sands of good quality.

Prograding barrier islands following the inland progress of a rising sea-level shoreline can leave a thin, transgressive sand sheet where sediment starvation is common. If there is an abundance of sand, thick barrier islands can step inland in an en-echelon pattern, leaving older shore face barrier islands and lagoon systems in increasingly more marine environments, especially where sea-level rise is intermittent or cyclic (Johannessen et al. 2015).

Paralic fluviatile-delta and barrier island systems will concentrate the coarser sediments in elongate, slope parallel deposits associated with temporarily stable basin water levels. These could also be associated fluviatile-delta systems and aeolian sediments of similar age upslope. The best case for the development for the highest-quality paralic systems consists of relatively slow flooding, with a temporary stall of flooding that allows substantial clastic sediment to form at particular seashores. Periods of rapid reflooding from the ocean linking with these basins have the potential to bury the temporary shoreline and subaerial sediments under the younger marine succession. The rate of burial is probably too fast for significant erosion of them to take place.

Fluviatile deposits may be highly erosive during low stands when reworked shallow water clastic sediments can be transported into deep water. The most extreme erosion will likely underlie channels, and channel fill deposits of fluviatile systems can be expected to have a high degree of coarse-grained clastic sediment. Deltaic sediments would fan out downslope from the one-time intersection of a fluviatile channel upslope. Fluviatile deposits will root in buried erosive fluviatile systems such as the Messinian Nile canyon (Mahmoudi and Gabr 2009).

Fluviatile systems may have the general appearance of marine turbidite systems but they would be associated with more deeply incised in V-shaped valleys. On slopes they may be deeply incisive and thickly sediment filled, but on lower gradient slopes, braided stream deposits over large areas could be the locations of large areas of coarse sediment deposits. While fluviatile systems may also cross cut geological structures conspicuously because of their greater erosive potential, turbidites that have a weaker erosive power downslope and will tend to flow around obstructions on the seafloor (Boswell et al. 2012). Fluviatile systems may also have less well-developed levee and overbank deposits than turbidites.

4.4 Aeolian–Sabkha Reservoirs

Aeolian systems are subaerial wind-driven deposits that often exhibit exceptional grain size sorting, and they can be quite thick. They are amongst the best reservoir sediments because of their characteristic high porosity and permeability and resistance to compaction and liquefaction. They typically have less predictable development with respect to existing sediment supply systems because wind can move the sands long distances from a sediment source without major erosional features such as channels. For instance, paleo-weather effect can initiate very different weathering and sediment transport regimes. In addition, erosion potential can also be very different at different topographic levels where higher altitude freezing can release fragmental and clastic material that may be highly mobile in fast running water at lower elevations.

A good example is the Norphlet Fm., an Upper Jurassic clastic unit that was deposited both conformably and unconformably upon the LouAnn Salt both onshore and offshore Alabama, Mississippi, and Louisiana (Tew et al. 1991). The salts were deposited in a Callovian (166.1–163.5 Ma) tectonic depression caused by crystal thinning in what is now the Gulf of Mexico (Lundin and Dorè 2016). The Norphlet Fm. is formed from reworked sediments in the basin that are immediately overlain by the Kimmerigian Smackover Fm. The Norphlet Fm. has been subdivided into three sub-units; all deposited under arid climatic conditions. The basal Norphlet Fm. is a thin black shale unit. This <20 m basal shale also underlies the fine silts that occupy interdunal areas of interspersed aeolian facies.

The Lower Norphlet is locally arkosic and has porosities up to about 20% along with high permeability. It appears to be locally derived from erosion of the Appalachian highlands, which were then located just north and east. The overlying aeolian Upper Norphlet Fm. is comprised of typical aeolian beds with features such as high-angle cross-bedding, frosted grains, and wind-textured geomorphologies. The greater part of the Norphlet Fm. is a series of medium to coarse-grained sand bodies up to over 120 m thick. They have been prolific gas reservoirs and have undergone substantial drilling. The Norphlet Fm. is interpreted as having formed in a depositional environment similar to the aeolian Navajo/Nugget sandstones in the southwestern region of the United States or the sub-Atlas Saharan Mountains. The Norphlet Fm. formed as the Gulf of Mexico opened slow enough that carbonates formed as the basin slowly sank.

The extensive drilling and sub-surface mapping of the Norphlet Fm. has been done in considerable detail as to continuity of individual dune deposits and their shape. Studies have shown that they were deposited, along with the inter-dune silts, on the flanks of a sediment-starved basin in which aeolian conditions were responsible for substantial winnowing and sediment redeposition. Sediments at the top of the Norphlet Fm. show reworking of sand in shallow marine conditions (referred to by local geologists as the Smackover Sea) that mark the onset of ocean progradation over this region of subsiding crust. These topmost reworked sands are hard, white, contain local iron mineralization, and have very low porosity. This is overlain by a dense limestone, which marks the onset of shallow marine conditions.

In addition to the Norphlet Fm. in the Northern Gulf of Mexico, an analog (the Pikit Fm.) has been identified in the southern Gulf of Mexico in the conjugate Campeche Escarpment Basin that margins the wide shelf north of the Yucatan Peninsula. In addition to having exciting paralic sediment possibilities, there are also a number of pre-salt exploration opportunities (Rodriguez et al. 2017). Although these may now be buried too deeply to host NGH deposits, its existence further demonstrates the potential for high quality reservoirs that are unrelated to marine turbidites that could host NGH concentrations in what are now thoroughly marine regions.

A lack of major fine-grained sediment inputs as the basin opened and the water level within the basin fell would enhance the likelihood of high quality sediments prior to the establishment of a fully marine environment. As water level in a drying basin goes down, increasingly large new areas of marine sediment are exposed to subaerial conditions where they are subject to erosion having the potential to sort and be redeposited further downslope, with sedimentation focused at or near temporarily depressed sea levels within the basin. Very large areas of easily eroded sediment on the exposed shelf and exposed slopes could readily provide the amount of sediment in high-quality reservoirs. The Norphlet Fm. is an excellent example of a narrow area of laterally, nearly continuous, aeolian sands above sabkha (salt flat) deposits that formed as the Gulf of Mexico first opened.

4.5 Contourites

Contourites are oceanic sedimentary deposits caused by the circulation of thermohaline oceanic currents that winnow and redeposit existing sediments. Reworking of marine sediments by contour-following ocean currents separate fine grained from coarser grained sediment. This results in banks, waves, and sheets of higher porosity and permeability sediment and deposition of fines down-current. The finer-grained evacuated material may be deposited in down-current sediment drift fans up to 1 km thick (Maldonado et al. 2003). This current-driven process characteristically creates depressions or furrow features (Fig. 4.3) that are elongate sub-parallel lineations that are somewhat regularly spaced, typically a few kilometers in length, a few tens of meters wide. In some case they are cut into coarse gravel and sand substrate or into fine-grained cohesive sediments (Robesco et al. 2014).

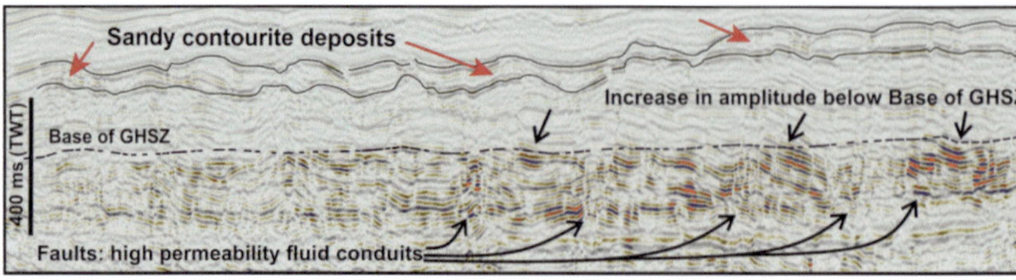

Fig. 4.3 Contourite horizon (between thin solid lines, red arrows) within the GHSZ in the Porcupine Basin in the western Ireland offshore. Calculated base of GHSZ coincident with top of free gas zone as described in detail by Roy et al. (2017). Faults not shown but small displacements similar to those associated with GHSZ in Chap. 2, Fig. 2.7 are pervasive from below GHSZ through the contourite horizon. Permission to publish from the Petroleum Affairs Division of the Department of Communications, Climate Action and Environment, Government of the Republic of Ireland

The single sub-horizontal series of bed containing the contourite bodies (red arrows, Fig. 4.3) is located about halfway between the base of the GHSZ and the seabed and shallower. The contourites are dominated by thick sandy sediments within the Mid Miocene to early Pliocene succession. The discontinuous nature of some of the reflectors within this zone suggests that individual sand bodies are laterally discontinuous, which would be expected in a reworked sediment with periodic contour currents. The contourite body lithologic succession is the equivalent of RTb seismic stratigraphic layer of Stoker et al. (2001). Sediment waves are predominant in the uppermost stratigraphic beds in the contourite horizon, which is Pliocene to Holocene in age (Fig. 4.4).

Free gas is marked by the distinct increase in amplitude and reflection coefficient below the base of the GHSZ. The widespread distribution of the acoustic signature is characteristic of a strong gas flux. The abrupt termination of the seismic response of gas at the base of the GHSZ suggest strongly that in the absence of any particular physical cap to upward gas passage, that the boundary is a phase transition any gas

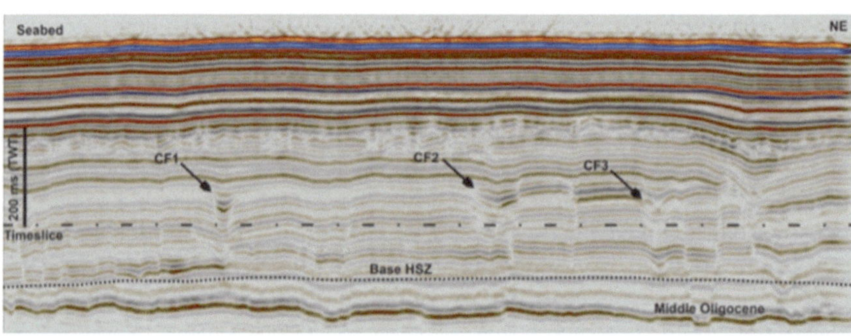

Fig. 4.4 Detail of contourite furrows in the sandy contourite horizon (above). Permission to publish from the Petroleum Affairs Division of the Department of Communications, Climate Action and Environment, Government of the Republic of Ireland

within the GHSZ is almost certainly in the form of NGH, similar to NGH shows in the Feni Drift contourite deposits (Stoker et al. 2001). Note that there is no well-defined BSR at the base of the GHSZ, which indicates that the sediments are not uniformly muddy or without good bed-differentiation, but contain beds that are well differentiated by sediment type. On these records, no discontinuous BSRs (Hillman et al. 2017) or 'string of pearls BSRs' (Max and Johnson 2014) across dipping permeable strata that could act as migration pathways for natural gas appear to occur. But this could be a function of the seismics, which were carried out to identify deeper hydrocarbon reservoirs. Bed differentiation, faults linking the free gas zone with the thick sandy beds in the contourite horizon, and a well-defined GHSZ that is thick enough to impart some mechanical stability to any large concentration of NGH that occurs in the contourite setting here, are promising for NGH exploration.

Sediment waves in which sediment is further winnowed commonly form as a result of the current activity. Courser grained deposits that remain behind may be best developed along only one margin of an erosional channel in a direct analog to fluviatile systems or along a continental slope. The are important as they may occur within a GHSZ (Fig. 4.3) and are not dependent on some influx of course grained sediments such as some turbidites and can be unrelated to sediment supply and winnowing processes on shallower continental shelves. Their early identification depends almost entirely on interpretation of reflection seismic data as there is no relation to existing marine depositional modeling.

Contourites are often associated with turbidite events because their sedimentation can cause local slope instability. and may also cause local gravity readjustment and limited mass movement These turbidites are formed by gravitational rather than direct water current processes and often initiate at the highest local seabed slopes, often at a bathymetric notch.

Where winnowing has been well developed, well-sorted sandy beds having excellent reservoir character may be developed. Fine grained drifts can locally and regionally develop large and thick sediment accumulations, which may have an economic importance for conventional hydrocarbon reservoirs (Viana et al. 2007). Contourites and associated turbidites within GHSZs are not a first order exploration target for conventional hydrocarbons because they are too close to the seafloor, although they may provide a reservoir host in deeper structural or stratigraphic traps. However, identification of their presence should be part of a thorough NGH evaluation (Roy et al. 2017). Although turbidites are the present primary exploration target (Chap. 3), contourites have the potential to be much more common potential hosts for NGH concentrations than has been previously suspected.

4.6 Sequence Stratigraphy-Related Marine Sequences

Snedden et al. (2003) also discuss the sequence stratigraphic approach to highstand, transgressive, and lowstand type conventional hydrocarbon reservoirs. Although these are fully marine, albeit often shallow marine, both clastics and carbonates may include high-quality reservoirs in the same range of reducibility as rift and immediately post-rift clastics. They are part of the marine sequence that indicates fully marine conditions have been reached offshore (Fig. 4.1). These beds may only be important as potential reservoirs or those that may be elevated into the GHSZ by tectonic action that could host NGH concentrations (Chap. 5).

Other thick marine successions, such as those around New Zealand, flank a cratonic core that was once part of Gondwana (Graham 2008). NGH has been observed in the thick sediments flanking New Zealand (Schwalenberg et al. 2010) but we could not find any record of preserved deeply subsided early rifts related to the segmentation of Gondwana. The potential of these sediments remains to be tested for NGH potential but most of the bottom simulating reflectors (BSRs) we have seen illustrated in publications appears to be well defined over long distances, which probably means that fine grain sands dominate sediment composition. This would limit the potential for NGH productivity.

4.7 The Special Case of High-Quality Reservoir Potential in the Mediterranean and Black Seas

The Mediterranean and Black seas (Fig. 4.5) were deep marine basins in the northern Tethyan system abutting Europe when tectonic activity related to northward impingement of Africa and the Arabian block of continental crust isolated

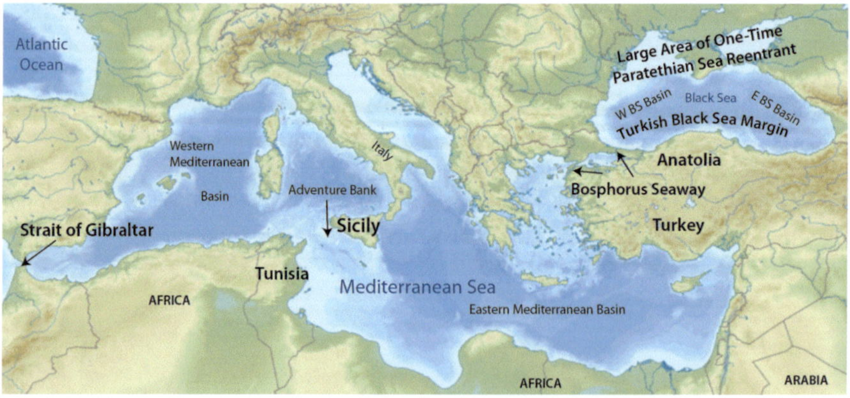

Fig. 4.5 General location map of the Mediterranean and Black Seas. After USGS and Max and Johnson (2015). Also see Chap. 3

them from the open Atlantic Ocean (Robertson et al. 2012; Stoica et al. 2013). Flexural uplift in the vicinity of the Straits of Gibraltar was responsible for the formation of a low land bridge that blocked Atlantic Ocean water from entering the Mediterranean Sea (Groves 2009). As a result, both the Mediterranean and Black seas and possibly part of the Caspian basin dried to different depths as evaporation exceeded water replenishment.

The region is structurally complex with the oceanic crust in the Eastern and Western Black Sea basins having formed at different times and apparently having had different structural histories and spreading center orientations that appear to be related to different subduction zones and regions of oceanic crust (Okay et al. 1994). The Mediterranean Sea is older and structurally more complex, having once been part of the Tethyan Ocean. Whereas the Arabian cratonic core essentially closed the Tethyan Ocean against the Eurasian craton, leaving various continental and oceanic terranes in Turkey and eastward, the African craton did not completely impinge upon Europe.

The Mediterranean and Black seas have had a Neogene and younger geoenvironmental history quite different from that of the open oceans. Their history has important implications for the presence of, and exploration for, NGH. The restricted deep basins of the Mediterranean and Black seas were cut off from the open ocean in the upper Neogene during the Messinian salinity crisis, during which their water levels decreased to the point where thick salt deposits formed from hypersaline residual water in the deepest parts of both the Eastern and Western Mediterranean basins. This was a rare geoenvironmental episode during which part of a large, complex deep oceanic basin nearly completely desiccated in the late Messinian age.

The OUGSME (2005) has summarized the event as: from 5.96 $\pm$ 0.02 Ma there was transition to regressive evaporite deposition through the entire Mediterranean. The first sign of hypersalinity was the deposition of the Lower Evaporite unit, which indicates a relative sea-level fall of between 200 and 1000 m at 5.8–5.5 Ma when complete isolation was established. An erosional surface/M-layer (lowstand) on land deeply incised fluvial channels. The onset of the transgressive Upper Evaporites marked the development of a 'Lago Mare' consisting of non-marine, brackish lakes that only occurred in the deepest basins that collected river drainage from only a few major rivers (Ebro/Spain, Rhone/France, Po/Italy, and Nile/Egypt). At 5.33 Ma there appears to have been an abrupt return to full marine conditions throughout the entire Mediterranean at about the Miocene/Pliocene boundary. (Krijgsman et al. 1999). The Zanclean Flood is the name given to the onset of Atlantic waters beginning to refill the Mediterranean Basin (Hsü 1983; Gillet et al. 2007; Gautier et al. 1994).

During this time of abnormally low water levels, conditions were suitable for the formation of aeolian and paralic reservoirs on the continental slopes of the present Mediterranean and Black seas. Whereas the Mediterranean was desiccated to the point where thick evaporites formed, the Black Sea desiccated but not to the point at which basin evaporite and sabkha conditions were established in basin floors.

4.8 Sea Level Low Stands in the Mediterranean and Black Seas

In open ocean conditions, sea level rise and fall relates most commonly to glacial episodes that sequester water in ice on land. This lowers sea level so that the upper 100–150 m of continental shelf and upper continental slope is exposed. Interglacial periods are associated with sea level rise. Restricted seas, such as the Mediterranean and Black Seas, may be cut off from the open ocean when sea level is lower than geological/bathymetric shallows such as the straits of Gibraltar and the Bosporus. This renders them special cases for NGH exploration because preexisting NGH will not survive dramatically low water levels.

4.8.1 The Mediterranean Sea

The Mediterranean and Black seas have not been considered to be optimal NGH exploration regions. At the end of the Messinian Salinity Crisis (Gautier et al. 1994) during which time the Mediterranean Sea nearly dried and there were only scattered saline lakes and thick salt pans of evaporites in the then-subaerial abyssal regions, the connection with the Atlantic reopened in the Gibraltar–Morocco region owing to tectonic movement. Since then the Eastern Mediterranean Basin has maintained an open interchange with the Atlantic (Gillet et al. 2007), most recently through the Straits of Gibraltar. The seawater connection between the Eastern and Western Mediterranean Basins, however, can only be maintained through a narrow shallow or spill point in the central Sicily Channel between Tunisia and the Adventure Bank to the SW of Sicily. The Strait of Messina that separates NE Sicily from mainland Italy, is narrow, very shallow and has been throughout the Pleistocene. It was not a marine connection between Eastern and Western Mediterranean basins during Glacial sea level low stands.

Although there is presently a complex seafloor bathymetry in the Sicily Channel that separates Tunisia and Sicily (Fig. 4.6a), with a spill point off the NW corner of the Adventure Bank (Fig. 4.6b) at between −300 and 400 m depth (UNEP 2015), we could not be certain how shallow this spill point was during the last glacial episode and how it would have affected water circulation in the region (Sorgente et al. 2003).

Plag et al. (1996) suggested that sea-level fluctuated episodically, with phases of rapid rise separated by phases of quasi-stability reflecting glacial and interglacial events during the Pleistocene. These sea level changes provide the opportunity for preservation of erosional and depositional features indicative of now submerged, relatively lower sea-levels, where lateral transitions from subaerial bank to unbroken marine successions occur. This area is in the far-field of northern European ice sheet loading effects (Lambeck 1993, 1996, 2014) and south of the

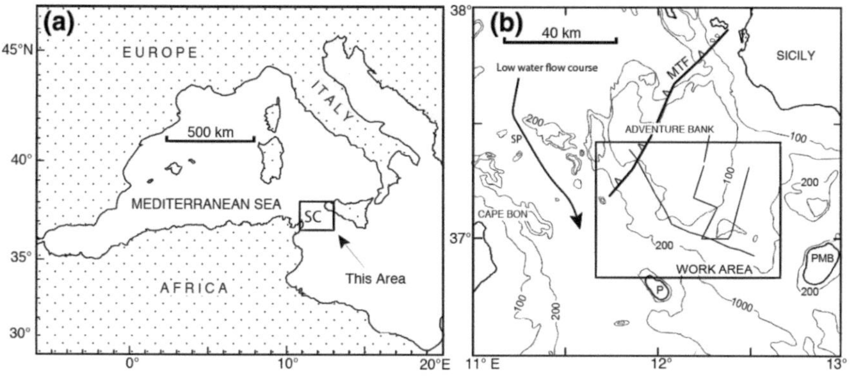

Fig. 4.6 Location map. **a** general location of area. SC, Sicily Channel, **b** generalized map of Sicily Channel between western and eastern Mediterranean basins. SP, spill point, arrow shows low water movement through Sicily Channel at low stand

outermost fore bulge limits, which, according to Wingfield (1995), produced extreme sea-level lowstand of ca. 160 m in the Celtic Sea south of Ireland.

Adventure Bank, which is an isolated, seafloor platform between 50 and 100 m depth SW of Sicily (Fig. 4.6) lies on the tectonic foreland of the African Plate south of the Maghrebian-Sicilian mountain belt, which passes along the northern coast of Algeria and Tunisia and crosses the Strait of Sicily to merge into the Cantabrian arc of the Italian Apennines (Max et al. 1993). Seafloor south of Sicily is now in a transtensional stress regime, which has resulted in grabens of both northeast and east-southeast trends (Jongsma et al. 1985).

Parts of Adventure Bank were emergent and subjected to erosion and near shore marine sedimentations during low sea level of the last interglacial. The bank is composed of mid-Mesozoic to Late Miocene marine sedimentary rocks, with isolated volcanic and intrusive igneous rocks. These are unconformably overlain by shallow Pliocene-Quaternary marine sediments (Jongsma et al. 1985; Max et al. 1993). Caiti et al. (1996) attribute compaction of the Pliocene-Quaternary sediments to subaerial exposure during oxygen isotope stage 2, corresponding to the maximum Late Weichselian (ca: ^{14}C, 25,000–17,000 yBP) glaciation of northwestern Europe.

The sea area between the Adventure Bank and Tunisia in the Sicily Channel (Fig. 4.6) is generally shallow, rocky, and rugged, or having an irregular, generally rocky seafloor, with geophysical evidence of repeated fault movements up to recent times. Active tectonism (Argnani 1993) in the area has caused both upraising and lowering of local areas, which will introduce small paleo-depth variation of temporary sea level features. Although, the history of erosion and sedimentation in the Adventure Bank is important to establishing whether the relative level of the bank bar during the lowstand was such that the Mediterranean was separated into a western basin connected to the Atlantic and an eastern basin that was not.

Surveys in 1994–1995 over the southern margins of Adventure Bank (Fig. 4.6b) were carried out on NATO's Research Vessel, *NRV Alliance* as site surveys to assist analysis of other work. Methods included acoustic reflection profiling, side-scan sonar and multibeam bathymetry. Concurrent high-frequency Uniboom and medium energy profiling allowed both ±0.25 m reflector resolution down to 20 m below seabed and lower resolution to ±5 m to about 150 m. Side-scan sonar yielded about 0.25 m resolution. These surveys took place on the Eastern side of the spill point in the Sicily Channel. The history of lower sea level low stands there would reflect the history of sea level change for the entire Eastern Mediterranean Basin and the Adriatic and Aegean seas.

Patchy Quaternary sediments on Adventure Bank form sand waves and fill small depressions. Continuous successions of Pliocene-Quaternary marine sediments are seen in deeper water on the flanks of the bank and in two grabens. Evidence of erosion on the bank shows local variations, which may be due to differential fault movements (Argnani 1993).

Colantoni (1988) noted that the northeast-trending graben incised into the southern Adventure Bank was filled with Plio-Quaternary marine sediments ranging to Holocene in age, and that the flat seafloor of the bank was probably subaerial during Quaternary low sea-level stands. The smooth sediment carpet of the bathymetric depressions along the grabens passes up-slope to rocky banks.

Reflection seismic records show parallel bedded, continuous sedimentary sequences in the grabens. No prominent erosion surfaces or channeling are present that would indicate intertidal or subaerial environments or alternating subsidence and upraising. Because the strata are now in a shallow marine environment, the entire history of graben sedimentation likely is one in which subsidence has kept pace with sedimentation. Sediment drapes along the shallower margins of the grabens, show downthrown subsidence along the grabens' marginal faults associated with periodic erosion up-slope. Three seismic-stratigraphic units were profiled on the shallow parts of Adventure Bank: acoustic basement, Pliocene-Quaternary sediments, and remnants of Quaternary deposits.

Acoustic Basement This oldest unit is formed of shallow marine sedimentary and volcanic rocks ranging in age from Mesozoic to Late Miocene. Sampled sediments are Upper Miocene calcarenites with similar lithologies to that of the coarse fraction of local seabed sediments (Tonarelli et al. 1993). Upright open folds are common, while acoustic velocities indicate pronounced lithification (Caiti et al. 1996). These rocks are truncated above by a sharply defined erosion surface. No igneous rocks were identified.

Plio-Quaternary Sediments Generally undeformed sediments of this age range are found on the flanks of Adventure Bank and in the grabens. They unconformably overlie acoustic basement sedimentary rocks and abut igneous constructs. Plio-Quaternary sediments are more than 100 m thick in a graben on south-east Adventure Bank (Colantoni 1988). On the graben's western margin this Pliocene to Holocene sequence reduces to a featheredge along the acoustic basement margin.

Substantial erosion of Pleistocene sediment on the largely sediment-free bank was probable during the Oxygen Isotope Stage 2 sea-level lowstand interval.

The two main basins of the Mediterranean Sea appear to have had similar sea level lows at from 120 to 140 m below modern levels as the Atlantic and the world ocean generally. This strongly suggests that seawater flow from the Atlantic would have entered the Western Mediterranean Basin through the deeper Strait of Gibraltar at >300 m depth and flowed into the Eastern Mediterranean Basin through a relatively narrow channel between Sicily and Tunisia after first filling the western Mediterranean basin up to that level (Garcia-Castellanos et al. 2009).

There are indications of sea-level low stands adjacent to the graben margins and to the peripheries of seabed rock outcrops. Thin, but prominent acoustic reflectors in the Plio-Quaternary sequence in the grabens are identified. These may represent individual beds formed by coarse grained, aeolian, or volcanoclastic detritus. Alternatively, they may represent lag deposits or erosional surfaces.

On the eastern side of the north-east trending graben on south-east Adventure Bank the off-bank Plio-Quaternary sediments thin across a marginal fault at the depressed western margin and pass into a series of onlapping feather-edged beds. The erosive base of the sediments is a rough slope into the graben with local down-steps (Fig. 4.8) suggestive of erosional benches on wave cut platforms. Two weak abrupt changes of slope in the erosion surface (N1, N2) are coincident with the tops and feather-edges of overlying sediment packages.

A distinct acoustic reflector 1B In the Plio-Quaternary sequence of the graben displays both a highly reflective top and strong scattering beneath (Fig. 4.9). In the graben, it appears to represent a generally horizontal bed in a parallel sequence of similar beds. However, this reflector slopes up towards the margin of the bank and assumes the character of an even more prominent reflector before becoming sub-horizontal (P1) again at the level of N1. The top of the buried terrace (P1), immediately graben-ward of N1, lies at mean sea level minus 131 m. Its form is suggestive of a nearshore terrace of sand and gravel, associated with apparent foresets sloping into the graben.

Above reflector 1B-slope-P1 (Fig. 4.8, to N1) there are in excess of 10 m of seismically laminated sediments in the graben with various reflectors at levels 1A and RS. These onlap towards the graben margin so only the uppermost, as sediment packet RS, appears to continue at some 3.5 m thickness above reflector P1, and the uppermost 1.25 m of these are those cored and described in Table 1 as an upward fining sequence. A second bench results as the sedimented seabed (P2) terminates against outcropping basement of the bank at N2. The form of the P2 terrace is dissimilar to the buried P1 terrace in that it has only a narrow inner section, higher and bank-ward of N1, and a broader, graben-ward, slope of lesser slope in its central part over the buried P1. The inner terrace P2 lies at mean sea level minus 121 m.

In contrast to the NE trending graben (Fig. 4.7), the ESE trending graben is asymmetric. In the south, sediments from the graben onlap onto the small basement high (Figs. 4.8 and 4.9), whereas in the north this east-south-east trending graben does not have outcropping acoustic basement at the seabed. Shallow marine

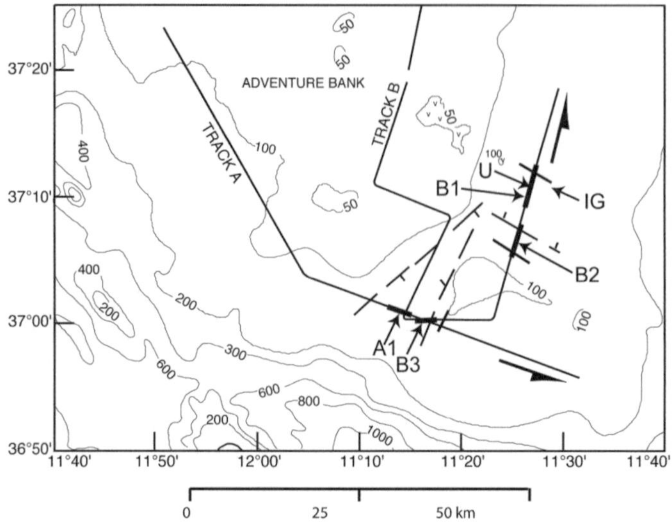

Fig. 4.7 Survey track A (1994) and B (1995). Arrows show ship's heading. Heavy lines on track segments are figure locations. Dashed lines indicate major faults along graben margins, ticks show downthrows. Cross-track short solid lines show palaeocliffs. U, basement ridge; IG, igneous intrusion

sediment terminates at N2 against a cliff base in rockhead at 132 m depth. Above this level is sediment is banked-up against the cliff with a steep 1-in-1, graben-ward slope 8 m high (distortion caused by unequal horizontal and vertical scale). The sediments here must be at least partly cemented to allow this very steep slope to be maintained. More deeply buried beds in this east-south-east trending graben have greater displacement against the margin than in the other graben. Neotectonic movement continues since the seabed sediments are affected.

Wingfield and Evans (1998) demonstrated over a 50 km length of Yorkshire coast that middle and late Holocene coastal erosion of virtually homogeneous glacial diamict sediments of Middle and Late Pleistocene ages have resulted in a striking set of planes of erosion, but post-erosion sediment is widely absent. An intertidal and subtidal "shoreface ramp" displaying a seaward slope of *circa 1-*in-100 between cliff-foot (mean sea level plus 3 m) and mean sea level minus 14 ± 1 m appears to be the focus of active coastal erosion (Tsimplis et al. 1995). However, minor seabed lowering continues down 40 m depths during the largest storms. Lateral cliff retreat is due to the lowering and consequent landward extension of the shoreface ramp. This type of analysis on raised lowstand provides a framework against which to assess the submerged features on Adventure Bank.

Wave erosion of lithified rock by wave effects occurs in a zone that extends from above mean sea level to, at most, minus 40 m depth, and effectively to as little as 14 m on the open sea coast of Yorkshire (Wingfield and Evans 1998). Almost identical depths of wave erosion effects of consolidated and lithified deposits were found by Trenhaile (1972, 1974, 1987) elsewhere in the United Kingdom and by Sunamura (1992) off Japan. It follows that each of the several eroded scarps and their lesser graben-ward slopes from the cliff-foot were incised by wave action.

Shoreface ramps may be evidenced at Adventure Bank across the upper surface of P1 to another cliff at N1 acoustic basement (Fig. 3), and across the terrace forming the inner continuation of Reflector 0 between mean sea level minus 134 and 131 m (Fig. 4.9). In the latter case, at present seabed, an abrupt change of slope at mean sea level minus 131 m marks the foot of a cliff of about 8 m height, but here developed in the cover sediments. If the latter cliff and both these suggested shoreface ramps are correctly identified, they were eroded in at the most semi-consolidated deposits and presumably at as much as an order of magnitude faster rate that the now buried former shoreface ramps and associated cliffs cut across acoustic basement (Wingfield 1995, Table 3).

It may be inferred that the sea-level fall necessary to produce these structures are suggestive of a drop to 131 m depth (P1, N1) and a slight rise to about 121 m (level P2, N2). As one or both of these indications of major sea-level low stands occur immediately below or at the seabed, they probably formed during the last Oxygen Isotope Stage 2 recognized by Fairbanks (1989) of about 120 m depth off Barbados and confirmed by Bard et al. (1990) there at a slightly greater depth. Barbados and the Adventure Bank are both active tectonic areas but indications of relatively low sea stands retain a local validity (Pirazzoli 1993). We do not correlate the lower levels of low stand benches with either pulses within the Late Pleistocene curve, or with, in the case of the N1/P1 level, an entirely separate low sea-level excursion (Stanley 1995; Plag et al. 1996) of possibly one of the end-Oxygen Isotope stages 6, 8, 10 or 12 (Bard et al. 1996).

In summary, we do not know the age of the rock head erosion surface underlying the onlapping sediment (Figs. 4.8 and 4.9). It is probably a composite erosion surface from one or more of the Pleistocene glacial episodes. An older beach terrace (BT on Fig. 4.8) may be associated with an older lowstand of the last glacial episode, or it could represent an older and temporary beach terrace related to the most recent glaciation. However, the simplicity of the sedimentation at and above the beach terrace makes it almost certain that this succession has undergone only one immersion, and that would have been the most recent. Thus, the most recent stable sea level lowstand at N2 (Figs. 4.8 and 4.9) is almost certainly related to the most recent sea level low stand associated with at least the later part of the most recent glacial episode at about 130 m below present sea level.

Seismic evidence from the SE Adventure Bank strongly suggests that both the eastern and western basins of the Mediterranean were part of one oceanographic system during at least the last glacial maximum as well as during our present interglacial with its higher sea levels. The low sea level connection was maintained through the Sicily Channel between Tunisia and Sicily. Our geophysical data indicates that there was seawater exchange between the Eastern and Western Mediterranean basins during the last glacial event when the open ocean sea level was about 130 m lower than today. This is important because synchronous sea level

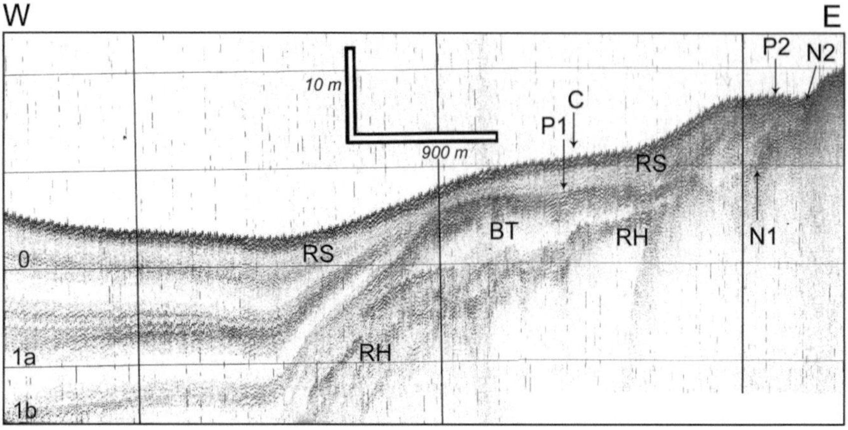

Fig. 4.8 Uniboom refection acoustic profile on Track B, (B3, Fig. 4.7). Plio-Quaternary sediments onlap acoustic basement across the eastern graben margin. Horizontal time lines at 25 ms correspond to some 10 m thicknesses in cover sediments and some 1.5% less in the seabed sediments (RS). C, Control core position. N1 and N2, lower and upper abrupt changes of slope in acoustic basement; P1 above the ST sediment package and P2 above the RS sediment package are sub-horizontal sediment terraces associated with onlap on the erosional surface; RH, rockhead is acoustic basement with no discernable bedding. Reflectors O, 1a and 1b. Vertical exaggeration *circa* × 40

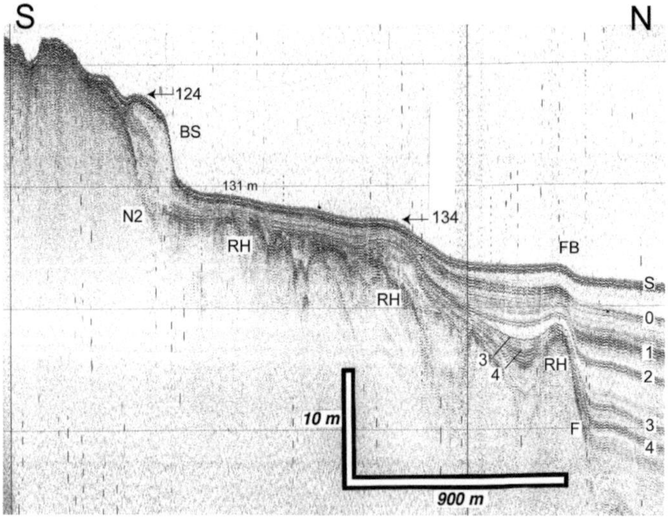

Fig. 4.9 Uniboom reflection profile on Track B, (B2, Fig. 4.7). The profile displays off-lap of the Pliocene-Quaternary sediments at the southern margin of the east-south-east trending graben into the graben. Reflectors 0, 1, 2, 3 and 4, N2, change in rockhead slope to small cliff. RH, rockhead of acoustic basement same surface as RH on Fig. 4.9; F, southern fault to graben; FB, seabed expression of fault movement; S, seabed reflector; BS, uppermost sediment bench. Vertical exaggeration *about* 50×

change means that the pressure effect on GHSZ thickness throughout the Mediterranean can be modeled as a single system and confirms the modeling of Praeg et al. (2011).

4.8.2 The Black Sea: Sand Transfer to the Slopes

During the Holocene, the Mediterranean maintained a marine connection with the Atlantic Ocean, whereas the Black Sea did not. The Bosporus sill depth, over which flooding of the Black Sea from the Mediterranean took place, is generally placed at between 30 and 35 mbsl., however, we propose that it was about 10 m shallower (at 23 $\pm$ 5 m) than the depth identified by Ryan et al. (1997) and Giosan et al. (2008). In either case, sea level low stands in the Black Sea indicate that the Bosporus acted as a dam to water in the eastern Mediterranean and the Aegean Seas during the Holocene sea level low stand. When the Black Sea reflooded at the onset of the present interglacial, the impact on NGH conditions altered substantially. According to Poort et al. (2005), influx of 8–10 °C warmer saline water, despite the rise in sea level by from 100 to 150 m would result in reduction of the GHSZ by from 15 to 62% of its expression during the glacial episode. Nonetheless, sufficient GHSZ thickness may host large NGH concentrations.

There are two main viewpoints as to the Holocene sea level lowstand (Nicholas and Chivas 2014), (1) that the sea level minimum was as low as 100–120 m below present sea level based on reflection seismic data (Ryan et al. 1997) and, (2) an alternate view based on geochemistry that the minimum level was only a few tens of meters below its present level (Aksu et al. 2002). Çağatay et al. (2009) found a lowstand at about 85 m below its present sea level. There is general agreement that reflooding of the Black Sea of about 8400 yBP Ryan et al. (2003), although slightly older than the original estimate by Ryan et al. (1997) of 7150 yBP.

Lericolais et al. (2007, 2013) showed substantial sub-surface data from the Danube and Dnieper/Dniester turbidite systems along the NW sector of the Black Sea as well as sediment influx from the Anatolia region of the central Turkish coastline of the Black Sea. Nicholas and Chivas (2014) identified aeolian depositional environments, and soils related to a sea level low that placed reconnection of the Black Sea with the Mediterranean at from 8200 to 8600 ^{14}C yBP, with a sea level rise from a low 107 m below present sea level. This is in general agreement with Ryan (2007), who in his summary of viewpoints on water level depression and timing also notes apparently aeolian dune fields between 65 and 80 m water depth with wave-truncated bedded sediment terraces with beach-like berms at depths of 90–100 m. Sediments related to the unconformity contain shell fragments dated between 9500 and 8500 yBP, suggesting that the younger unconformity represents a post-Younger Dryas regression that depressed the Black Sea's water level.

Soulet et al. (2011) have used high resolution geochemical and AMS radiocarbon data sets and records going back 32,000 yBP to establish as quantitative reservoir age record for the last glacial history of the Black Sea "Lake". During the last glacial maximum, the Black Sea was strictly isolated, with reconnection and water exchange with the Mediterranean Sea at about 9000 yBP. More precisely, reliable published radiocarbon ages, Soulet et al. (2011) proposed an atmospheric radiocarbon age for a complete reconnection at 8995 ± 145 yBP. Complete flooding may have taken place on the order of about 900 years during which mixed lacustrine biota gave way to marine biota (Soulet et al. 2011). These dates are in general agreement with the previous estimates of Ryan et al. (2003), and Ryan (2007) of about 8400 yBP, although slightly older than the original estimate by Ryan et al. (1997) of 7150 yBP.

A younger date for flooding of 5600 yBP that is often referred to in summary literature is no longer supported. Thus, complete flooding may have taken place on the order of about 900 years during which mixed lacustrine biota gave way to marine biota. Lister et al. (2015), however, concluded that there was a transgression of the SW shelf of the Black Sea by at least about 10300 yBP and perhaps as early as 12,000 yBP. Whatever the minimum sea level was, it would have had much less impact on the likelihood of sandy turbidites on the Turkish Black Sea slopes than to the north where shallow continental shelves are broader. Sand delivery to the continental slope by turbidites across the narrow continental shelves probably took place on a semi-continuous basis. Whereas on wide continental shelves more coarse-grained sediments are more likely to be deposited on the slopes during sea level lows, with strong erosion only a short distance from marine depositional areas, the presence of sand bodies on the slopes is likely.

Anatolian rivers appear to deliver more than 33% of the total sediment input into the Black Sea (Popescu 2002), even if they represent only 8% of the total freshwater discharge. Because of the narrow continental shelves along the Turkish Black Sea coast, almost all of the sediment, including the coarser fraction, would have reached the continental slope. This high sediment load may be related to the proximity of the North Anatolian Fault system and associated tectonism in northern Turkey and its Black Sea continental slope. A similar sediment delivery system probably operated during earlier sea level lows, which implies that considerable coarse sediment was delivered to the continental slope that could be formed into sand deposits. Identification by Cifci et al. (2002) and Dondurur et al. (2013) of clustered slump-scars on the upper continental slope suggest that submarine landslides are a long-standing mechanism for triggering mass flows resulting in turbidite deposition along the margin. In addition, the tectonic activity here may be a progenitor of high gas flux into and through the upper sediments along this margin.

Off the mouths of the Sakarya and Filyos rivers in northern Turkey, small deltas occur at depth on the upper shelf (e.g. Algan et al. 2007). These could have been built by recurrent hyperpycnal currents (Lamb and Mohrig 2009) in the semi-freshwater Black Sea during sea level low-stands. Sediment overloading on the steep slope, which appears to be common on the Turkish shelf (Cifci et al. 2002) constitute evidence of additional triggering mechanisms which would relate to distal turbidites also recovered in deeper-water cores.

Late Glacial warming that led also to the early melting of Anatolian glaciers (Akçar et al. 2008; Zahno et al. 2010), beginning at around ~18.3 kyBP. This implies a slight increase in the sediment supply onto the Turkish margin, and enhanced sediment failures and transport to the deep basin. The water brought to the Black Sea after this meltwater pulse event led to a rise of the Black Sea level to −40/−20 m (Lericolais et al. 2007, 2013).

These Black Sea Holocene deposits are mainly composed of silt and clay, and may indicate the progradation of deltas off the main Anatolian rivers after the transgression. Mid-Holocene debris flows appear as an anomaly recorded in the distal basin, in respect with the continuous record of gravity processes on MD04-2762 since the LGM. They occurred during sea level highstand conditions, suggesting that they might not be triggered directly by gas-hydrate release, as the increase in hydrostatic pressure would have stabilized the hydrate deposits (Maslin and Thomas 2003). Major earthquakes could have been another possible triggering mechanism.

Redistribution of sediment supply from shelves onto upper slopes during sea level rises can lead to an instability with sediment overloading generating shelf-edge/upper slope instabilities. Mass transport deposits have also been observed in the Nile Fan under sea level rise and linked to enhanced pluvial conditions and sediment mobilization (Garziglia et al. 2007).

Plio-Quaternary sedimentation along the Turkish margin of the Black Sea since the Messinian drying event has covered the boundary and any one-time sedimentary units formed before the flooding. Because of the relatively thick sedimentation along this margin and into the ocean basin of up to 4 s TWT (Finetti et al. 1988), only thinner depositional areas or areas that have been thinned by mass flow evacuation of sediments into deeper water may have potential high-quality reservoir sediment along the Messinian–Pliocene boundary.

4.8.3 GHSZ and NGH Prospectability in the Mediterranean and Black Seas

The Mediterranean Sea is sufficiently warm at depth that the GHSZ is relatively thin and encountered at greater depths than in the open oceans. Because the Mediterranean does not receive the deep cold water of the open world ocean, its seafloors are relatively warm. In the open ocean, NGH is stable from a water depth of about 500 m depth. In contrast to the open ocean, seawater in the Mediterranean warms toward the bottom, with temperatures ranging from about 12.5 °C at 1 km depth to about 14 °C at abyssal depths greater than 3 km (Praeg et al. 2011). Figure 4.10 shows the intersection points for two arbitrary geothermal gradients. If a low enough geothermal gradient underlay sediment containing NGH at a water depth of about 3.4 km, for instance, the GHSZ could be as thick as is commonly found in the open ocean. Higher heat flow, for instance 50 °C/km, yields a thinner GHSZ, whose base would lie along a horizontal line passing through the intersection of the arrow head and the NGH phase boundary.

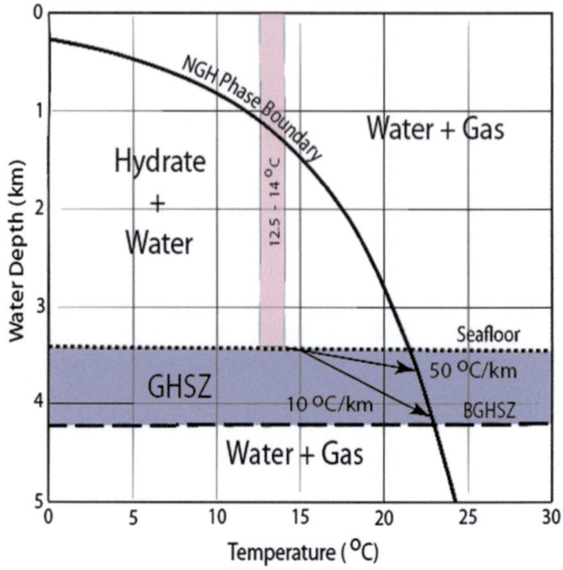

Fig. 4.10 Generalized phase boundary for methane hydrate in the Mediterranean (Methane hydrate = NGH). Redrawn from Praeg et al. (2011)

The Eastern Mediterranean region, including sediments on land bordering the Mediterranean Sea, is a proven natural gas province (Bruneton et al. 2012). In the abyssal regions, gas and fluid venting, mud volcanoes, and shallowly buried NGH that may also be exposed on the seafloor, are common (Praeg et al. 2011). NGH has only been positively identified in the eastern part of the Eastern Mediterranean basin, for instance by Aloisi et al. (2000), Dahlmann et al. (2005), Lykousisa et al. (2009), and Perissoratis et al. (2011), amongst others, in which different aspects of NGH recovered from the abyssal seafloor in the Eastern Mediterranean and its potential for extensive development in the Eastern Mediterranean Basin are discussed. Although a number of BSRs have been observed (personal observations), a strong BSR that probably indicates muddy sediment hosts has been observed in the Nile Fan (Praeg et al. 2011).

Praeg et al. (2011, Fig. 2), using more detailed data, modeled thicker GHSZ in the Mediterranean than Wood and Jung (2008). In the Eastern Mediterranean basin, the GHSZ ranges in thickness from 200 to 400+ m close to the Turkish coast where there is only a narrow continental shelf to greater than 300 m in the North African continental slope. The largest region of GHSZ thickness over 400 m thick lies in the sea area SE of Sicily to the approaches to the Adriatic Sea and Greece. Elsewhere in both the eastern and western parts of the Eastern Mediterranean Basin, although not along the eastern Mediterranean slopes, the GHSZ is commonly over 200 m in thickness. In the western Mediterranean Basin, the GHSZ is less than 100 m thick in continental slopes.

Because the GHSZ is thicker in the eastern Mediterranean Basin, we would regard the continental slopes of Turkey, from north of the western end of Cyprus to the shallower margin of the Aegean Sea, and the slopes of the Italian and Greek Sea areas, as well as off SE Crete and the Nile Fan, especially as high-quality sand reservoirs may exist in the GHSZ (Max and Johnson 2015) to be prospective for NGH.

The Black Sea is divided into two tectonic basins, the Western and Eastern that are separated by a median ridge that passes from about the SW prolongation of Crimea SE toward Turkey (Barka and Reilinger 1997). The floor of the Black Sea is almost flat; sediment fill has obscured the deeper crustal structure. The Black Sea has a strong natural gas petroleum system with many methane seeps, mud volcanism, and NGH (Haeckel et al. 2015) and new discoveries of large gas fields (Offshore 10/15/15).

Wood and Jung (2008) show a widespread GHSZ in virtually the entire Black sea, which is an artifact of their data point spread in a map of the entire world oceans, while Vassilev and Dimitrov (2003) show more detail contouring a sediment-heat flow model. BSRs indicating the presence of a GHSZ with underlying gas have been identified by Popescu et al. (2007). NGH identifications in the proto-delta of the Dnieper River continental slope of the Black Sea (Lüdmann et al. 2004; Naudts et al. 2005) suggest that organic rich sediments were provided to the Black Sea Basin. Because the deep basin sedimentary facies are liable to be predominantly muddy sediments that are unsuitable hosts for NGH (Chap. 2), we identify five NGH prospect zones along the slope-rise sediments (Fig. 4.11) that could contain suitable sandy sediment host beds for NGH concentrations.

In addition to the large Mediterranean and Black Sea marine basins, the Sea of Marmara (SoM) has a number of deeps below 1 km water depths that could also be prospective for NGH. Bourry et al. (2009) have sampled thermogenic compound

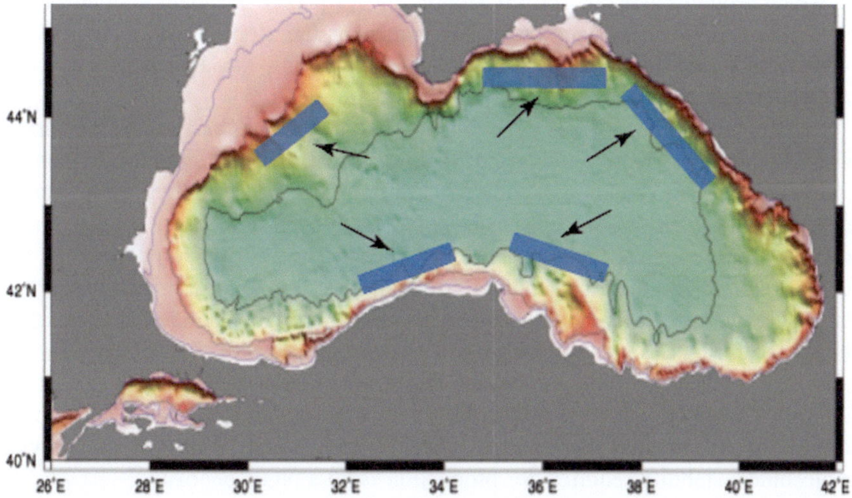

Fig. 4.11 Diagrammatic slope prospective zones (arrows) using the slopeward margin of the GHSZ with thickness >300 m as shown for the whole Black Sea by Vassilev and Dimitrov (2003). Base map (unattributed) at http://cmtt.tori.org.tw/data/App_map/maplist.htm.3.7

NGH from about 670 m in the western part of the SoM, which is about at the feather edge of the local GHSZ. Although free gas vents having gas mixtures similar to those found in the NGH have been observed at about 1200 m, no other NGH has yet been sampled. At the western high the GHSZ is calculated to be about 100 m thick and depending on geothermal gradient variations, it should be thicker in the lower slopes of the SoM basins. The prospectivity of the SoM also remains to be evaluated.

Mediterranean Sea The Mediterranean Sea was initially part of the open Tethyan Ocean. It was initially cut off from the Indian Ocean to the east and the Atlantic to the west by impingement of Africa northward into Eurasia. The region is a nexus of collisional activity containing areas of oceanic crust preserved in the Eastern and Western Mediterranean basins. The Mediterranean comprises a unique example of a large sea that had dried almost completely.

Reflooding generally began when seawater entered through a spillway from the Atlantic Ocean about 4 million years ago over what has now resolved into the shallow Strait of Gibraltar. The western Mediterranean flooded until the shallow sill between Sicily and Tunisia was overtopped by the sea level in the Atlantic Ocean and the Mediterranean began to fill again (Garcia-Castellanos et al. 2009). The Eastern Paratethys Sea, including the Black Sea and a number of other shallower basins, were then flooded when the precursor to the shallow Bosporus seaway was itself overtopped.

The continental slopes of the Mediterranean would have loomed as steep sided mountains during the Messinian Salinity Crisis from the perspective of the dry plateaus and super saline lakes that now form the abyssal depths (Hsü 1983). Subaerial conditions on what had been marine slopes would have provided surfaces on which aeolian sands could form. Conditions for paralic sediment systems would have been provided by temporary sea levels on the slopes related to periodic balances between water inflow and evaporation during the drying and possibly irregular water supply and drying during reflooding.

Similar geological histories of drying and partially drying basins also may apply elsewhere, but the Mediterranean and Black seas are the best examples of which we know. In addition to the dramatic lowering of sea level in the Mediterranean, glacial episodes during the Pleistocene each would have been associated with sea level lows. In the Mediterranean Sea, the older glacial episodes may have lowered sea level enough so that the levels in the Eastern and Western Mediterranean basins and the Black Sea were different and possibly much lower on the continental slopes than they were during the most recent glacial episode. Although sea level lows for the most recent glacial episode are known, the levels of older sea level low stands are not well understood. It is possible that older high-quality sand beds could have formed on continental slopes and covered by enough sediment to provide geotechnical stability required for production from a NGH deposit.

From the perspective of the presence of one-time NGH in marine basins, Pierre et al. (2002) consider that sedimentary structures in Tortonian marls reflect sediment deformation and brecciation caused by NGH destabilization related to falling sea levels and possibly warming as the water level lowered. During the Messinian low-stand it is virtually certain that any pre-existing NGH was destabilized and converted to its constituent gas and water. All NGH now present in the Mediterranean marine sediments must have formed since sea levels increased enough to provide adequate pressure for NGH stability.

Black Sea The Black Sea is considerably smaller than the Mediterranean. From the mid-Cretaceous, its marine depositional areas have been close to tectonically active mountain building related to plate collision. Considerable coarse sediment could have been supplied from time to time, even to slope rise sediments. The southern margin of the Black Sea is a tectonically active sector of the Alpine-Himalayan orogen (Tari et al. 2000). The southern margin of the Black Sea is located close to the 'sand machine' from which significant clastic sediments could have been supplied. Large rivers (Danube and Volga) can supply significant sediment to the Black Sea, particularly during transition from Glacial to interglacial periods.

The Paratethyan region to the north that encompassed the Black Sea (Chap. 3) did not appear to dry to the same extent, as no comparable evaporite precipitation took place, but major shifts in the deuterium enrichment suggests a strongly negative water budget coinciding with the Messinian Salinity Crisis (Vasiliev et al. 2012, 2013). Leever et al. (2009) suggested from seismic analysis that the connectivity of basins in a drainage area showed that the degree of interconnectivity of basins in a drainage area determined the amount of relative drying. For instance, they suggest that the Pannonian basin in the NW Paratethys was essentially full but the Dacic Basin that was 'upstream' from the Black Sea dried less than did the Black Sea.

Gillet et al. (2007) concluded that sea level in the Black Sea had dropped by about 1600 m from its present level based on deep erosional features between the Messinian and Plio-Pleistocene sequences. This interpretation of a deep lowering of the Black Sea water level was in agreement with the conclusions of Hsü and Grovanoli (1979), which were based on drilling. Konerding et al. (2010) found the results of a sequence stratigraphic analysis from the Romanian Black Sea sediments that conformed to this deep drying interpretation but did not verify it unequivocally.

However, Tari et al. (2015) disagree with the identification of the sediment horizon known as the 'pebbly breccia' as shallow water sediment. The regard it as a conformal mass flow deposit of shallow water facies from a distant area that has been transported. Although Tari et al. (2015, 2016) regard the sea level drop to be significantly less than the suggested $\sim$1600 m, and discount what appear to be erosional notches at various depths along the continental slope, they do not identify any shallower sea level low stand. Grothe et al. (2014) have used new biostratigraphic data from two key sedimentary successions located in a deep and marginal setting of the Black Sea Basin to demonstrate that the Pebbly Breccia probably pre-dates Mediterranean Sea level drop during the Messinian Salinity Crisis and thus is unrelated to that event. Krezsek et al. (2016), however, suggest that Messinian sea level fall was 500–600 m offshore Romania based on the depth of erosional valleys cut into the slope and the Messinian erosional surface. Because of the lack of definitive data suggesting a deeply dried Black Sea, it is almost certain that it did not dry to the same extent as the Mediterranean basin. However, depositional environments at and above temporary shorelines based on the most recent Messinian sea level evidence indicates that restricted formation of high-quality reservoirs could have taken place on the upper continental slopes of the Black Sea.

Gas Flux along the Turkish Margin of the Black Sea NGH formation must be related to a strong existing and relatively recent gas flux (Chap. 2). NGH concentrations reflect recent conditions of gas movement into sediments in the GHSZ that sequester the natural gas in NGH concentrations (Max and Johnson 2014). The Turkish Black Sea margin currently appears to be a major gas province (White 2013). Focused flow (Chap. 2) into a permeable horizon within the GHSZ has the potential to form large concentrations of NGH (Max et al. 2006). Faults capable of delivering high gas flux (Plaza-Faverola et al. 2015) may be an important mechanism for bringing gas into or near the base of a GHSZ.

Persistence of NGH In open ocean regions, NGH can be expected to persist because marine conditions were maintained, although the location and thickness of the GHSZ can be expected to vary according to sea level (pressure) and seafloor temperature changes. NGH is strongly responsive to environmental change. During the Messinian Salinity Crisis, virtually all pre-existing NGH would have dissociated because pressure would have been lowered to the point where it would have been unstable. The natural gas would have escaped into the sea, from which it would either escape into the atmosphere directly by venting if the water was shallow enough and individual gas vents large enough or it would have been absorbed by the seawater where it would be a primary energy source at the base of the food chain.

Formation of High-Quality Reservoirs Long-standing Anatolian rivers appear to deliver more than 33% of the total sediment input into the Black Sea (Popescu 2002; Popescu et al. 2004), even if they represent only 8% of the total freshwater discharge. Because of the narrow continental shelves along the Turkish Black Sea coast, almost all of the sediment, including the coarser fraction, would have reached the continental slope. This high sediment load may be related to the proximity of the North Anatolian Fault system and associated tectonism in northern Turkey and its Black Sea continental slope (Şengör et al. 2005). A similar sediment delivery system probably existed during earlier sea level lows, which implies that considerable coarse-grained sediment was delivered to the continental slope that could be formed into sand deposits. Identification by Cifci et al. (2003) and Dondurur and Cifci (2007, 2009) of clustered slump-scars on the upper continental slope suggest that submarine landslides are a long-standing mechanism for triggering mass flows resulting in turbidite deposition along the margin. In addition, the tectonic activity here may be a progenitor of high gas flux into and through the upper sediments along this margin, and as NTH dissociation (Nixon and Grozic 2007; Sultan et al. 2004).

Plio-Quaternary sedimentation along the Turkish margin of the Black Sea since the Messinian drying event has covered the unconformable boundary with underlying sediments. The sedimentation along this margin and into the ocean basin was relatively thick, up to 4 s TWT (Finetti et al. 1988). As a result, only thinner sedimented areas or areas that have been thinned by mass wasting of sediments into deeper water may have potential high-quality reservoir sediment along the Messinian–Pliocene boundary.

We regard the Turkish margin of the Black and Mediterranean seas as prospective for NGH as well as for conventional hydrocarbons. There should be abundant sandy beds within the GHSZ, which have escaped scrutiny because they are too shallow for conventional hydrocarbons. Also, because the important North Anatolian Fault Zone passes along the E-W Turkish Black Sea margin faulting related to this intercontinental transpressive shear zone could have excavated once more deeply buried high-quality sand reservoirs structurally high enough so that they are now in the GHSZ.

Exploration Potential During low sea level stands another group of potential NGH reservoirs may have formed on the continental slopes of in the present Mediterranean and Black seas, as the underpinning to younger marine sedimentation. The entire slope region that was then above the temporarily lowered sea level would have hosted fluviatile and aeolian deposits at a higher relative elevation to near-shore marine and subaerial longshore and beach related sand systems and organic-rich lagoon muds. On the slope above, fluvial and aeolian sediments will have formed. Both of these sediment systems form large, well sorted sand bodies that would be coarser grained than typical marine turbidite system lithologies. These sands would potentially constitute very high-quality reservoir host beds.

Although we have not carried out a detailed analysis, it would appear that the Messinian erosion surface approaching the steep slopes of the deep basin margins lies within the GHSZ (Praeg et al. 2011). The margin of the Mutruh Basin offshore Egypt (Tari et al. 2012), the higher elevations of the irregular top of the incised Lago Mare (Messinian) sequence offshore the Rirt Basin, Libya (Bowman 2012), and sediments approaching the flanks of the Levantine Basin offshore Cyprus and Lebanon (Skiple et al. 2012) all display only thin sediment cover on top of the Messinian Lago Mare or Messinian erosional surface at water depths deep enough to have a GHSZ that appears to extend to below the Messinian unconformity, atop which paralic and aeolian sediments can be expected to have formed.

We regard the Mediterranean Sea including the Turkish continental margin as prospective for NGH in high-quality sand reservoirs related to subaerial deposition on temporarily exposed continental slopes during the Messinian Salinity Crisis when substantial coarse sedimentation from northern Turkey could be expected. There are a number of areas in the Mediterranean where sedimentation has not been pronounced since the reflooding at the end of the Messinian and some high-quality reservoir sediments deposited during the salinity crisis could still reside within the GHSZ. The Black Sea appears to be a major gas province but because the same extreme Messinian drying probably did not take place, the potential for non-marine high-quality reservoirs is not as great as in the Mediterranean (Max and Johnson 2014). Nonetheless, high-quality sands deposited into deepwater during the glacial maximum lowstand could provide reservoir hosts for NGH concentrations that have been accessed by the high gas flux in the Black Sea.

4.9 Exploration for High-Quality Reservoirs

All sand bodies associated with subsided clastic rift sediments would lie stratigraphically below sediments deposited by advancing marine conditions. These could be present in a GHSZ if marine sedimentation above them was thin or eroded, or if faulting had raised these blocks of the sediment into the GHSZ. Normally thick marine sedimentation is likely to have more source beds (Piggottt and Pulham 1993), and these can be buried deeply beneath host sediments within the GHSZ.

Aeolian activity, fluviatile activity, or temporary shoreline deposition would lie on or root in an erosional surface and underlie prograding marine sediments marking upslope migration of sea level. One-time barrier island, point bar, and more complex sand bodies such as off-beach sand wedges, and beach and back-beach dunes would tend to be lensoid and discontinuous. Individually and as a group these would be disposed laterally along a slope at successively prograding shorelines. Any deepwater continental slope, especially along a passive continental margin, having a substantial natural gas flux that has undergone significant subsidence, tectonic activity, or sea level rise may have had paralic or aeolian sand deposition and drowned fluviatile systems that are unrelated to the present marine depositional environment. These could provide host sediments for significant concentrations of NGH.

Aeolian sands are subject to rapid erosion because there is usually little or no cementation to provide strength. If aeolian deposits were not flooded rapidly enough, they would likely be incorporated in paralic sands along strand lines owing to beach erosive process that would redeposit the sands.

Fluviatile deposits would be excellent reservoirs not only because they could consist largely of coarse to medium grained sands with high porosity and permeability. One-time seacoast sediments would pass upslope into fluviatile and aeolian sediments while seaward from temporary shorelines, deltaic sediment would be the result of river mouth sedimentation. NGH occurs in thermodynamic traps in which NGH mineralization solutions pass upward into a GHSZ (Chap. 2) so there may not need to be a stratigraphic or structural trap up dip. This would provide a similar geomorphology to marine turbidites deposits (Chap. 3), where the base of the GHSZ intersects sands in the system. Suitable mineralizing solutions passing along the sand body will crystallize NGH when conditions of NGH stability are reached. An ideal NGH reservoir would consist of fluviatile deltas having a high clastic sediment delivery system that would form a large area of braided stream/deltaic sediments. These would constitute a single large reservoir host. Any such deposit or just above the base of a GHSZ would be a prime exploration target for NGH concentrations.

It is possible that other similar depositional environments, now buried in the continental slope, occur at different depths within the continental slope sediments. Identifying temporary lowstand levels within basins with drying and subsequent flooding histories should be an important facet of basin geological analysis. Near-shore depositional conditions related to temporary strand lines, aeolian deposits and fluviatile deposits further upslope can be expected to have formed and would constitute sand deposits having high-quality reservoir characteristics.

The nature of exploration targets commonly determines exploration and very importantly, seismic survey design. For instance, close-spaced slope parallel 2D seismic lines or seismic attributes extracted along multiple surfaces along the slope created from 3D datasets are optimal for identifying downslope features such as turbidite systems, but they may miss slope-parallel paralic depositional environments and aeolian targets that may be at any angle to one-time paleoslope or present slope orientation. Fluviatile systems, in contrast, may be predominantly downslope but may be more sinuous in form and difficult to distinguish from turbidite systems. Whereas turbidite systems will have a settling delta that may be related to slope, deltas associated with paralic systems related to temporary shoreline, which may not be related to the steepness of the paleoslope. Thus, designing seismic surveys and processing workflows for NGH exploration should take into account slope parallel and variable orientation of potential reservoirs in addition to marine turbidite systems.

References

Boswell, R., Frye, M., Shelander, D., Shedd, W., McConnell, D. R., & Cook, A. (2012). Architecture of gas-hydrate-bearing sands from Walker Ridge 313, Green Canyon 955, and Alaminos Canyon 21: Northern deepwater Gulf of Mexico. *Marine and Petroleum Geology, 34,* 134–149.

Bott, M. H. P. (1995). Mechanisms of rifting: Geodynamic modeling of continental rift systems. In K. H. Olsen (Ed.), *Continental rifts: Evolution, structure, tectonics. Developments in Geotectonics* (Vol. 25, pp. 27–43). Amsterdam: Elsevier.

Bowman, S. A. (2012). A comprehensive review of the MSC facies and their origins in the offshore Sirt Basin, Libya. *Petroleum Geosciences, 18,* 457–469.

Brownfield, M. E., & Charpentier, R. R. (2003). Assessment of the undiscovered oil and gas of the Senegal Province, Mauritania, Senegal, The Gambia, and Guinea-Bissau, Northwest Africa. *U. S. Geological Survey Bulletin 2207-A*, 29.

Bryant, I., Herbst, N., Dailly, P., Dribus, J. R., Fainstein, R., Harvey, N., et al. (2012). Basin to basin: Plate tectonics in exploration. *Oilfield Review (Schlumberger) Autumn, 24*(3), 38–57.

Çağatay, M. N., Eriş, K., Ryan, W. B. F., Sancar, Ü., Polonia, A., Akçer, S., et al. (2009). Late Pleistocene-Holocene evolution of the northern shelf of the Sea of Marmara. *Marine Geology, 265,* 87–100. https://doi.org/10.1016/j.margeo.2009.06.011.

Cifci, G., Dondurur, D., & Ergun, M. (2003). Deep and shallow structures of large pockmarks in the Turkish shelf, eastern Black Sea. In *Contributions from the Combined 7th International Conference on Gas in Marine Sediments and the NATO Advanced Research Workshop on Seafloor Hydrocarbon Seeps. Geo-Marine Letters, 23*, 311–322.

Condie, K. C. (1997). *Plate tectonics and crustal evolution* (4th ed., 293 pp). Oxford, New York: Butterworth Heinemann. ISBN 0 7506 3386 7.

Crager, B. (2014). *An overview of subsea systems*. SUT Subsea Awareness—Texas A&M University March 21, 2014 (Endeavor), 48 pp. http://sut.tamu.edu/sites/default/files/Session%206.%2021-Mar-14%20SUT-TAMU%20Subsea%20Overview.pdf. Accessed February 20, 2016.

Davison, I., & Steel, I. (2018). Geology and hydrocarbon potential of the East African continental margin: A review. In D. Macgregor, J. Argent & P. Sansom (Eds.), *Thematic set: Tectonics and petroleum systems of East Africa—Part 1* (pp. 57–91).

Dondurur, D., & Cifci, G. (2007). Acoustic structure and recent sediment transport processes on the continental slope of Yesilirmak River fan, eastern Black Sea. *Marine Geology, 237*(1–2), 37–53.

Dondurur, D., & Cifci, G. (2009). Anomalous strong reflections on high resolution seismic data from the Turkish shelf of the eastern Black Sea: Possible indicators of shallow hydrogen sulphide-rich gas hydrate layers. *Turkish Journal of Earth Sciences, 18*(2), 299–313.

Finetti, I., Bricchi, G., Del Ben, A., Pipan, M., & Zuan, Z. (1988). Geophysical study of the Black Sea. *Bolletino di Geofisica Teorica ed Applicata, 30*(Plate 2, 117–118), 197–324.

Frye, M., Shedd, W., & Schuenemeyer, J. (2013). *Gas hydrate resource assessment, Atlantic outer continental shelf* (Bureau of Ocean Energy Management (BOEM) Report RED 2013-01, 57 pp).

Garcia-Castellanos, D., Estrada, F., Jinénez-Munt, I., Gorini, C., Fernàndez, M., Vergés, J., et al. (2009). Catestrophic flood of the Mediterranean after the Messinian salinity crisis. *Nature Letters, 462*(10), 778–781.

Gautier, F., Clauzon, G., Suc, J. P., Cravatte, J., & Violanti, D. (1994). Age and duration of the Messinian salinity crisis. *C.R. Academy of Science Paris (IIA), 318,* 1103–1109.

Gibson, G. M., Roure, F., & Manatschal, G. (2015). *Sedimentary basins and crustal processes at continental margins: From modern hyper-extended margins to deformed ancient analogues* (Vol. 413, 338 pp). Geological Society of London Special Publication. ISBN 978-1-86239-720-0.

Gillet, H., Lericolais, G., & Rehault, J.-P. (2007). Messinian event in the black sea: Evidence of a Messinian erosional surface. *Marine Geology, 244*(1–4), 142–165. Publisher's official version: http://dx.doi.org/10.1016/j.margeo.2007.06.004, Open Access version: http://archimer.ifremer.fr/doc/00000/3324/. Accessed August 6, 2015.

Graham, I. J. (2008). *A continent on the move: New Zealand geoscience into the 21st century* (386 pp). Geological Society of New Zealand. ISBN 978-1-877480-00-3.

Grothe, A., Sangiorgi, F., Mulders, Y. R., Vasiliev, I., Feichart, G.-J., Brinkhuis, H., et al. (2014). Black Sea dessication during the messinian salinity crisis: Fact or fiction? *Geology, 42*, 5.

Groves, R. (2009). Choking the Mediterranean to dehydration: The Messinian salinity crisis. *Geology, 37*(2), 167–170. https://doi.org/10.1130/G25141A.1.

Hillman, J. I. T., Cook, A. E., Sawyer, D. E., Küçük, H. M., & Goldberg, D. S. (2017). The character and amplitude of 'discontinuous' bottom-simulating reflections in marine seismic data. *Earth and Planetary Science Letters, 459*, 157–169.

Hsü, K. J. (1983). *The Mediterranean was a desert: A voyage of the glomar challenger*. Princeton, New Jersey: Princeton University Press.

Hsü, K. J., & Grovanoli, F. (1979). Messinian event in the Black Sea. *Palaeogeography, Paleoclimatology, Palaeocology, 29,* 75–93.

Inagaki, F., & 45 others. (2015). Exploring deep microbial life in coal-bearing sediment down to ~2.5 km below the ocean floor. *Science, 349*, 420–424.

Intawong, A., & Nodgson, N. (2017). Deepwater turbidites offshore Namibia shown to provide high-quality reservoir sands. *Offshore Magazine*, 6 pp. http://www.offshore-mag.com/articles/print/volume-77/issue-4/geology-geophysics/deepwater-turbidites-offshore-namibia-shown-to-provide-high-quality-reservoir-sands.html. Accessed May 16, 2017.

Johannessen, P. N., Nielsen, L. H., Nielsen, L., Møller, I. M., Pejrup, M., Fruergaard, M., et al. (2015). How can a 15 m thick, 14 km long and 5 km wide marine barrier island reservoir sandstone be developed?—A case study from the holocene to recent Rømø Barrier Island, Danish Wadden Sea. Sedimentology of paralic reservoirs. In *Proceedings on Recent Advances and their Applications*, May 18–19, 2015. Geological Society of London, 62.

Kelly, M. (2012, November). Transforming East Africa's exploration successes into value. *Offshore Magazine, 40*, 42.

Konerding, C., Dinu, C., & Wong, H. K. (2010). Seismic sequence stratigraphy, structure and subsidence history of the Romanian Black Sea shelf. In M. Kaymakci, R. A. Stephenson, F. Bergerat & V. Stabostenko (Eds.), *Sedimentary basin tectonics from the Black Sea and Caucasus to the Arabian platform* (Vol. 340, pp. 159–180). Geological Society of London Special Publication.

Krezsek, C., Schleder, Z., Bega, Z., Ionescu, G., & Tari, G. (2016). The Messinian sea-level fall in the western Black Sea: Small or large? Insights from offshore Romania. *Petroleum Geoscience, 22*(4), 392–399.

Krijgsman, W., Hilgen, F. J., Raffi, I., Sierro, F. J., & Wilson, D. S. (1999). Chronology, causes and progression of the Messinian salinity crisis. *Nature, 400,* 652–655.

Leever, K. A., Matenco, L., Rabagia, T., Cloetingh, S., Krijgsman, W., & Stoica, M. (2009). Messinian sea level fall in the Dacic Basin (Eastern Paratethys): Palaeogeographical implications from seismic sequence stratigraphy. *Terra Nova, 22*(1), 12–17. https://doi.org/10.1111/j.1365-3121.2009.00910.x.

Lundin, E. R., & Dorè, A. G. (2016). The Gulf of Mexico and Canada Basin: Genetic siblings on either side of North America. *GSA Today, 27*(1), 8.

Mahmoudi, A. El., & Gabr, A. (2009). Geophysical surveys to investigate the relation between the quaternary Nile channels and the Messinian Nile canyon at east Nile delta, Egypt. *Arab Journal of Geoscience, 2*, 53–67. https://doi.org/10.1007/s12517-008-0018-9.

Maldonado, A., Barnolas, A., Bohoyo, F., Galindo-Zaldivar, J., Hernâândez-Molina, J., Lobo, F., et al. (2003). Contourite deposits in the central Scotia Sea: The importance of the Antarctic circumpolar current and the Weddell Gyre flows. *Palaeogeography, Palaeoclimatology, Palaeocology, 198,* 187–221.

Max, M. D., & Johnson, A. H. (2014). Hydrate petroleum system approach to hydrate natural gas exploration. *Petroleum Geoscience, 20*(2), 187–199. Geological Society of London (Published in Online First March 21, 2014). https://doi.org/10.1144/petgeo2012-049.

Max, M. D., & Johnson, A. H. (2015). A fresh look at the Mediterranean and Black Sea potential for high quality natural gas hydrate sand reservoirs. U.S. Department of Energy, National Energy Technology Laboratory, Morgantown WV. *Fire in the Ice, 15*(1), 15–18.

Max, M. D., Johnson, A., & Dillon, W. P. (2006). *Economic geology of natural gas hydrate* (p. 341). Berlin, Dordrecht: Springer.

Max, M. D., Johnson, A. H., & Dillon, W. P. (2013). *Natural gas hydrate Arctic Ocean deepwater resource potential* (113 pp). SpringerBriefs in Energy.

Mohriak, W. U., & Leroy, S. (2012). *Architecture of rifted continental margins and break- up evolution: Insights from the South Atlantic, North Atlantic and Red Sea-Gulf of Aden conjugate margins* (Vol. 369, pp. 1–40). Special Publication—Geological Society of London. https://doi.org/10.1144/sp369.17. hal-00730844.

NAS. (1989). *Margins: A research initiative for interdisciplinary studies of the processes attending lithospheric extension and convergence* (289 pp). Continental Margins Committee, Ocean Studies Board, National Research Council. ISBN 0-309-54327-4.

Nixon, M. F., & Grozic, J. L. (2007). Submarine slope failure due to gas hydrate dissociation: A preliminary quantification. *Canadian Geotechnical Journal, 44*(3), 314–325. https://doi.org/10.1139/t06-121.

Offshore 1/12/16. (2016). *Seismic review highlights AGC Profond play potential offshore northwest Africa.* Offshore http://www.offshore-mag.com/articles/2016/01/seismic-review-highlights-agc-profond-play-potential-offshore-northwest-africa.html. Accessed April 23, 2016.

Offshore 2/9/16. (2016). *CGG to survey offshore South Africa.* Offshore http://www.offshore-mag.com/articles/2016/01/cgg-to-survey-offshore-south-africa.html. Accessed April 23, 2016.

Offshore 2/12/16. (2016). *Woodside discovers more deepwater gas offshore Myanmar.* Offshore http://www.offshore-mag.com/articles/2016/02/woodside-discovers-more-deepwater-gas-offshore-myanmar.html. Accessed April 23, 2016.

Offshore 10/15/15. (2015). *Lukoil discovers potential giant gas field offshore Romania*. http://www.offshore-mag.com/articles/2015/10/lukoil-discovers-potential-giant-gas-field-offshore-romania.html. Accessed April 23, 2016.

Okay, A. I., Celal Şengör, A. M., & Görür, N. (1994). Kinematic history of the opening of the Black Sea and its effect on surrounding regions. *Geology, 22*, 267–270.

OUGSME. (2005). *The Messinian salinity crisis-Nov 2005*. Open University Geological Society Mainland Europe Branch. http://www.ougseurope.org/index.php?id=28. Accessed May 1, 2015.

PG. 2014. *Reservoir quality of clastic and carbonate rocks: Analysis, modelling and prediction*. Meeting of the Petroleum Group of the Geological Society of London May 28–30, 2014, 232 pp. https://www.geolsoc.org.uk/~/media/shared/documents/specialist%20and%20regional%20groups/petroleum/Reservoir%20Quality%20Abstract%20Book.pdf?la=en. Accessed July 13, 2016.

Pierre, C., Rouchy, J.-M., & Blanc-Valleron, M.-M. (2002). Gas hydrate dissociation in the Lorca Basin (SE Spain) during the Mediterranean Messinian salinity crisis. *Sedimentary Geology, 147*, 247–252.

Piggott, N., & Pulham, A. (1993). Sedimentation rate as the control on hydrocarbon sourcing, generation, and migration in the deepwater Gulf of Mexico. In *Proceedings, Gulf Coast Section SEPM 14th Annual Research Conference* (pp. 179–191).

Plaza-Faverola, A., Bünz, S., Johnson, J. E., Chand, S., Knies, J., Mienert, J., et al. (2015). Role of tectonic stress in seepage evolution along the gas hydrate-charged Vestnesa Ridge, Fram Strait. *Geophysical Research Letters, 42*, 10. https://doi.org/10.1002/2014gl062474.

Popescu, I. (2002). Analyse desprocessusse, dimentairesre, centsdansl' e ventail profond du Danube (MerNoire) (PhD thesis). Universite de Bretagne Occidentale, 282 pp.

Popescu, I., Lericolais, G., Panin, N., Normand, A., Dinu, C., & Drezen, E. (2004). The Danube submarine canyon (Black Sea): Morphology and sedimentary processes. *Marine Geology, 206* (2004), 249–265. https://doi.org/10.1016/j.margeo.2004.03.003.

Praeg, D., Geletti, R., Wardell, N., Unnithan, V., Mascle, J., & Camerlenghi, A. (2011). The Mediterranean Sea: A natural laboratory to study gas hydrate dynamics? In *Proceedings of the 7th International Conference on Gas Hydrates (ICGH 2011)*, Edinburgh, Scotland, United Kingdom, 8 pp, July 17–21, 2011.

Ravnås, R., Nøttvedt, A., Steel, R. J., & Windelstad, J. (2000). Syn-rift sedimentary architectures in the Northern North Sea. In *Dynamics of the Norwegian margin* (Vol. 167, pp. 133–177). Geological Society of London, Special Publications. ISBN 9781862390560.

Rebesco, M., Hernández-Molina, F. J., Van Rooij, D., & Wåhlin, A. (2014). Contourites and associated sediments controlled by deep-water circulation processes: State-of-the-art and future considerations. *Marine Geology, 352*, 111–154. https://doi.org/10.1016/j.margeo.2014.03.011.

Reynolds, A. D. (1999). Dimensions of paralic sandstone bodies. *American Association of Petroleum Geologists Bulletin, 83*(2), 211–229.

Reynolds, T. (2015). Paralic reservoirs. (Abs) Keynote speech: Sedimentology of paralic reservoirs. In *Proceedings on Recent Advances and Their Applications*, May 18–19, 2015. Geological Society of London.

Robertson, A. H. F., Parlak, O., & Ustaömer, T. (2012). Overview of the Palaeozoic-Neogene evolution of Neothethys in the eastern Mediterranean region (southern Turkey, Cyprus, Syria). *Petroleum Geoscience, 18*, 381–404.

Rodriguez, K., Hodgson, N., Saunder, M., & Geiger, L. (2017). Recent survey illuminates possible potsalt Aeolian sandstone play fairway. *Offshore, 77*(2), 33–34.

Roy, S., Moreno, H. M., & Max, M. (2017). Natural gas hydrates and fluid flow: Implications from Irish offshore. In *Proceedings of the 9th International Conference on Gas Hydrates*, Denver CO, 2 pp, June 25–30.

Ryan, W. B. F. (2007). Status of the Black Sea flood hypothesis. In V. Yanko-Hombach, A. S. Gilbert, N. Panin, & P. M. Dolukhanov (Eds.), *The Black Sea Flood question: Changes in coastline, climate and human settlement* (pp. 63–88). Dordrecht: Springer.

Ryan, W. B. F., Jamor, C. O., Lericolais, G., & Goldstein, S. L. (2003). Catestrophic flooding of the Black Sea. *Annual Review Earth & Planetary Sciences, 31,* 525–554. https://doi.org/10.1146/annurev.earth.31.100901.141249.

Ryan, W. B. F., Pitman, W. C., Major, C. O., Shimkus, K., Moskalenko, V., Jones, G. A., et al. (1997). An abrupt drowning of the Black Sea shelf. *Marine Geology, 138,* 119–126.

Schwalenberg, K., Wood, W., Pecher, I., Hamdan, L., Henrys, S., Jegen, M., et al. (2010). Prelimary interpretation of electromagnetic, heat flow, seismic, and geochemical data for gas hydrate distribution across the Porangahau Ridge, Nes Zealand. *Marine Geology, 272,* 89–98.

Scrutton, R. A. (Ed.) (1982). Dynamics of passive margins. In *Geodynamics series* (Vol. 6, pp. 1–200). American Geophysical Union.

Şengör, A. M. C., Tüysüz, O., İmren, C., Sakinç, M., Eyidoğan, H., Görür, N., et al. (2005). The north Anatolian fault: A new look. *Annual Reviews Earth and Planetary Sciences, 33,* 37–112.

Skiple, D., Anderson, E., & Fürstenau, J. (2012). Seismic interpretation and attribute analysis of the Herodotus and the Levantine basin, offshore Cyprus and Lebanon. *Petroleum Geoscience, 18,* 433–552.

Smelror, M., Key, R., Daudi, E., & Njange, F. (2006). Frontier with high expectations. *GeoExPro.* http://assets.geoexpro.com/legacy-files/articles/Mozambique%20Frontier%20with%20High%20Expectations.pdf. Accessed June 31, 2015.

Snedden, J. W., Sarg, J. F., & Ying, X. (2003). *Exploration play analysis from a sequence stratigraphic perspective.* http://www.searchanddiscovery.com/documents/snedden/images/snedden.pdf.

Stoica, M., Lazăr, I., Krijgsman, W., Vasiliev, I., Jipa, D., & Floroiu, A. (2013). Paleoenvironmental evolution of the east Carpathian foredeep during the late Miocene–early Pliocene (Dacian basin; Romania). *Global and Planetary Change, 103,* 135–148.

Stoker, M. S., Van Weering, T. C. E., & Svaerdborg, T. (2001). *A mid- to late cenozoic tectonostratigraphic framework for the rockall trough* (Vol. 188, no. 1, pp. 411–438). Geological Society, London, Special Publications.

Sultan, N., Cochonat, P., Foucher, J. P., & Mienert, J. (2004). Effect of gas hydrates melting on seafloor slope instability. *Marine Geology, 213*(1–4), 379–401. https://doi.org/10.1016/j.margeo.2004.10.015.

Tari, E., Sahin, M., Burka, A., Reilinger, R., King, R. W., McClusky, S., et al. (2000). Active tectonids of the Black Sea with GPS. *Earth Planetary Space, 52,* 747–751.

Tari, G., Hussein, H., Novotny, B., Hannke, K., & Kohazy, R. (2012). Play types of the deep-water Matruh and Herodotus basins, NW Egypt. *Petroleum Geoscience, 18,* 443–455.

Tari, G., Fallaha, M., Kosia, W., Floodpageb, J., Baurb, J., Batic, Z., et al. (2015). Is the impact of the Messinian salinity crisis in the Black Sea comparable to that of the Mediterranean? *Marine and Petroleum Geology, 66*(1), 135–148.

Tari, G., Fallash, M., Schell, C., Kosi, W., Bati, Z., Sipahioğlu, N. Ö., et al. (2016). Why are there no Messinian evaporites in the Black Sea? *Petroleum Geoscience, 22*(4), 381–391.

Tew, B. H., Mink, R. M., Mann, S. D., Bearden, B. L., & Mancini, E. A. (1991). Geologic framework of Norphlet and Pre-Norphlet strata of the onshore and offshore eastern Gulf of Mexico area. *Gulf Coast Association of Geological Societies Transactions, 41,* 590–600.

Vasiliev, I., Reichart, G. J., van Roij, L., Sangiorgi, F., & Krijgsman, W. (2012). Drying up of the Black Sea during the Messinian salinity crisis of the Mediterranean. *Geophysical Research Abstracts,14,* 1. EGU2012-9321, 2012 EGU General Assembly 2012.

Vasiliev, I., Gert-Jan Reichart, J., & Krijgsman, W. (2013). Impact of the Messinian salinity crisis on Black Sea hydrology—Insights from hydrogen isotopes analysis on biomarkers. *Earth and Planetary Science Letters, 362,* 272–282.

Viana, A. R., Almeida, W., Nunes, M. C. V., & Bulhões, E. M. (2007). The economic importance of contourites. *Geological Society, London, Special Publications, 276*(1), 1–23. https://doi.org/10.1144/gsl.sp.2007.276.01.01.

White, K. (2013). The Black Sea hots up. *GeoExPro, 10*(1). http://www.geoexpro.com/articles/2013/06/the-black-sea-hots-up. Accessed July 13, 2016.

Chapter 5
Valuation of NGH Deposits

Abstract Valuation of a natural gas hydrate (NGH) deposit is concerned with determining the amount of natural gas that can be technically extracted from a deposit. NGH will rarely occupy all porosity, probably topping out at 80%, or slightly greater. NGH may also be patchy because of both inconsistent mineralization and variable porosity and permeability in the reservoir. There are different ways to value a NGH deposit: (1) reservoir analysis, which is an analog of valuing a conventional gas deposit, (2) 3D body analysis, which is based on high resolution seismic data, and (3) cell analysis, which is an analog of conventional economic-geology exploration. In addition to an estimate of gas-in-place, estimates for technical recovery must be made. NGH in place may not be technically recoverable gas. It is likely that a number of the NGH deposits, which may be discovered, will occur in discontinuous volumes that are not completely hydraulically linked through their pore water. For example, discontinuous impermeable shale beds and lenses can disrupt continuity. Where production-well bores cannot be expected to link some NGH volumes, the gas valuation figure must be discounted.

5.1 Petrogenesis

Valuation of a natural gas hydrate (NGH) deposit differs from valuing conventional gas deposits, which essentially fill all porosity at the apex of a geological reservoir, having separated from oil that may be below and subjacent water that provides drive to pressurize the gas head. Gas, oil, and water constitute a single, stratified, hydraulic system. As conventional oil and gas are produced, pressure will drop at a rate proportional to the size of the deposit throughout the producing system. Secondary recovery techniques of pumping gas or water injection to maintain production pressure is often a first step to maintaining production. An NGH deposit will also be hydraulically connected by remaining pore water, but when depressurization begins, converted water will be derived along with the gas, and pressures throughout the region in which NGH is dissociating may develop considerable

M. D. Max and A. H. Johnson, *Exploration and Production of Oceanic Natural Gas Hydrate*, https://doi.org/10.1007/978-3-030-00401-9_5

variations in pressure swings that must be managed, particularly in mid-life when areas having smaller NGH domains fully dissociate before larger domain NGH regions.

The first step of valuation is to determine the volume of NGH in place in a geologically—geographically-defined reservoir. The second step is to design a drilling plan that will compensate for any hydraulically connected NGH that would result in incomplete conversion of the NGH to its constituent natural gas and water, and its subsequent drainage, with due consideration to the rate of extraction (Max and Johnson 2011). The NGH-in-place may not be directly related to technically recoverable gas. The aim of a production plan should be to make NGH-in-place directly equivalent to technically recoverable gas.

Commercial valuation involves many financial factors that we do not discuss in detail because as these external factors alter, the level of commerciality may change, irrespective of technical recoverability. From an extractive point of view, a producing NGH reservoir is one in which stimulation conversion of NGH is on-going and extraction of natural gas can be balanced against the rate of conversion to yield smooth operation. This will have the effect of reducing stress on production and processing equipment.

A conventional gas deposit forms by the migration of free and dissolved gas from subjacent source beds along secondary permeability (faults, fractures) into a permeable horizon in a geological trap where the gas can accumulate. Gas is produced, draining it from the trap, under natural pressure, initially providing much if not all of the drive. In contrast, in a NGH accumulation gas and dissolved gas also migrated upward into the GHSZ from source beds in the same manner (Chap. 3), but its ascent is trapped when it reaches conditions of stability in the GHSZ and crystallizes in place (Chap. 1). The NGH system extracts some natural gas from the biogeochemical cycle as gas produced at all levels of a sediment pile, or in association with subduction zones is sequestered in the GHSZ on its upward passage to the oceans and atmosphere.

Within the thermodynamic stability of its reservoir conditions, NGH is a solid crystalline material (Chap. 2) with petrogenetic and provenance similarities to the formation of metallic stratabound mineral ore deposits. Just as the mineralizing fluids form stratabound sulfides, for instance, crystallize when they reach a zone of physical chemical stability, gas and dissolved gas rising into the GHSZ (Chap. 2) spontaneously begin to crystallize solid NGH when stability conditions for the solid form of water and natural gas are reached. Formation of NGH satisfies the definition of a chemical reaction (when two or more substances react to form a third substance which has a definite stability field). Some comments about NGH not being formed by a chemical reaction because the weak electric bonding may not be recognized as being 'proper', have been made to us (personal experience). But NGH is clearly not a solid solution or any other chemical arrangement.

As in the valuing of traditional mineral deposits, there are different methods for estimating the grade each metal or material in a deposit can be applied, depending on circumstances. Many of the methods commonly used by economic geologists offer perspective to describing and valuing NGH deposits in addition to approaches normally used by conventional gas methods.

5.1.1 Mineralization Grade

Because NGH does not fill a reservoir in the same way as conventional gas, but only partly fills porosity, similar to some metallic mineral deposits, we prefer to use the term 'grade' instead of pore fill for purposes of valuation. This is so that the link with traditional mineral deposits is clear and so that all permeability, including unfilled primary intergrain volume and secondary porosity that may also carry NGH as part of a single exploitable deposit, can be described and valued using standard economic geology methods.

In general, inorganic mineral deposits can be divided into high and low-grade deposits (Lindgren 1932). High-grade NGH deposits have high concentration and range from filling most of the available pore space to also replacing rock so that the "ore" percentage may be much higher than original porosity. High-grade deposits usually occupy relatively small volumes, which means that the value of the extracted material may be very much higher than the cost of extraction and processing.

In addition to the requirement for a suitable host rock or strata, high-grade deposits must have a sufficient supply of constituents and a suitable transport mechanism to carry a very great deal of the constituents to the concentration site. In mineral deposition, there is usually a combination of parameters that cause a localized chemical reaction that fixes the ore. In a NGH deposit, the chemical reaction for NGH nucleation and growth is attained when the natural gas-bearing solutions arrive into the GHSZ.

Low-grade deposits, in contrast, are composed of NGH dispersed as grains, nodules, veins, within in huge volumes of host sediment that is poorly bed-differentiated and has little or no permeable host in which high-grade accumulations might form. The pattern of NGH concentration usually consists of the highest percentages of NGH in total sediment being greatest nearer the base of the GHSZ, diminishes irregularly upward, as was seen in drilling the Blake Ridge (Holbrook et al. 1996; Paull and Ussler 2001). Low-grade NGH percentages are low, rising to no more than thin units of 8–10% but are more commonly less than 1% and traces throughout their 'reservoir'. Natural gas migration is generally accomplished by diffusion and can be associated with fluid migration into veins and faults, which often may follow beds.

5.2 Valuation

According to NGH petroleum system analysis, the first order offshore NGH exploration targets, which are considered to be exploitable with existing technology, are turbidite sands that reside within GHSZ (Boswell et al. 2012; Max and Johnson 2014). Ascending groundwater pore fluids carried the mineralizing solutions from elsewhere and have driven the crystallization of concentrated NGH in a focused flow situation (Chaps. 2 and 3). Crystallization can be patchy and estimating the amount of NGH-in-place for a large deposit is subject to many assumptions that must be narrowed to reduce evaluation risk. There is a substantial difference in how the definition of conventional oil and gas and NGH cells and their values will be determined. As in developing estimates for natural gas in conventional deposits, it is always prudent to develop three estimates for low, best, and high values of NGH-in-place and technically recoverable gas. Constraining geology factors governing reservoir potential can be carried out with the direct application of methods for estimating hydrocarbons in place such as that of Imrie and Macrae (2016), although these require further refinement for NGH valuation. Only after production has begun will it be possible to narrow the estimates and quantify a proven resource (SPE 2016).

5.2.1 *Regional Estimates: Shelf or Basin Analysis*

The first stage of exploration is to localize promising regions in which NGH concentrations may occur. This begins with a regional or basin analysis. Good examples of this first approach have been published by Frye (2008) and Frye et al. (2013) for the southern and eastern continental margins of the United States (Chap. 3). First, heat flow data is used to establish geothermal gradient and water temperature data are used to determine the variable thickness of GHSZ for an entire continental slope region. Second, 2D seismic and energy company reports including analysis of seismic sections are compiled to identify the aerial distribution of suitable turbidite sands residing in the GHSZ. Third, areas that appear to have promising NGH exploration targets are located so that more detailed seismic surveys can be targeted upon them (Thakur and Rajput 2011).

Once NGH concentrations have been mapped by seismic survey and other means (Chap. 3), they can be superficially classified by apparent size. This allows ranking of their overall potential for long-term natural gas production. Location is also taken into consideration because of its effect on producibility (Chap. 9). For instance, it may be more desirable from the viewpoint of conserving CAPEX to develop relatively smaller deposits that are closer to a possible landing as well as having better logistics. In this way, an infrastructure network can be most cost efficiently built out from shore to the larger, more distant deposits.

It then becomes important to value deposits to determine reasonable estimates of gas-in-place. A system of classification for gas-in-place may either rank NGH deposits on the same scale as conventional gas deposits, or it could be a special scale in which the valuation is discounted by some figure thought to represent a cost factor additional to valuing a conventional deposit. We prefer a straight comparison because the development costs of conventional gas deposits can vary substantially.

5.2.2 Reservoir Analysis

This method applies traditional conventional gas valuation in which the reservoir that is considered to constitute the geological 'trap' is estimated volumetrically as a function of height and area. Typically, interpretation is begun using either individual 2D seismic lines or lines extracted from a 3D survey (in-line, cross-line, or diagonal). For conventional gas evaluation, the gas trap will be restricted, for example in an anticlinal or stratigraphic closure. Bounding impermeable strata and at its base, by either water or oil, which will form a relatively sharp nearly horizontal boundary (in a closure without strong asymmetrical water drive). In NGH evaluation, the concentration will occur at the base of the GHSZ or at some distance within it where the combination of pressure and temperature reach stability conditions for the concentration of dissolved natural gas (Chap. 1). The most important aspect of a conventional deposit is that the volume of the reservoir from the base of the gas column will fill all porosity.

The shape of the NGH body will be constrained by its bounding impermeable strata, similar to a conventional deposit but up dip the NGH will not necessarily be terminated by a geological trap. NGH deposits may terminate at the structurally highest part of the lowest NGH concentration, even if permeability and porosity of the NGH host bed remains the same. Alternatively, structurally higher NGH concentrations may occur (Max et al. 2006). NGH deposits may be much more complicated in form than conventional deposits. As with all high-grade deposits whose distribution within the boundaries of a deposit may vary, a tight integration of seismic analysis and NGH petroleum system modeling will reduce valuation risk. Such integration will closely follow similar procedures for conventional gas and oil in which Sylta (2008) details procedures for the reduction of uncertainty by using tight integration of seismic and hydrocarbon migration modeling. In this method, areas of greatest uncertainty are identified and examined in greater detail. Reiteration using the more detailed restricted data can allow substantial reevaluation of the entire prospect. Feedback built into the analysis procedure means that reassessment is a function of narrowing geological risk.

In a conventional valuation, it can be assumed that only a fraction of the available porosity linked into permeability would not be occupied by gas, and that it essentially constitutes a single, hydraulically connected body of gas. For a NGH deposit, a numerical estimate for NGH in place can be made by applying a pore fill factor for an estimated porosity for similar sandy host beds, which can then be multiplied by the (conservative) compression factor of 160. Estimates for pore fill can vary from knowledge of other deposits that are similar and much better known through an estimate based on a geophysical estimate from seismic properties. These include interval velocity determined from seismic survey and broad-body estimates based on a few research wells with either accurate logging or analysis of pressure cores.

Stress modeling has been used to successfully identify geological and geomechanical "sweet spots" in tight sand wells (Quenes 2015). Although not directly applicable to valuation of NGH deposits, the concept of using geophysical and geologic data in geomechanical simulations (3D workflow) may have some derivative value in developing accurate valuation methodology for NGH deposits. These can be expected to vary in space in both a regular pattern (generally diminishing upward away from the base of the GHSZ) (Chap. 3) and/or irregularly, as in a complex sedimentary host sediment that affected migration of mineralizing solutions during the principal phase of mineralization.

A NGH deposit in permeable strata (Chap. 3) may be bound at its margins by impermeable strata and at its base by either gas or water. A deposit's up-dip boundary and possibly its lateral boundary may be gradational or patchy and not easy to characterize (Dai et al. 2012). But Kurihara et al. (2010) concludes that small scale inhomogeneities do not appear to affect production predictions. Estimation of NGH should include all NGH that will convert when depressurization dissociation is applied, so long as the gas will flow to the collector system.

5.2.3 3D Body Analysis

This method will be a special application of seismic interpretation in which a limited number of individual seismic lines are used to build up a 3D picture of the deposit as a whole and its immediate surroundings. This especially includes features reflecting the groundwater system which has delivered the mineralizing solutions. The shape of the deposit is referred to as a 'body'. Algorithm interpolation is used to fill in the likely variation of NGH within the body.

The geotechnical parameters derived from Vp, Vs, attenuation and other acoustic properties (Chap. 3) can be used to estimate NGH grade within the body. The main issue with production modeling using a 3D seismic approach is that it is dependent on seismic interpretation based on other data in a general model relating the relative percent of NGH in a volume to its effect on acoustic response. Nonetheless, for preliminary valuing, the analysis will narrowly bracket an accurate valuation of NGH-in-place.

Time slice definition of the NGH pay zone through a deposit can also be carried out using standard workstation analysis of seismic data. The shape of the body can be estimated by interpolation between time slices and by splitting and re-splitting time slices as a sensitivity test. When volume changes become insignificant, the time slices already processed define the shape and pore filling estimated using velocity analysis and other geotechnical properties can be used to define a good estimate of total NGH in place (Hunter and Digert 2014).

Pseudo-3D body analysis has been carried out using 2D data from the Walker Ridge location in the northern Gulf of Mexico (Gaddipati et al. 2011). Myshakin et al. (2012) may be the best production resolution for the Walker Ridge NGH deposit in the northern Gulf of Mexico (Chap. 3). The current iteration of the TOUGH thermodynamic NGH production model is able to input 3D data from which it can produce 2D sections showing NGH conversion progressing (Reagan et al. 2015), which also updates references to model updates, as also can the Japanese process model (Kurihara et al. 2010).

5.2.4 Cell Analysis

Cell valuation is a process of establishing grade by direct observation of core or detailed well logging and then defining one cell parameter by defining the distance along the core in which relative mineral grade(s) vary only by some defined percentage. This allows for absolute values with a statistical range to be established along the course of each well. A meshwork of data points among a number of wells can be interpolated in much the same way that a far greater amount of data is processed for 3D analysis of the concentration.

Each cell along a logged wellbore or core is a data point. With enough data points having conservative estimates for NGH grade and quality of recovery, the solution to valuation will become statistically significant because anomalies or bad data points will be normalized within the much larger data set. These values are then projected toward the margins of the cells with interpolation between values of adjacent cores made between adjoining cells using statistical techniques. Although there will be far fewer data points and no internal seismic data, these direct observations can be used to calibrate seismic surveys.

In an economic mineral deposit, the volume to be mined, or from which material is to be removed by dissolution or related processes, is defined as a function of the limits of the economic mineralization plus the volume of waste that must be extracted to expose the economic mineralization to efficient mining methods. The cost of mining this entire volume is weighed against the value of the recoverable minerals within the volume.

In the case of an open pit mine, this volume would include 100% of the mineralized material with an ore grade above a calculated cutoff value plus an additional volume of overburden or waste, which must be removed to mine the ore. The overburden may contain subgrade mineralization, which may become economic simply because it must be mined to reach the ore.

In the case of underground mines, volumes, or total tonnage, and values of recoverable mineralization are calculated in a similar fashion. Designing safe and stable underground workings may require the preservation or erection of support pillars that contain economic mineralization. This results in a reduction of mineralization value from the volume mined.

The same concept of a NGH deposit defined by grade (i.e., percentage of pore fill) does not result in a well-defined margin for valuation. When a deposit is subjected to depressurization dissociation, the pressure drop will almost certainly be felt in outer areas where NGH grade is lower. This will result in dissociation of all of the NGH subjected to an appropriate pressure drop, regardless of grade. In fact, where there is a higher surface area of NGH in lower grade volumes, more gas will be produced initially because dissociation is a diffusion-driven surface effect process (Chap. 11).

Valuation drilling With inorganic mineral deposits, narrow drill cores, generally about 5 cm in diameter are used to establish grade. The drilling, even in crystalline rocks, is relatively rapid compared with larger diameter drilling and materials requirements are much less. The small cores can be split and for many elements and compounds, rapidly chemically analyzed using automated spectroscopy. This allows a digital dataset to be produced directly in the field. Depending on the scale and nature of the deposit, different drill hole spacing can be statistically calculated and implemented to give an accurate estimation of deposit value. We are not sure what consideration has been given to NGH valuation drill spacing, which in any case need not be as close spaced in NGH valuation because the results will most likely be used to calibrate detailed seismic data, rather than providing all valuation data. If more than a single well is necessary, however, they will almost certainly not be as closed spaced as they commonly are in evaluation of a metallic mineral deposit.

Following a seismic analysis of a NGH deposit (Chap. 3), drilling will be used to ground-truth the deposit in much the same way that exploration drilling is used to proof conventional deposits. Logging while drilling (LWD) and pressure core sampling would normally be carried out at this time. In addition, a much more detailed seismic analysis would be carried out using sensor-embedded optic cable laid out behind the drill assembly that could be abandoned following hole completion, leaving the sensor net behind it. For narrow-hole evaluation, sensors will have to be miniaturized to operate in a narrow bore configuration.

With a number of wells containing these sensor nets, complex, fine-scale seismic analysis and acoustic performance modeling of a volume should be obtainable using a number of methods. Jaiswal (2016) has used a rock physics model and has integrated full-waveform inversion, seismic attributes and rock physics to estimate NGH pore filling based on NGH relationships in a nearby well. Once wellbore data is available along a seismic line, direct parameterization should result in a well-constrained 3D reservoir model. Wang et al. (2016) predicted NGH saturation using a constrained sparse spike inversion (CSSI) to generate an acoustic impedance model integrating seismic reflection and well-log data. Tamaki et al. (2016) used well data and 3D seismic data from the Nankai, Japan NGH deposit and applied a geostatistical approach using seismic data to constrain NGH pore fill estimates.

Narrow-bore drilling for preliminary valuation of a NGH deposit from a seafloor drilling rig can reach the full distance to below the base of a GHSZ (Sect. 7.4.8); a full valuation section can be obtained for each bore. Automated narrow bore drilling from the seafloor should be much faster than wider bore drilling and deeper depths may be obtained in part because the same size core magazine and well frames should be able to hold a much longer compiled drill stem than wider-bore. Multiple cutting capability, such as high-pressure water jets and abrasive cutting (Kolle 1999) can be used to keep the drill face cool and prevent unwanted conversion of NGH.

The Badger Tool (Sect. 7.4.6) for rigless drilling (Schmidt 2013), is a particularly innovative narrow bore drilling concept that may prove to be extremely inexpensive and useful for providing direct observational and measured data through a NGH deposit. This is an open-hole drill that has a drilling head similar to a bottom hole assembly (BHA) of a coiled tube drilling apparatus (Sect. 7.4.8). The drilling tool trails a cable containing power and communications and both are disposable. After the drilling target has been reached, the cable is cut and the drilling and logging tools and the cable are abandoned in place.

5.2.5 Water in the NGH Reservoir

Pore water within immediate contact of NGH, including lower NGH values within and in the margins of the deposit, will have a concentration equal or slightly greater to the concentration necessary to provide a driving force of crystallization (Max et al. 2006). Upon depressurization, pore water from outside the deposit with too low a concentration of dissolved natural gas that comes into contact with NGH will result in dissolution. Exsolution of the dissolved gas will occur upon further depressurization as in the March 2013 Nankai technical production test (Fujii et al. 2015).

Technically recoverable natural gas associated with complete NGH conversion and a fully communicating hydraulic system draining a reservoir is not simply the amount of gas sequestered in the NGH. In an accreting NGH deposit with a strong positive gas flux, which may be reflected by free gas below NGH at the base of the GHSZ, high dissolved gas concentrations in pore fluids can be expected in the lower GHSZ. Not only is there gas compressed within the NGH, but, depending on pressure-depth, there can be considerable gas dissolved in the pore water. The amount of dissolved gas increases with increasing pressure and depth in the NGH prospect zone (Max et al. 2006, Fig. 6.8). Paull and Ussler (2001), for instance, about 1 L of methane-saturated water at 3 km water depth will evolve 4.24 L of methane at surface conditions. At the surface, the same water contains only 0.02 L of methane at full saturation and STP (at 20 °C).

Thus, in addition to the gas within NGH-in-place that is determined by valuation, there may be considerable dissolved gas in the pore water, especially in ultra-deepwater where pressure is high. In addition to the NGH gas and the dissolved gas, free gas associated with the deposit that occurs in continuity within a permeable bed will also be part of the hydraulic system and when flow of NGH converted gas begins. Free gas can also be expected to be recovered. Not all gas will be recovered, however. Some gas will be lost as it will be dissolved in the water that is produced from NGH conversion. Additional gas will be dissolved in the pre-existing pore water, which will absorb more gas as free gas forms generally within the dissociation domain and tends to find equilibrium with the water. The amount of gas that will be recovered from produced water will depend on the lowest pressure it is subjected to in the production process (Max et al. 2006).

Interaction between gas and water in NGH production is substantially different from that of a conventional gas well. In NGH production, pressure and temperature will be lower than pre-production formation conditions in the well and in the dissociation envelope while in a conventional reservoir pressure may be both much higher, and very hot. A conventional gas well, in the best case, will contain little water in the gas reservoir although some water may be dissolved in the gas. In contrast, a NGH well will always begin with gas and water intimately mixed as the NGH is converted.

NGH occupies the same physical space and, once converted, its gas and water will act similar to conventional gas and oil. All gas and water will migrate down pressure gradients in sediment porosity where there is a high enough permeability to allow flow. In addition to the gas from converted NGH and pre-conversion pore water, recoverable free gas below NGH below the GHSZ, will also flow.

By the simple expedient of allowing the gas-enriched water to rise in a contained environment such as a pipe that rises to some intermediate water depth, more free gas will effuse from the water and will add to the total gas produced. Production systems will involve water pumping to decrease reservoir pressure triggering de-pressurization dissociation. Gas separation and collection systems should take into account a flexible system for recovering dissolved gas from produced water that can respond to variations in the gas price. NGH deposits in shallower water depths will have less gas dissolved in the water. In some deposits, it might be commercially and environmentally responsible to separate and recover at least some of the dissolved gas from produced water and at other times it might not.

Technically recoverable natural gas in a NGH deposit can be expressed as:

$$\text{(NGH gas} + \text{dissolved pore water gas} + \text{free gas)} - \text{(initial dissolved pore water gas and gas in converted water lost)} = \text{technically recoverable gas}$$

Valuation steps for a NGH deposit can be summarized as:

- Determination of NGH-in-place
- Evaluation of reservoir quality and potential productivity
- Determination of free and dissolved gas-in-place
- Determination of the cut off point for dissolved gas recovery (if any)
- Determination of technically recoverable gas for that deposit

- Geohazard evaluation
- Production plan for maximum production with cost and safety considerations
- Valuation of technically recoverable gas.

Commercially recoverable gas is, of course, some fraction of the technically recoverable gas (Boswell and Collett 2011). If development costs can be kept low and gas prices are high, the fraction of the technically recoverable gas can be high. Although we can do nothing about controlling the price of gas, we feel that by leveraging NGH-optimized existing technology substantial reductions in the cost of natural gas production from NGH deposits can be realized. This technology is specified only to the requirements necessary for the NGH resource and by applying new technology and practices.

5.3 Geophysical Characterization of NGH Deposit Settings

Seismic response in a NGH deposit varies directly with small variations in NGH percentage pore fill, which alters the geotechnical properties, (Shelander et al. 2012). Following this seismic interpretation, drilling can be carried out to ground-truth or calibrate the seismic interpretation. This procedure was followed in valuing the NGH-in-place at the Walker Ridge deposit in the northern Gulf of Mexico (Boswell et al. 2012). In order to correlate the acoustic response within the NGH body, detailed knowledge of the acoustic velocity structure of the sediment within which the NGH body has developed is desired. Gross estimates of the geoacoustic profile (a curved line representing a trace of Vp with depths) have been based on compaction of different sediments under their own weight (Hamilton 1979, 1980), but this method has now been superseded.

Wood et al. (2016) developed a technique to invert vertical sound speed profiles obtained from reflection seismic data, for grain and pore fluid properties. They have applied this process to seismic data from the Bering Sea to better constrain gas and NGH concentrations. The inversion is based on iterative forward modeling of the sediment constituents and pressure–temperature (PT) regime to match the observed sound speed profile. Inversion input can be either interval or stacking velocities, and they have avoided the assumption that stacking velocities are the same as root mean square average velocities. The method uses a series of constituent sediment physics models whose inputs are mainly porosity, gas saturation, temperature, pressure, effective pressure and grain type (for calculation of effective elastic moduli). The value of this approach is that every model run in the forward algorithm is geologically consistent. The model runs relatively quickly because vast portions of model space in the seismic record below the GHSZ are eliminated from searching. This is because NGH cannot exist outside its PT stability zone. With a more accurate Vp structure of the region of sediments in which the NGH deposit occurs, a more accurate determination from seismic analysis of the NGH deposit can be made.

Estimations of NGH in place from 3D seismic interpretation (Shellander et al. 2012) provide a sufficient basis for a global estimation of gas in place using estimates of host strata in the GHSZ and a range for NGH pore fill in a mineralized zone, although refinements will improve estimates. For reservoir engineering and for planning the 3D layout of wellbore a 3D valuation of the potential reservoir, including defining cut-offs, a more detailed seismic analysis is necessary. This will require either a higher frequency seismic analysis, which in deep water can only be achieved by source and receiver being closer to the seafloor. This could be done from a specially fitted AUV (rather than the less precise and more difficult to manage deep-tow seismic technology (DETAGS), or seafloor OBS and/or geophone cabling, or both, with overlain survey paths so as to achieve both Vp and Vs at higher frequencies that will resolve thinner beds and a much higher resolution of NGH saturation. For a full evaluation, a 3D digital model will provide the basis for an optimal well bore configuration to define the 3D nature of the NGH concentration.

5.4 The Creaming Curve

NGH is a new natural gas resource, with no production data as of mid-2016 on which to project its development (including gas production), the time over which the gas can be produced, and the investment and operating potential. Virtually all other gas resources once had the same pre-data character until markets could be established to drive resource identification and exploitation. These aspects could not be initially projected. By the time of development of unconventional gas resources, however, a methodology for carrying out these analyses had been developed using conventional gas data and could be applied early to coalbed methane and shale gas/oil (Chap. 10).

The 'creaming curve' is a graphical method that describes the relationship between the history of discovery, the sequence of the exploration wells drilled, and the cumulative volume of reserves discovered in any defined region from a single basin to much larger areas. It was introduced by Meisner and Demirmen (1981) after they found an empirical relationship in exploration data. It is a graphic representation used to analyze past exploration trends and to provide insight for forecasting remaining oil and gas resources potential. WHUT (2008) provides a brief mathematical discussion of the curve projections and Premier Oil (2012) provides a detailed illustration of the use of creaming curve projects as an aid to exploration in SE Asia.

Reserves are plotted as a function of time in a field determined by the number of discoveries as exploration proceeds and the cumulative reserves (Fig. 5.1). Plotting the two variables shows the exploration trend. The first stage of a graph shows the first discoveries, with significant increase in field reserve size. The second stage shows the gradual decrease of the mean field size, and the third stage is an interpolated trend line continued from a smoothed discovery/reserves line (Bohorquez

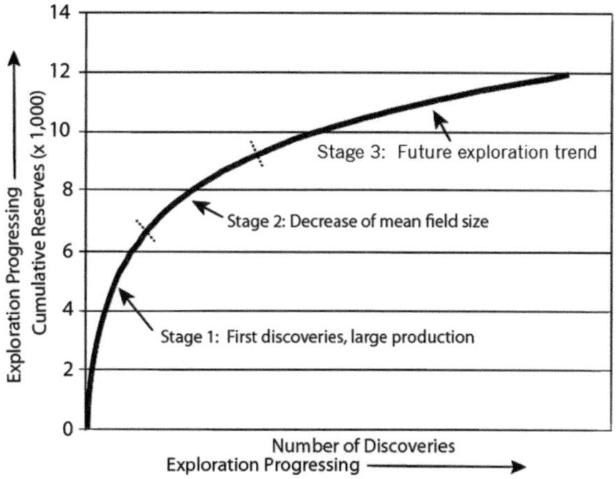

Fig. 5.1 Basic elements of a creaming curve drawn after Bohorquez (2014, Fig. 1)

2014). Laherrere (2009) gives many examples of plays, fields, regions, and physical properties of oil, expanding the use of creaming curve diagrams.

The 'conventional' discovery curve has been used in an idealized form (Fig. 5.2) by ExxonMobil to determine when the best discoveries in an area have been found so as to provide guidance for continuing in the area or moving to a new area in which the first part of its future operational history is liable to be the most rewarding (Snedden et al. 2003). Creaming curves can also provide guidance for the possible existence of a second or third 'tier' in a particular exploration play. Ideally, in their view, when the different tiers of a sequence stratigraphic play appear to have been largely exhausted and the size of each newly defined reserve diminishes to a certain point, the profit margin on the investment will decrease below the level at which they prefer to operate. At this point, after they have benefitted from the larger early discoveries and they have skimmed the 'cream' of the producible product of the area, the graph provides an indication of a point of diminishing returns after which a

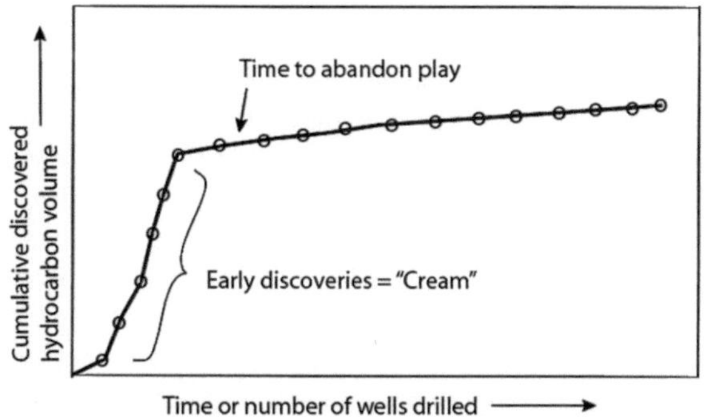

Fig. 5.2 Idealized diagram of a conventional discovery curve showing early high production period followed by mature production and smaller discoveries after Snedden et al. (2003). The break to advice to abandon the play is rarely this clear

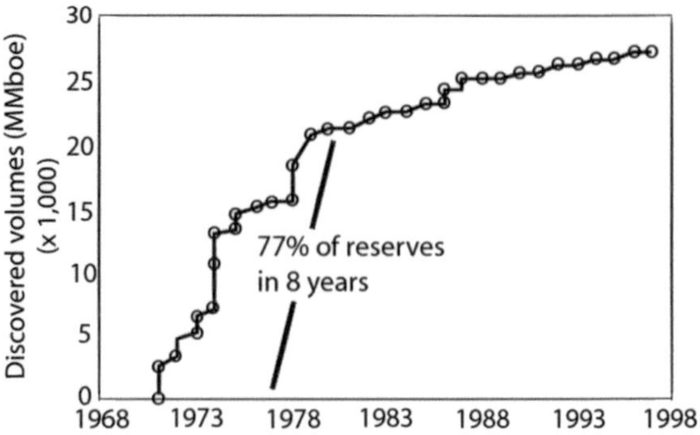

Fig. 5.3 Example of an actual creaming curve from Snedden et al. (2003, Fig. 2), the Middle Jurassic gas play, UK, Norway and Denmark. MMBOE = million barrels oil equivalent

play may be abandoned. Snedden et al. (2003) also discuss the relationship of highstand, transgressive, and lowstand type conventional hydrocarbon reservoirs related to sequence stratigraphy (Chap. 4).

There are now hundreds of creaming curve diagrams published for virtually all fields and regions, and for different types of hydrocarbon. Most show approximately the same relationship with a relatively steep early history reflecting high production, and a gradual flattening of the curve as the rate of aggregate increases increasingly falls off. Major plays have a dramatically steepening early production. In contrast, smaller fields discovered and produced at greater intervals, such as those to the west of Ireland, do not show a pronounced early strong production (Caulfield and Corr 2015). Creaming curve diagrams can be used early in the history of a play development to indicate not only whether it is liable to be a major profit center, but also to indicate that some regions are not high-profit prospective. Figure 5.3 shows an example of a strong gas play from the North Sea that conforms to the idealized pattern shown in Fig. 5.2, where most of the reserve identification took place in the early phase of development.

In assessing the early development of NGH, it will be important to use creaming curve diagrams to indicate potential value. The nature of the general play is the same worldwide in that all NGH deposits will occur only in the restricted GHSZ (Chap. 3), regardless of the geological age or nature of the NGH host strata (Chap. 4). Although host beds for the primary NGH target identified to date consist of deepwater turbidite sands, other NGH plays may occur. Max et al. (2013), for instance, have identified deep troughs in Arctic and other continental shelves as possibly even more attractive plays, with outlier plays requiring much more assessment.

Thus, valuation of deposits can be done in a number of manners, with actual production or a long-term production test necessary to complete the valuation process. Once production begins, play assessment using creaming curve analysis and other tools will be used to value the overall resource.

References

Bohorquez, M. O. (2014). Hydrocarbon discovery potential in Columbian basins: Creaming curve analysis. In *Adapted from Oral Presentation given at AAPG 2014 European Regional Conference & Exhibition*, Barcelona, Spain, May 13–15. AAPG Datapages. http://www.searchanddiscovery.com/pdfz/documents/2014/10613bohorquez/ndx_bohorquez.pdf.html. Accessed September 21, 2015.

Boswell, R., & Collett, T. (2011). Current perspectives on gas hydrate resources. *Energy and Environmental Sciences, 4,* 1206–1215.

Boswell, R., Collett, T. S., Frye, M., Shedd, W., McConnell, D. R., & Shelander, D. (2012). Architecture of gas-hydrate-bearing sands from Walker Ridge 313, Green Canyon 955, and Alaminos Canyon 21: Northern deepwater Gulf of Mexico. *Marine and Petroleum Geology, 34,* 134–149.

Caulfield, C., & Corr, H. (2015, August). Industry reassessing potential of Irish Atlantic Margin. *Offshore Magazine*, 42–45.

Dai, S., Santamarina, J. C., Waite, W. F., & Kneafsey, T. J. (2012). Hydrate morphology: Physical properties of sands with patchy hydrate saturation. *Journal of Geophysical Research, 117* (B11205), 12. https://doi.org/10.1029/2012jb009667.

Frye, M. (2008). *Preliminary evaluation of in-place gas hydrate resources: Gulf of Mexico outer continental shelf* (OCS Report MMS, 2008-0004, 136 pp). US Department of the Interior Minerals Management Service Resource Evaluation Division.

Frye, M., Schuenemeyer, J., Shedd, W., Piper, K., & Herman, B. (2013). Gas hydrate resource assessment on the United States outer continental shelf: A mass balance model. In *Proceedings of the 7th International Conference on Gas Hydrates (ICGH 2011)*, Edinburgh, Scotland, United Kingdom (13 pp), July 17–21.

Fujii, T., Suzuki, K., Takayama, T., Tamaki, M., Komatsu, Y., Konno, Y., et al. (2015). Geological setting and characterization of a methane hydrate reservoir distributed at the first offshore production test site on the Daini-Atsumi Knoll in the eastern Nankai trough, Japan. *Marine and Petroleum Geology, 66*(1), 310–322. https://doi.org/10.1016/j.marpetgeo.2015.02.037.

Gaddipati, M., Myshakin, E. M., Boswell, R., & Anderson, B. J. (2011). Gas production modeling from a complex 3-D description of marine hydrate deposits. In *Proceedings of the 7th International Conference on Gas Hydrates (ICGH 2011)*, Edinburgh, Scotland, United Kingdom (11 pp), July 17–21.

Hamilton, E. L. (1979). Sound velocity gradients in marine sediments. *Journal of the Acoustical Society of America, 65,* 909–922.

Hamilton, E. L. (1980). Geoacoustic modeling of the sea floor. *Journal of the Acoustical Society of America, 68,* 1313–1340.

Holbrook, W. S., Hoskins, H., Wood, W. T., Stephen, R. A., Lizarralde, D., & Leg 164 Science Party. (1996). Methane hydrate and free gas on the Blake Ridge from vertical seismic profiling. *Science, 273*, 1840–1843.

Hunter, R., & Digert, S. (2014). *Resource characterization and quantification of natural gas hydrate and associated free-gas accumulations in the Prudhoe Bay–Kuparuk river area on the north slope of Alaska*. Final Technical Report September 30, 2001–March 31, 2014. United States Department of Energy National Energy Technology Laboratory, Office of Fossil Energy, U.S. Department of Energy, 370 pp.

Imrie, C. E., & Macrae, F. J. (2016). Application of experimental design to estimate hydrocarbons initially in place. *Petroleum Geoscience, 22,* 11–19.

Jaiswal, P. (2016). Hydrate quantification: Integrating full-waveform inversion, seismic attributes, and rock physics. *Special Section: Exploration and Characterization of Gas Hydrates. Interpretation, 4*(1), SA55–SA71.

Kolle, J. J. (1999). *A comparison of water jet, abrasive jet and rotary diamond drilling in hard rock* (8 pp). Tempress Technologies Report. http://ttinc10.qwestoffice.net/papers/comparison.pdf. Accessed April 23, 2016.

Kurihara, M., Sato, A., Ouchi, H., Ebinuma, T., Suzuki, K., Masuda, et al. (2010). Prediction of production test performances in eastern Nankai trough methane hydrate reservoirs using 3D reservoir model. OTC 20737. In *Offshore Technology Conference held in Houston*, Texas, USA (22 pp), May 3–6, 2010.

Laherrere, J. (2009). *Creaming curves & cumulative discovery at end 2007 by continents*. http://aspofrance.viabloga.com/files/JL_cream_end2007.pdf. Accessed February 11, 2016.

Lindgren, W. (1932). *Mineral deposits* (p. 930). New York and London: McGraw-Hill Book Company, Inc.

Max, M. D., & Johnson, A. H. (2011). Methane hydrate/clathrate conversion. In M. R. Khan (Ed.), *Clean hydrocarbon fuel conversion technology, Woodhead publishing series in energy* (Vol. 19, pp. 413–434). Cambridge, U.K.: Woodhead Publishing Ltd. ISBN 1 84569 727 8, ISBN-13: 978 1 84569 727 3.

Max, M. D., & Johnson, A. H. (2014). Hydrate petroleum system approach to hydrate natural gas exploration. *Petroleum Geoscience, 20*(2), 187–199. Geological Society of London (Published in Online First March 21, 2014). https://doi.org/10.1144/petgeo2012-049.

Max, M. D., Johnson, A. H., & Dillon, W. P. (2006). *Economic geology of natural gas hydrate* (p. 341). Berlin, Dordrecht: Springer.

Max, M. D., Johnson, A. H., & Dillon, W. P. (2013). *Natural gas hydrate Arctic Ocean deepwater resource potential* (113 pp). Springer Briefs in Energy.

Meisner, J., & Demirmen, F. (1981). The creaming method: A Bayesian procedure to forecast future oil and gas discoveries in mature exploration provinces. *Journal of the Royal Statistical Society A, 144*(1), 1–31.

Myshakin, E. M., Gaddipati, M., Rose, K., & Anderson, B. J. (2012). Numerical simulations of depressurization-induced gas production from gas hydrate reservoirs at the Walker Ridge 313 site, northern Gulf of Mexico. *Marine and Petroleum Geology, 34*, 169–185.

Paull, C. K., & Ussler, W. (2001). History and significance of gas sampling during DSDP and ODP drilling associated with gas hydrates. In C. A. Paull & W. P. Dillon (Eds.), *Natural gas hydrates occurrence, distributions, and detection* (Vol. 124, pp. 53–65). American Geophysical Union, Geophysical Monograph.

Premier Oil. (2012). *Where are the big fields hiding in SE Asia? A premier persepctive*. Presentation at Geological Society Meeting, Burlington House, London, GB, February 21, 2012. http://c299791.r91.cf1.rackcdn.com/andrewlodge.pdf. Accessed February 11, 2016.

Quenes, A. (2015). Stress modeling, '3-G' workflow pinpoints shale sweet spots. *Oilpro*. http://oilpro.com/links/detail/21689/3-g-workflow-pinpoints-shale-sweet-spots?utm_source=DailyNewsletter&utm_medium=email&utm_campaign=newsletter&utm_term=2015–09-04&utm_content=Article_15_txt. Accessed September 4, 2015.

Reagan, M. T., Moridis, G. J., Johnson, J. N., Pan, L., Freeman, C. M., Boyle, K. L., et al. (2015). Field-scale simulation of production from oceanic gas hydrate deposits. *Transport in Porous Media, 108*, 151–169. https://doi.org/10.1007/s11242-014-0330-7.

Schmidt, V. (2013, April). Rigless drilling enters the pilot stage. *OE Digital*, 47–48.

Shelander, D., Dai, J., Bunge, G., Singh, S., Eissa, M., & Fisher, K. (2012). Estimating saturation of gas hydrates using conventional 3D seismic data, Gulf of Mexico joint industry project leg II. *Marine and Petroleum Geology, 34*, 96–110.

Snedden, J. W., Sarg, J. F., & Ying, X. (2003). *Exploration play analysis from a sequence stratigraphic perspective*. http://www.searchanddiscovery.com/documents/snedden/images/snedden.pdf. Accessed February 11, 2016.

SPE. (2016). Estimating prospective resources and reserves. *Oilfield development*. http://www.petroleumonline.com/modules/m005/hl_005_012.asp. Accessed September 21, 2015.

Sylta, Ø. (2008). Analysing exploration uncertainties by tight integration of seismic and hydrocarbon migration modelling. *Petroleum Geoscience, 14*, 281–289.

Tamaki, M., Suzuki, K., Fujii, T., & Sato, A. (2016). Prediction and validation of gas hydrate saturation distribution in the eastern Nankai trough, Japan: Geostatistical approach integrating well-log and 3D seismic data. *Special Section: Exploration and Characterization of Gas Hydrates. Interpretation, 4*(1), SA83–SA94.

Thakur, N. K., & Rajput, S. (2011). *Exploration of gas hydrates, geophysical techniques* (281 pp). Heidelberg, Dordrecht, London, New York: Springer. ISBN 978–3-642-14233-8.

Wang, X., Qian, J., Collett, T. S., Shi, H., Yang, S., Yan, C., et al. (2016). Characterization of gas hydrate distribution using conventional Ed seismic data in the Pearl River Mouth Basin, South China Sea. *Special Section: Exploration and Characterization of Gas Hydrates. Interpretation, 4*(1), SA25–SA37.

WHUT. (2008). *Creaming curves and dispersive discovery*. http://mobjectivist.blogspot.com/2008/03/creaming-curves-and-dispersive.html. Accessed February 11, 2016.

Wood, W., Martin, K., Barth, G., & Scholl, D. (2016). *New constraints on gas and gas hydrate estimates in the Bering Sea using an automated sediment physics modeling approach (Abs)*. American Geological Union Fall Meeting, San Francisco, December 14–18, 2015. https://agu.confex.com/agu/fm15/webprogram/Paper62806.html. Accessed January 18, 2016.

Chapter 6
Deepwater Natural Gas Hydrate Innovation Opportunities

Abstract Natural gas hydrate (NGH) is unique among conventional and unconventional gas resources. It is a stable, solid crystalline material in its reservoir. All NGH deposits worldwide will be ideally found in partially consolidated marine sediments and sand or coarser beds within about 1 km or less from the seafloor, making them easily accessible from the seafloor. Most of these will be turbidite sands similar to those that are older and more deeply buried, which may host conventional oil and gas deposits. NGH deposits are generally not associated with oil or hazardous chemicals. When NGH is inexpensively converted to its constituent gas and water, the resulting natural gas and water facilitate unique drilling opportunities. NGH deposits have very low environmental risk, even in remote regions such as the environmentally sensitive Polar Regions. The unique characteristics of NGH concentrations will potentially allow inexpensive exploration, reservoir characterization, and production opportunities. In particular, there are opportunities in drilling and reservoir wellbore plans and new technologies for NGH exploitation that could provide an innovative pulse to offshore energy operations.

6.1 NGH Technology Opportunities

It can be easy for an experienced oil or gas explorationist to misunderstand natural gas hydrate (NGH) because its paragenesis, the manner in which concentrations form, the bounding elements of those concentrations, and the secondary recovery techniques that have to be applied to produce the natural gas are very different from conventional oil and gas reservoirs. But at the production end, it is just ordinary natural gas. And the nature of the relatively pure natural gas produced from NGH also offers opportunities for cost-cutting innovations. NGH offers a number of technology opportunities that should allow the development of new methodologies and technologies for the NGH resource. These could have dramatic impact upon commercialization of NGH in the same manner as, for instance, horizontal drilling and new fracking methods were applied to exploitation of the shale resource,

M. D. Max and A. H. Johnson, *Exploration and Production of Oceanic Natural Gas Hydrate*, https://doi.org/10.1007/978-3-030-00401-9_6

although we do not envisage fracking as being necessary for NGH production. We attempt here to illustrate the particular aspects of NGH and its occurrence from the point of view of opportunities for innovation in technology and methods that could have a dramatic effect in reducing the overall cost of developing the NGH resource.

Like the other unconventional gas deposits, coalbed methane and shale gas, NGH will not spontaneously flow as conventional gas and oil deposits do. The depressurization method for conversion of the NGH to its constituent natural gas and water is superficially similar, but because all exploration and production activities will take place in deepwater and ultra-deepwater where issues such as flow assurance exist, the NGH production strategy will have to combine both unconventional and conventional technology approaches from the outset. In addition to differences with existing 'conventional' hydrocarbon E&P, the NGH petroleum system (Max and Johnson 2014) differs from conventional petroleum systems in a number of ways that reduce exploration requirements (Chap. 3). These differences (Max et al. 2006; Max and Johnson 2011a) offer opportunities for lowering the cost of development to well below that for conventional hydrocarbons.

The first and most prominent general opportunity is the probable scale of the resource (Chap. 1). It is unlikely that NGH innovation will be dead-ended in only a few fields. Although it is not yet known how much NGH exists in richer deposits in which commercialization of the resource will begin, conservative estimates based on petroleum system analysis are very large (Johnson 2011). If a very small percentage of it could be commercialized (i.e., produced at cost and profit margins that would allow it to compete with other gas resources), the resulting production would be of great significance on a global scale and especially to major energy importing countries that have NGH opportunities in their coastal areas (Chap. 3). The NGH resource that we discuss throughout is comprised of sandy marine turbidite systems for which exploration and production techniques are currently sufficient to initiate production without major innovation. It is not known how the development of new technology and methodology for the turbidite play will increase the commerciality of progressively lower grade NGH concentrations.

A typical high-grade NGH reservoir, which almost certainly will be either a Class 1 or 2 deposit (Chap. 2), and will consist of permeable, sandy strata with a high percentage of the permeability filled by NGH. A more complete discussion of the different types of NGH development can be found in Max et al. (2006). NGH is not distributed uniformly within the GHSZ (Chap. 5). NGH occurrences range from high-grade deposits with significant concentrations of NGH in sandy strata in strongly bed-differentiated sedimentary sequences to dispersed NGH in poorly bed-differentiated finer grained, muddy sediments. One of the general characteristics of NGH deposits is that converted gas from NGH may not concentrate for easy recovery, especially in oceanic NGH. In conventional gas deposits, structural traps such as anticlinal closures, which seem to be a dominant deepwater geological trap, can have spill points associated with a bounding synclinal closure or a fault that can act as a migration pathway (Hermanrud et al. 2014). In contrast, a NGH trap does not have a spill point as the up-dip termination of a typical sand bed deposit will be the level at which crystallization of NGH becomes unstable due to changes in pressure and temperature.

NGH deposits have been divided into four major classes, first based on whether they are concentrated or dispersed, and second depending on what geological strata or structure is in contact with the NGH and whether the groundwater system is open or closed to the passage of pore water. Where gas is in contact with overlying NGH at the base of the GHSZ (Class 1), as at Nankai (Takahashi and Tsuji 2005; Noguchi et al. 2011) and the northern Gulf of Mexico (Boswell et al. 2012a, b) the optimum conditions for hydrate conversion by artificial means exist because decompression and/or heating dissociation can be used. Ideally, NGH that is converted near the base of the GHSZ may be concentrated, although as the conversion front migrates further up dip into strata in which NGH saturation can be expected to be less, gas may concentrate in a NGH-topped physical trap. Class 2 high-grade hydrate deposits may not provide a good situation for converted natural gas to concentrate if water is mobile within the mineralized strata and trapped gas is relatively thin. This would be due to the permeability not being open enough in the mineralized formation to allow gas to pass freely into it and water to percolate through it.

The cost of NGH development is currently modeled as the cost of conventional resource development plus the cost of converting NGH to its constituent gas and water, plus the costs of other common issues such as flow assurance. To our knowledge most of current cost projections for NGH (to our knowledge) do not involve the application of any innovative new technology or approaches. We propose that there are a number of technical fields that offer considerable scope of development for driving down costs. These include drilling, wellbore lining, and preparing a deposit for production, as well as some new production methodology. If these new technologies and practices can be realized, there is substantial scope for making NGH much more competitive than may be presently modeled. We have counseled in our publications over the years that the particular attributes of NGH and the sediment within which it occurs would allow development of new methodologies and technologies that could have dramatic impact upon commercialization of the potential NGH resource. We outline some of those here. There are a number of technical fields, particularly in drilling and preparing a deposit for production, as well as some new production methodology, that offer considerable scope for reducing costs.

6.2 Exploration Opportunities

The basic exploration method for conventional hydrocarbons, which is high resolution 2D and 3D seismic surveys whose digital datasets are analyzed using graphic workstations can be applied to NGH exploration with virtually no change in operational approach. The geotechnical effect of different NGH pore fill of sandy reservoir hosts on seismic performance (Chap. 2), which has been verified by exploration drilling, is now known in enough detail for close valuation (Frye et al. 2012). An opportunity exists for refining the geotechnical approach to both identification and characterization (Chap. 3) and valuation (Chap. 5) so as to narrow the present estimation range. There is also scope for designing more frequency-specific seismic surveys to better expose the full acoustic response character of the NGH prospect zone (Chap. 2). Finally, it is possible that improvement could be made in the resolution of electro-magnetic (EM) methods so as to reveal fine-scale NGH values in a deposit.

6.3 Drilling

As the hydrocarbon energy industry has proceeded from land across shallow water and into deepwater and ultra-deepwater (Chap. 3), the vessels, drilling equipment, and innovation have become more robust and costly in order to drill through sediments and rocks having different geomechanical character to reach the drilling targets in deep and ultradeep water. Conventional exploration and production is now taking place in a range of water depths in which the GHSZ is thick enough to host substantial NGH concentrations. There is no difference in the water depth requirement for NGH that has not already been encountered by drilling activities for conventional hydrocarbons. Thus, there is no technical reason why exploration and production of NGH cannot take place using conventional equipment. However, this deepwater capability is expensive. NGH production will be entirely gas, although it may not be entirely methane. Higher density natural gas may exist deeper than the GHSZ, which is a phenomenon of methane hydrate in areas with strong thermogenic gas systems. Introducing drilling technology and methods that could reduce the cost of exploration and production for NGH is considered by us to be the key to commercialization of the resource. Drilling and support activities seem to us to be the segment of the development spectrum in which the greatest cost savings can be made.

6.3.1 Material Requirements

Drilling requirements are established primarily by the geomechanical properties of the lithologies that have to be penetrated to reach the drilling target identified through geophysical exploration (Table 6.1). The scale of the drilling requirement, including access to the drill site at sea, is directly related to drilling vessel cost. Major cost factors involve the length of time the drilling unit has to be deployed to complete exploration and production preparation operations. These costs, as with the earlier exploration costs, all need to be made as part of capital expense (CAPEX) to be recovered later. If a prospect is abandoned, these are all sunk costs.

Table 6.1 Material drilling requirements

Attribute, issue	Reservoir/characteristics	Opportunity for NGH-specific requirements
Material	Semi-consolidated sandy sediments worldwide	Relatively easy drilling
Reservoir integrity	NGH stable in reservoir. NGH helps stabilize reservoir	Keeping conditions the same during drilling prevents gas leakage
Pressure	Formation pressure	During production below reservoir pressure

Keeping exploration and drilling costs as inexpensive as possible is desirable, especially if the costs have to be borne through overall company income, as in the case of abandonment.

Comparison with conventional hydrocarbon drilling requirements shows immediately that the materials that have to be penetrated to reach and drill within the NGH reservoir are going to be much less demanding than for conventional hydrocarbons. Because the material drilling requirements are at the low end of the drilling requirement spectrum, the power of the drill, the amount of materials, and the size of the drilling platform can also reflect the low-cost end of drilling and access requirements.

6.3.2 Geotechnical Attributes and Reservoir Stability

Most conventional hydrocarbons occur in rock formations having considerable mechanical strength whereas all first-order economic oceanic NGH targets will occur within porous and permeable unconsolidated and semi-consolidated marine sediments of the GHSZ. Thus, the mechanical properties of NGH targets and their immediate geological environs are substantially different from those of conventional deposits (Table 6.2). This both offers opportunities and introduces some restricted conditions for drilling and exploration, reservoir preparation, and production from NGH reservoirs. NGH formation increases mechanical strength, but recovery of natural gas from NGH inevitably involves some weakening of the sediment host that comprises the reservoir as the solid crystalline NGH dissociates and bulk modulus drops. The geotechnical attributes are a key item to be considered in the production plan for NGH deposits in order to maintain reservoir stability.

In general, it appears that more attention needs to be focused on reservoir stability of NGH deposits, but because pressures can be managed within the NGH producing reservoir, there are degrees of control that do not exist for highly pressurized conventional deposits. Being able to manage the pressure within the reservoir and potentially to control the movement of gas and water as independent issues offers opportunity for innovation.

Table 6.2 Comparison of geological and physical characteristics of conventional and NGH deposits

Issue	Conventional gas	NGH deposits
Material	Rock	Semi-compacted sediment
Reservoir strength	Strong, unlikely to deform. Decompression compaction possible. Well contained	Yes, differential compaction and sediment deformation
Strength/time factor	Reservoir/trap maintains strength over time	Reservoir weakens over time, bulk modulus diminishes as NGH is converted
Mass flow possible during production	Unlikely	Yes, differential compaction and sediment deformation
Reservoir maintenance	Decompression and compaction possible. Water flood?	Weak, requires plan to compensate for weakening reservoir
Pressure differential with seafloor	Very high	Low
Temperature	Very high	Low
Special materials required for borehole casing	Yes	No

6.3.3 Wellbore Stability

Wellbore stability is a factor from the seafloor all the way down to a NGH deposit of interest as well as possibly below it. The distance from the seafloor to depths at which NGH deposits are liable to be found are about the same as the open hole drilling that conventional wells usually drill before surface casing is run and cemented to form the foundation for the wellhead. This could be used to suggest that a considerable proportion of the drilling for NGH deposits could be done also using open hole methods with seawater used as the drilling fluid. Using seafloor water as the drilling fluid would have the added benefit of cooling the well bottom, where temperatures will be somewhat elevated owing to the effects of the local geothermal gradient and friction from the bit. Thus, there may be little difference in overall drilling practice between NGH and conventional drilling, at least to the NGH depth of interest.

Wells themselves will have to accommodate the enclosing effect of casing in a conventional well. But because of the lower temperatures and pressure variations in a NGH well and the weak geomechanical sediments, wellbore walls need to be supported as soon after drilling as possible. Because of the relatively soft and particulate nature of the geological strata, care has to be taken to carefully control drilling fluid movement to prevent washouts until the well can be lined. Again, because of the relatively benign nature of the NGH and the ambient surroundings of the well, considerable innovation is possible as the harsh conditions associated with conventional oil and gas wells do not have to be compensated. Materials in the wellbore liner can be considerably more flexible, for instance, the corrosion resistant, mechanically very strong casing that has to be used on conventional deepwater wells is not required for NGH production preparation drilling.

6.3.4 Drilling Depths, Pressures and Temperatures

The depth to which drilling must reach beneath the deepwater seafloor is much shallower for NGH deposits than for conventional hydrocarbons. Deepwater production often entails reservoirs over 2000 m beneath the seafloor. It would now appear that in ultra-deepwater regions the conventional oil and gas drilling targets are most commonly located about 2500–3000 m below the seafloor. In contrast to conventional oil and gas, NGH is much shallower (Table 6.3).

Thus, although the water depths in the ultra-deepwater may be about the same, the depth below sea level that must be reached for conventional oil and gas drilling targets is very different and much deeper (Fig. 6.1). The greater depth difference means that the high pressures and temperatures anticipated in deep conventional wells are not present in NGH deposits and this factor pertains worldwide. This difference in depth to drilling target is crucial to the innovative methods that might be applied to both exploration and production activities for NGH.

The complex drill strings and decreasing casing size with depth necessary for deep conventional wells will not be necessary for NGH wells. Most of the NGH deposits will be in the 200–500 m depth range below the seafloor (Fig. 6.2), although in deeper water or in regions having very low geothermal gradients, NGH deposits could be found deeper. It is unlikely that most NGH deposits will be found at sub-seafloor depths of as much as 1–1.2 km. However, it may be necessary to drill deeper wells, especially in cases where it is desired to intersect an NGH deposit laterally and from below. The pressure differential between NGH deposits and the seafloor and between well pressure and the ambient formation pressures within the GHSZ is much less than for conventional targets. Hence, NGH wellbore lining or casing does not have to be as strong or as resistant to high temperatures and corrosive materials as it has to be in conventional wells.

Table 6.3 Examples of deep oil and gas wells

Resource	Max depth	References
NGH	<1200 m beneath seafloor	Max et al. (2006)
Conventional East Coast India ONGC	Water depth 3174 m	ONGC (2013)
Conventional Brazil drilling record	Water depth 3078 m well to 2988 m below seafloor	MP (2015)

Note that the Challenger Deep in the Mariana Trench is the deepest ocean seafloor on Earth at 36,201 ft. (11,304 m)

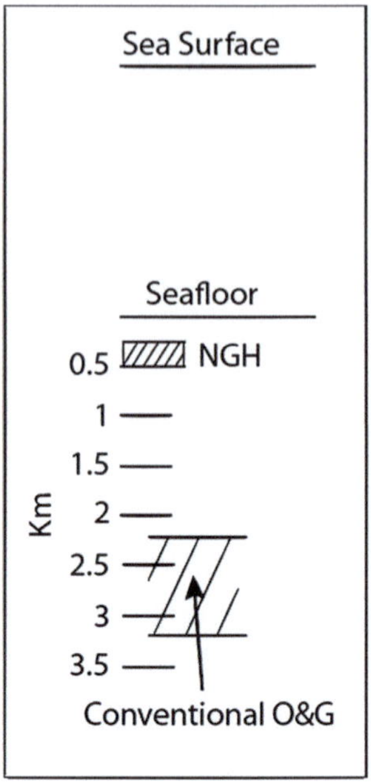

Fig. 6.1 Drilling depths of NGH and conventional deepwater oil and gas deposits, and also some non-deepwater conventional deposits. A seafloor depth of 2500 m is assumed

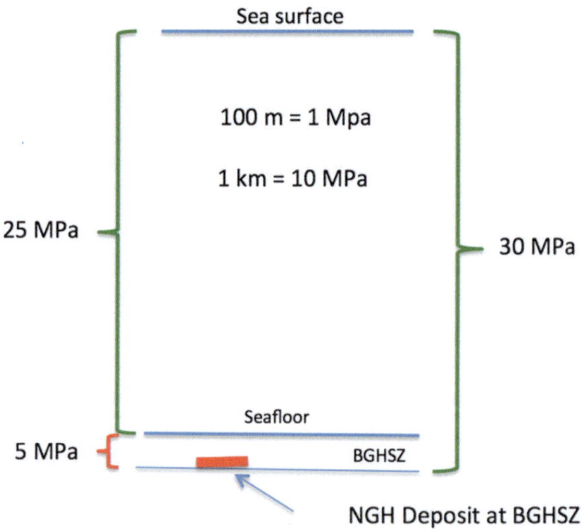

Fig. 6.2 Diagrammatic example of pressure depth relations for a NGH deposit in ~2500 m water depth with a 500 m thick GHSZ

The considerable difference in the depth below seafloor and the stability of the NGH reservoir, which has locked up the natural gas in solid form, offers many opportunities for innovation in drilling methods. For instance, drilling will largely be in mechanically weak semi-consolidated sediments to the level of the NGH deposit and below. Where lithification has taken place, for instance in carbonate sediments or some earlier diagenetic crystallization has taken place, greater mechanical strengths can be anticipated. It is likely these stronger zones will be rare. These do not require the same heavy-duty drilling capability used for conventional deepwater oil and gas drilling. This opens up possibilities for different approaches, which may springboard off the seafloor drilling capability that is presently being developed for autonomous site investigation and shallow exploration.

6.4 Production Opportunities

Conventional deep-water hydrocarbons are commonly a mixture of oil and gas, with a high probability of water production. In contrast, NGH is almost always a pure methane-water compound dominated by methane. NGH in sediments along passive continental margins is not normally associated with oil or gas condensates that are derived from thermogenic sources. In active margins with active petroleum systems, there is often a trace of oil associated with the NGH. NGH usually has little N_2, SO_x, CO_2, and other contaminants that are often found in conventional deposits, and almost pure water is produced when hydrate dissociates.

Until NGH is artificially converted to its constituent gas and water as the first step of production, there will be little or no free gas that could result in uncontrolled venting. Dissociation is a surface-effect phenomenon—that is, the dissociation takes place only where diffusion drives NGH-forming gas from the NGH crystalline structure and allows the water cage structure to collapse. Thus, the surface area of the NGH deposit is crucial to reaching high rates of dissociation. Ideally, substantial permeability will exist and dissociation rates will be high. If the surface area of NGH exposed to pore water is too small, it may be necessary to create higher exposed surface area.

A relatively small drop in pressure or rise of temperature near the phase boundary at the base of the GHSZ will cause the NGH to become unstable and it will dissociate into its component gas and water (Chap. 10). Hence, we consider that the most practical method to initiate dissociation is to induce a drop in pressure such that the NGH becomes unstable. A relatively small drop in pressure or rising of temperature near the phase boundary at the base of the GHSZ (i.e., the phase boundary) will cause the NGH to become unstable and it will dissociate into its component gas and water (Chap. 10). Thus, deposits at or near the base of the GHSZ will involve the lowest energy to convert the NGH.

The natural dynamic of conventional hydrocarbon deposits, however, presents a continuing problem, particularly during exploration and early production. Because of their high pressures, conventional gas and oil deposits have the potential to uncontrollably vent if a breach occurs in an exploratory or production drilling system, pipeline, or other infrastructure. And because of their high temperatures, they can be difficult and dangerous to handle. If a conventional deepwater hydrocarbon would be exposed to the overlying seafloor water, it would spontaneously blow out. Careful enforcement of regulations, development of best practices, expensive deepwater equipment, and simply the high costs of dealing with blowouts in deepwater conventional deposits may reduce blowout risk but the danger remains.

The exception to the inherent stability of a NGH deposit is in those circumstances where free gas occurs below the NGH in a sand layer. Although generally small amounts of gas occur within the free gas zone (FGZ of Hornbach et al. 2004), the thickness of the FGZ and the pressure of the gas in it can approach critical levels at which a blowout through the GHSZ may occur (with or without triggering, for instance from slope failure) as has been discussed by Dillon et al. (2001). As part of a production scenario, it may be advisable to first extract the free gas in the FGZ before initiating conversion of NGH.

Free gas deposits that often underlie or may occur within a GHSZ constitute the only persistent threat of overpressuring that may occur. Drilling should be refrigerated so that NGH stability temperature for any particular depth in the drillhole are not exceeded. Free gas zones may occur within a permafrost GHSZ where the sediments were dry before being frozen. Gas migration and could find residence in the porosity of dry, highly porous strata. This is in contrast to a marine environment where all sediments could be expected to be water saturated. Free gas zones will be rarer, but strong gas injection could expel water and then reside in the porosity.

6.4.1 Temperature and Pressure: Production Hazard Potential

Temperature, pressure, chemical constituents strongly influence the corrosion of e metal parts used in wells. The high partial pressures in conventional production wells and apparatus accelerate chemical reactions and dramatically increase the requirements for special metals, composites, and polymers to resist corrosion. Because of the strength of materials necessary to deal with conventional hydrocarbons, metals will almost certainly have to be used. In addition, high temperatures and pressures increase requirements for everything in a well that will come into contact with the hydrocarbons. Well casing cementing, which provides for well stability and barriers to fluid or gas escape from the well are vital because of the driving force of reservoir pressure.

As deepwater discoveries are made and production is implemented, both higher pressure and temperature are increasingly an issue that resorting to heavier materials alone may not completely solve. For instance, Griffin and Thethi (2016) regard ultra-deepwater riser designs as reaching the limit of pipe manufacturing and weld fabrication, which is causing significant engineering revision to be undertaken. This is expensive because completion equipment has to be redesigned with new higher strength, low-corrosion metals and elastomers to counter the increased corrosion potential, as do coatings (Offshore 2015). A similar redesign process is under way for cementing because of the extreme conditions.

In general, with increasing depth both temperature and pressure tend to increase but geothermal gradients differ considerably so that there is no absolute relationship that applies everywhere. There can be a very large spread of temperatures at the same depths, in particular in the heavy to light oil range (Hosseini et al. 2011) in temperature and pressure as a three-tiered system, there is almost a complete transition amongst field and well examples from shallower to deep and from low to high temperature and pressure (Shadravan and Amani 2012 of Figs. 3 and 5). In addition, different companies, e.g., Halliburton, Baker Hughes, Schlumberger, etc. use different divisions and graphical representations of the three zones. In any case, the important take-away is that deepwater natural gas deposits are characteristically both very hot and very high pressure, which is relatively expensive to produce.

NGH is clearly very different from deepwater conventional resources as both its temperatures and pressures are much lower (Table 6.4), especially with reference to seafloor pressures and temperatures. Thus, the requirements for materials, and potentially also the manner of their deployment, can be considerably lower with respect to strength and resistance to chemical attack.

The pressure in the production systems of conventional and NGH wells reflect their formation pressures or pressure in the reservoir. Overpressures exist in conventional gas deposits, but the differences between conventional and NGH development will be even greater. Conventional gas deposits can be choked down to reduce pressure at the wellhead but the factor of beginning the extraction process at

Table 6.4 Representative pressures and temperatures from deepwater wells contrasted with NGH reservoir conditions

Depth	Pressure/depth at well base	Temp in reservoir	References
Natural gas hydrate NGH	Maximum 1.0 km (10 MPa), probable depth limit of <0.8 km (8 MPa) below seafloor	Maximum ~35 °C in reservoir, max ~20 °C in production	Max et al. (2006)
Conventional, GoM deepwater (tertiary play)	Total depth 20,000 psi (138 MPa)	400 °F (204 °C)	Offshore (1/1/10)
Conventional, GoM Wilcox trend	7000–30,000 psi (48–207 MPa) Water depths 4000–10,000 ft. (1220–3050 m) Total well depths 10,000–35,000 ft. (10,688 m)	130–300 °F (54–149 °C)	Chavez et al. (2014)
Conventional, world deepwater	Very high	>300 °F (149 °C)	SPE (2016)
Conventional, ultra-deepwater North Sea	35,000 psi (241 MPa)	Up to 450 °F (232 °C)	OST (2012)
HP/HT	69–103 MPa	150–175 °C	Shadravan and Amani (2012)
Extreme HP/HT	103–138 MPa	175–200 °C	
Ultra HP/HT	>138 MPa	>200 °C	

Generalized pressure–temperature tier system last three rows of table pressures much higher than at wellhead are unavoidable. NGH, in contrast, will have lower pressures and probably lower temperatures in the product gas flow than its formation where depressurization dissociation is used to convert NGH to its constituent gas and water.

6.4.2 Production Containment: Leak-Proof Production from NGH

In conventional production, the pressure differential between the well and the bounding rocks and sediments increases toward the seafloor and outward pressure on the well increases. Pressure differentials are well are highest at the seafloor. Any breach of the well, may lead to uncontrolled blowout. Thus, great care is taken by industry to prevent well leakage at any water depth. This involves costly apparatus and materials, all of which is implemented in a highly regulated environment. Mechanical failure will result in catastrophe.

In strong contrast, NGH production in which depressurization dissociation will be used to drive NGH conversion, which means lower pressure in the well than in the surrounding sediments when depressurization conversion of NGH is initiated. Pressure will be lower from the reservoir itself to the seafloor wellhead. Locating pumps in the well from the outset to control pressure within the production well system as a whole, offers opportunities for innovation. Even where thermal conversion is initiated, pressures will rise above reservoir formation pressure but never to the extent of that in conventional deepwater gas. From the seafloor the gas must be lifted by pressurization, which also provides opportunities for innovation, particularly with respect to flow assurance.

Location and relative depressurization will be carefully planned and implemented as one of the engineering opportunities that would provide the greatest opportunity to control depressurization. Depending on the scale of the dissociation envelope and the placement of well bores, it may be possible to initiate gas–water separation within the reservoir. Thus, whereas conventional wells invariably contain a strong potential to vent and blow out, in a NGH production well, if there is any breach in the well liner, leaks will be into the well (Fig. 6.3). Leaks must be prevented as it would result in more water within the well that would have to be removed, either by separation within the well or in processing equipment on the seafloor. Leakage is less likely in a NGH well because the differential pressure inside pipes or casing and in the bounding sediments, is much lower than in a conventional well. Therefore, potential for innovation exists in wellbore liners and other equipment that does not have to be so robust.

As part of the depressurization dissociation used for NGH conversion, pressure within a NGH well virtually rules out gas overpressure-driven blowout danger of inducing mass flow or large-scale sediment deformation. Thus, a NGH well will have a very high level of geosafety and low environmental risk. A systemized

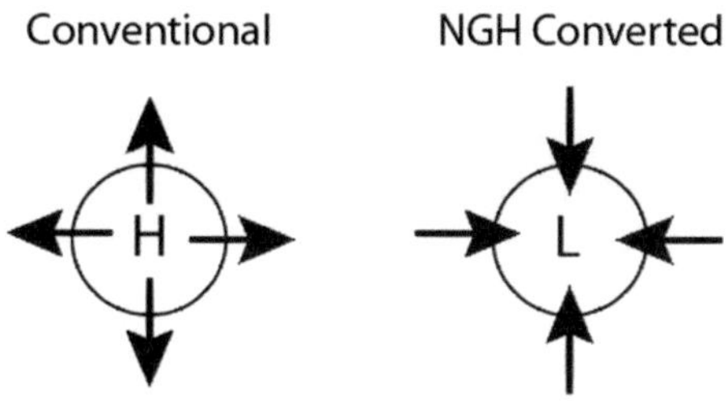

Fig. 6.3 Pressure differentials in conventional and NGH wellbores in which the gas has been converted by depressurization. Arrows show relative pressure differential and direction of pressure gradient and fluid movement toward lower pressure

control system would be utilized to provide the desired levels of gas extraction while managing water and gas flow, their separation and maintaining safety. Such a system also offers potential for innovation that could spill back into conventional hydrocarbon production.

6.5 Operations on the Seafloor

Industry usually chooses the least expensive safe drilling options available for any particular targets and water depth, but the use of larger, more capable, and more expensive equipment in the ultra-deepwater is unavoidable for conventional hydrocarbons unless substantial innovation for seafloor drilling is achieved. It is entirely likely that development of seafloor industrial apparatus (pumps, processing, media separation, etc.) for NGH, which is a much easier drilling target, will assist the development of seafloor drilling for conventional hydrocarbons. Moving all operations to the seafloor and making them autonomous or semi-autonomous seems to us to be a next step in offshore energy industrial plant.

Moving deep and ultra-deepwater production facility to the seafloor is already underway (Chaps. 7 and 8). Not only are water, sand, gas, and oil separation increasingly being engineered to operate on the seafloor for long periods of time but power systems are also being moved to the seafloor. Siemens, among other companies developing equipment to be operated on the seafloor, is developing and integrating the four main components of a seafloor power grid (SPG) required to move operations completely to the seafloor (Beckman 2014). Each SPG consists of a pressure-compensated transformer, medium-voltage switchgear, variable speed electric drives, and power control and communications systems. It is intended that these will be integrated with an array of new equipment for subsea production, processing and boosting using industry-standard underwater electrical connectors. Cooling will be natural with no moving parts and is designed to work at all depths without any special configuring.

Forster et al. (2015) note that not only processing technologies that have been hosted on floating platforms are now being brought to the seafloor, but industry is also beginning to standardize deepwater and seafloor technology. The most critical factors in interoperability are common power levels, connectors, computer control, and data management systems that can be integrated. Work is continually underway to harden seafloor facilities for greater durability in harsh environment. For instance, initiatives to improve coating and cathodic protection for thermally cycled risers and subsea pipelines are the subject of joint industrial activity amongst major companies (Offshore 2015-1). The generation of environment-specific designs that will provide increased longer-term reliability of coatings for long-term use without maintenance in the difficult of access, deep water. Reducing operational and commercial risk through longer design life with reduced maintenance costs for facilities in remote locations should result in significant reduction of the anode mass and electric erosion or deposition on subsea structures.

Drilling will follow a possibly higher level of shallow hazard studies that will allow avoidance of gas pockets and other problematic zones. Because of the low NGH reservoir pressure differential with the seafloor, it is simply unnecessary to have substantial drilling mud in a riser as there is no need to hold high pressure conventional gas in its reservoir. Light pumps or other methods can maintain pressure. For instance, advanced riser development work underwritten by the U.S. Department of Energy (Head and Long 2016) identify as technology gaps risers that have to sustain pressures up to 20,000 psi, temperatures up to 350 °C, and low assurance inhibitor performance in these conditions. This calls for very robust and potential expensive technology that is completely unnecessary to handle natural gas produced from NGH. Deepwater risers for NGH gas using composite materials should be significantly less expensive.

It may well be that a mixed NGH-free gas zone will occupy a relatively wide metastable zone at the base of the GHSZ, which is up to 80 m thick at Blake Ridge and a sub-GHSZ free gas zone up to 300 m thick will be found in association with potentially commercial NGH concentrations. In this case, artificially induced well pressure will have to counter positive gas pressure. The pressure of gas in closed-system permafrost NGH traps is not much higher than hydrostatic pressure. There is little data relating to gas pressure at and below GHSZs in hydraulically open oceanic NGH systems as drilling has avoided them. It would appear that the trapped gas pressure is not very high as high pressure would cause local pull-down in the base of the GHSZ, which has not been observed yet.

Drilling programs during the past decade have begun to confirm the magnitude of the recoverable NGH resource potential. These have been conducted in Arctic locations in Canada, Alaska, and Norway, and in offshore India, South Korea, China and especially Japan in the Nankai Field SE of Tokyo, and in the northern Gulf of Mexico of United States. These programs have proven the effectiveness of remote and subsurface exploration tools, especially seismic analysis, and have validated exploration models. While technical challenges remain, the field programs have provided insights that will allow these challenges to be addressed.

Carrying out drilling operations directly from the sea floor means that in weather and climate-restricted areas such as the Arctic Ocean, weather-independent operations can be carried out. Resupply can be made by submersible supply vessels, building on available Navy technology and experience. A number of scenarios are currently being developed to deal with the issue of stranded (conventional) gas and all will apply directly to natural gas produced from remote NGH deposits. Much of the technology that will support exploration and production activities in the NGH deepwater prospect region (Chaps. 1 and 2) developed for conventional hydrocarbons now exists in a more robust form than necessary for NGH. Further technology developments for NGH will be highly NGH-specific. Logistical and support equipment, including relatively small and less expensive ships than are now used are required to support cost-cutting equipment.

Subsea industrial sites on the seabed make it possible to utilize remote-controlled transport of hydrocarbons at any offshore facility. This would completely remove the requirement for huge, costly topside facilities and is the lower-cost alternative to the exploitation of the challenging deep-water oil and gas reserves worldwide. Advanced subsea drilling and processing technology could bring about a cataclysmic change and introduce a disruptive process that usually accompanies major technology shifts. The oil and gas sector would be significantly transformed. Opportunities for innovation range across all processes.

Subsea Processing is liable to be slowly accepted by industry but is a huge growth market (Crager 2014). The major elements of seafloor industrial sites that have to be implemented are:

- Purpose built vessels and engineered solutions including horizontal/vertical trees
- Potential new ways to set seafloor casing or its equivalent
- Low specification blowout preventers that can isolate wells
- Boosting, including pumping and suction (BHAI, GE, One Subsea, FMC, Ake, etc.)
- Product water disposal after separation and processing (GE, One Subsea, FMC, Cameron, Twister, etc.)
- Gas compression
- Carry out all processing on seafloor so that only gas is transported.

6.6 Environmental Security

This is one of the most misunderstood NGH issues. NGH has commonly been characterized as a hazard to global warming by overexcited environmental enthusiasts because of the fear that large amounts of NGH will suddenly begin to dissociate and flood the atmosphere with methane, which is a powerful greenhouse gas. Most of the methane leaking from continental shelves may never have been sequestered temporarily in the form of NGH. We are not aware of any substantive evidence that climate-driven dissociation of NGH is a root cause of the bubble plumes such as those observed in the natural gas rich province underlying the East Siberian Sea as well as the Barents Sea. Both are shallow water areas situated in the environmentally sensitive region of the Arctic. Further, the solid evidence for deepwater gas venting is that, for instance in the case of the Macondo blowout in the northern Gulf of Mexico, that none of it reach the atmosphere as it was dissolved in the ocean (Chap. 10). We agree with Ruppel and Kessler (2017) that the risk of enhancement of the global greenhouse with methane from NGH as extremely low for the foreseeable future, compared with, for instance, methane from rotting vegetation (biogas).

NGH appears to have about the same purity worldwide, with few exceptions. In addition to the basic hydrate-forming component, any chemical or dissolved ionic material migrating with the dissolved gas that is not a hydrate-former is rejected from the crystallizing NGH into the pore water where it will dissolve and be carried away. For instance, salt is dissolved in the pore water. Low salinity zones were one of the primary indicators that dispersed NGH had dissociated in cores recovered in the Blake Ridge (Paull et al. 2000). Formation of NGH will increase the salinity of surrounding pore water. If the rejected salt is not dispersed by diffusion or pore water percolation, NGH conversion sill simply restore the original salinity of the pore water. However, drill tests have shown that NGH dissociation dilutes the salinity of pore water to less than that of the original marine pore water (Ussler and Paull 2001). This strongly suggests equilibration of the rejected salt in the pore water and suggests a dynamic mixing environment during NGH formation. Without mixing, there would be no low salinity zones marking the presence of dissociated NGH. Oil is also rejected by growing NGH as are other chemicals (personal laboratory observation). The purity of NGH has implications for risk management and environmental impact, particularly in more environmentally fragile regions such as the Arctic Ocean, whose sediments may contain very large amounts of NGH (Max et al. 2013).

NGH has very low environmental risk for drilling and production. It is probably the lowest risk of all gas resources (Chap. 10). Conventional hydrocarbons are at high pressure and when connected to the surface by a drill stem/riser or to a lesser extent to the seafloor have the potential to blow out or leak uncontrollably. This natural process is referred to as primary recovery. When primary recovery processes diminish and the flow of hydrocarbons from the reservoir falls below a certain point, artificial stimulation (EOR) is applied to increase hydrocarbon flow and recovery. This stimulation includes (but is not limited to) acid application to increase porosity and permeability, hydraulic fracturing (fracking) that increases permeability, and gas injection to increase pressure and lower viscosity of liquid hydrocarbons.

These techniques are referred to as 'secondary recovery methods'. NGH, like a number of other 'unconventional' gas resources, must first be stimulated through one of the secondary recovery techniques. In the case of coalbed methane, pressure is lowered by pumping out pore water so that the gas can exsolve and form free gas that will flow. In the case of shale gas (and oil), fracking increases permeability and proppants keep fractures open. NGH also has a unique secondary recovery technique. It must be converted from its stable, solid form to its constituent gas and water (Max and Johnson 2011b).

Conventional deepwater hydrocarbon deposits have a maximum volume of naturally dynamic material during exploration and the early phases of production, which is when the risk of uncontrolled leakage is greatest. In contrast, the gas in NGH is naturally contained as a very stable solid form in its reservoir. Gas must first be produced by conversion before it can be recovered or released. This renders NGH inherently resistant to uncontrolled venting during gas extraction operations. Where a gas phase is generated in the reservoir through hydrate conversion, the volume of gas in the reservoir can be kept small and its pressure controlled by matching conversion to extraction.

Without large volumes of overpressured gas, significant blowouts cannot take place, even if there is a serious engineering failure. Hydrate conversion to gas is controllable and must be controlled for production. If a breach should take place, stimulation can be stopped rapidly and the naturally buffered hydrate conversion will cease, even if no remedial measures are taken. The converted gas in the reservoir and extraction system becomes produced gas only when it is safely transported to the surface (Max and Johnson 2011b).

Thus, there is a marked contrast between conventional hydrocarbon and NGH deposits, the former having measurably higher exploration and production risks. This difference could be extremely important for the nature of regulations governing NGH recovery. The environmental risk factors are much lower than for conventional hydrocarbons. Early production of gas from NGH poses an infinitesimally small risk to the producer and the environment. Once the gas has been produced, however, it is subject to the same type of risk as other produced gas although its pressure and temperature will be less and it will have fewer pollutants. We propose that accelerated development of the NGH resource can take place in the near term, even in environmentally fragile regions. As a practical matter, NGH production may be considered to be essentially fail-safe.

6.7 Lightweight Exploration and Production

The conventional exploration and production capability currently being used can be referred to as a 'heavyweight' character. Huge, expensive drilling ships and semi-submersible platforms with massive drilling rigs have been developed for deepwater energy exploration and production activities. They have been developed to reach commercial targets many miles beneath the sea floor through strata having a wide range of physical properties that require very complex drilling systems. The penetrated rocks often exhibit variable pressure gradients and high temperatures.

Deep-water drill ships and semi-submersible platforms that are capable of operating over deep continental margins, where the GHSZ and its potential subjacent free gas deposits are located, are highly complex and costly. A semi-submersible deepwater drilling rig costs up to half a million dollars a day to lease and it can take three months or more to drill a complicated deepwater conventional well. Amortizing all expenses can result in overall costs for conventional deepwater drilling of upwards of a million dollars a day. These costs are high due to the drilling capabilities that are necessary because of the two main ocean drilling requirements for safe conventional exploration and development capability: (1) to drill in a wide variety of geological materials, and, (2) to drill to a considerable depth below the seafloor. Currently available deep-water vertical drill strings, composed of steel pipe, can be up to 8–10 km in length. In contrast, drilling for NGH in the GHSZ will involve drilling into relatively mechanically weak sediments with a narrow composition range in no more than 1.2 km vertical depth into the seafloor, although inclined wells may go much farther.

Conventional deepwater targets are often beneath complex salt layers or salt welds. The pressure changes can be dramatic and difficult to control. In some cases, the pressure gradient may actually decrease with depth over certain intervals so that additional casing has to be set in order to insure that fracture gradients are not exceeded. Salt diapirism can strongly affect depositional environments, for instance in the Gulf of Mexico, where diapirs can create seafloor morphology. In addition, minerals in addition to halite (common salt) can be dissolved in fluids that percolate higher in the sedimentary pile. Chemical as well as mechanical complexities can be developed very locally, which makes prediction of sediment characteristics difficult.

Additional evaporite issues are especially important for the subsalt drilling targets offshore Brazil where the evaporite minerals include carnallite, tachyhydrite, and bischofite. Drilling into pressured strata presents significant blowout risks. We refer to these deepwater conventional drilling conditions, equipment, and cost structure as the conventional 'heavy-duty paradigm'. The NGH resource is amenable to application of innovative, 'lightweight', methodology and technology that is possible and that can be more efficient and much less costly than the existing technology of the heavy-duty paradigm. This is because the new lightweight requirements are matched to the nature of the material in which drilling takes place and the physical conditions encountered in production activities. NGH-specific lightweight technologies are not simply engineering optimization of conventional technology. They involve new approaches that generally cannot be retrofitted directly into conventional hydrocarbon exploration and development.

In the first instance, lighter duty vessels from which drilling operations can be carried out or controlled may be used, in contrast to the heavy drilling capability currently in practice. To some extent this movement to lighter and less expensive vessels has been underway for some time, along with the application of coiled tubing for at least well maintenance, if not the primary drilling capability. However, lighter drilling tools and support apparatus offer considerable scope for innovation. Lightweight drilling capability is particularly appropriate for the shallow NGH drilling targets.

The essential differences between present deep-water energy exploration and production activity and the much less rigorous requirements necessary for NGH does not yet appear to be factored into the methodology and technology of deepwater energy-related activities. This has resulted in the cost of recovering gas from NGH being considered from the viewpoint of the application of existing expensive equipment and their costs.

Little or no drilling below 1.2 km below the seafloor will result in relatively low-pressure differentials between the seafloor and the NGH pay zone (Fig. 6.1) and much less powerful drilling apparatus can be used. Smaller drill ships and lighter-weight drilling capability and hook loads can be used. The generally soft semi-consolidated sediment containing the NGH drilling targets is essentially similar to semi-consolidated sediments worldwide. This allows less expensive and less costly approaches to be used. Smaller and more lightweight seafloor completion apparatus including much smaller blowout preventers and other links to riser pipe can be used. In addition, because the reservoir and production pressures of the gas will be relatively low by industry standards, less robust gas transmission processing facilities and pipelines will be required. This will reduce the overall cost of exploiting the NGH resource while increasing the commerciality of the resource. We refer to this mix of equipment and the practices that arise from using it for NGH exploration and production as the NGH 'lightweight paradigm'.

The background to developing and implementing lightweight exploration, development, and production technology for the commercial production of natural gas from NGH is a matter of continuing speculation and modeling. NGH exploration has been largely carried out with ships and survey equipment designed and fabricated for conventional hydrocarbon exploration, although recently, and especially in the Arctic, smaller vessels of opportunity have been used for surveying and drilling. Advances in reducing the size, weight and cost of equipment, as well as enhanced computational power for the acquisition, processing, modeling, and simulation of seismic and other data, that are suitable for NGH exploration and possibly production, are currently in the early stages of development. But the opportunities presented by a lightweight approach have not been previously described as part of a related series of technology and methodology opportunities.

An aspect of the trend in deepwater hydrocarbon exploration, development and production is the fact that specialist contractors are increasingly narrowing the range of their products. This is illustrated by the way in which the major contractors continue to purchase smaller contractors who have developed specialist technology, especially if it constitutes Intellectual Property. For instance, Schlumberger, one of the leading oilfield services companies, recently purchased Smith International, an American drill-bit company for $11 billion in order to acquire its technology.

Many people are not intimately associated with the recent changes in the energy industry in which individual small companies may control critical technology. For instance, there are hundreds of types of drill bits, each for different rocks and sediments that are being drilled through, and the purchase of Smith International may have been for only a restricted range of technology, but one that was very important to Schlumberger, due to increasing demand of a particular technology. That same level of complexity flows through all aspects of the drilling and completion process: drill collars, drilling fluids, casing and liners, cements, packers, perforating technology, completion fluids (and kill muds), flow assurance, manifolds, trees, valves, and more. In each of these technical fields, solutions have not yet been developed that are uniquely focused on NGH.

By implementing a NGH-specific set of technologies and methodologies, to which the NGH-specific costs are added, a much lower overall cost should be achieved, and this will have a strong impact upon the perception of commercialization of NGH resources. It may be possible to keep capital expenditure and operating costs well below those currently envisaged by the current conventional model of pricing of NGH commerciality. By implementing new methodologies and technologies for maintaining high production rates, safety, and reliability, NGH should prove to have a cost profile that will render it highly competitive, even after taking into account conversion costs. Low internal CAPEX costs will provide a lower base for further internal CAPEX reduction not available to conventional hydrocarbon exploration and production.

Offshore service companies have carried out substantial improvements in their conventional drilling technique over the past fifteen years as major energy companies have tended to outsource their one-time dominance if the field. This period of development roughly corresponds with developing technology and practices for deepwater exploration and production, although much of the new technology has also been applied in non-deepwater energy situations. Examples include multiple offset well bores from a single pad that are now commonly being used for shale gas and oil recovery, and accurately inclined and horizontal well boring. For brevity, we use the term 'deepwater technology' for all drilling, support and production technology and activities that may be useful for oceanic NGH development, regardless of their mode of development or history of application.

Lighter duty vessels will not be required to be on station full time, although operations are envisaged that will be all weather and full time. All drilling, as well as processing, which is largely already on the seafloor, will be moved to the seafloor. An autonomous or robotic approach will be used, including innovative approaches for both rotary, CTD, and autonomous drilling including the use of a new class of semi-autonomous bottom hole assemblies and semi-autonomous service vehicles/apparatus. Shallow, light drilling platforms only will be required owing to the geomechanical character of the sediment to be encountered. Finally, lower materials requirements will also lower cost curves.

Because NGH is a relatively shallow drilling target below the seafloor and the strata to be drilled are liable to be similar worldwide, a lighter drilling platform than is currently used for deeper conventional hydrocarbon deposits is possible. We suggest that new NGH-specific drilling practices be employed to reduce costs by using existing technology, consisting of a light drilling capability and using a combination of a mud boat or light drilling ship and coiled tube or some other lighter drilling method. Lighter drilling capability may be used because the NGH is a solid, crystalline material that is stable in its ambient environment and is very unlikely to cause blowouts when appropriate drilling practices are followed.

The cost of NGH exploration and recovery could be less than conventional hydrocarbons because only relatively shallow geophysical methods are required, shallow drilling targets and conditions are likely to be similar worldwide—and no more than 1 km beneath the seafloor. In addition, NGH gas production-system temperatures will be relatively low, almost certainly below 40 °C (Table 6.3).

This is in contrast to conventional hydrocarbons that may have elevated temperatures requiring substantial and costly handling that induce higher levels of risk. The temperature and pressure differentials will be lower in NGH recovery than found in deep conventional hydrocarbons, so flow assurance will probably be easier to manage and less costly.

6.8 Summary of NGH Opportunity Issues and Conclusions

NGH has the lowest environmental risk of any natural gas deposit.

- Occurs at shallow depths into the sediment; relatively easy to drill to
- Not being highly pressurized, there is no potential for blowout, as in conventional gas deposits
- The pressure in the reservoir during conversion for gas recovery has to be controlled for NGH conversion
- In case of a malfunction during production, gas conversion can cease, and existing gas and water in the reservoir will recrystallize to NGH and again become stable
- NGH concentrations potentially suitable for production are almost 100% pure water and natural gas
- There is rarely associated oil, particularly on passive continental margins.

After production, the natural gas is a better product than conventional gas

- NGH produces only fresh water, which will dilute pore water
- Generally, no dangerous chemicals; not corrosive
- Pore water becomes less saline due to dilution by NGH conversion water
- Temperatures of the produced water and gas will not rise above 40 °C and will usually be about 20 °C
- Materials can have less rigorous requirements and lower cost than required for potentially very high pressure and temperature conventional natural gas reservoirs that may have considerable saline water and dangerous corrosive chemical.

Outlook

- Seafloor drilling for NGH is less expensive and operationally easier because of the low-pressure differential between the deposit formation pressure and the seafloor.
- NGH deposits have relatively benign pressures, temperatures, and chemical characteristics, so methods that are considerably less expensive can be developed and employed than those which are used for conventional gas deposits.
- The existing industrial base already has developed much of the technology required for NGH exploration and production, although miniaturization and new interoperability will be required.

- New technology and methodology could lower both CAPEX and operational costs.
- It is envisaged that application of hazardous chemicals and fracking will be unnecessary for NGH production.
- Lower operational costs should be obtainable through new technology development.
- Apparatus developed for conventional hydrocarbons will operate with a higher safety factor in NGH deposits.
- NGH-specific technology development may have some application in non-NGH deepwater spinoffs.

References

Beckman, J. (2014). Subsea power grid accelerating transfer of topsides functions to the seafloor. *Offshore, 74*(10), 84–85.

Boswell, R., Collett, T. S., Frye, M., Shedd, W., McConnell, D. R., & Shelander, D. (2012a). Subsurface gas hydrates in the northern Gulf of Mexico. *Marine and Petroleum Geology, 34,* 4–30. https://doi.org/10.1016/j.marpetgeo.2011.10.003.

Boswell, R., Collett, T. S., Frye, M., Shedd, W., McConnell, D. R., & Shelander, D. (2012b). Architecture of gas-hydrate-bearing sands from Walker Ridge 313, Green Canyon 955, and Alaminos Canyon 21: Northern deepwater Gulf of Mexico. *Marine and Petroleum Geology, 34,* 134–149.

Chavez, M., Garcia, G., Pogoson, O., Li, L. Y., Cardona, A., & Nelson, R. (2014). Study aims to enhance ultra-deepwater well design. *Offshore* 1/22/114. http://www.offshore-mag.com/articles/print/volume-74/issue-1/drilling-and-completion/study-aims-to-enhance-ultra-deepwater-well-design.html. Accessed February 21, 2016.

Crager, B. (2014, December 12). Subsea production systems an enabling component of offshore production. *Rice Global E&C Forum (Endeavor Management). Figure 2, taken from Offshore Magazine*, 53 pp. http://www.forum.rice.edu/wp-content/uploads/2011/06/RT-141212-Crager.pdf. Accessed February 20, 2016.

Dillon, W. P., Nealon, J. W., Taylor, M. H., Lee, M. W., Drury, R. M., & Anton, C. H. (2001). Seafloor collapse and methane venting associated with gas hydrate on the Blake Ridge—Causes and implications to seafloor stability and methane release. In C. K. Paull & W. P. Dillon (Eds.), *Natural gas hydrates occurrence, distribution, and detection*. Geophysical Monograph 124 (pp. 211–233). American Geophysical Union.

Forster, L., Paes, T. M., & Baker, R. J. (2015). Industry moves subsea processing toward standardization, consistency. *Offshore, 75*(3), 64.

Frye, M., Shedd, W., & Boswell, R. (2012). Gas hydrate resource potential in the Terrebonne Basin, Northern Gulf of Mexico. *Marine and Petroleum Geology, 34,* 150–168.

Griffin, C., & Thethi, R. (2016). HP/HT wells forcing rethink on riser designs, materials. *Offshore*. http://www.offshore-mag.com/articles/print/volume-76/issue-5/subsea/hp-ht-wells-forcing-rethink-on-riser-designs-materials.html?cmpid=enl_Offshore_OffshoreDailyNewsletter_2016-06-01&email_address=%25%25EmailAddress%25%25. Accessed June 1, 2016.

Head, B., & Long, R. (2016). *Offshore Magazine*. 3/10/16. http://www.offshore-mag.com/articles/print/volume-76/issue-3/engineering-construction-installation/research-project-seeks-to-deliver-ultra-deepwater-riser-concepts.html?cmpid=EnlOSDailyMarch302016&eid=288179069&bid=1353203. Accessed March 30, 2016.

Hermanrud, C., Halkjelsvik, M. E., Kristiansen, K., Bernal, A., & Strömbäck, A. C. (2014). Petroleum column-height controls in the western Hammerfest Basin, Barents Sea. *Petroleum Geoscience, 20,* 227–240.

Hornbach, M. J., Saffer, D. M., & Holbrook, W. S. (2004). Critically pressured free-gas reservoirs below gas-hydrate provinces. *Nature, 427,* 142–144.

Hosseini, B. K., Ardali, M., Chalaturnyk, R. J., & Mamora, D. D. (2011, November 15–17). *A new analytical approach to investigate heated area in thermal recovery techniques*. Paper SPE-148836-MS presented at Canadian Unconventional Resources Conference, Alberta, Canada.

Johnson, A. H. (2011). Global resource potential of gas hydrate—A new calculation. *Fire in the Ice. NETL, U.S. Department of Energy, 11*(2), 1–4.

Kvenvolden, K. A., Golan-Bac, M., McDonald, T. J., Pflaum, R. C., & Brooks, J. M. (1989). 15. Hydrocarbon gases in sediment of the vøring plateau, Norwegian Sea. In O. Eldholm, J. Thiede, & E. Taylor, et al. (Eds.), *Proceeding of the Ocean Drilling Program, Scientific Results 104*, 319-326.

Max, M. D., & Johnson, A. H. (2011a, July 17–21). Hydrate petroleum approach to natural gas hydrate exploration. In *Proceedings of the 7th International Conference on Gas Hydrates (ICGH 2011)*, CD, Paper 637. Edinburgh, Scotland, United Kingdom, 12 pp.

Max, M. D., & Johnson, A. H. (2011b). Methane hydrate/clathrate conversion. In M. R. Khan (Ed.), *Clean hydrocarbon fuel conversion technology*. Woodhead Publishing Series in Energy No. 19 (pp. 413–434). Cambridge, UK: Woodhead Publishing Ltd. ISBN 1 84569 727 8, ISBN-13: 978 1 84569 727 3.

Max, M. D., & Johnson, A. H. (2014). Hydrate petroleum system approach to hydrate natural gas exploration. *Petroleum Geoscience, 20*(2), 187–199. Geological Society of London (Published in Online First March 21, 2014). https://doi.org/10.1144/petgeo2012-049.

Max, M. D., Johnson, A. H., & Dillon, W. P. (2006). *Economic geology of natural gas hydrate* (p. 341). Berlin, Dordrecht: Springer.

Max, M. D., Johnson, A. H., & Dillon, W. P. (2013). *Natural gas hydrate arctic ocean deepwater resource potential*. Springer Briefs in Energy, 113 pp.

MP. (2015). http://en.mercopress.com/2015/05/05/petrobras-sets-record-for-exploratory-drilling-in-water-depth-of-3.000-meters. Accessed February 21, 2016.

Noguchi, S., Furukawa, T., Aung, T. T., & Oikawa, N. (2011, July 17–21). Reservoir architecture of methane hydrate bearing turbidite channels in the eastern Nankai Trough, Japan. In *Proceedings of the 7th International Conference on Gas Hydrates (ICGH 2011)*, Edinburgh, Scotland, United Kingdom, 9 pp.

Offshore 1/1/10. (2010). *Offshore Magazine*. http://www.offshore-mag.com/articles/print/volume-70/issue-1/gulf-of_mexico/lower-tertiary-play.html. Accessed February 21, 2016.

Offshore 2015-1. (2015). *Offshore Magazine. Subsea hydraulic power unit widens field*. http://www.offshore-mag.com/articles/2015/04/subsea-hydraulic-power-unit-widens-field-development-options.html. Accessed February 27, 2016.

Offshore. (2015). *Offshore Magazine. Study examines options for protecting harsh environment subsea facilities*. 5/28/15. http://www.offshore-mag.com/articles/2015/05/study-examines-options-for-protecting-harsh-environment-subsea-facilities.html. Accessed April 23, 2016.

ONGC. (2013). *Transocean sets world record for deepwater drilling*. http://www.rigzone.com/news/oil_gas/a/127610/Transocean_Sets_World_Record_for_Deepwater_Drilling#sthash.gvjqrhQB.dpuf. Accessed April 23, 2016.

OST. (2012). http://www.offshore-technology.com/features/featurerisky-business-deepwater-drilling-north-sea/. Accessed February 21, 2016.

Paull, C. K., Matsumoto, R., Wallace, P. J., & Dillon, W. P. (Eds.). (2000). In *Proceedings of the Ocean Drilling Program, Scientific Results*, 164 (pp. 3–10).

Ruppel, C. D., & Kessler, J. D. (2017). The interaction of climate change and methane hydrates. *Reviews of Geophysics, 55*(1), 80 pp. Wiley Online Library.

Shadravan, A., & Amani, M. (2012). HPHT 101—What petroleum engineers and geoscientists should know about high pressure high temperature wells environment. *CSCanada Energy Science and Technology, 4*(2), 36–50. ISSN 1923-8460 [PRINT]. https://doi.org/10.3968/j.est.1923847920120402.635.

SPE, 2016. (2016). SPE petrowiki deepwater drilling. http://petrowiki.org/Deepwater_drilling.

Takahashi, H., & Tsuji, Y. (2005, May 2–5). Multi-well exploration program in 2004 for natural hydrate in the Nankai-Trough offshore Japan. In *OTC 17162, 2005 Offshore Technology Conference*, Houston, TX, USA, 10 pp.

Ussler, W., & Paull, C. K. (2001). Ion exclusion associated with marine gas hydrate deposits. In C. K. Paull & W. P. Dillon (Eds.), *Natural gas hydrates occurrence, distribution, and detection*. Geophysical Monograph 124 (pp. 41–65). American Geophysical Union.

Chapter 7
Leveraging Technology for NGH Development and Production

Abstract From the earliest days of industrial production of oil and gas, assumptions about limits to the ability to recover hydrocarbon energy consistently have been proven wrong through technology development. Migration of exploration and production technology to frontier regions, deeper water, and harsher conditions more remote from land is necessary for the oil and gas industry to create the largest value of potential new resources and diversity supply. With advances, production technology increasingly is being moved from surface platforms to the seafloor to decrease costs and improve operations. Seafloor installations are becoming progressively more robotic. Some conventional oil and gas drilling, particularly that not having to deal directly with very high reservoir pressures, is being moved to the seafloor. All of this investment in technology development is being borne by industry for conventional oil and gas resources, but most of it can be optimized for development of NGH resources. This means the amortized cost of NGH development can be relatively small from the outset. In addition to leveraging existing technology developments, the physical properties of NGH and its reservoir sediments will allow seafloor processing to be carried out substantially less expensively than convention gas, which means seafloor processing installations for NGH gas could be inherently less expensive. NGH seafloor installations will be assembled, operated, and serviced entirely underwater, with both site provision and gas production located remote from the seafloor installation. Virtually all exploration and production technology for NGH resources can be moved to, and operated from, the seafloor.

7.1 The Curve of Technology and Innovation

Apart from a general trend of moving equipment to the seafloor and widespread automation, we identify drilling as the critical opportunity area that offers the greatest potential for reducing NGH development costs. Due to of the more benign nature of NGH and associated production issues, considerable cost savings over conventional natural gas can also be made. It is important to note that there is scope

M. D. Max and A. H. Johnson, *Exploration and Production of Oceanic Natural Gas Hydrate*, https://doi.org/10.1007/978-3-030-00401-9_7

for inexpensive experimentation owing to the very low risk of blowout or pollution. The unique physical and chemical characteristics of the NGH resource will allow much less expensive and robust equipment to be used that is more closely matched to the lower NGH requirements (Chaps. 2 and 7).

NGH is the last natural gas resource on Earth to be developed. It can be developed using conventional technology in deepwater, but in our view that would be like trying to drive a small nail with a pile-driver. The particularities of NGH (Chaps. 2 and 3) offer many opportunities for applying modified and new technologies and practices that will have the effect of dramatically lowering the cost of NGH exploration and production. It could prove to be the least expensive and most environmentally low-risk of all the various gas resources. The U.S. Department of Energy (DOE 1998, 2015) regards NGH as a producible natural gas resource in which new or modified deepwater technology is necessary for development as a source of natural gas, and we agree. The governments of Japan, South Korea and India also have energy-related NGH programs that are facing the same set of new NGH issues. About 30 technology gaps have been identified for the Norwegian sector of the Barents Sea; these focus on drilling technologies including large well bores, optimizing reservoir performance, gas compression and pumping, subsea facilities and power supply, and a number of ice and hazardous weather issues (Offshore 11/25/15).

Current technology developments for conventional oil and gas that can be almost directly applied to producing from NGH are subsea electrical power and control systems, separation and water treatment systems, and subsea gas compression. In addition to their development, industry is also concerned with standardizing deep-water technology (Forster et al. 2015a; Offshore 3/9/16). The main improvements in exploration will probably be confined to designing dedicated seismic surveys specifically for the GHSZ depth range and in better geotechnical data processing (Chap. 2). These improvements will be largely leveraging of techniques that are underway. Similarly, production of NGH will generally leverage technology that is already under development, although it may be applied in a very different fashion. NGH separation will generally involve only relatively unpolluted water and natural gas and will not have costs associated with complex separation systems that not only have to be carried out afloat but also on nearby land (Offshore 5/1/08).

NGH exploration, especially drilling, has been carried out using heavy-duty conventional equipment and equipment rated for conditions that make it strongly over-specified for NGH applications. It is thus more expensive than necessary. There is very little information available about the extent to which companies and national NGH programs have recognized that new NGH-specific technology could dramatically lower the cost of exploration and production. The leading broad NGH development program is the German program SUGAR (GEO 2015), which was initiated as part of a biological—climate program. This was reorganized in 2014 to include a second phase that is divided into six sub-projects: (1) Detection and monitoring of NGH deposits, (2) Exploration and characterization of NGH deposits, (3) Simulation of NGH accumulations, (4) Simulation of NGH exploitation, (5) Optimization of NGH production technologies, and, (6) Drilling technologies for NGH deposits.

In this chapter we discuss both the leveraging of technology developments that are already being made as they might apply to the NGH exploration zone on deep continental shelves and continental slopes (Chaps. 2 and 3), and innovation or step-changes where breakthrough is possible in the way NGH exploration and production may be carried out that constitute truly new practices and hardware for NGH. To some extent this chapter is a ‘shopping list’ of technology that we consider can be used in one form or another for NGH exploration and production. We summarize these technologies and the degree to which they could be applied as an aid to readers not familiar with the field as a whole.

In theory, everyone wants to be able to implement new technology to overcome technological and operational barriers, or to accomplish tasks in less expensive, safer, and more efficient manners. New technology, from the ragged inception of the industrial revolution, has improved processes, and enabled huge wealth to be created. In practice, however, the technology innovations that appear to be widely demanded are often more difficult to bring to practice because of cost and risk. All development comes off the bottom line and achieving funding for developments that may or may not work are rarely justified in small companies. The offshore industry has a tendency not to assume new development risks. Also, the first use of any really new technology always faces higher risks. Companies like to be early adaptors, not innovators, because it lowers their risk. The energy industry has few venture capitalists within it (Chap. 10).

There is little doubt that it is only because of new technology developments that working in deepwater and ultra-deepwater has been enabled, not just by entrepreneurial companies but by groups of companies in joint industry projects (Offshore 9/15/15). We expect that deepwater technology will follow a development similar to shale in which fracking and inclined drilling were key. These technologies enabled shale oil and gas production at its huge commercial scale. Offshore, development is more difficult and costly, but deepwater is where large discoveries will be made. The constant hardening of underwater equipment, to make use in ever deeper waters and more hostile conditions, is often less a matter of new technology than it is improvements or modifications of older technology and practices rather than a step change in the way things are done (FMC 2015).

For operators of major companies, particular problems have to be solved, as for instance being able to work safely in increasingly deeper water. Initially, this drive into new regions often leads to more expensive solutions being used, simply because that particular technology is available. Adaptation and then innovation follows as a natural means of reducing costs, increasing safety, and reducing environmental risk. Hopefully, lessons learned from other new resource developments (Holditch 2013) will allow for less troubled development for NGH.

Leveraging current technology: Offshore, improvements to existing technology or borrowing technology developed for other purposes has been important in adapting to ever harsher conditions for hydrocarbon exploration as seafloor water depths have increased. There are also many examples of technology development achieved through taking advantage of existing technology designed originally for other fields. Technology often can be improved or used in ways that were originally not foreseen when it was developed, even though it may be used widely in the field for which it was originally developed. Innovation can also consist of bringing together certain equipment and practices that can literally add up to be worth more than the sum of its parts when applied to a new field, such as NGH exploration and production. However, simply applying existing technology to another purpose will almost never be patentable.

Possibly the best example of technology 'stretching' has been the combination of horizontal or 'extended-reach' drilling (ERD) and a particular form of hydraulically fracturing of oil and gas-rich shale beds so that the trapped hydrocarbons will flow and be produced. Fracking itself was not a new technology, but applied successfully to oil and gas shale reservoirs that had the initial permeability of construction cement, it was a dramatic development. It was an old technology for increasing porosity in existing reservoirs as part a secondary technique for enhanced production. A small company, Mitchell Energy, which could be regarded as the key developer, did not even cover the cost of fracking and controlled horizontal well drilling on shale tests until its 36th well (Begos 2012). Further, it is likely that the technologies and their application would not have advanced as rapidly as they did without U.S. Department of Energy funding, which overall exceeded 100 million dollars. 3D location of drill bits for lateral drilling may have been borrowed from Navy acoustic location techniques and GIS software has been adapted for use in military rapid area assessment and command and control situations (Max et al. 1998).

Further 'stretching' involves refracking of an already fracked well in which production has declined using Schlumberger's broadband sequence fracturing technique. Initial horizontal wellbore fracking, which usually costs from $6 to 8 million, is normally done by using perforating guns and fracturing plugs to isolate about a few hundred feet of wellbore in turn. Refracturing, in contrast, is usually done to the entire wellbore at once without mechanical isolation using substituting chemical plugs between each stage that dissolve or react with the water after fracking. Refracking costs in 2015 had dropped to between $750,000 and $1.5 million (WO 1/8/16). In addition, gas production in one example was increased from 100 Mcfd to about 2.7 MMcfd for the first 30 days of production, raising the well's overall production by an estimated 1.0 Bcf. It might be possible to consider using the emerging refracking techniques for NGH production preparation, especially where wellbores already had pre-perforated sections in place. A number of background technologies that offer promise of being leveraged and applied to NGH development are discussed in Sects. 7.3, 7.4, and 7.5. Long-reach inclined or 'horizontal' drilling techniques have also been adapted to subsea extreme conditions (ultra-deepwater) because market forces can only be met with new approaches (ABB 2017).

Technology innovation: Breakthrough innovation is at the other end of the technology development scale from incremental improvements to existing technology. This involves re-imagining within an application area and approaching it differently. Valuing technology innovation, the somewhat risky process of picking winners and losers, can be done in a number of ways. Because it is often difficult to

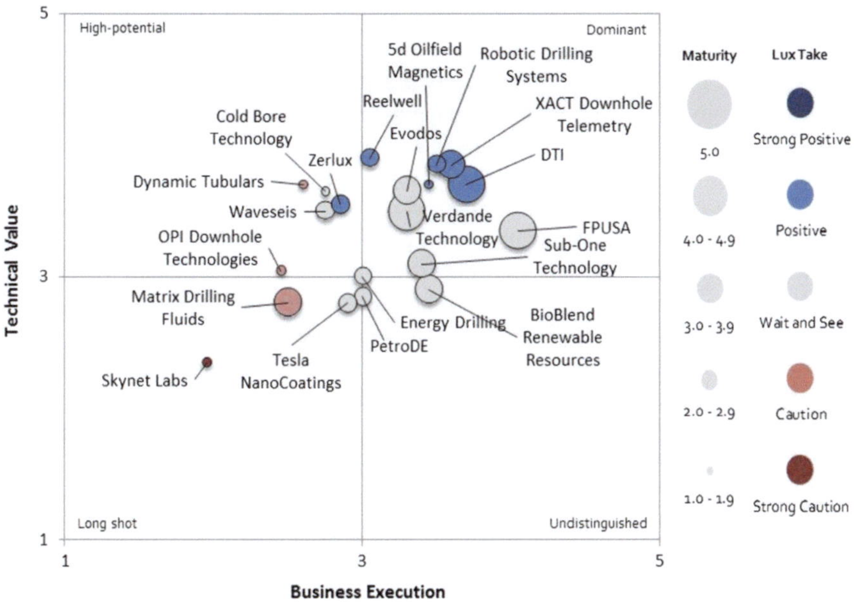

Fig. 7.1 Lux Research, Inc. Innovation grid for drilling technologies. By permission of Lux Research (Lux 2015), which has a full explanation of their methodology for identifying risks

understand how valuable a particular technology innovation will be, evaluation of existing companies can be made on a business execution basis (Fig. 7.1).

According to Lux Research, 'dominant' companies were first to the market with disruptive technologies and are poised for growth. 'High-potential' companies have attractive technologies but struggle with business aspects. 'Undistinguished' and 'long-shot' companies lack the best in-class technologies although they might make adequate partners, but technology development being made by companies that are not yet on the development-investment map may have the potential for breakthrough, with potentially disruptive technologies.

The drilling rig, drill string, fluid system, casing, cement, and completions are factors for all drilling operations, and each offers particular opportunities for improved cost performance. For instance, adaptive drill bits (BH 2017) adjust to downhole conditions autonomously.

Efficiencies can be made in three ways. (1) Improved or new technology can cut costs by replacing less well-performing technology, (2) Increasing automation decreases personnel requirements, and, (3) Cutting non-productive time (NPT), particularly during drilling, when NPT can average 10–15% of ultra-deep wells but can be up to 30% in harsh climate environments. Drilling from the surface inherently involves potential for weather disruption, particularly in regions having hurricane seasons or in polar regimes in which ice predictions prove incorrect.

Salizar (2015), in an elegant discussion, notes that the demand for energy must result in the development of new technologies to meet the requirements for industry to successfully explore and produce oil and natural gas, simply because new energy frontiers—deep and ultra-deepwater and the polar regions—require new capabilities. The pace of innovation in the energy industry has increased, in part because it has been possible to incorporate the enabling technologies in electronics, robotics, materials, and autonomous capabilities that have been driven by other requirements in other fields. Innovation in the offshore energy industry can leverage innovation in a broad range of fields. Innovation may sometimes consist of using existing technology differently, or in conjunction with other technology for which it was originally not foreseen to be advantageously compatible. Improvements in existing technology and development of new technology are a constant theme, as is innovation, which involves step-changes in the way things may be done.

The map of investment potential (Fig. 7.1) shows only a small number of Lux-selected companies bringing particular technologies. But we note that there is a clear preference for drilling and drilling related issues such as downhole technology and telemetry, robotization of seafloor drilling capability (Spagnoli and Freudenthal, 2013), and other remote stand-alone technologies. Robotization of seafloor activities is currently underway (Banks 2018) and is prime area in which new and innovative technologies for NGH development will take place.

7.2 Moving to the Seafloor: Subsea Industrial Sites

Exploration and commercial activity at sea has been an era of ship-based expeditions for over 500 years. This heroic age, however, is coming to a close. Technology has now brought us to the tipping point of a sweeping transformation in the way work in the oceans and especially the deep sea will be carried out. The historical focus on expeditionary science is shifting due to new technology that has the potential to produce orders of magnitude more and better data that can be more rapidly recorded and communicated, all while vastly increasing efficiency and reducing costs through having fewer research vessels available and fewer scientists at sea (Kintisch 2013). The seafloor will be the new base of commercial operations in deepwater and ultra-deepwater. In particular, substantial processing that commonly includes separation of materials, is increasingly taking place on the seafloor (Dabade et al. 2017) and has been recently summarized (Roth et al. 2018).

There are strong commercial reasons for moving to the seafloor to maximize deepwater development (DW 2015). Subsea processing technologies in particular, are coming of age (Perry 2013). Abili et al. (2014) note that expenditure on subsea processing systems is expected to exceed \$3.4 billion, with deepwater expenditures expected to increase by 130% to \$260 billion by 2018 in both new fields and older brownfields. According to Musarra (2015), subsea factories are business critical for deepwater development. Operations are also becoming more mobile as issues related to new classes of support vessels and stranded gas and oil are in the process of being resolved (Chaps. 7 and 11). Seafloor processing equipment and field layout options are possible with utilization of increasingly capable advanced technology (Forster and McKee 2016). Much of this advanced equipment for subsea processing, including separation, compression, pumping systems including boosting and gas compression, power systems, connectors, and seawater treatment systems, is already available from commercial suppliers (I-O 2016). Use for NGH would likely involve a lower-stress application than the more robust operating conditions of conventional hydrocarbon work for which it was designed.

Furlow (2003) suggested that subsea processing was slow to take off, despite what he saw as the many advantages, with technology development times reduced by leveraging on existing technology. As he noted, the technology was in an incipient stage and the economics appeared to be extremely favorable. We have probably now reached the tipping point for moving to the seafloor in that the pilot demonstrations in working situations, the reduction of the size of equipment, increases in the dependability of deep sea equipment, and the economic and operational advantages are now becoming clear enough to convince even the conservative skeptics in the offshore industries. Because of technology development, it is now possible to envisage seafloor industrial sites for oil and gas production (Offshore 1/25/13).

In addition to commercial benefit, the new marine technology also has the potential to open new fields of research and dramatically increase our understanding of our planet and its biosphere. There are important aspects of being able to increase our knowledge base in the ocean, especially the deep ocean. Much of the technology developed by the energy industry in its pursuit of deep ocean natural resources will be applied to better understand the ocean-atmosphere system control of climate, our food supply base, new biochemical discoveries, and other non-energy related issues.

AUV and robotic technology is a relatively recent development in marine research and working capabilities (Kezina 2015), but is already being commercialized, judging from Kongsberg Hydroid advertisements in Sea Technology December 2015 for “Intelligent marine robots you can rely on”. Moreover, the intrinsic difficulty of the multidisciplinary integration required to design and operate full ocean depth AUVs has resulted in relatively few research institutions or companies having succeeded in producing operational, full ocean depth AUVs. Moreover, practically none of them has displayed a real capability to host effective sampling tools for both water column and seafloor. Most of the commercial devices on the market are limited to seafloor survey and sonar mapping. New technology for establishing semi- and fully-autonomous commercial activities on the seafloor constitutes the next step in energy ocean commerce (Ponce 2018).

Over the course of a century and a half, oil and gas exploration and production activities have transitioned from land, across coastal regions, and into increasingly deeper water on continental shelves to finally reach the deeper waters of continental slopes and basin floors. Industry has continually demonstrated an ability to provide new technology and practices for the increasingly more remote and difficult conditions and operational situations. Existing technology has been upgraded, modified, strengthened and used in new ways. New technology with step improvements in capabilities has been added to the technology mix as necessary and required to allow cost-efficient commercial delivery of gas and oil to consumers driving the international economies.

More robust solutions are required by increased distances from shore and greater water and drilling depths. The first offshore developments usually followed geologically-related resource trends into deeper water where identification of new targets was based on new geological domains. For instance, in the northern Gulf of Mexico, offshore Nigeria and elsewhere, deepwater geology is not directly traceable into older onshore geological trends. Current deepwater exploration for resource targets both reservoirs associated with the early formation of passive continental margins and deep marine sedimentation. There is now extensive exploration activity and discoveries on all continental margins, including the promising Arctic Ocean (Tippee 2015). Production will follow when economic cost factors make it viable. The oil and gas industry will probably continue to move into ever deeper water to create the largest value of potential new resources, and not only replace existing production but increase it. Deepwater resources have the potential to grow production by close to 80% by 2020 (Müller 2014; Sheehan 2015).

Although production in deep water must eventually be brought to the sea surface so that it can be transported to consumer markets, industry is finding that investment is lower and it is more cost and operationally effective to locate more of the processing technology on the seafloor rather than on a platform floating kilometres above. This is the reason why conventional oil and gas subsea solutions are being developed and implemented for the future. Subsea processing can be broadly considered to be any technology such as downhole pumps that are placed on the seafloor, moving equipment such as for injection, separation, power handling and other processing to the seafloor; and placing gas compression and pumping equipment on the seafloor. "Most floating production systems require subsea-tiebacks consisting principally of umbilicals, flowlines, and communication systems, which will steadily increase from 2016", and, "For aging fields, moving activity to the seafloor can lead to substantial cost savings, while many marginal greenfield products can become economic with the aid of subsea processing" (Crager 2014, 2015). As companies achieve experience and confidence in low risk levels, tiebacks are getting longer. Major companies are already leading the way to the seafloor. These development trends are already under development, with the entire cost being supported by conventional gas E&P and can be leveraged for NGH development.

Because of the special physical characteristics of NGH and its anticipated host sediments in the GHSZ (Chaps. 2 and 7), it should be possible to move not only processing to the seafloor, which is presently underway for conventional gas, but all activities, including drilling. In order to dramatically reduce the cost of exploration and production of NGH, it is optimal to move virtually everything to the seafloor except a production gas line to the surface via seafloor pipeline to shore to a surface production system that would be overhead in deep water, or to a production platform in shallow water. Adapting new seafloor technology allows NGH-specific equipment development to piggyback on conventional development costs and experience, which can be optimized for the less demanding pressure, temperature, and chemical conditions encountered with NGH. The entire cost of these development trends means that the cost of moving NGH exploration and production activities to the seafloor will have largely been borne by the conventional industry.

Discussions led by industry attendees at the first International National Gas Hydrate meeting at the U.S. Geological Survey Center in Reston, VA in 1991, focused on the cost of developing technology with which to explore for and produce NGH (Max et al. 1991a, b). The cost was regarded as prohibitive because (1) NGH only occurred in deepwater and ultra-deepwater for which the industry at that time did not have adequate drilling or production technology, and (2) seismic identification of NGH was incompletely understood. Much of the cost of developing deep water technology for NGH at that time was regarded as prohibitive for its development. In a relatively short 25 years, the energy industry has moved operations into waters of sufficient depth to identify and extract natural gas from NGH deposits.

Almost all of the deepwater development has been carried out by industry, although governments, which may act through national oil companies, provide some support. For instance, Equinor is pioneering the concept of 'underwater factories' rather than just wellheads and is on track to initiate its first full installation by 2020 (Dupre 2015). Governments have also supported technology development. The U.S. Department of Energy, for instance, is supporting deepwater technology development through the 'Research Partnership to Secure Energy for America' (RPSEA) (Kliewer 2014; Long 2014). Although established to support conventional oil and gas development, for instance in reverse-circulation cementing, new ram actuators, nanotube umbilicals, downhole fiber optics, air gun replacements, drilling safety and environmental issues, amongst other initiatives, many of these could have direct application to NGH development.

Technology solutions for the development of the NGH resource are now largely in place and routine. ROV/AUV pipeline inspection, seafloor repairs, and project management work is now common (Bibby 2015). Industry is commonly drilling and producing conventional oil and gas in water depths that include the NGH prospect zone below 1000 m (Chap. 3), and is already moving production technology to the seafloor. Not only is water, sand, gas, and oil separation increasingly being engineered to operate on the seafloor for long periods of time, but power systems and all the components of technology are being developed for the move to the seafloor. Forster et al. (2015a) note that processing technologies that have been hosted on floating platforms now being brought to the seafloor, and that industry is also moving to standardize seafloor technology across the industry. The most critical factors in interoperability are common power levels and connectors, and computer control and data management systems (Barnes and Tunnicliffe 2008; Pirenne and Guillemot 2009; Li et al. 2015) that can be integrated. Moving an increasing amount of processing equipment to the seafloor is driving requirements for subsea power systems (Ball 2014). Connectors that previously required manual connection using saturation diving are being replaced by ROV assembly (Offshore 3/23/16), which has the potential to transition into autonomous assembly.

Work is continually underway to harden seafloor facilities for greater durability in harsh environments. Initiatives to improve coating and cathodic protection for thermally cycled risers and subsea pipelines are the subject of joint industrial activity amongst major companies (Offshore 5/28/15a). Individual companies are reaching out to establish joint subsea technology development activities. For example, GE Oil and Gas is partnering with Equinor on seafloor power systems (Offshore 3/22/16). This complex is intended to house testing equipment for flexible pipelines and large-bore risers. It provides a test site to simulate some of the world's harshest underwater conditions and will focus on the expected high-pressure/high-temperature (HP/HT) reservoirs that are being encountered in deepwater. GE's focus will be on R&D for the next generation of flexibles for ultra-deepwater (Offshore 6/24/15).

Joint industry projects are focusing on production equipment (Segård et al. 2015). Also, benefits should be the generation of environment-specific design data that will provide increased longer-term reliability of coatings in often inaccessible, deepwater service. Reducing operational and commercial risk through longer design life with reduced maintenance costs for facilities in remote locations should result in significant reduction of the anode mass and electric erosion or deposition on subsea structures.

7.3 Background Technology Trends

Technology and methodology improvements and innovations are already taking place across the spectrum of deepwater and ultra-deepwater (Chap. 2) conventional exploration and production. The time is right for moving to the seafloor (Forster et al. 2015b). Subsea gas compression, seawater treatment, ultra-deep technology standardization, electrical connectors, hydraulic system, acoustic releases, navigation systems, hydrophones and transducers, amongst other equipment, are ready for use for any purpose in the ultra-deep environment.

These are the background to development of NGH-specific exploration and production. Most of the current developments can be applied to improving the efficiency of any activity on the seafloor, such as gathering scientific data or seafloor mapping. Many of these can be leveraged or borrowed for NGH exploration and production with little modification. Some of these technologies are useful simply because they work in the water depths that NGH deposits are anticipated. Some are directly applicable for NGH while others are indirectly applicable in that modifications are necessary, or they will be used in a manner that was not fully envisaged during its development for conventional exploration and production.

Operators often focus on the development of particular components that they need to improve a particular system and work with a particular developer to produce a de-risked competitive technology application. These developments often lead to method or apparatus patents, from which further value can be obtained either by sale, license, royalty, or joint venturing to the wider industry. In theory, operators want standardization so that equipment acquisition can be optimized for products across a range of manufacturers. In most cases standardization is accomplished through industry standards specifications, requirements, performance, materials, components, etc., promulgated by institutions, for instance such as the International Organization for Standardization (ISO) the Institute of Electrical and Electronics Engineers-Standards Association (IEEE-SA), and governmental agencies and other international organizations.

Much of the technology advances that are currently underway can be used in a less robust and less expensive manner for NGH because of the special physical character of the NGH resource (Chap. 2). The topics of the subsections in this section are illustrative of development trends whose costs are being borne entirely by conventional deep- and ultra-deepwater developments.

7.3.1 Convergence of AUVs, ROVs and Robotization of the Seafloor

The terms "robotics" and "automation" have a precise technical meaning. Robotics focuses on systems incorporating sensors and actuators that operate autonomously or semi-autonomously in cooperation with humans. Robotics emphasizes intelligence and adaptability to cope with unstructured environments. Automation deals with efficiency, productivity, quality, and reliability of systems that operate without manual intervention (RTC 2013). This is usually applied in a field of repetitive activity. Not only will autonomous underwater vehicles (AUVs) and remotely operated vehicles (ROVs) converge into a new composite technology but robotics and automation will also.

Automation has been one of the drivers of increased efficiency in the energy industry on land as well as at sea. Automation and reduction of the number of personnel at sea has always been a naval interest as well as an economic factor of working at sea. Until electronic control systems became practical, however, most work and control systems at sea were manual. The SCADA (supervisory control and data acquisition) system has been in place in the Gulf of Mexico for at least two decades. Now operators in an offshore office using touch screens to not only monitor sensor data but also control production from every well on a platform. Offshore platforms may now have fewer than 10 people on them, although the platform may initially have had 50 or more personnel at any one time. Older platforms had galleys that were big enough to serve 60 at a time.

The use of completely unmanned wellhead platforms on fields offshore Norway is now being studied and will be likely to be implemented now that remote control rooms can exert direct control and monitoring of automatic systems (Musarra, 2013; Larmon 2015; Offshore 10/23/15b). Remote operations extend to ships as well as seafloor and ocean systems. The 80 m Highland Chieftain supply vessel was driven through a series of maneuvers in the late summer of 2017 operating in the Scottish sector of the North Sea was controlled for over 4 h directly from an office in San Diego, CA, about 8000 miles away (Offshore 17-9-5). On 9 November 2017, the Valemon control room on land at Bergen, Norway, initiated the first remote operating of a production platform (PennEnergy 17-11-9; Wright 2017). The Norwegians appear to be leading the way in remote platform operation because the logistics inherent in the large area of Norwegian offshore operations and the hostile environment that make human servicing of platforms often difficult. The new Norwegian approach is geared to smaller and deeper-water fields. Unmanned wellhead platforms have been the norm in the Gulf of Mexico for at least 50 years.

Unmanned wellhead platforms off Norway are now practical because most of the actual production equipment is being placed on the seafloor and less of the critical production machinery will be on the platforms.

Many of the deepwater GOM facilities are developing multiple fields from single facilities for which wellheads are unmanned. The Jack/St. Malo platform is between the two fields that are 25 miles apart. The Na Kika platform serves 6 fields and uses 90 miles of deepwater flowlines. All manifolds are on the seafloor. Na Kika started production 2003. Automated production facilities are now becoming the norm.

The main argument in favor of unmanned wellhead platforms, which are regarded as having the long-term potential to be functional and robust, is that this could be an efficient development solution in terms of both cost and production. Normally unattended platforms offer substantial savings and are particularly adapted to needs of 'stranded' fields (Beckman 2016). In addition, platforms will remain more accessible for direct human inspection and maintenance than seafloor installations. The Norwegian Petroleum Directorate (NPD) has commissioned studies of the effects of widespread 'demanning' of platforms and has already called for increased adoption of dry-tree, unmanned wellhead platforms on fields offshore Norway. There will be different types of unmanned wellhead platforms and some process equipment may be retained on the platform rather than being put on the seafloor. All unmanned platforms or those manned by very small crews will retain helicopter and ship access.

As exploration has increasingly moved away from shore and into deeper water, technology improvement and adaptation of industrial processes have progressed from land to offshore ships and platforms. This has now reached the extreme of liquefying natural gas to LNG on ships and moving industrial systems to the seafloor. As part of a technology redistribution process, some other operational components, such as remote control systems and electrical supply, are increasingly being taken off rigs and relocated to the shore.

Exploration drilling is still largely done from ships and floating platforms but production capacity is now increasingly being moved to the seafloor. Placing vital production equipment on the seafloor places increased demands upon equipment and process reliability. Seafloor industrial sites have to function safely and reliably for decades. In addition, active or preventative maintenance to replace parts will be used to minimize repair, which may require a more time-sensitive intervention.

Two types of marine technology are developing toward a common goal that will evolve to new technical levels. Robotization of seafloor industrial sites is in the process of taking place. The mobile technology that will attend the immobile equipment fixed to the seafloor will enable the process equipment, such as electrical power switching and conditioning systems, gas compression, and separation, to be maintained and repaired in situ without being brought to the surface during its entire design life, except for purposeful replacement with updated equipment.

At the same time that autonomous marine technology has been a center of innovation and improvement, marine technology has been consolidating around new sensors, batteries, control and communication systems, and replacement of direct human intervention and control for advanced automation. For instance, integration of fiber-optic sensors to enable tactile robotic manipulation (Kampmann and Kirchner 2012, 2014) is already underway. Robotization of certain aspects of underwater engineering, such as inspections, are already underway (Cathx 2015; Offshore 6/25/15). Engineering solutions for turnkey seafloor intervention and subsea robotic systems work are available (Cellula Robotics 2016), particularly in geotechnical and geophysical applications. And automated drilling is gaining momentum in offshore operations (Minyaeva 2015).

ROVs: Remotely Operated Vehicles (ROV) are connected with the surface where an operator controls the movement, attitude, and tools aboard, usually using real-time visual imagery from cameras carried on the ROV. ROVs are tools for carrying out specific work underwater at depth greater than divers can normally work, where the work is inappropriate for a diver, or where it can be done better by an ROV. Even in shallow water where divers can work, ROVs are usually superior because they are not affected by time at depth limitations, and usually require less material and operational support. ROVs can carry out much more intensive power-demand work, for instance in lifting heavy loads or operating equipment with large motors because a power cord is commonly part of the cable between the ROV and the surface or other support system. ROVs can also be operated from larger submersibles, such as JASON, which was operated from a manned submersible in underwater archeology work (OET 2015), or where small ROVs are deployed from large ROVs in order to enter small spaces in a semi-autonomous manner. Multi-system interactive vehicles are in the forefront of underwater robotics (Phoenix 2015).

Although manipulator ROVs have been dominated by hydraulic power systems, electric ROVs are proving to be able to do the same work, except for very heavy load requirement vehicles. Electric ROVs are lighter weight and between 1/4 and 1/2 the cost, have no hydraulic fluid to spill (which makes them inherently less of an environmental risk), do not require heavy-duty, high pressure plumbing systems, and are generally easier to maneuver (Lumsden 2015). Being electric also means that components for carrying out different types of work are facilitated because different working modules can be attached very quickly, even underwater using wet-pin connectors. This electrical technology is more widely applicable to seafloor industrial installations, although it is presently being developed for ROV.

AUVs: Intelligent machines are replacing human activity in the energy industry. The number of people at sea carrying out exploration activities and work on production platforms has fallen dramatically as technology has been automated.

Technology for working underwater without direct continuous human control or intervention is now becoming dominant below water. Technology has now brought us to a tipping point of working at sea.

Operations at sea have generated new technology that is vastly increasing efficiency and reducing costs through having fewer workers at sea. AUV technology is a relatively recent development in marine research and working capabilities (Moline and Benoit-Bird 2016), but the burst of innovation for automation of survey and repetitive work even at full ocean depth has succeeded in producing operational, full-ocean depth AUVs. Moreover, practically none of them have displayed a real capability to host effective sampling tools for both water column and seafloor. AUVs with military applications, however, include both pre-programmed and course-seeking with, in the case of mines counter measures, autonomous detection, classification, localization, and resolution (CMRE 2015). There is considerable software engineering available for rapidly upgrading AUV activities required for supporting seafloor industrial sites.

Commercial AUVs are being used today mainly for mapping or survey. They can be programmed to operate at certain depths above the seafloor or at a particular depth and to carry out a pre-planned survey that is programmed into their control system. There are many advantages to this mode of survey over a surface ship survey, particularly in cost and particularly for understanding details of the seafloor and with seismic, magnetic, and gravity tools, for sub-seafloor survey. AUVs have transitioned from vehicles tied to the surface for recharging, reprogramming, and data downloading to long term submerged vehicles operating from underwater 'garages' in which they can connect for power and digital function transmission (Offshore 6/23/17). SAAB Seaeye, for instance (SASE 2015) is one of a number of commercial systems that can carry out long-term work programs near the seafloor without surfacing for long periods of time. Digital underwater communications developed for military applications allow for constant wireless communication at ranges of about 30 km at a carrier frequency of 9.5 kHz with a bit rate of 50 bps to 12 km at a carrier frequency of 6.5–19 kHz and a bit rate of 1000 bps (Conrad and Schreiber 2017). In a deepwater seafloor industrial installation in which acoustic noise is low, it is likely that digital underwater communication can have higher band widths and data rates between all components of a fixed drilling or production installation than systems developed for military applications.

AUVs are becoming more 'self-sufficient' as they are increasingly equipped with automatic navigation systems. According to Inzartsev et al. (2010), both internally-referenced guidance systems that can be recalibrated at intervals by external manual or automatic update will transition into a new type of AUV having an inertial + mixed control system. Russian offshore technology producers are planning to create Russia's first innovation cluster of companies for designing underwater robots for civilian and military use. This is in part because their sources for Western equipment of this type have been made unavailable to them by Western sanctions. Nonetheless, it is projected that working, underwater robots could soon appear in Russia's Far East Kezima (Inzartsev et al. 2010) by early 2016.

Technology allowing AUVs to self-navigate along a pipeline using optical pattern recognition for an untethered survey is an example of a step toward a greater scale of robotization (Cathx 2015).

The workhorses of the offshore industry, ROVs, are well into a process of evolving into underwater robots (Francis 2014), with increasing programming for independent actions. This preceded hydrocarbon production activities being moved to the seafloor as soon as can be engineered. Reducing costs is a main driver of AUV/ROV development because it would appear that operators have reached the tipping point of water depth in which it will be less expensive and more reliable to place infrastructure on the seafloor than to operate from the surface. Continuous operating of processing equipment on the seafloor, including maintenance and servicing that may include replacement of parts, will require a new class of long-term submerged robotic underwater vehicles that can operate both in remotely controlled and autonomous mode, depending on use.

We envisage that the traditional AUV shape will morph into much more complex forms resembling ROVs with manipulators that will allow for complex movement and operands within a seafloor industrial site. These robotic underwater vehicles (RUV) will include many mechanical functions that are achieved today using 1-atmosphere diving suits. As more autonomous underwater capability is required, development of entirely new ranges of 'compound' robotic AUVs will be brought into operation. For instance, Kongsberg Maritime and Equinor have signed an agreement with Eelume, a NTNU spin-off company to develop a new class of swimming robots suitable for carrying out maintenance, inspection, and repair and replacement activities while submerged for years (PennEnergy 4/18/16). In situ testing of this new breed of active robot (having a high degree of autonomous function, Chap. 8) is scheduled for late 2017 (Hydro 2016). We envisage seafloor industrial equipment will be instrumented for close range interaction with service vehicles based on our technology vision of communicating collaborative RUVs.

7.3.2 Preparation of Seafloor Sites

Subsea structures are typically supported on mud mat-type foundations that spread load over an area. Occasionally short piles might be used, along with cementing around a wellbore at the seafloor. Foundation stability analysis and mediation is governed by American Petroleum Institute codes for Planning, Designing and Constructing Fixed Offshore Platforms, for instance API RP 2A and API RP 2GEO, 'Recommended Practice for Geotechnical and Foundation Design Considerations' (API 2014). As more industrial equipment is placed on the seafloor over larger areas for longer periods of time, and more heat and possibly waste products are generated locally, new practices and regulations will be required. Preparation of seafloor sites for NGH equipment will follow current developments, with respect to different activities (Chap. 8).

7.3.3 Power Systems

With the move into deeper water that is requiring more equipment to be placed on the seafloor, supply of power is a major issue that is already being resolved, at least for production equipment. Thus, we anticipate that the power solutions for underwater industrial sites will be largely achieved soon to the levels that are required for NGH. The generators on a platform that powered all equipment on the rig now can be directed to the seafloor via umbilicals. However, the Equinor Troll-A platform, which is 70 km offshore from Norway, is now receiving power from shore utilizing a subsea cable (Offshore 10/23/15a), which makes the power supply more dependable and lowers logistical supply costs to the rig. Longer undersea power cables are feasible, along with power distribution systems within seafloor industrial sites.

As more equipment is being placed on the seafloor, electrical control and switching systems are also being placed on the seafloor rather than bundling wire in an umbilical that allows control from the surface. Siemens, among other companies who will be competing for seafloor technology business, is developing and integrating the four main components of a seafloor power grid (SPG) required for operations to move from topside to the seafloor (Beckman 2014). Equinor is a principal partner in this particular activity (Dupre 2015) and is currently working with Siemens in a $100 million subsea power JIP to put 100 MW of long-term (30-year life) power on the deepwater seabed (to depths of 3000 m) (Dykes 2017).

The power-handling components each consist of a pressure-compensated transformer, a medium-voltage switchgear, variable speed electric drives, and power control and communications systems. It is intended that these will be integrated with an array of new equipment for subsea production, processing and boosting using industry-standard electrical connectors. Cooling is provided by the surrounding cold seawater and with an oil-bath pressure-compensating housing around the device so that it will work at all depths without heavy pressure chambers.

The most recent consideration for bringing power to the seafloor is to utilize tidal current electrical generators to generate the power on-site or nearby where current conditions are better (Dykes 2017).

7.3.4 Data Acquisition and Management

Data acquisition depends on accurate and well-located sensors having sufficient data rates, high-speed communications, and reduction of data that can be input through automatically authored instructions to automated equipment. Increasing numbers and new types of sensors are being installed on subsea equipment with the aim of creating large data sets for better well and field management. This provides a working basis for NGH application. As a result, increasing amounts of data from many more sensor nodes have to be transmitted and managed, especially in largely robotic systems in which automatic control is dependent on real-time sensor data. Sensors are dramatically shrinking in size while gaining durability through the use of new materials.

Laser sensors are used for spectroscopy, usually in the infrared, to monitor composition and detect potential pollutants. Lasers are grown to be specific for narrow wavelengths, are solid state, have high survivability at elevated temperature with high sensitivity and are selective to individual substances, are non-destructive, fast, robust, reliable, and durable. Infrared spectroscopy requires gas samples, however, owing to interference with water spectral lines. Thus, gas has to be extracted from the water using membranes. Sensors can be very tiny but the membrane assemblies are currently too small to send down hole, so analyses would have to be done above the wellhead during exploration or during processing during production. Sensors that scan across multiple frequencies are less common a group of very tiny lasers emitting light at specific wavelengths. Quantum cascade lasers are now about size of a fingernail. Laser spectroscopy can also be adapted for individual isotopes to indicate, for example in the case of methane, biogenic and thermogenic components.

The application of laser spectroscopy to the deep sea mainly concerns dissolved gases and fluids, bubbles, and NGH. Cruises normally have everything needed for full quantitative work at sea including reference gas standards and full computer processing. Atmospheric sensors adapted to deep-sea work carried by AUVs usually have a silicon membrane backed by a frit for support to produce the gas sample for spectroscopy.

It is also possible that cross-field sensor development will have application in NGH and conventional wells. For instance, sensors developed for measuring blood flow in stents in the medical field (C&EN 2015), points the way to new sensor development for monitoring flow in wells (Schlumberger 2016). Sensors in wells can be based on optimizing MWD/LWD technology, which is being continuously improved for conventional wells (Ghiselin 2014, 2015). Downhole sensor/power supply systems are now available for coiled tubing drilling operations (Livescu et al. 2016). Abili et al. (2014) note that development of several large oil and gas fields is incorporating multi-phase meters and wet-gas meters. These instruments provide essential data for optimizing production, measuring oil, gas, water fractions, and flowrates. Development work has already begun to focus on NGH. For instance, Delta Subsea and VEGA have signed an agreement to develop a gamma ray NGH localization tool (Davis 2015).

Drilling data may display a highly variable quality. Well conditions, conductivity of drilling fluid, sensor sensitivity, electrical connections and fields, and signal degradation, amongst other causes, can render quality variable over time within a well and between wells. This complicates the jobs of real-time understanding, as well as look-back analysis. In order for automation of drilling systems to proceed accurately and safely, acquisition, compiling, and analysis of real-time data will be necessary to carry out operations; for instance, for automatically steering a drilling tool in a 3D orientation matrix according to a development plan.

Processing control systems will require a high degree of self-auditing in order to automate processes.

Major companies are embracing ‘big data’. BP, for instance, has stated that it will have at least 650 wells feeding into a single large dataset, collecting at least two million data points per minute into proprietary software (Fehrenbacher 2015). Many locations will be controlled by a small number of control rooms, all connected by high speed communications. Consolidation of real-time information and remote management are all part of increasing efficiencies that should reduce costs while increasing safety. As part of this consolidation, more of the data analysis and control system modules are tending to be more automatic with proportionally less human involvement and oversight.

7.3.5 Long Range Communications

Major companies, such as BP, are engaged in bringing their exploration and production activities into an industrial internet context in which sensors monitoring vibrations, temperature, pressure, and other well properties will store, contextualize, and allow the data to be visualized in order to give BP real-time insights into over 4000 wells world-wide (Offshore 7/9/15). This will involve upgrading of existing systems. Supervisory control and data acquisition (for instance, SCADA) systems with radio and hardwire communication links have been used by some oil companies for over 20 years. Integrated operations centers (IOC) bring enough data from one or more offshore rigs to offices on land so that operational functions that used to be carried out offshore can now be concentrated in a small number of localities (Musarra 2015). This centralized data gathering is particularly good for maintaining logistics and supply, as well as actually monitoring and actively controlling some functions of production and support operations. Supervisors and technicians can monitor wells and pipelines, including opening and closing valves using a touch screen from a shore-based office. Although most communications within a seafloor industrial site will be wired with wet connectors, long range communication will require different fiber optic and underwater modem and radio links. The objective of real time communications is to improve well performance and production while minimizing downtime and to consolidate control for a number of wells and installations from only a few data management centers.

Communication within a seafloor drilling/production installation will use a combination of wired and wireless. Within each device and between them, hard-wired systems will be most practical and secure. For advanced AUV and other robotic systems, however, short range wireless communication may be necessary because the swimming robots have to enter restricted spaces and exit from a different access than the entry point. Wireless communication can be supported by underwater acoustic modems, which have the advantage of working about as well in muddy water as in clear; or by nearfield light modems, which have the advantage of having much higher-speed data transfer but require relatively clear water.

Research in underwater modem technology is strongly supported by the Office of Naval Research (http://www.onr.navy.mil). Some of the areas of research are: code divided multiple access signals, spread spectrum, Doppler, MDAPSK (Multi-level Differential Amplitude and Phase-shift Keying), OFDM (Orthogonal Frequency Division Multiplexing that may also be either non-contiguous or differential), electronically-phased acoustic arrays (Smith et al. 2013), acoustic multiple-input and output systems, synthetic array beamforming, and very compact solid-state transducers and arrays. Baud rates for short distance communications will involve very little lag time but attenuation of signal over distance is will remain a difficult issue to overcome.

Because deepwater sites are liable to be far from land and mobile phone technology, satellite links will probably be necessary (Farrell 2015). In any case, satellite links can be made more secure. In addition to communications, managing dataflow particularly during drilling operations and monitoring pressure and gas/water flux during production can best be done using a GIS platform (Noll and Hogeweg 2015). As these systems are already in development and in some cases already installed, we do not envisage any NGH-specific equipment will be necessary.

For sites nearer to land, all control activities for seafloor installations can be based on land. For instance, the Corrib gas field off the NW of Ireland (Sect. 10.3) has an underwater umbilical for electric and hydraulic power for the seafloor processing system for which the control room is on land about 90 km away. Thus, the basic elements for remote physical monitoring of a seafloor site are already part of the industry toolkit. Application to NGH deposits at greater distance from shore is only a matter of scale and implementing different control systems.

Increasing complexity of the work required by long-duration robotic systems implies a need for multi-modal feedback sensing of the environment to the control system as human intervention will be rare. Control systems that react to their environment will be able to take into account short-term excursions to fixed programming such as fish, lose connections requiring fixing, sediment movement, and other unpredictable environmental factors. Emerging multi-modal tactile sensory systems will combine static and dynamic force sensor arrays together with relative and absolute force measurement and autonomous reaction (Meng et al. 2006; Kampmann and Kirchner 2012, 2014).

7.3.6 Conventional Drilling from Surface Vessels

The most obvious cost element in deepwater exploration is ship and semisubmersible drilling and support. We regard the existing new support and smaller drill ships to be perfectly adequate to support NGH seafloor industrial sites. Major conventional hydrocarbon drill ship costs are the single greatest cost offshore (Table 7.1). The drillships and semi-submersibles are huge because they have to carry a large amount of supplies, such as drill pipe, wellbore liners, risers, power

Table 7.1 Example of deepwater drillship costs as of spring 2015

WD = working depth	Day cost
Drillship <4000′ WD	\$258,000
Drillship >4000′ WD	\$521,000
Semisubmersible <1500′ WD	\$272,000
Semisubmersible >1500′ WD	\$344,000
Semisubmersible >4000′ WD	\$437,000

Costs have dropped with the post-2014 low price environment for hydrocarbons but remain significant

systems, and drilling mud, although supply vessels moored to them and standing off nearby often can provide certain services, such as storage of drilling muds. The deeper the water and the deeper drilled, the more supplies must be brought to the drilling platform. Because they have to drill in very deep water, and then drill to tens of thousands of feet into the seafloor, often through hard rock such as igneous sills in a sedimentary succession, the drilling equipment has to be among the most powerful (and expensive).

The deepwater drilling costs other than the cost of the drilling rig (referred to as spread costs) are approximately equal to the drilling rig day rate. That is, if the drillship costs \$400,000/day, the assumption is that the total cost (including casing, logging, transportation, etc.) would be about \$800,000/day. This relationship may vary with the day rate, however. Spread costs may be difficult to reduce but the day rate can vary considerably. For instance, at its peak in 2014, the day rate for new contracts was on the order of \$600,000–\$650,000 per day. As of August 2015, because of excess capacity the day rate had fallen to less than \$250,000, which is forcing companies to dramatically reduce and even scrap drilling platforms in order to survive (Triepke 2015). In addition are the ‘hard’ additional costs of drill bits, logging, drill mud, casing, cement, transportation, etc. These will vary a lot depending on the TD of the well, number of casing strings, number of log runs, and more. There are also issues of pre-drill expenses for permits, site surveys that would be the same for conventional and NGH wells.

So long as conventional drilling technology is used from a surface platform, a certain high level of cost is involved. When a full, heavy-duty drilling capability is installed on the platforms a very high drilling cost is inevitable. An obvious cost saving initiative is to reduce the size and capability of the drilling platform, something that is not possible for conventional ultra-deepwater. New classes of light weight service vessels are now appearing offshore (Lloyds 2016). These commonly have large open deck spaces, often with provision for through hull drilling moon pools or over the side drilling using coiled tube drilling. One or more large cranes are also a feature about mid-ships, which are capable of working in ultra-deepwater. There is usually a helicopter deck forward of the wheelhouse and cabin space that is also well forward. Although this is not the place to discuss the various special vessels of this type that are now beginning to work offshore, they are generally configured for different purposes and more than one vessel can work together at the same site at the same time. The important factors are their seaworthiness for their size, and their ability to work with a variety of conventional drilling and seafloor exploration and production activities. Platform supply vessels (PSV) of this class with very large stern to amidships unencumbered deck space can be configured to supply equipment to NGH seafloor sites and service them (with the reduced NGH requirements) both during exploration and production at a small fraction of conventional drill and production vessels including semi-submersible platforms.

7.4 Drilling

Drilling is the most expensive component of ultra-deep offshore exploration and production. It is traditionally carried out from the surface using a mobile offshore drilling unit (MODU). These can be anything from a jack-up to a drillship. Also, FPSOs used in support of a production platform may store, process, and offload oil to smaller tankers. In an environment such as the post-2014 low cost energy paradigm, developing and promoting technologies to reduce drilling cost may be the first step to survival of many businesses. Although increased automation has been a constant factor on drilling platforms that has over the years significantly reduced the number of hands on the drilling floor, improvements in technology and the need to reduce costs is accelerating the trend toward more automated drilling (Minyaeva 2015).

The principal danger in shallow seafloor drilling is intersecting an over-pressured gas sand or a gas concentration in a permeable bed beneath NGH. Methodology for evaluating NGH drilling hazards that is based largely on seismic survey has already been brought into practice as part of a NGH drilling program in the northern Gulf of Mexico (McConnell et al. 2012). Conventional 'top-hole' drilling is done with a hole open to the seafloor to a depth below which casing can be safely set so that a wellhead with blowout preventers can be put in place. Drilling can then safely encounter high pressures in the well such that flow into the wellbore that would cause blowouts will be prevented. Conventional top-hole drilling may span the depth of an entire GHSZ. It is well understood that simply keeping temperature at or below its ambient temperature in the GHSZ (which increases downward from the seafloor), so refrigeration using seafloor water directly or with heat exchange provides sufficient in situ means for reducing the risk of unintentional conversion and production of free gas to essentially zero. Well cuttings can be left on the seafloor without any possibility of dissociation of NGH taking place.

In general, oil industry people have concerns about drilling safety, as over-pressured gas pockets have been encountered within GHSZ and have been an issue in some conventional deepwater wells. However, shallow hazard studies using both seismic and electrical survey can identify problematic zones and allow them to be avoided, preserving a safety factor. Similarly, drilling into a free gas sand beneath the base GHSZ will result in drilling fluids being cut and well control becoming an issue. At the very least, this is a concern expressed by drillers. But it has also been taken seriously by NGH exploration programs and, thus far, drilling safety has been maintained. It is likely that if the procedures developed are followed, drilling safety will continue to be well managed. During production, the greatest safety advantage is the controlled shutdown with the re-formation of hydrate that will follow the ceasing of NGH conversion should anything get out of control.

Drilling from the seafloor is a technology concept that has been in development for some time. Kolle (1989) and Kolle and Bibee (1996), proposed to implement an autonomous seafloor drill was designed for deploying seismometers at depths of 30–100 m, and built a working prototype of an autonomous seafloor drill rig with only 50 m design depth capability in soft sediments that was tested at sea. The most advanced versions today are drilling-capable ROVs that are lowered to the seafloor and controlled and powered from a floating platform. For instance, the MeBo200 is a portable seafloor drill developed by BAUER (Bauer 2015) that can be deployed from research vessels such as the RV Meteor, which is much smaller and less expensive to operate than conventional drillships. Other research vessels with the same characteristics, such as being just short of 100 m in length and having cranes on the fore and stern decks, can safely carry and deploy the MeBo200. They include the RV Alliance, which is operated by NATO, and a majority of the larger research vessels operated by many countries (Bourillet et al. 2007; Gutenberg 2015).

The MeBo200 is designed to drill in both hard and soft rock to depths of 80 m in 2000 m of water but because it is not able to take pressure cores it is regarded by some to have limited use for NGH exploration drilling. The MeBo200 was originally intended for shallow drilling for geotechnical sampling of the upper 125–200 m of marine sediments. It can recover cores suitable for engineering geology to a depth of about 160 m. Drilling from the seafloor is much less weather dependent than drilling from a vessel, even when the drill is controlled and powered from the surface. It is estimated that on a time-basis, seafloor drilling is 3–5 times more efficient than shipboard geotechnical drilling in water depths of 1300 m and greater in soft sediment conditions.

The MeBo500, which is in a design/fabrication stage in the autumn of 2015 will be larger than the BeMo200 and designed to drilling to about 500 m (Pers. Comm., Wallman 2015). It will be a 3-piece modular system that is will be assembled on the seafloor. To be used across the worldwide range of NGH targets, however, a seafloor drill should be capable of drilling through entire GHSZ depths and some distance below so that it may become a primary tool for making relatively inexpensive direct assessments or ground truthing of seismic identifications of NGH deposits. Other seafloor drilling systems such as the Cellula Robotics CRD100 Geotechnical drill (Jackson et al. 2016) are also moving from development and laboratory to commercial operations. We suggest a working drilling depth of about 1200 m below sea floor (mbsf) would be sufficient for NGH drilling worldwide.

A very shallow (3 m) low cost mineral exploration drilling system that is ROV-mounted offers another perspective to seafloor drilling (Søereide et al. 2017). Lightweight and low cost, in its present configuration it is not suitable for NGH prospecting but the basic engineering and operating procedures should be useful for development of NGH seafloor mounted drilling systems.

Drilling automation, which is already a development issue for cost reduction (Minyaeva 2015), is one of the critical technologies that will enable seafloor drilling. A benefit of drilling automation for ship drilling is that it can minimize human exposure to harsh offshore weather and sea condition environments, including Polar Regions. However, automation drilling using a seafloor rig controlled from the surface will still suffer from weather related issues. Only a fully automated seafloor drilling rig (e.g., a robotic system) has the potential to reduce drilling downtime to near zero, while implementing superior precision and execution. Drilling from the seafloor will require a much more sophisticated automated control system that may be only rarely be remotely controlled.

In addition to modifying support vessels for NGH exploration and production work that may not involve drilling directly from the sea surface, a new class of smaller drillship designed for economical deepwater operation in remote locations is becoming available for less rigorous drilling requirements (Pinkstone et al. 2017). These smaller ships are highly amenable to automation and the first 'digital' Rig is now operational (NC 2018). This automation is a giant step toward autonomous drilling systems that mark a step-function in asset performance improvements. In addition, these vessels are idea for well intervention and seafloor servicing work that does not require long-term sea keeping. Although designed to support conventional hydrocarbon development, they appear to be ideal, with little modification, to support NGH E&P activities.

7.4.1 Riserless Drilling

The technical and operational challenges of using a riser in deep and ultra-deepwater grow with increasing depth, requiring longer pipelines and communications lags. Risers are subject to high tensions, extremely variable internal and external pressure differentials with depth, and high levels of potential fatigue damage, all of which can create drilling emergencies in relatively short times. In addition, using a riser in deepwater can affect well planning and operations as downhole pressures may fluctuate and compensate. Top-hole drilling is normally carried out in an uncased, open hole, after a shallow hazard survey for geological factors such as faults, gas pockets that can 'blow out', and shallow NGH, which can convert to gas and cause sediment fluidization and mass flow. It is common to use a Lightweight Intervention (LWI) for riserless tophole drilling of the shallowest part of the well from the seafloor down to some depth selected through a shallow hazard study in which no drilling problems or petroleum concentrations are expected. Tophole drilling can often be as deep as 1+ km, which is almost coincident with the depth section in which NGH will be found in the GHSZ (Chap. 2). Technology for riserless drilling may be applied directly to NGH seafloor industrial installations.

7.4.2 Steerable Drilling Systems

Weatherford International, National Oilwell Varco, Haliburton/Baker Hughes Inc., and Schlumberger all provide proprietary steerable rotary drilling systems that offer precision directional drilling. Ghiselin (2014) summarized the state of the technology. In recent years, the technology has taken a leap from (passive) geometrical steering, where the driller follows a borehole plan to a new active method where integration of real time LWD data tool string is used to steer the drill head. The prediction that steering systems would improve efficiency (Downton et al. 2000) is now being put into practice (Holme 2016).

In rare emergency situations, such as at the Macondo well where a relief well had to intersect the original hole at a precise location, directional drilling technology is essential. Less critical situations, such as sidetracking around an obstruction in a wellbore, also benefit from the ability to control the wellbore trajectory. Without advanced directional drilling technology, it would not be possible to drill some wells. Steerable systems allow complex wellbore geometries to be achieved. Such complex wellbore geometries could be used to optimize NGH conversion and gas flow/separation.

7.4.3 Dual Gradient Drilling/Managed Pressure Drilling

Using selectable weight drilling mud is commonly necessary to counteract the buoyancy effect of oil and gas encountered during drilling in order to prevent blowouts. In general, at the seafloor the density balance for drilling mud is ~9.0 lbs/gallon (abbreviated 9.0#). If a system maintained only hydrostatic pressures, 9# mud would be sufficient to drill to any depth. Some level of lithostatic pressure will always be present so some increase in drilling fluid weight is always required. The problem is that the pressure regime in deepwater sediments can be quite variable due to differential compaction, overpressured sands, geotechnically unstable strata, etc. While it is relatively rare to have a well start flowing within in the first kilometer into the seafloor, where it will encompass the GHSZ, this has happened in the Gulf of Mexico and elsewhere. Riserless drilling technology or riserless mud removal (RMR) systems are now covered by regulations written primarily about fracture gradients in deeper strata (Parish 2015). These could be applied directly to the relatively weak reservoirs anticipated for NGH deposits without drafting new regulations.

RMR is a dual-gradient technology used to drill the top or open-hole sections of a subsea well. Systems are available off the shelf (Weatherford 2017). A system that recovered and reused drilling fluid was implemented in 2006 by Total E&P UK on its Jura West 3/15-10 North Sea exploration (Offshore 11/1/15). The system used a suction module to collect the mud and cuttings at the seabed. These were then pumped back to the surface along an 8-inch or medium diameter flexible hose, enabling the cuttings to be separated at the surface and the drilling fluid to be reused. Too much down pressure can be a problem with drilling ahead after setting casing as mud weights are increased. RMR is a process that eliminates the downward pressure that would be exerted by the mud column between the seafloor and the surface. The RMR system also allows a rig to move off station without having to recover all the drilling fluid.

The RMR system is useful in shallow geohazard and unstable areas containing soft clays, shallow water flow, and shallow gas where specialized drilling fluids are needed for borehole stability. The density of the drilling fluid has to be high enough to hold back the formation pore pressure but not so high as to fracture or deform the formation. The system provides for gas mitigation, in which gas can be separated and diverted from the mud stream at the surface. The set-up and operation of the system is much more flexible than riser drilling as most of the installation is carried out on the seafloor and the hose is simpler to handle than pipe.

In deepwater and ultra-deepwater in which the NGH prospect zone occurs (Chap. 3), drilling muds in risers are becoming more difficult to manage. This has led to the development of dual gradient drilling, which is a type of managed pressure drilling (Schubert et al. 2006), in which the riser is full of seawater and is pressure equilibrated with the ambient seawater over its full length while heavy muds are used in the wellbore. Managed pressure drilling (MPD) technology has grown from basic closed-loop circulation systems on fixed structures to full riser-integrated system on drilling platforms (Wuest et al. 2015). When the mud is circulating in the wellbore, bottom hole pressure is a function of riser hydrostatic pressure and mud-system induced pressure. Dual gradient drilling enables operators to reach very deep reservoirs without having the problems of overpressured risers.

Although originally developed for very deep drilling (over 3000 m), the method could be applied to NGH drilling from the seafloor in which it has been desirable to have higher wellbore pressures than if seawater were used. Whereas in single gradient drilling the drilling mud column extends from the base of the wellbore to the drillship, in dual gradient drilling, the mud return does not go through a large-diameter conventional riser (DC 2012). A separate mud-lift system uses a feed/return line powered by seafloor pumps and mud can be circulated in the hole without direct return to the surface. In addition, some additional pressure can be added to the hole by the pumps.

7.4.4 Seafloor Hydraulics

Siemens and Equinor have developed and qualified subsea hydraulic power units designed to provide hydraulic power at an offshore well site in water depths to 500 m (Offshore 4/15/15). The system takes electrical power from switchgear at the well site. The hydraulic system has an onboard control module that operates hydraulic valves, sleeves, and safety equipment, as well as providing hydraulic pressure. The main advantage of siting the unit on the seafloor is to remove the requirement for long hydraulic hoses and the fluid from these surface lines, along with the hydraulic friction losses in long hoses from a floating platform or ship to the seafloor. Siting on the seafloor also reduces the hydraulic fluid storage requirements on the surface, as well as reducing requirements for fluid volume. There is also improved hydraulic performance and lowering energy cost because the electrical system is more compact. This is a cost-effective option, replacing complex and long umbilicals, especially in ultra-deepwater.

Seafloor industrial sites will require power to be supplied to them. Thus long, subsea power cables, either from a surface platform or from nearby land via islands is required, and industry is already meeting requirements (Offshore 6/30/15). For instance, a 46 km long combined communications and 230 kW power supply cable has been installed by Aramco to its T-20 tie-in platform in the Arabian Gulf (OE 2015). Supplying power directly to a seafloor industrial site and then distributing it is a more cost-effective option than using complex and long umbilicals, especially in ultra-deepwater.

7.4.5 Advanced Drilling Tools

Rotary drilling, in which the entire drill string is rotated from the drilling floor, and coiled tube drilling in which the rotary drive motor is located at the end of the drill string, are the two drilling most commonly used drilling methods. Changing a drill bit involves pulling the entire drill string or the tubing that has been inserted into the well out of the hole so that the bit can be replaced. Because of the length of time it takes to replace a drill bit in ultra-deep water—which has the longest drill strings, considerable efforts to improve the longevity of rotary drill bits (Ghiselin 2015) are of interest across the energy community. Increasing bit longevity is key to maintaining drilling rates, particularly in hard rocks where a high rotary speed and a low weight on the bit yields longevity and high rate of penetration (ROP). NGH drilling is anticipated to have very low drill bit attrition because of the generally softer, more yielding nature of the formations in which NGH will be found, in comparison with conventional drilling targets.

Non-rotary drilling tools have been developed and proven that may prove to have application for NGH. High pressure water jet drilling tools can be effective in hard rock to the point where pressure reduction may be necessary for drilling in softer rocks or only partially consolidated sediments, such as those marine sediment in which NGH will be found (Kolle 1999; Kolle and Theimer 2004; Theimer and Kolle 2007).

7.4.6 Narrow Bore and Rigless Drilling

Industry is adopting narrower well bores to reduce drilling and construction costs, as narrow holes reduce the amount of materials necessary and usually increase drilling rates. Narrow bores are particularly expedient for exploration wells. Narrower or slimmer holes will use less consumable materials such as casing, drilling fluids, cement, etc., produce fewer cuttings, reduce transportation, logistics, and weight and materials handling requirements, reduce the potential for leaks at wellbore reduction junctions, among other benefits. Narrow bore drilling for preliminary valuation of a NGH deposit (Chap. 5), would be advantageous, especially if a seafloor drilling rig is used, because it should be much faster than wider bore drilling. Also, it should be possible to retrieve more core sections into existing seafloor well boring frames using higher capacity small-tube magazines to hold core pipe.

Another innovative drilling method is through-tubing rotary drilling (TTRD) that is designed to perform both sidetrack and intervention drilling on existing wells without changing components (David 2006). It appears to be a cost-efficient infill drilling technique that presently has its greatest use in mature fields. The objective is to prolong the use of a completion by being able to intersect and drain isolated marginal reserves. Because each main well in a NGH field will likely have multiple wellbores in a complex geometry, TTRD may be used in drilling the initial well spread or to access NGH patches that were not affected by the main conversion episode during production. Because NGH fields will be 4-D monitored with respect to the changing acoustic structure in a deposit during production, identification of orphaned NGH zones can be identified and TTRD drilled to expand the coverage of the conversion method even during production.

Additional exploration drilling technology that is being brought to market is the Badger Tool for rigless drilling (http://www.bxpl.com). This uses a relatively narrow bore in an open hole in which seawater may be circulated. Quick evaluation of the strata can be made by lowering an expendable tool to the seabed, drill into it while evaluating the geology using logging while drilling (LWD) methods. In addition to logging, it is now also possible to carry out seismic-while-drilling (Vieitez and Cox 2017), which could be immensely useful in revealing the small-scale details of sediment NGH distribution in the reservoir. When the drilling plan is achieved, abandon the tool and leave the entire equipment, including the umbilical, in the wellbore. The electric or electro-hydraulic drill tool is fed power through an umbilical while the tool descends, pulling the umbilical behind it (Schmidt 2013). This application could also be useful for valuing a NGH deposit (Chap. 5). This and other rigless drilling are currently re-shaping the subsea well intervention market and much of this technology may be applied directly to NGH drilling (Armitage and Hernandez 2017).

The sediments that must be drilled, and in which NGH will be concentrated, will be relatively soft. Where NGH is developed, conventional drilling will be difficult because solid concentrated NGH is very hard and often tenacious (Durham et al. 2003).

We suggest that a NGH-specific cutting method could utilize a programmed angled scanning laser to cut the NGH using narrow dissociation channels to produce cuttings that could be mechanically separated from the drilling face much more easily than by using mechanical cutting tools alone. Thus, the size and power of drilling motors can be dramatically reduced for NGH drilling. Reduced specification drilling motors already under development for narrow-hole drilling could be used (Scott et al. 1991). Drilling tool development is one of the NGH technology development issues in which economies can be found by matching equipment to NGH requirements.

7.4.7 *Inclined and Horizontal Well Bores*

Horizontal drilling from sidetracks is common for enhance recovery in conventional gas wells, especially in cases where more than one producing structure or zone in different strata is desired, and as a result widespread lateral drilling is has become predominant (Fig. 7.2), for development of shale resources (EIA 2011). In addition, conventional exploratory or wildcat wells are usually (though not always) vertical. Offshore development wells are typically directionally drilled, and it is not uncommon for the bottom-hole locations of development wells to be a mile or more laterally from the surface location. Even in conventional resources, local geology might dictate deviated well trajectories, such as drilling around salt bodies. Reservoir drainage or production from a particular well also might be improved if a well could be oriented to penetrate multiple reservoirs or fault blocks, or was constructed at a low angle to intersect fractures or to maximize wellbore surface area within a low permeability reservoir. A multilateral well has more than one lateral wellbore and will typically provide better sweep of a reservoir and provide better drainage, and also to be able to drain more than one reservoir compartment. Multilateral well bores probably will have substantial application to NGH, especially in a reservoir where sands and thin shale partings have the potential to retard hydraulic transmission of pressure drop and drainage of the reservoir as a whole.

In general, conventional wells produce well enough from vertical and low-angle directional wellbores, especially where there is an extensive producing horizon and

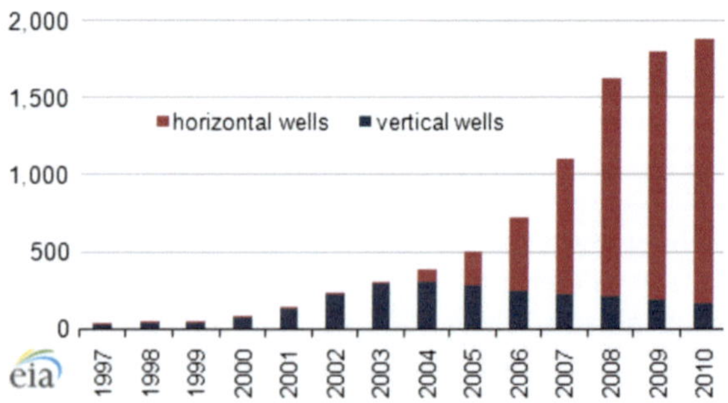

Fig. 7.2 Annual Barnett shale natural gas production by well type (Bcf). 'Horizontal' includes inclined. *Source* U.S. Energy Information Administration based on HPDI, LLC

there is no need to carry out lateral drilling. In shale, lateral drilling is vital to increase the fracked volume. There appears to be no general pattern that can be applied to most shale wells because no two shale locations are the same, often even in the same play and relatively close by. Shales also often vary vertically and laterally, which requires a basic understanding of the geological and geotechnical properties in order to optimize the fracking effect upon which the productivity of the well will depend. Shale and conventional lateral wellbores generally toe up or down or are geometrically undulating or irregular in both plan and/or profile, but they cannot have more complex geometries, such as turning corners at more than 90° angles using conventional drilling and wellbore casing technique.

Successful horizontal drilling, which was the first such attempt in the world in NGH, was reported from work-up drilling in the semi-consolidated marine sediments and gas-enriched zones of the Japanese Nankai deposit W of Tokyo (Takahashi and Tsuji 2005). This proof of concept for horizontal drilling in mechanically weak marine sediments hosting NGH demonstrates that conventional drilling techniques are adequate for NGH deposits. Considerable improvement in drilling rate, directionality, lateral reach wellbore complexity, and other developments using essentially existing technology have substantially increased drill pipe technologies in the rotary drilling toolbox (Teelken et al. 2016).

7.4.7.1 Coiled Tubing Drilling

Coiled Tubing Drilling (CTD) technologies replace the traditional rigid, jointed drill pipe with a long, flexible coiled pipe fabricated of flexible steel and/or composite tubing (Williams et al. 2001). It is delivered to the well site coiled on a reel and unreeled during the drilling to follow the drill head as the pipe is pushed or slid into the wellbore. Available coiled tubing ranges from 1″ (microbore) to 5″ in diameter and larger. Normal practice on land is for the tubing to be inserted into producing oil or gas well through an existing wellhead initiated by rotary drilling. CTD is regarded as a 'high maintenance' technology (LEA 2002), although the greater part of the maintenance is concerned with the actual drilling and down-hole sensor capability. CTD rigs are generally small and can be truck mounted or mounted on a small vessel for work at sea (Duque et al. 2012). For instance, Beckman (2017), regards an inexpensively deployed coiled-tube drilling system as being beneficial for downhole maintenance as part of conventional drilling operations. The expensive and time-consuming aspects of rotary drilling are already being superseded by rapidly lowering a coiled tube with a bottom hole device designed for the particular purpose for which it is being deployed.

There are only a few basic components of an upright CTD in which the tubing is unreeled from a coil of pipe (CT) and lowered vertically into a wellbore (Fig. 7.3) over an arch that is designed not to exceed the flexibility limit of the pipe. A curved guide beam termed a gooseneck guide conveys the pipe from the reel in a slightly greater radius of arc into a vertical position at the wellbore mouth and passes the tubing into the CT injector, which is a friction engine that grips the tubing and

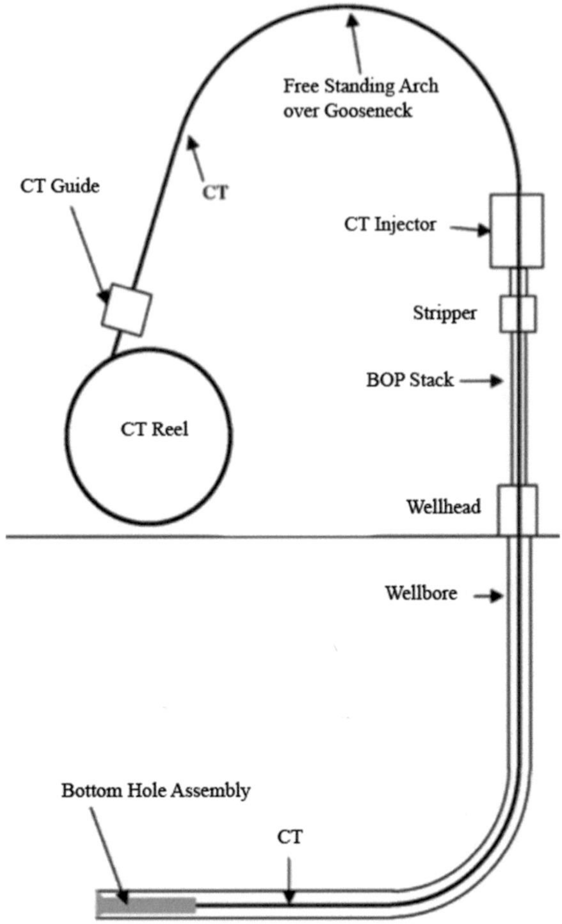

Fig. 7.3 Generalized diagram of a CTD drilling rig. CT—Coiled tube on reel and deployed into wellbore. Gooseneck guide for free standing arch not shown

pushes or pulls it into or out of the well. Below the injector head is the stripper containing packing elements that provide a seal around the tubing in order to isolate the well's pressure from that at the wellhead (not shown). The stripper is the primary means of preventing blowout under normal operating conditions.

Below the stripper is the blowout preventer stack (BOP), which provides the ability to cut the coiled tubing pipe and seal the well bore (shear-blind) and hold and seal around the pipe (pipe-slip). Older quad-BOPs have a different ram for each of these functions (blind, shear, pipe, slip). Newer dual-BOPs combine some of these functions together to need just two distinct rams (shear-blind, pipe-slip). The BOP is affixed to the top of the well at the wellhead, which provides the pressurized tunnel down to the top of the wellhead. Between the tree and the riser is the final pressure barrier, the shear-seal BOP, which can cut and seal the pipe (not shown).

Drilling is carried out using a motor or a hydraulic assembly and bit (along with other components) near the head of the drill 'stem'. At the end of the drill string is the bottomhole assembly (BHA). It can range from something as simple as a jetting nozzle, for jobs involving pumping chemicals or cement, to a larger string of logging and steering tools and but usually has a motorized, low power rotating drill. Pressure on the drilling tool, where a mechanical force is required, is achieved by rams at the wellhead that advance the tube into the drill hole. This may result in helical buckling of the tube in the wider diameter wellbore, thus close monitoring of the drill face pressure may be used to control insertion pressure. High speed drilling motors and special cutting bits yield drilling rates comparable with rotary drilling.

Because of the less demanding temperature, pressure, and chemical conditions in NGH wellbores, there is considerable scope for long-term operation of technology designed for much more demanding conditions, as well as the possibility of implementing the technology less expensively. Thus, refining of a seismically determined NGH geometry and the geometry of its sediment host by implementing LWD services in a bottom hole assembly (BHA), real-time updating of the 3D reservoir mode can be used to optimize the NGH borehole plans during drilling (Ghiselin 2014), which are likely to be much more complicated geometrically than reservoir penetrations of conventional gas deposits. The system can deliver both power and real-time data, switching between applications is a simple as changing out the Bottom Hole Assembly (BHA). Offshore interventions requiring both CT and wireline options are: Perforating, stimulation, and gas lifting; milling and cleanout operations; logging operations; mechanical intervention, and well abandonment. Integrated high-pull wireline conveyance systems can accelerate access to wells with complex geometries (WO 2016).

Vessel-based CTD is already being deployed, as well as seafloor wellhead installations designed to provide for proper torsion and well control (Johnston 2017). Much of this equipment may have direct application or application with slight modification to efficiently operated entirely within a seafloor drilling/industrial installation.

7.4.8 Multi-pad and 'Octopus' Drilling

A single wellhead can be used for multiple vertical, inclined, and 'horizontal' wellbores from a single drilling vessel in much the same way that multiple production wells are now drilled in shale beds from a single drill pad. In conventional developments, the multiplicity of production wellbores is a function of providing the best drainage, especially where reservoirs may not be of the highest quality. In shale developments, wells are geometrically located so that as little source rock in the reservoir escapes fracking as possible, followed by having a drainage pattern that magnifies the initial production rate. Thus, the basic technology and drilling experience is available to be applied to the well geometry we envisage for pressure control, gas and water separation, and gas drainage from NGH reservoirs (Chap. 8).

Drilling from the same pad facilitates high density drilling similar to integrated fracture systems in thick shale horizons (Zeits 2013). Following shale terminology,

"High-density drilling" means up to 32 wells per single drilling pad in full development mode, with laterals landed in a number of stacked intervals, a truly staggering density relative to the 7–8 well patterns that have been considered "dense" until recently.

7.5 Production Issues

A number of production technologies are in advanced stages of development for assisting production in ultra-deepwater environments. Seafloor production systems mainly consist of subsea templates, wellheads and production trees that meter and control the flow of fluids both into and out of oil and gas wells, wellheads, manifolds, subsea and topsides or remote control facilities, and flowlines and their connection systems (Offshore 5/28/15b).

By and large, these technologies will be directly applicable to natural gas derived from NGH without any significant or costly modification. In fact, because the water and gas stream from converted NGH is liable to be purer and less chemically hazardous or polluting, less robust (and less costly) equipment may be used for NGH production than for conventional. For instance, automated communication cable laying (Choi et al. 2015) can be used to connect equipment on the seafloor robotically and upgraded capability could lay pipes. Some technologies, on the other hand, will be greatly over-specified for NGH production. For instance, Subsea Services Alliance plans to produce a deepwater 15,000 psi intervention riser suitable for handling high temperature and high pressure conventional gas and oil (SSA 2015). DOE is supporting work under the auspices of the Research Partnership to Secure Energy for America to develop deepwater riser technology that at the wellhead will be safe at a maximum operating pressure of 12,000 psi and maximum operating temperature of 350 °F (Head and Long 2016). Produced NGH natural gas will have pressures between formation pressure and seafloor pressure (only slightly higher than ambient pressure before lift compression) and in its temperature range of about 10–30 °C. This means that much less costly riser systems can be implemented for NGH natural gas.

7.5.1 Gas Scrubbing, Separation, and Compression/ Artificial Lift

Compression accelerates gas flow and accelerates lift to a production platform or into a transmission pipeline. Compression compensates for decreasing pressure in the reservoir or in a reservoir in which pressure is abnormally low. This type of lift is not to be confused with 'artificial lift' that includes a variety of methods for increasing pressure in the reservoir to boost production where natural reservoir pressure is not enough to provide an acceptable production rate unaided (Kammerzell 2015).

From its beginning in 1994 at a water depth of 50 m, subsea boosting projects, which increase the pressure in the well stream, have been placed in gradually increasing water depths. Conventional wells are now being drilled in the lower part of the NGH Prospect Zone (Chap. 3). Transocean Ltd. reported that the ultra-deepwater drillship *Dhirubhai Deepwater KG1* has set a new world record for the deepest water depth by an offshore drilling rig (Offshore 7/8/13). The rig recently spudded a well in 3174 m of water while working for ONGC off the east coast of India. In 2001, the first seafloor separation system was installed at Equinor Troll C platform in 340 m water depth while the deepest system so far is Shell's Perdido field at about 2500 m (Müller 2015). Subsea water injection, which is used to increase recovery, can now be placed on the seafloor, requiring only the supply of electricity to operate rather than extensive infrastructure connecting the seafloor with topside.

Equinor has the vision of establishing 'seafloor factories' for gas and oil processing. It has taken the lead in developing and installing this technology and considers that subsea gas compression is a future technology that has arrived. Moving the gas compression from the platform to the wellhead substantially increases the recovery rate and life of the fields. Prior to gas compression, gas and liquids are separated out, and then after pressure boosting are recombined and sent through pipelines to production platforms and trunk pipelines. In addition to improving recovery, subsea gas compression will be more energy efficient than the traditional topside solution. The technology significantly cuts energy consumption and CO_2 emissions over the field's life to about half of the cost of compression from topside. The Gullfaks and Åsgard fields are world firsts in subsea compression (Davies and Ramberg 2016). The technology has a low environmental footprint compared to a compression platform although a significant preparation of the seafloor was required. This is all part of a learning process of moving equipment to the seafloor.

Gullfaks: The first subsea wet gas compressor installation was successfully installed at the end of June, 2015 at the Equinor Gullfaks C platform at 135 m depth about 100 km west of Bergen, Norway (Kliewer 2007). This involves two 5 MW compressors running in parallel with capacity for 10 MMm^2/d. Because the boosting equipment draws from the gas supply, there is some question whether separate depressurization pumps and compression booster pumps will be necessary or whether one in-line pump could do the job of reservoir depressurization and boosting. Subsea gas compression equipment is already being installed and its use will widen to increase production from conventional fields by applying suction to the reservoir. This subsea compression facility, which is estimated to add 22 million barrels of oil equivalent, and extend plateau production by approximately two years (Reed 2015; Offshore 7/7/15) came on-stream in September 2015 (Equinor 11/3/15) and into routine production on 7 October 2015.

FMC Technologies Inc. was contracted by Equinor to develop the processing system. FMC called on Framo Engineering for contributions to the subsea pump and boosting, and CDS Engineering for its gas-bypass design to keep the seabed separator as compact as possible. Assembly took place in March at Tønsberg in southern Norway where the 500-ton (453-metric ton) foundation base structure and some of the modules were built. Because of the restricted crane capacity available, the base structure was rolled onto a barge and the two heaviest modules—the 230-metric-ton (253-ton) manifold and 170-metric-ton (187-ton) separator vessel—were lifted onto it by a floating crane. The partly assembled station then was rolled back into the construction hall for installation of the desander, multiphase and water flow modules, and the two 2.3 MW pumps. Considerable care and expenditure on preparing proper seafloor foundations for the equipment may be cut in the future as innovations are made.

Åsgard: Testing of the Siemens compressor at Equinor's Åsgard Minimum Flow Project is ongoing. Equinor's goal is to have two 10 MW gas compressor systems running in parallel in about 250–300 m of water powered from and controlled from the Åsgard B platform to boost gas production from two nearby fields. Gas is piped 40 km to the Åsgard B semisubmersible production platform (Paganie 2015). Boosting technology is expected to spread rapidly.

For instance, OneSubsea, a Cameron and Schlumberger company, has installed a smaller subsea production system including wet gas pumping systems for a 13 well gas project offshore North Africa. This will include subsea production equipment, tooling, and installation and commissioning services with deliveries beginning in Q3 2016. In the Norwegian Sea, Equinor has contracted for 22 seafloor gas compression modules in the Åsgard Field, the largest of which will weigh 70 metric ton (Offshore 6/9/15). Other subsea power systems are currently being planned for new fields and retrofitted to producing fields to improve production performance (ABB 2017).

FMC Technologies and Siemens Industrial Turbomachinery are promoting industry interest and support for the development and marketing of an electrically-driven, centrifugal gas compressor for operation in water depths up to 3000 m that will be capable of operating for several years without requiring maintenance.

Adapting existing boosting and lift equipment to operate efficiently will be a NGH development issue. It is likely that little adaptation of existing gas compression equipment will be necessary to apply it directly to NGH production, especially if deepwater and ultra-deepwater units have already been developed. Subsea gas compression and boosted lift to the surface from the wellhead will be a vital part of production from NGH deposits. The gas pressure within the collector pipes in the reservoir and leading to the wellhead will be lower than formation pressure, and possibly lower than the ambient seafloor pressure. As a result, the production rate of gas from the wellhead would be much too slow to support a commercial extraction rate without management of gas pressures within the well and boosting away from the wellhead.

7.5.2 Sand Control

Sand control is regarded as potentially a more serious production issue with NGH than with conventional gas because the NGH reservoir host sediments will be much shallower. Buried, semi-consolidated turbidite sands (Egawa et al. 2013) that are mechanically weaker than the geological reservoirs of more deeply buried conventional gas deposits. Indeed, the slightly early shutdown of the March 2013 JOGMEC technical production test in the Nankai NGH field was reportedly due to sand flow blocking the wellbore and preventing gas flow (JOGMEC 2013). However, sand control issues have regularly plagued recovery of conventional gas. Each of the currently used technologies for sand control (substantial gravel packs, sand screens, frack-packs, and other technologies) were all once new when introduced. The need to control sand produced along with the gas will be ongoing, as is the development of new technology to improve performance.

Hydraulic screens (Offshore 9/10/15) are a recent alternative to gravel packing and can be used to supplement existing mechanical sand control technologies. Another non-mechanical method is chemical sand control. This allows intervention without workover and placing new mechanical sand technology is also now possible (Piemontese et al. 2014a). Chemical treatment can create an ionic attraction between sand and fines and to agglomerate and reduce or stop fines migration. This modifies the relative permeability of the reservoir host sediment. A quasi-chemical method being improved for conventional open-hole wells, which may have a close similarity to NGH drilling methods, involves using swell packers that can conform to the rough, irregular shapes of boreholes where there is no casing (Simonds and Sladic 2015). The swelling initiates when the material carried in collars is exposed to water in the well and isolates sections of the sand screen so that asymmetrically high production does not take place further up the hole, to which water and gas in the reservoir is liable to flow without packers. Swell packers have a proven track record in conventional oil and gas well preparation, but they may take up to two weeks to swell fully, which in high cost deepwater wells can be prohibitive. Considerable effort is underway to improve swell packer performance for deepwater, and some of the technology that will be developed for NGH wellbores, may be back-fitted for conventional hydrocarbons production.

Accordingly, for sand control, existing technology and new developments in place for conventional gas may only require modest adaptation, although new technology shows promise in some situations (Piemontese et al. 2014b). There should be little difference between NGH and conventional sand control issues, except that possibly more sand may be expected to be produced with NGH conversion because of the lower induration of the host sediments. On the other hand, because reservoir pressure can be controlled, an innovative wide-area collector system of complex well bores could be used to help control sand production (Chap. 8). Reservoir engineering based on production simulations (Worthington and Hattingh 2014) and fluid flow production simulations (Segura et al. 2011) will be needed to optimize both gas production and sand production.

7.5.3 Flow Assurance

The science and practice of assuring flow in pipelines and equipment is critical. Different classes of flow obstruction materials such as tars, asphaltenes, waxes, and other materials may form in clumps, lumps, or large masses across entire flow section, in addition to NGH. Although unwanted hydrate may form and block flow, they may also be the easiest to remediate because their formation and dissociation can be controlled by modifying pressure or temperature conditions to imitate the highly reversible hydrate chemical reaction (Chap. 2). Since the 1930s, the unwanted formation of NGH has been recognized as an industrial issue because very stable blockages of crystalized unwanted hydrate can form in oil and gas pipelines and obstruct or completely block flow. Flow assurance is important from a commercial point of view because blockages can stop oil and gas production, which conventionally cannot be restarted until the blockage has been cleared. Also, gas leakage at wellheads can encrust the critical components of blowout preventers and processing equipment and interfere with their operation and reduce overall safety.

Flow assurance has two main objectives; the first is to stop or retard the formation of unwanted NGH and the second is to mitigate or clear a blockage after it has formed or started to form. The use of both kinetic and thermodynamic inhibitors that slow down NGH nucleation and growth or cause the NGH to form small agglomerations that will flow in oil (cold flow) can be used to mitigate the potential of NGH to block flow. These inhibitors take the form of methanol and chemicals that interfere with crystallization. In addition, there are non-chemical flow assurance solutions (CRM 2016). These are not discussed here because they will be largely unnecessary for flow assurance of NGH, especially when depressurization is used as the primary NGH conversion technique, and gas flows at a certain level can be maintained (Chap. 8).

Clearing NGH blockages can be done in a number of manners. The most disruptive is to recover the pipe to the surface and clear the blockage physically, and then relay the pipeline. There are a number of images of this operation available. This remediation (Offshore 6/10/15), along with physical and chemical cleaning methods inside the pipe are most common for the non-NGH materials such as tar, creosote, and other semi-solid hydrocarbons that tend to be chemically stable, it has also been used for major NGH blockages. Unwanted NGH formation can be treated in a number of ways. Because NGH is susceptible to changes in temperature and pressure, it can be induced to dissociate within a pipe or equipment without opening or entering the pipe. Depressurization may be difficult to effect or it might cause some safety issue that could result in the collapse or rupture of a pipe. Heating tends to be the best treatment, and it is increasingly being administered by ROVs carrying heating equipment (Hernandez 2015). In addition, well vacuuming can remove flowline NGH and other materials (Duque et al. 2012). Active heating of seafloor processing systems removes the need for chemical injection and other thermal techniques (Duke et al. 2017).

There are a number of chemical approaches including kinetic inhibitors, which generally retard nucleation and thermodynamic inhibitors, which retard growth. Methanol injection has been a proven technique used on land and extended into the offshore but long pipeline lengths requiring huge infrastructure and materials in the ultra-deepwater are proving to be an expensive problem. Many non-chemical solutions also exist (C-Ray 2014). We do not discuss conventional methods for unwanted NGH remediation because they are generally not required for NGH production.

Considerable research and definition engineering are ongoing on subsea heating options to prevent formation of hydrate or cause their dissociation. For instance, in the 30 km gas pipeline which runs from the KME field in the South China Sea, ~130 km offshore Sabah, East Malaysia in 750 m of water to the new shelf edge Kebabangan platform, NGH remediation is an important issue (Offshore 6/17/15). Both microwave and laser heating from outside the pipe cause dissociation of NGH inside the pipe. Microwave heating is technically feasible but development emphasis appears to be on laser-based NGH remediation technology. Both are directly applicable to NGH-produced natural gas, as they can treat unwanted NGH formation both within equipment and in the ambient seafloor industrial installation without physical intervention.

Laser heating is being driven by the evolution of fiber and diode laser efficiency, reliability, solid state operation. Current (2016) versions will operate in up to 3000 m, but their scope is for any water depth. Decreased costs of off-the-shelf lasers that will work underwater with little modification are also a driver for laser NGH mitigation. Laser-based hydrate remediation technology uses focused warming to create a relief path for pressure equalization and chemical flow across the hydrate plug. The technology also can create a path for chemical flow across a hydrate plug. This enhances the effectiveness of the hydrate remediation methods based on chemical inhibitors, allowing the media to attack the plug from multiple directions in the pipe. The tool is based on a string of laser heads ranging from short sections or multiple sections that apply thermal energy to the subsea structure. The laser string is mounted on a buoyancy-compensated ROV. As the lasers heats the pipe surface directly, and up to the entire circumference of the pipe that is not resting on the seafloor can be affected, the water and gas that dissociates from the plug is diluted in the flow mixture. If dilution lowers the concentration of the hydrate-forming components enough, NGH will not reform (Sect. 2.1.2).

7.5.4 *Floating Gas Compression and Transport for Stranded Gas*

Remote oil and gas that cannot be transported by pipeline must be shipped. In order to have the most efficient transport, gas must be compressed. The offshore oil and gas business has been traditionally focused on oil, but discovery of large gas deposits in ultra-deepwater remote areas with limited or no access to local gas markets are driving new transport requirements. In the future, because of CO_2 considerations, gas may also become the hydrocarbon fuel of choice. Also, largely because of climate concerns, flaring of incidental gas produced along with oil is increasingly prohibited or heavily regulated to reduce methane and CO_2 emissions to the atmosphere, which is the result of large-scale flaring. Traditionally, LNG plants are onshore and the gas is brought to them, usually by short pipelines from nearby offshore or overland, where gas can be gathered from huge areas. In the case of stranded gas, there is no easy access to natural gas because ultra-deepwater conventional oil and gas deposits are being found at increasing distances from the nearest shore and in more remote regions. In order to capture stranded gas and bring it to market, the gas must first be compressed into a higher energy-density form on ships. Responding to the increased demand for delivered natural gas, new vessels are being brought into service and more are being planned (Offshore 11/3/15).

Aspirations for large conventional oil and gas discoveries are pushing exploration into deepwater, which can only be serviced by combining the new technology and vessels to effectively service deepwater (Khurana 2015). LNG plants are now being brought to the gas supply so that once stranded natural gas can be added to the world gas supply (Guttulsrod 2015). Floating LNG (FLNG) 'ships' developed and currently under construction show how industry is adapting to the drive for increased world gas supplies (Sect. 8.5.4). FLNG ships have to be very large, over 300 × 60 m, with a topside weight of ~35,000 metric ton in order to make the project financially viable (Loose et al. 2015). Large capacity FLNG could have dimensions of up to 500 × 75 m and topside weight much greater than 50,000 metric ton. Shell's Prelude FLNG at 488 × 74 m is the largest FLNG currently planned or under construction. Because it can take between 12 and 24 h to restart the liquefaction process, leaving station to deliver LNG will be less common than at-sea LNG transfer.

These vessels, and all of the forms of compression that are being designed for conventional natural gas can service production from NGH without any modification, and are a classic example of how an investment for a particular gas resource in the deepwater environment can service all without requiring special modifications. Once the natural gas from a NGH deposit has been brought to the surface or to land it does not differ in any significant way from conventional natural gas, except perhaps that is would be purer and much lower in pollutants than conventional natural gas. The emerging response to far offshore and remote conventional natural gas, which is to build floating gas compression facilities, can be directly leveraged for NGH transport.

The predominant form of gas compression for transport in the energy industry today is liquefied natural gas (LNG), which is produced through energy—intensive refrigeration. Other forms of compression are compressed natural gas (CNG), in which the gas is simply compressed and refrigerated to reduce the heat of compression, gas to liquids (GTL) , in which a synthetic near-diesel liquid fuel is created, and as artificial gas hydrate (AGH) , and synthetic methane hydrate fuel (SMHF) (Max et al. 1997), which could be produced by pressurization and refrigeration (Sugar 2011). Both AGH and SMHF involve compression of about $\times 164$ through recrystallization that forces the gas molecules close together in the NGH crystal lattice (Chap. 2). These are tabulated with technical parameters and discussed in Max et al. (2013, Sect. 13). LNG is the only actively practiced gas transport method currently being used commercially. CNG and AGH, as well as GTL are all in various phases of development. For large and mid-size gas fields, converting the gas to LNG on a special vessel (floating LNG ship or FLNG) provides a commercial avenue to direct export worldwide (Offshore 7/29/15). To service an enlargement of the LNG fleet to provide transport necessary to underpin a worldwide LNG business there is an increasing need for natural gas transport shipping. FLNG is currently only being constructed to service very large natural gas fields.

Although some of the FLNG vessels are very large and are intended to process the entire production from major fields for decades, small modular LNG plants that are built in a factory and then transported to a site are currently available for sale (GE 2014). These smaller modular plants with dramatically lower CAPEX could be installed on smaller vessels, perhaps even disused container vessels that have large hold and deck space. Modules could be configured to container or multiple container sizes so that they could easily be installed and uninstalled without significant modification to the vessel itself. Utilizing modified vessels in this way could be much less expensive than new very large designs. More than one of these smaller vessels could service a single location, providing redundancy.

The economics of compressing stranded gas constitute a special case for determining commerciality. In contrast to a normal calculation for revenue and costs as would be the case when the feedstock natural gas would be equally available to all forms of compression, 'rescuing' stranded gas is key to bringing it to market. Because some of the produced natural gas that could otherwise not be brought to market can be used to provide the energy necessary for the compression, the cost of carrying out each form of compression may be discounted. Although land LNG plants have the lead for large cargos and virtually indefinite gas supply, for stranded gas there will be a complex formula involving CAPEX of each compression/transport system and the energy density for each (Btu/shipload) that will determine the Btu value of each load. When load income is taken into account, LNG may not prove to be the most efficient or profitable form of transport for all stranded gas, especially for smaller fields (Table 7.2). LNG is also now regularly being used as a fuel for seagoing vessels such as offshore supply vessels (OSV) (Offshore 11/4/15).

FLNG designs are considered to be "novel" in comparison to floating production, storage and offloading facilities (FPSOs), which is very similar to a large drilling platform. New mooring technology is also required for such large, station-keeping vessels (Kliewer 2014). The first offshore LPG storage and offloading unit was classed by ABS (2016) in the mid-1970s, the first purpose-built LNG storage and offloading unit was delivered in 1997, and the first FPSO which was capable of producing LNG (FPSO units) was put into service in 2005. The 2007–2014 period saw the development of multiple FLNG designs. There are a

Table 7.2 Modified from Max and Johnson (2014). Gas transport options for high Arctic stranded gas

Fuel type	Transport conditions	Energy density (Btu/ft^3)
Natural gas (STP)	Not transported at STP	1000 +[!]
Pipeline	Compressed, rate of transport adds time factor	Depends on pressure and velocity
CNG	Compressed	235,000 @ 205 Atm
LNG	Compressed and cooled 25 kPa–162 °C	563,000
AGH/SMHF	−20 C 1 Atm	164,000
Gas to liquids	Liquid ambient P-T	~800,000–900,000 (est.)
Diesel	Liquid ambient P-T	939,000
Gasoline (petrol)	Liquid ambient P-T	805,000

Diesel and petrol (gasoline) are shown for comparison. Rounded to nearest 1000. AGH/SMHF assumes variable occupation of entire volume, actual for pelletized material will be less. Natural gas energy density is for methane, all other natural gas mixtures have higher energy density. Other fuels energy densities (DOE 2016) are not direct competitors. CNG depends on tank pressure; higher-pressure tanks cost more initially. Natural gas (at STP) is about 0.1% energy density of oil while NGL (Natural Gas Liquids) can be as much as 65% of oil depending on their mixture. In addition to LNG, provision for separation of condensate liquid petroleum gas (LPG) (mainly propane or butane) may be required. LPG and condensate add considerable value and improve overall economics of thermogenic natural gas deposits

number of FLNGs now under construction and to be brought into service beginning in 2016. This sector is one that will continue to see growth, and possibly more advanced designs such as autonomous or unmanned FLNG (Maslin 2015). Total expenditure for the 2014–2020 period for FLNG-related hardware could total as much as $64.4 billion (Guttulsrod 2015).

7.5.5 *Water Injection/Extraction*

Injecting and extracting water from ultra-deepwater wells may be necessary for the efficient operation of 'subsea factories' with oil and gas production facilities located directly on the seabed. If reservoir pressure is too low or drops as a function of production or longer step-outs, compressing gas on the seafloor to expedite transport of the gas to the surface is one of the next frontiers in offshore engineering. Gas compression equipment, having been placed on the seafloor, is expected to operate throughout the life of the field. This requires long-term reliability and maintainability a critical part of every component. Design for seafloor industrial sites should include redundant parts for back up, along with plug-and-play design for remote replacement of important parts such as water pumps.

The principal method for conversion of NGH to its constituent natural gas and water is emerging from a number of options (Max and Johnson 2011) as depressurization. JOGMEC's 2013 technical production test on the Japanese Nankai NGH concentration (JOGMEC 2013) all but confirmed that downhole pumping for depressurization was practical and efficient. Downhole electric submersible pumps (EPS) are gaining in flexibility and dependability, and most important, in their ease of swapping out or replacement (Algeroy and Nutter 2017) for conventional wells. Controlled depressurization using downhole pumping is regarded by us as one of the keys to NGH production. Thus, modeling how to best incorporate ESPs in any production system will be fundamental to reservoir management models.

7.5.6 Realtime Monitoring of Reservoir Conditions

Real-time monitoring of reservoirs using acoustic arrays on the seafloor to monitor 4D changes in a reservoir are critical to well and reservoir control. For instance, the Equinor Sverdrup field in the Norwegian sector of the North Sea is to feature permanent reservoir monitoring from start-up. The objective is to achieve a goal of 70% recovery (Offshore 1/17/18). Sverdrup has contracted with Alcatel Submarine Networks to install 380 km of fiber-optic seismic cables on the seafloor and more than 6500 acoustic sensors covering an area of more than 120 km^2 permanent reservoir monitoring (PRM). This will allow for continuous recording of changes in the subsurface. Maintaining a complete picture of fluid movement and compaction and minor structural adjustment (possibly small-scale faulting) will allow fine adjustments to optimize production. The full-time monitoring should also serve as a basis for future recovery interventions in the field such as infill wells, water and gas injection, and 3D and 4D visualization of the reservoir throughout its projected 50 year life. In addition to monitoring requirements for conventional gas and oil wells, sensors within the reservoir will also be necessary. NGH production alters mechanical strength of the reservoir and heat control and inducing fluid movement within the reservoir will also be important.

NGH significantly stiffen or increase the mechanical strength of strata with NGH concentrations because water is incorporated in the solid NGH or displaced during formation. Bulk modulus is increased as NGH replaces water in pore fill and if enough NGH forms it becomes linked generally within porosity, shear modulus also increases. Thus, the mechanical strength of any NGH concentration will be greatest at the initiation of NGH conversion and natural gas production. As NGH conversion proceeds, mechanical strength will decrease. Although conventional reservoirs under production are already being monitored so as to be able to respond to changing conditions in a reservoir, advanced monitoring is required for NGH production because conversion procedures will have to be altered in nearly real time to maintain mechanical coherence of the reservoir as well as to purposefully alter pressure within a reservoir from place to place to achieve objectives that can only be done in NGH reservoirs. At present, shooting of conventional high-resolution surveys is achieved mainly using towed surface seismic arrays.

Full-time, rapid-response monitoring or 4D reservoir surveillance of production-induced changes in a NGH reservoir can most accurately be carried out remotely entirely from the seafloor using acoustics, from which mechanical strength calculations can be made based on the changes in Vp within the reservoir. Deepwater node systems based on ocean bottom seismometers (OBS) use high quality 3D seismic data in many oil fields to support field development (Hayes 2017). Optimization of software for the shallower GHSZ Monitoring wells should allow existing hardware and the experience of establishing and operating systems to be directly applied to monitoring NGH conversion and also the movement of gas and water relative to remaining NGH. Incipient potential failure zones within the reservoir may be controlled by NGH conversion management wherein conversion conditions from place to place in the reservoir may be controlled.

Monitoring of pressure and temperature, which can be used to indicate rate and degree of NGH conversion, can be made using sensors within production and narrow bore observation wells (7.4.6). An advantage of a full seafloor system is that acoustic sources such as Helmholtz projectors for pressure wave generation and shear wave generators in which wave form and propagation direction can also be placed on the seafloor so that acoustic paths can be known with very high confidence. This will allow a step increase in reservoir resolution and a realtime reservoir monitoring solution to NGH production.

7.6 Modularization of Equipment

Traditionally, individual ships and semi-submersibles being used for exploration and production had purpose-built equipment installed prior to being placed on site. We anticipate that this design philosophy will follow through to NGH seafloor industrial sites. More innovative design and manufacturing techniques for breaking the equipment down into modules that can be easily transported and assembled on site allow for whole-system testing prior to installation just as effectively as a purpose-built installation. This aspect of system design has many implications for the establishment of remote or seafloor industrial sites, particularly for reducing costs and reducing assembly time using very cost-efficient designs. Many modular offshore systems are now pre-assembled at the factory for deployment to a marine platform in port or at sea. Whole production systems, for instance, can be fabricated in modules including electronics, controls, and production equipment so that it can be factory-tested before it is disassembled and transported for installation.

Although modularization is currently cutting costs in equipping floating platforms, sometimes at sea and at its anchored site, modularization is particularly appropriate for equipping a seafloor industrial site. Each module can be of a size and weight such that they can be transported to a site following full system testing under controlled conditions. Relatively small vessels, in relation to the standard floating exploration and production platforms, can transport modules, lower them to the seafloor (Millar 2015; SB 2015), and then assemble them on the seafloor using ROVs and AUVs that have been purpose built.

7.7 Leveraging of Conventional Technology

Transitioning production equipment and some drilling equipment to the seafloor is already taking place and 'lightweight', cost-effective operational infrastructure is also becoming available to service it. Many of these developments can be leveraged for NGH exploration and production. Many, such as electrical connections, electronic control systems, pipes and pipe fittings designed for ROV assembly, amongst other technologies, can be used more or less off the shelf. Others may need some modifications, which in many cases need not be as robust as those used for conventional deepwater hydrocarbons. Thus, because the operational requirements of NGH are less demanding, existing technology that can be leveraged can probably be made less robust and less expensive while achieving necessary design life.

Improvements in technology can be incremental or step-like, in safety, efficiency, and cost. Very few technologies cannot be improved and many technologies used today were introduced over a 100 years ago and are still being improved. Modifications of technology for new circumstances such as new environmental and operational conditions, as the opportunities in NGH offer to virtually every conventional technology (Chap. 8), indicate to us that NGH-specific operations may be undertaken very rapidly with only little modification to existing technology. Also, the greater proportion of technology development and improvement in the offshore energy industry are in the water depth range where potentially exploitable concentrations of NGH occur, which means that NGH technology development can piggy-back financially on conventional deepwater opportunities.

References

ABB. (2017). *Long step-out systems for subsea pump and compressor applications economic hydrocarbon recovery under extreme conditions.* ABB white paper (20 pp). http://www.lead-central.com/AssetManager/02427e68-6f15-4f3a-9749-d37abf613741/Documents/Long_Step_Out_Power_System_9AKK107045A2327%20(1).pdf. Accessed October 23, 2017.

Abili, N., Citi, G., & Gassert, M. (2014, December 12). Fluid sampling, subsea processing help maximize deepwater development. *Offshore Magazine*. http://www.offshore-mag.com/articles/print/volume-74/issue-12/subsea/fluid-sampling-subsea-processing-help-maximize-deepwater-development-p1.html. Accessed April 23, 2016.

ABS. (2016). *Floating liquefied natural gas*. American Bureau of Shipping. http://ww2.eagle.org/en/what-we-offer/gas-solutions/floating-liquefied-natural-gas.html. Accessed March 22, 2016.

Algeroy, J., & Nutter, G. (2017). Offshore rigless conveyance system: AccessESP enables efficient electric submersible pump replacement. *Sea Technology, 58*(4), 15–116, 118.

API. (2014). *Recommended practice 2A-WSD planning, designing, and constructing fixed offshore platforms—Working stress design (Outline of issues)* (7 pp). American Petroleum Institute. Accessed November 1, 2015.

Armitage, P., & Hernandez, F. C. (2017). Rigless technologies are re-shaping the subsea well intervention market. *Offshore, 77*(7), 36–37.

Ball, E. (2014). Seafloor factory concept drives subsea power technology development. *SPE News* (6 pp). http://www.spe.org/news/article/seafloor-factory-concept-drives-subsea-power-technology-development Accessed May 31, 2016.

Banks, M. (2018, February 24–25). OGTC backs P&A progress. *InnovOil* 60.

Barnes, C. R., & Tunnicliffe, V. (2008). Building the world's first multi-node cabled ocean observatories (NEPTUNE Canada and VENUS, Canada): Science, realities, challenges and opportunities. In *Marine Technology Society OCEANS 2008 Proceedings* (pp. 1213–1220).

Bauer. (2015). Sea floor drilling. http://www.bauer.de/export/sites/www.bauer.de/pdf/bma/products/bmt_maritime_technologies/Seafloor_drilling.pdf.

Beckman, J. (2014). Subsea power grid accelerating transfer of topsides functions to the seafloor. *Offshore, 74*(10), 84–85.

Beckman, J. (2016, April). Normally unattended platforms offer opex, capex reductions. *Offshore Magazine, 76*(40), 65–66.

Beckman, J. (2017). Vessel-deployed coil tubing system offers benefits for downhole maintenance. *Offshore, 77*(3), 46–47.

Begos, K. (2012). Fracking developed with decades of government investment. *Huffington Post.* http://www.huffingtonpost.com/2012/09/23/fracking-developed-government_n_1907178.html. Accessed November 3, 2015.

BH. (2017). *OilPro. Baker Hughes announces the industry's first adaptive drill bit* (2 pp). http://oilpro.com/announcement/4612/baker-hughes-announces-industrys-first-adaptive-drill-bit?utm_source=DailyNewsletter&utm_medium=email&utm_campaign=newsletter&utm_term=2017-03-14&utm_content=Article_5_txt. Accessed March 14, 2017.

Bibby. (2015). *PennEnergy. Bibby Offshore Singapore delivers subsea projects to Asia Pacific.* http://www.pennenergy.com/articles/pennenergy/2015/07/bibby-offshore-singapore-delivers-subsea-projects-to-asia-pacific.html?cmpid=EnlDailyPetroJuly72015&eid=288179069&bid=1113517. Accessed April 23, 2015.

Bourillet, J. F. (Coord.), Augris, C., Cirac, P., Mazé, J. P., Normand, A., Loubrieu, B., et al. (2007). *European ocean research fleets. Towards a common strategy and enhanced use* (64 pp). Position Paper 10, European Science Foundation Marine Board.

Cathx. (2015). *Cathx ocean.* http://www.cathxocean.com. Accessed April 23, 2016.

Cellula. (2016). *Company website.* http://www.cellula.com.

Choi, J.-K., Yokobiki, T., & Kawaguchi, K. (2015). Automated cable-laying system. *Sea Technology, 56*(12), 43–45.

CMRE. (2015). *Autonomous naval mine countermeasures.* Status page. http://www.cmre.nato.int.

Conrad, M., & Schreiber, S. (2017). Military digital underwater communication. *Sea Technology, 58*(5), 10–14.

Crager, B. (2014, March 21). *An overview of subsea systems. SUT subsea awareness.* Texas A&M University, Endeavor Management ©2014, Posted to Internet. http://oilpro.com/post/4421/green-field-subsea-boosting-life-of-field-boosting-unconstrained. Accessed April 23, 2016.

Crager, B. (2015). *Benchmarking of subsea decommissioning, part 1.* http://oilpro.com/post/4421/green-field-subsea-boosting-life-of-field-boosting-unconstrained.

C-Ray. (2014, April). *2014 survey of offshore non-chemical flow assurance solutions.* C-Ray Media, Poster. Sea Technology One sheet inset.

CRM. (2016, April). *2016 survey of offshore non-chemical flow assurance solutions.* Poster compiled by C-Ray Media, Inc. Offshore Magazine. One sheet.

C&EN. (2015, June). Electronic stent monitors blood vessels. *Chemical & Engineering News. American Chemical Society*, 29.

Dabade, T., Subramanian, S., Roth, R., Ponagandia, V., McKee, M., Paulin, M., et al. (2017, March). Worldwide survey of subsea processing and active heating system. *Poster, Intercsea, Inc. & Offshore Magazine, 77*(3).

David, O. O. (2006). Limit of through-tubing rotary drilling (TTRD) well completion. *Journal of Applied Sciences, 6*(15), 3116–3121. ISBN 1812-5654.

Davies, S., & Ramberg, R. M. (2016). Gullfacks, Åsgard represent world-firsts in subsea compression. *Offshore Magazine, 76*(1), 52–53.

Davis, D. (2015). *Delta SubSea and VEGA sign agreement to develop subsea gamma ray hydrate localization tool ("EUREKA") delta subsea news*. http://www.deltasubsea-rov.com/media-gallery/press-releases/entry/delta-subsea-and-vega-sign-agreement-to-develop-subsea-gamma-ray-hydrate-localization-tool-eureka. Accessed May 1, 2015.

DC. (2012). Dual-gradient system evaluation highlights key high-risk issues. *Drilling Contractor, Innovating While Drilling, July/August, The Offshore Frontier*. http://www.drillingcontractor.org/dual-gradient-system-evaluation-highlights-key-high-risk-issues-17111. Accessed April 23, 2015.

DOE. (1998). *A strategy for methane hydrates research & development* (58 pp). DOE/FE-0378. U.S. Department of Energy Office of Fossil Energy. Accessed November 27, 2015.

DOE. (2015, February). *Publications and presentations of DOE supported methane hydrate R&D 1999–2015* (282 pp). The U.S. DOE Methane Hydrate R&D Program. Accessed November 19, 2015.

DOE. (2016). *Energy density chart* (4 pp). Alternative Energy Data Center, U.S. Department of energy. http://www.afdc.energy.gov/fuels/fuel_comparison_chart.pdf.

Downton, G., Kausen, T. S., Hendricks, A., & Pafitis, D. (2000). New directions in rotary steerable drilling. *Oilfield Review* (Spring), 12 pp.

Duke, D., Dabade, T., Subramanian, S., & Roth, R. (2017). Active heating technologies enhance benefits of subsea processing systems. *Offshore, 77*(3), 53–53.

Dupre, R. (2015). Statoil's subsea factory tackles the power source hurdle. *Offshore, 75*(1), 76–77.

Duque, L., Guimaraes, Z., Chagas, J. V., Barros, R., Fonseca, P., & Siqueira, N. (2012, November). Concentric coiled tubing well vacuuming effectively removes flowline hydrate. *Offshore,* 80, 82, 84.

Durham, W. B., Kirby, S. H., & Stern, L. A. (2003). The strength and rheology of methane clathrate hydrate. *Journal Geophysical Research, 108*(B4), 2182. https://doi.org/10.1029/2002jb001872.

DW. (2015, January 2). Lower-cost subsea processing must be a priority, says Douglas-Westwood (Offshore Staff). *Offshore Digital Magazine*. http://www.offshore-mag.com/articles/2015/01/lower-cost-subsea-processing-must-be-a-priority-says-douglas-westwood.html. Accessed Oct ober 31, 2015.

Dykes, A. (2017). *100 Megawatts InnovOil December annual for 2017*, 46–47. http://edition.innovoil.co.uk/html5/reader/production/default.aspx?pubname=&pubid=e8e37904-9f19-4ca5-a43d-3b547fc6a3ce. Accessed.

Egawa, K., Furukawa, T., Saeki, T., Suzuki, K., & Narita, H. (2013). Three-dimensional paleomorphologic reconstruction and turbidite distribution prediction revealing a Pleistocene con- fined basin system in the northeast Nankai Trough area. *AAPG Bulletin, 97*(5), 781–798. https://doi.org/10.1306/10161212014.

EIA. (2011). *Technology drives natural gas production growth from shale gas formations*. U.S. Energy Information Administration. http://www.eia.gov/todayinenergy/detail.cfm?id=2170#.

Farrell, N. (2015, May). Small data: Applications for scaled-down satcom, using satcom to connect to monitoring networks. *Sea Technology,* 27–30.

Fehrenbacher, K. (2015). Big oil turns to big data as oil prices plummet. *Fortune*. http://fortune.com/2015/09/08/big-oil-big-data/. Accessed September 13, 2015.

FMC 6/2/15. (2015). *FMC Technologies and Technip complete Forsys Subsea*. http://ir.fmctechnologies.com/releasedetail.cfm?ReleaseID=915800. Accessed March 11, 2016.

Forster, L., & McKee, M. (2016). Advances in subsea processing offer greater design flexibility. *Offshore Magazine, 76*(3), 44–45. http://www.offshore-mag.com/articles/print/volume-76/

issue-3/subsea/advances-in-subsea-processing-offer-greater-design-flexibility.html. Accessed March 25, 2016.

Forster, L., Paes, T. M., & Baker, R. J. (2015a). Industry moves subsea processing toward standardization, consistency. *Offshore, 75*(3), 64.

Forster, L., Paes, T. M., & Baker, R. J. (2015b). Industry moves subsea processing toward standardization, consistency. *Offshore*. http://www.offshore-mag.com/articles/print/volume-75/issue-3/subsea/industry-moves-subsea-processing-toward-standardization-consistency.html. Accessed March 11, 2015.

Francis, D. (2014). ROVs evolve into underwater robots. *Offshore, 75*(5). http://www.offshore-mag.com/articles/print/volume-74/issue-5/subsea/rovs-evolve-into-underwater-robots.html. Accessed June 22, 2015.

Furlow, W. (2003). Subsea processing slow to take off despite advantages, maturing technology. *Offshore Magazine*. http://www.offshore-mag.com/articles/print/volume-63/issue-12/technology/subsea-processing-slow-to-take-off-despite-advantages-maturing-technology.html. Accessed July 9, 2015.

GE. (2014). The definitive guide to small-scale liquefied natural gas (LNG) plants. *General Electric*, 8 pp.

GEO. (2015). *The SUGAR project*. GEOMAR Helmholtz Centre for Ocean Research Kiel. http://www.geomar.de/en/research/fb2/fb2-mg/projects/sugar-2-phase/. Accessed April 26, 2016.

Ghiselin, D. (2014). Steering for the sweet spots. *Offshore, 74*(4), 86, 88.

Ghiselin, D. (2015). Advances in drill bits improve efficiency, performance. *Offshore, 75*(5), 90, 92.

Gutenberg. (2015). *List of research vessels by country*. Guttenberg Press. Website. http://www.self.gutenberg.org/articles/list_of_research_vessels_by_country. Accessed April 26, 2016.

Guttulsrod, T. -I. (2015). FLNG design adapts to an evolving energy mix. *Offshore, 75*(5), 96, 98. http://www.offshore-mag.com/articles/print/volume-75/issue-5/engineering-construction-installation/flng-design-adapts-to-an-evolving-energy-mix.html. Accessed May 25, 2015.

Hayes, D. (2017). Semipermanent reservoir monitoring with ocean bottom nodes, 4D seismic improves understanding reservoirs. *Sea Technology, 58*(4), 10–14.

Head, B., & Long, R. (2016). Research project seeks to deliver ultra-deepwater riser concepts. *Offshore* 3/10/16. http://www.offshore-mag.com/articles/print/volume-76/issue-3/engineering-construction-installation/research-project-seeks-to-deliver-ultra-deepwater-riser-concepts.html?cmpid=EnlOSDailyMay42016&eid=288179069&bid=1394652. Accessed May 10, 2016.

Hernandez, F. C. (2015). Fit-for-purpose ROVs can help remediate flow assurance issues. *Offshore Magazine*. http://www.offshore-mag.com/articles/print/volume-75/issue-10/flowlines-and-pipelines/fit-for-purpose-rovs-can-help-remediate-flow-assurance-issues.html?cmpid=EnlOSDailyOctober272015&eid=288179069&bid=1217137. Accessed October 27, 2015.

Holditch, S. A. (2013). Unconventional oil and gas resource development—Let's do it right. *Journal of Unconventional Oil and Gas Resources, 1–2,* 2–8.

Holme, L. P. (2016). Change in drilling strategy brings string of efficiency milestones to troll. *Offshore*. http://www.offshore-mag.com/articles/print/volume-76/issue-10/drilling-and-completion/change-in-drilling-strategy-brings-string-of-efficiency-milestones-to-troll.html. Accessed October 20, 2016.

Hydro. (2016). Eelume the snake robots, pros and cons. *Hydrographic Catalogue*. http://hydrographic-catalogue.com. Accessed May 15, 2016.

I-O. (2016). *Intecsea—Offshore Magazine Poster, March. 2016 Worldwide survey of subsea processing: separation, compression, and pumping systems*. Prepared by: L. Forster, T. M. Paes, R. J. Baker, D. Thati, D. Antilley, K. Hawthorn, J. Burke, A. Mendoza, J. Allen, M. McKee, E. K. Albough, & D. Davis, with poster assembled by C. Jones & S. Aguirre, 1 Sheet.

Inzartsev, I., Kamorniy, A., Kiselyov, L., Matviyenko, Y., Rylov, N., Rylov, R., et al. (2010). Integrated positioning system of autonomous underwater robot and its application in high latitudes of Arctic zone. In F. Casolo (Ed.), *Motion control* (580 pp), 229–242. INTECH. ISBN 978-953-7619-55-8.

JOGMEC. (2013). *News release: Gas production from methane hydrate layers confirmed* (3 pp). Tokyo: Japan Oil, Gas and Metals National Corporation. http://www.jogmec.go.jp/english/news/release/release0110.html. Accessed April 26, 2016.

Jackson, E., White, D., & Spencer, A. (2016). CRD100 seafloor geotechnical drill. *Sea Technology, 57*(9), 10–13.

Johnston, C. (2017). Three emerging technologies are expected to advance subsea well operations. *Offshore, 77*(7), 34–35.

Kammerzell, J. (2015). Artificial lift—An essential technology for unconventional oil plays. *OilPro*. http://oilpro.com/post/10561/unconventional-artificial-lift?utm_campaign=newsletter378&utm_source=dailyNewsletter&utm_medium=email. Accessed April 26, 2016.

Kampmann, P., & Kirchner, F. (2012). A tactile sensing system for underwater manipulation. In *Proceedings of the Workshop on Advances in Tactile Sensing and Touch based Human-Robot Interaction. 7th ACM/IEEE International Conference on Human-Robot Interaction (HRI 2012)*, Boston, MA, 3 pp.

Kampmann, P., & Kirchner, F. (2014). Integration of fiber-optic sensor arrays into a multi-modal tactile sensor processing system for robotic end-effectors. *Sensors, 2014*(14), 6854–6876. https://doi.org/10.3390/s140406854.

Kezina, D. (2015). *Underwater robots could soon appear in Russia's Far East*. http://asia.rbth.com/science_and_tech/2015/06/09/underwater_robots_could_soon_appear_in_russias_far_east_46773.html. Accessed June 21, 2015.

Khurana, S. (2015). Online exclusive: New strategy needed for deepwater development, unconventional plays. *OilPro*. http://www.offshore-mag.com/articles/2015/01/online-exclusive-new-strategy-needed-for-deepwater-development-unconventional-plays.html. Accessed February 3, 2016.

Kintisch, E. (2013). A sea change for U.S. oceanography. *Science, 339*, 1138–1143.

Kliewer, G. (2007). Tordis becomes world's first subsea processing installation. *Offshore Magazine*. http://www.offshore-mag.com/articles/print/volume-67/issue-12/top-five-projects/tordis-becomes-worldrsquos-first-subsea-processing-installation.html. Accessed April 26, 2016.

Kliewer, G. (2014). RPSEA continues technology advance with ultra-deepwater R&D projects. *Offshore, 74*(10), 50, 53.

Kolle, J. J. (1989). *Deep-ocean seismometer implantation system. Phase 1*. http://oai.dtic.mil/oai/oai?verb=getRecord&metadataPrefix=html&identifier=ADA212153. Accessed October 3, 2015.

Kolle, J. J. (1999). *A comparison of water jet, abrasive jet and rotary diamond drilling in hard rock* (8 pp). Tempress Technologies Report. http://ttinc10.qwestoffice.net/papers/comparison.pdf.

Kolle, J. J., & Bibee, L. D. (1996). Deep ocean sensor implantation system. *International Journal of Offshore and Polar Engineering, 6*(4), 255–261. Accessed November 16, 2015.

Kolle, J. J., & Theimer, K. (2004, February 23–25). *Seismic-while-drilling using a swept impulse source* (9 pp). Paper: 92114 SPE/IADC Drilling Conference held in Amsterdam, The Netherlands.

Larmon, D. (2015). Unmanned oil rigs—Are they the way of the future? *The OGM Energy and Culture*. http://theogm.com/2015/06/12/unmanned-oil-rigs-are-they-the-way-of-the-future/. Accessed June 17, 2015.

LEA. (2002). *Introduction to underbalanced drilling* (51 pp). Leading Edge Advantage (www.lealtd.com).

Li, D. J., Wang, J., Zhang, Z. F., Yang, C. J., Jin, B., & Chen, Y. H. (2015). Research and implementation of an IEEE 1588 PTP-based time synchronization system for Chinese experimental ocean observatory network. *Marine Technology Society Journal, 49*(1), 47–58.

Livescu, S., Blanco, D., & Vacik, L. (2016). New coiled tubing system offers real-time monitoring capability. *Offshore Magazine, 76*(3), 39–40.

Lloyds. (2016, January). *Rules and regulations for the classification of ships* (1657 pp). Lloyds Register. http://www.lr.org/en/RulesandRegulations/ships.aspx. Accessed July 14, 2016.

Long, R. (2014). Industry collaboration works to advance subsea processing technology. *Offshore, 74*(11), 106, 108, 110.

Loose, A., Caswell, C., Bentley, M., LeBoeuf, R., & D'Souza, R. (2015). Early FLNG projects deal with technical and execution challenges. *Offshore, 74*(5), 94, 97–98.

Lumsden, R. (2015, December). The evolution of ROVs. *Sea Technology, 56*(12), 29–32.

Lux. (2015). *Lux innovation grid: Comparing drilling cost optimizing technologies*. From Lux Populi Newsletter—September 27, 2015, with kind permission of Christie Oliver and Brent Giles.

Maslin, E. (2015). Unmanned floating FLNG—The next frontier. *OE Digital*. http://www.oedigital.com/component/k2/item/9112-unmanned-floating-flng-the-next-frontier?tmpl=component&print=1. Accessed April 26, 2016.

Max, M. D., Dillon, W. P., & Malone, R. D. (1991a, April 23 and 24). In *Proceedings: Report on National Workshop on Gas Hydrates* (38 pp), U.S. Geological Survey. Reston, VA: U.S. Department of Energy DOE/METC—91/6/24, DE 91016654.

Max, M. D., Dillon, W. P., Malone, R. D., & Kvenvolden, K. (1991b, October 29). *National workshop on gas hydrates. (abs)*. EOS, Transactions American Geophysical Union. 72, 44, 476–477.

Max, M. D., & Johnson, A. H. (2011). Methane hydrate/clathrate conversion. In M. R. Khan (Ed.), *Clean hydrocarbon fuel conversion technology* (pp. 413–434). Woodhead Publishing Series in Energy No. 19. Cambridge, UK: Woodhead Publishing Ltd. ISBN 1 84569 727 8, ISBN-13: 978 1 84569 727 3.

Max, M. D., & Johnson, A. H. (2014). Hydrate petroleum system approach to hydrate natural gas exploration. *Petroleum Geoscience, 20*(2), 187–199. Geological Society of London (Published in Online First March 21, 2014). https://doi.org/10.1144/petgeo2012-049.

Max, M. D., Johnson, A. H., & Dillon, W. P. (2013). *Natural gas hydrate arctic ocean deepwater resource potential* (113 pp). Springer Briefs in Energy.

Max, M. D., Pellenbarg, R. E., & Hurdle, B. G. (1997). *Methane hydrate, a special clathrate: Its attributes and potential* (74 pp). Naval Research Laboratory Report NRL/MR/6101-97-7926.

Max, M. D., Spina, F., Bovio, E., & Vasseur, N. (1998, March 10–13). The E-Map, Innovative COTS software for data fusion and interactive communication for rapid environmental assessment. In E. Pouliquen, A. D. Kirwan Jr., & R. T. Pearson (Eds.), *Rapid Environmental Assessment, SACLANTCEN Conference Proceedings CP-44* (pp. 269–278). Lerici (La Spezia), Italy.

McConnell, D. R., Zhang, Z., & Boswell, R. (2012). Review of progress in evaluating gas hydrate drilling hazards. *Marine and Petroleum Geology, 34,* 209–223.

Meng, Q., Wang, H., Li, P., Wang, L., & He, Z. (2006). Dexterous underwater robot hand: HEU hand II. In *2006 IEEE International Conference on Mechatronics and Automation* (pp. 1477–1482).

Millar, J. (2015, June). 2015 worldwide survey of subsea construction vessels. *Offshore Magazine*. Poster.

Minyaeva, E. (2015). Automated drilling gains momentum in offshore operations. *Offshore*. http://www.offshore-mag.com/articles/print/volume-75/issue-10/drilling-and-completion/automated-drilling-gains-momentum-in-offshore-operations.html?cmpid=EnlOSDailyOctober272015&eid=288179069&bid=1217137. Accessed February 18, 2015.

Moline, M. A., & Benoit-Bird, K. (2016). Sensor fusion and autonomy as a powerful combination for biological assessment in the marine environment. *Robotics, 5*(1), 13 pp. https://doi.org/10.3390/robotics5010004.

Müller, J. F. (2014). *Global deepwater developments: Real supply additions or just fighting decline. Deepwater has the potential to grow production by close to 80% by 2020.*

Müller, J. F. (2015). *Subsea processing moves forward with challenges in tow*. http://www.offshore-mag.com/articles/print/volume-75/issue-4/subsea/subsea-processing-moves-forward-with-challenges-in-tow-p1.html. Accessed 29 Jun 2015.

Musarra, S. P. (2013). Statoil EVP: Subsea factories are business critical. *OE Digital*. http://www.oedigital.com/component/k2/item/4478-statoil-evp-subsea-factories-are-business-critical?tmpl=component&print=1. Accessed December 25, 2013.

Musarra, S. P. (2015, July). Integrated operations center keeps shell in sync with deepwater GoM assets. *Offshore*, 62–63.

NC. (2018). GE and Noble link up the world's first digital rig. *PennEnergy*. http://www.pennenergy.com/articles/pennenergy/2018/02/oil-and-gas-ge-and-noble-link-up-the-world-s-first-digital-rig-moving-towards-digitizing-marine-operations-excellence-and-analytics-based-on-predictive-models.html?cmpid=enl_pennenergy_pennenergy_research_2018-02-23&pwhid=7f771bc841f5df66ebcac04dbeb4928e7780bf3f90641c677bb3f8cc8d4941fc147ac159c501f5878031f6ba7f9e82feb9b8c395d6a3e9927d723448ebe1b263&eid=288179069&bid=2013609. Accessed February 23, 2018.

Noll, G., & Hogeweg, M. (2015, May). Bit data management at Port of Rotterdam; Using a GIS platform to streamline IT at growing maritime hub. *Sea Technology*, 31–32, 34.

OE. (2015). Saudi Aramco reaches subsea installation milestone. *Offshore Energy Today.com*. http://www.offshoreenergytoday.com/saudi-aramco-reaches-subsea-installation-milestone/. Accessed March 11, 2016.

OET. (2015). *Ocean Expedition Trust*. http://www.oceanexplorationtrust.org/#!2014-expeditions/cpn8. Accessed April 25, 2016.

Offshore 5/1/08. West African reservoir fluids intensify demands on separation design. *Offshore*. http://www.offshore-mag.com/articles/print/volume-68/issue-5/france/west-african-reservoir-fluids-intensify-demands-on-separation-design.html. Accessed September 1, 2016.

Offshore 1/25/13. *Ormen Lange: the beginning of entirely subsea oil and gas production?* http://www.offshore-technology.com/features/featureormen-lange-beginning-subsea-oil-gas-production/?WT.mc_id=WN_Feat. Accessed November 19, 2015.

Offshore 7/8/13. (2013). *Transocean sets new offshore water depth drilling record*. Accessed November 25, 2015.

Offshore 6/1/15. (2015). *Ultra-deepwater production set for steady growth, report claims*. http://www.offshore-mag.com/articles/2015/06/ultra-deepwater-production-set-for-steady-growth-report-claims.html. Accessed April 26, 2016.

Offshore 4/15/15. (2015). *Subsea hydraulic power unit widens field*. http://www.offshore-mag.com/articles/2015/04/subsea-hydraulic-power-unit-widens-field-development-options.html. Accessed April 26, 2016.

Offshore 5/28/15a. (2015a). *Study examines options for protecting harsh environment subsea facilities*. http://www.offshore-mag.com/articles/2015/05/study-examines-options-for-protecting-harsh-environment-subsea-facilities.html. Accessed June 16, 2015.

Offshore 5/28/15b. (2015b). *OneSubsea tapped for North African subsea production system*. http://www.offshore-mag.com/articles/2015/05/onesubsea-tapped-for-north-african-subsea-production-system.html. Accessed April 26, 2016.

Offshore 6/9/15. (2015). First Åsgard subsea compression kit heads to seafloor. *Offshore Magazine*. http://www.offshore-mag.com/articles/2015/06/first-sgard-subsea-compression-kit-heads-to-seafloor.html. Accessed April 26, 2016.

Offshore 6/10/15. (2015). *Delta SubSea, Zerlux collaborate on hydrate remediation technology*. http://www.offshore-mag.com/articles/2015/06/delta-subsea-zerlux-collaborate-on-hydrate-remediation-technology.html. Accessed November 14, 2015.

Offshore 6/17/15. (2015). *KPOC commissions deepwater Malaysia pipeline heating studies*. http://www.offshore-mag.com/articles/2015/06/kpoc-commissions-deepwater-malaysia-pipeline-heating-studies.html. Accessed December 24, 2015.

Offshore 6/24/15. (2015). *Offshore, GE unveils next-generation flexibles complex*. http://www.offshore-mag.com/articles/2015/06/ge-unveils-next-generation-flexibles-complex.html. Accessed January 25, 2016.

Offshore 6/25/15. (2015). *UK center targets new generation of robots for offshore inspection.* http://www.offshore-mag.com/articles/2015/06/uk-center-targets-new-generation-of-robots-for-offshore-inspection.html. Accessed June 26, 2015.

Offshore 6/30/15. (2015). *Subsea cables deliver power to offshore Safaniya upgrade project.* http://www.offshore-mag.com/articles/2015/06/subsea-cables-deliver-power-to-offshore-safaniya-upgrade-project.html. Accessed December 1, 2015.

Offshore 7/7/15. (2015). *Gullfaks subsea compressor nears testing phase.* http://www.offshore-mag.com/articles/2015/07/gullfaks-subsea-compressor-nears-testing-phase.html. Accessed November 4, 2015.

Offshore 7/9/15. (2015). *GE to connect BP wells worldwide to industrial internet.* http://www.offshore-mag.com/articles/2015/07/ge-to-connect-bp-wells-worldwide-to-industrial-internet.html. Accessed November 4, 2015.

Offshore 7/29/15. (2015). ABS gives approval in principle to Jensen Maritime LNG vessel design. http://www.offshore-mag.com/articles/2015/07/abs-gives-approval-in-principle-to-jensen-maritime-lng-vessel-design.html. *Offshore*, July 27, 2015.

Offshore 9/10/15. (2015). *OE 2015: Darcy features use of hydraulic screens for sand control.* http://www.offshore-mag.com/articles/2015/09/oe-2015-darcy-features-use-of-hydraulic-screens-for-sand-control.html. Accessed September 10, 2015.

Offshore 9/15/15. (2015). *Wood group outlines subsea JIPs.* http://www.offshore-mag.com/articles/2015/09/wood-group-outlines-subsea-jips.html. Accessed September 15, 2015.

Offshore 10/23/15. (2015a). *Statoil starts up shore-powered Troll A compressors.* http://www.offshore-mag.com/articles/2015/10/statoil-starts-up-shore-powered-troll-a-compressors.html. Accessed October 23, 2015.

Offshore 10/23/15. (2015b). *Rambøll examines potential for unmanned platforms offshore Norway.* http://www.offshore-mag.com/articles/2015/10/ramb-ll-examines-potential-for-unmanned-platforms-offshore-norway.html. Accessed October 23, 2015.

Offshore 11/1/15. (2015). Riserless mud system solves North Sea drilling problem. *Offshore, 66* (11).

Offshore 11/3/15. (2015). *FPS spending to increase, says Douglas-Westwood.* http://www.offshore-mag.com/articles/2015/11/fps-spending-to-increase-says-douglas-westwood.html. Accessed December 4, 2015.

Offshore 11/4/15. (2015). *Shell's new LNG-powered OSV enters GoM.* http://www.offshore-mag.com/articles/2015/11/shell-s-new-lng-powered-osv-enters-gom.html. Accessed February 23, 2016.

Offshore 11/25/15. (2015). *Barents Sea production study identifies technology gaps.* http://www.offshore-mag.com/articles/2015/11/barents-sea-production-study-identifies-technology-gaps.html. Accessed November 25, 2015.

Offshore 3/9/16. (2016). New JIP focuses on subsea processing standardization. *Offshore Magazine.* http://www.offshore-mag.com/articles/2016/03/new-joint-industry-project-focuses-on-subsea-processing-standardization.html. Accessed March 9, 2016.

Offshore 3/22/16. (2016). *GE oil & gas signs agreement with statoil for future subsea projects.* http://www.genewsroom.com/press-releases/ge-oil-gas-signs-agreement-statoil-future-subsea-projects-282665. Accessed March 22, 2016.

Offshore 3/23/16. (2016). *SSTB wo16: Freudenberg highlights recent Anadarko contract wins.* Accessed March 23, 2016.

Offshore 6/23/17. (2017). Shore-controlled ROV starts maneuvers in the North Sea. *Offshore staff.* http://www.offshore-mag.com/articles/2017/06/shore-controlled-rov-starts-maneuvers-in-the-north-sea.html?cmpid=enl_offshore_offshoredailynewsletter_2017-06-26&email_address=michaelmax1@mac.com&eid=288179069&bid=1795944. Accessed June 26, 2017.

Offshore 17-9-5. (2017). North Sea supply vessel steered remotely from California control center. *Offshore*, 09/01/2017. http://www.offshore-mag.com/articles/2017/09/north-sea-supply-vessel-steered-remotely-from-california-control-center.html. Accessed September 5, 2017.

Offshore 1/17/18. (2018). *Sverdrup filed to feature permanent reservoir monitoring from start-up.* http://www.offshore-mag.com/articles/2018/01/sverdrup-field-to-feature-permanent-reservoir-mon itoring-from-start-up.html?cmpid=enl_offshore_offshore_north_sea_report_2018-02-06&pwhid=7f771bc841f5df66ebcac04dbeb4928e7780bf3f90641c677bb3f8cc8d4941fc147ac159c501f5878031f6ba7f9e82feb9b8c395d6a3e9927d723448ebe1b263&eid=288179069&bid=1995078. Accessed February 6, 2018.

Paganie, D. (2015). Making the case for an all-subsea development. *Offshore*. 10/14/15. http://www.offshore-mag.com/articles/print/volume-75/issue-10/departments/comment/making-the-case-for-an-all-subsea-development.html. Accessed October 14, 2015.

Parish, J. R. (2015). Proposed regulations clarify the meaning of a safe drilling margin in the OCS. *Offshore*, 7/14/15. http://www.offshore-mag.com/articles/print/volume-75/issue-7/departments/regulatory-perspectives/proposed-regulations-clarify-the-meaning-of-a-safe-drilling-margin-in-the-ocs.html. Accessed November 14, 2015.

PennEnergy 4/18/16. (2016). Collaboration on swimming robots for subsea maintenance. *PennEnergy*. http://www.pennenergy.com/articles/pennenergy/2016/04/oil-gas-news-collaboration-on-swimming-robots-for-subsea-maintenance.html?cmpid=EnlDailyPetroApril192016&eid=288179069&bid=1377404. Accessed April 19, 2016.

PennEnergy 17-11-9. (2017). Norway's first platform to be remote-cooperated from land. *PennEnergy staff (Statoil)*. http://www.pennenergy.com/articles/pennenergy/2017/11/norway-s-first-oil-and-gas-platform-to-be-remote-operated-from-land.html?cmpid=enl_pennenergy_pennenergy_research_2017-11-24&email_address=michaelmax1@mac.com&eid=288179069&bid=1932942. Accessed November 24, 2017.

Pers. Com., Wallman. (2015). Personal communication from Prof. Klaus Wallman to Arthur Johnson at the Arctic Petroleum Conference, Stavanger, Norway, September 29–October 2, 2015.

Perry, R. (2013). Subsea processing technologies are coming of age. *Offshore*. http://www.offshore-mag.com/articles/print/volume-72/issue-4/subsea/subsea-processing-technologies-are-coming-of-age._printArticle.html. Accessed April 26, 2016.

Phoenix. (2015). *Underwater robotics*. https://www.phnx-international.com/robotics-rov.html. Accessed June 22, 2015.

Piemontese, M., Colombo, M., Pelicano, D., Singh, P., & Murphy, J. (2014a). Chemical remedial sand control using the subsea flowline succeeds. *Offshore Magazine*. http://www.offshore-mag.com/articles/print/volume-74/issue-12/africa/chemical-remedial-sand-control-using-the-subsea-flowline-succeeds-p1.html. Accessed January 5, 2015.

Piemontese, M., Colombo, M., Pipa, G., Pellicano, D., Singh, P., & Murphy, J. (2014b). Chemical remedial sand control using the subsea flowline succeeds. *Offshore, 75*(12), 52, 54.

Pinkstone, H., McCluskey, T., Nilsen, J. H., Klepsvik, J., & Lambregts, B. (2017). New drillship design combines advanced drilling and early production capabilities, concept applicable to marginal, remote fields. *Offshore, 77*(2), 35–38.

Pirenne, B., & Guillemot, E. (2009). The data management system for the VENUS and NEPTUNE cabled observatories. In *Marine Technology Society OCEANS 2009 Proceedings* (pp. 1–4).

Ponce, R. (2018). The maritime world enters the fourth industrial revolution. *Sea Technology, 59* (2), 10–13.

Reed. (2015). *The world's first subsea wet gas compressor has been installed*. http://oilpro.com/post/16263/world-first-subsea-wet-gas-compressor-has-been-installed-video?utm_source=DailyNewsletter&utm_medium=email&utm_campaign=newsletter&utm_term=2015-07-07&utm_content=Article_19_txt.

Roth, R., McLaurin, D., Paulin, M., McKee, M., Subramanian, S., Hume, C., & Ponagandia, V. (INTECSEA); Albaugh, E. K., & Davis, D. (Offshore Magazine). (2018, March). 2018 worldwide survey of subsea processing. Status of technologies and systems. *Offshore Magazine*, 1 Sheet Poster.

RTC. (2013). A roadmap for U.S. robotics—From internet to robotics. Chapter 5 in *U.S. Department of Defense Unmanned Systems Integrated Roadmap, FY2011-2036. Robotics Technology Consortium Workshop* (137 pp), Washington, DC, December 4, 2012.

Salizar, R. (2015). Diffusion of new O&G technologies: How beliefs & norms shape acceptance. http://oilpro.com/post/15221/diffusion-new-og-technologies-beliefs-norms-shape-acceptance?

utm_source=DailyNewsletter&utm_medium=email&utm_campaign=newsletter&utm_term=2015-07-08&utm_content=Article_4_txt. Accessed November 25, 2015.
SASE. (2015). *SAAB seaeye*. http://www.seaeye.com/pdfs/Seaeye_Falcon.pdf. Accessed June 23, 2015.
SB. (2015). *About the M/V Shelia Bordelon*. http://shelia.bordelonmarine.com.
Schlumberger. (2016). *Schlumberger introduces realtime downhole flow measurement service*. http://www.offshore-mag.com/articles/2016/03/schlumberger-introduces-real-time-downhole-flow-measurement-service.html. Accessed March 11, 2016.
Schmidt, V. (2013, April). Rigless drilling enters the pilot stage. *OE Digital,* 47–48.
Schubert, J. J., Juvkam-Wold, H. C., & Choe, J. (2006). Well control procedures for dual gradient drilling as compared to conventional riser drilling. *SPE Drilling & Completion, 21*(4), 287–295.
Scott, R., Bosio, J., & Boyington, B. (1991, July). Slimhole drilling, the story so far. *Oilfield Review,* 46–54.
Segård, B., Kuhnie, T. I., & Leinum, B. (2015, November). Joint industry project seeks to advance subsea standardization. *Offshore,* 66, 68.
Segura, J. M., Fisher, Q. J., Crook, A. J. I., Dutko, M., Uu, J. G., Skachkov, S., et al. (2011). Reservoir stress path characterization and its implications for fluid-flow production simulations. *Petroleum Geosciences, 17,* 335–344.
Sheehan, J. (2015). *Subsea factory dream nears reality for Statoil*. http://www.epmag.com/subsea-factory-dream-nears-reality-statoil-803556. Accessed June 19, 2015.
Simonds, R., & Sladic, J. (2015). Cup-seal packer enhances sand control in open-hole wells; Innovative design increases reliability, improves field economics. *Offshore, 75*(1), 72–73.
Smith, M. L., Roddewig, M. R., Strovink, K. M., & Scales, J. A. (2013, January). A simple electronically-phased acoustic array. *Acoustics Today,* 22–28.
Søereide, F., Ludvigsen, M., & Williamson, M. (2017). Low-cost drilling for deep-sea minerals exploration. *SeaTechnology, 58*(8), 10–14.
Spagnoli, G., & Freudenthal, T. (2013, December). Underwater drilling rig for offshore geotechnical explorations for oil & gas structures. *Oil Gas European Magazine*, 4 pp. http://www.researchgate.net/publication/259848572. Accessed November 15, 2015.
SSA. (2015). Subsea services alliance plans to produce 15,000 psi intervention riser system. *Offshore*. http://www.offshore-mag.com/articles/2015/04/subsea-services-alliance-plans-to-produce-15-000-psi-intervention-riser-system.html. Accessed May 1, 2015.
Statoil 11/3/15. (2015). *First subsea gas compression plant in the world on line—a step change in subsea technology*. http://www.statoil.com/en/NewsAndMedia/News/2015/Pages/17Sep_Aasgard_subsea.aspx. Accessed June 16, 2015.
Sugar. (2011). Transport of gas hydrate towards the shore. In *Quarterly Newsletter of the EU Sugar Submarine Gas Hydrate Reservoirs Program* (7 pp). http://www.elac-nautik.de/_uploads/images/pdf/Newsletter_no5_120811klein.pdf. Accessed April 26, 2016.
Takahashi, H., & Tsuji, Y. (2005). Multi-well exploration program in 2004 for natural hydrate in the Nankai-Trough Offshore Japan. In *OTC 17162, 2005 Offshore Technology Conference* (8 pp), Houston, TX, USA, May 2–5.
Teelken, R., Rossa, O. J., & Edwards, N. K. (2016). Smart drillpipe technologies for the driller's toolbox (Ove, Baker-Hughs). *E&P*. http://www.epmag.com/smart-drillpipe-technologies-drillers-toolbox-1449966?mkt_tok=eyJpIjoiWkdKaU56WTVOV1kwTTJOaSIsInQiOiJKRkkrR0liQ3RKVEpUN2pqZDVhK1ZVV1U2UEpaYjZsVWYyQjg3ZlEyMTJ1VEF2V0xuTDhjWHQyblA4ZlZYTk9ibHUyZW9ESVphUXk2eGtjdlJPM29XSHJEUjM5WXo1WnVSZmw1cnhoV0FZOD0ifQ%3D%3D#p=full. Accessed December 1, 2016.
Theimer, K., & Kolle, J. (2007). *Microhole high-pressure jet drill for coiled tubing* (Final Report TR-117 for DOE Award Number: DE-FC26-05NT15484) (51 pp). http://www.netl.doe.gov/kmd/cds/disk17/B%20-%20Microhole/NT15484%20Final%20Feb%201,%202005%20-%20Jun%2030,%202007.pdf. Accessed November 16, 2015.
Tippee, J. (2015). Arctic E&P activity on the rise. *Offshore, 75*(5), 52–53.

Triepke, J. (2015). *Everything will bottom in 2016*. http://oilpro.com/post/20940/everything-bottom-2016?utm_source=DailyNewsletter&utm_medium=email&utm_campaign=newsletter&utm_term=2015-12-17&utm_content=Article_10_txt. Accessed April 26, 2016.

Vieitez, C., & Cox, M. (2017). Walkasway seismic-while-drilling operation delineates salt dome. *Offshore, 77*(7), 48–49.

Weatherford. (2017). *Managed pressure drilling* (5 pp). Weatherford International. http://www.weatherford.com/en/products-services/drilling-formation-evaluation/secure-drilling-services/managed-pressure-drilling. Accessed March 25, 2017.

Williams, T., Deskins, G., Ward, S. L., & Hightower, M. (2001). *Sound coiled-tubing drilling practices* (Final Report). U.S. Department of Energy DOE/NETL-2002/1170.

WO. (2016). Systematic candidate selection improves Haynesville refracturing economics. *World Oil Magazine in OilPro*. http://oilpro.com/announcement/2657/systematic-candidate-selection-improves-haynesville-refracturing?utm_source=DailyNewsletter&utm_medium=email&utm_campaign=newsletter&utm_term=2016-01-08&utm_content=Feature_1_img. Accessed Jan uary 8, 2016.

WO. (2016). Integrated high-pull wireline conveyance systems accelerate access to well profiles. *WorldOil, 237*(12), 3 pp. http://www.worldoil.com/magazine/2016/december-2016/features/integrated-high-pull-wireline-conveyance-systems-accelerate-access-to-well-profiles. Accessed January 19, 2017.

Worthington, P. F., & Hattingh, S. K. F. (2014). Optimizing the value of reservoir simulation through quality-assured initiatlization. *Petroleum Geoscience, 20,* 283–301.

Wright, S. (2017). Statoil activates Valemon automated rig control room. *InnovOil, 59*(12), 30–31.

Wuest, C., Riddoch, J., McHardy, M., Karnugroho, A., Sosa, J. C. L., Bremner, S., & Toralde, J. S. S. (2015). Asia/Pacific spurs demand for integrated deepwater MPD systems. *Offshore Magazine* downloaded 3/22/16. http://www.offshore-mag.com/1/volume-75/issue-3/drill-comp/asia-pacific-spurs/asia-pacific-spurs-demand-for-integrated-deepwater-mpd-systems-full.html. Accessed April 26, 2016.

Zeits, R. (2013). Bakken: The downspacing bounty and birth of 'array fracking'. *Seeking Alpha*. http://seekingalpha.com/article/1248431-bakken-the-downspacing-bounty-and-birth-of-array-fracking. Accessed April 26, 2016.

Chapter 8
New Technology for NGH Development and Production

Abstract The main area in which new technology and approaches has the potential to dramatically reduce the cost of natural gas hydrate (NGH) development is in drilling and reservoir planning, and preparation for production. Substantial existing technology and emerging methods being developed for ultra-deepwater, particularly those based on operating processing equipment on the seafloor rather than in vessels overhead, can be used in a re-specified form. New approaches to drilling and reservoir planning are made possible by matching technology to the unique characteristics of NGH in its reservoir, including the shallow depth below the seafloor of potential pay zones, the additional benefits of depressurization-dissociation conversion (Max and Johnson in Advances in clean hydrocarbon fuel processing: science and technology. Woodhead Publishing, Cambridge, pp. 413–434, 2011), and the fact that the converted NGH product (which consists almost entirely of relatively pure natural gas and very low salinity water) will be at substantially lower pressures within the reservoir. Not only can pressure in the reservoir be controlled, but it is possible to maintain different pressures in different parts of the reservoir to better manage dissociation and water and gas movement. A completely new, integrated approach to drilling NGH deposits is intended to optimize the opportunities presented by NGH deposits.

8.1 Time for the big step down to the seafloor for everything NGH

In our opinion, NGH exploration issues have largely been resolved and depressurization-dissociation production has been essentially proven in principle. We concern ourselves here with new NGH-specific exploration and production technology and approaches. There would appear to be no significant technical roadblock to resource extraction.

The history of technology is that as costs increase, only new technology and the use of existing technology in ways that it may not have been intended originally have the capability to reduce costs. The resource target is very large—the median

M. D. Max and A. H. Johnson, *Exploration and Production of Oceanic Natural Gas Hydrate*, https://doi.org/10.1007/978-3-030-00401-9_8

value for gas in place in sand units of 43,311 Tcf (Johnson 2012). If it could be determined that approximately half of this total would be technically recoverable the percentage that would be economically recoverable may depend on dramatically lowering the costs to below that of using existing deepwater technology. Although we are confident that existing or lightly modified technology can be applied to production of NGH deposits, we suggest that a step-change in production technology and practices based on approaches that respond to the particular set of NGH characteristics introduces the possibility of driving down development costs materially. At this point deepwater NGH could become the major source of natural gas on a producer cost basis. We focus here on new technology that along with leveraged existing technology, may enable development of the natural gas hydrate (NGH) resource at competitive prices.

New technology or newly perfected technology has been the key to commercializing coalbed methane and tight and shale gas unconventional gas resources. These new technologies, however, were rooted solidly in using existing technology for different purposes or in different ways from which the equipment was originally used by adopting it to uses that enabled commercialization of the resources. Assessment of NGH characteristics and requirements suggests that there are a number of operational ways, lower specification, and less expensive equipment and vessels can be implemented for both exploration and production.

NGH may be the last of the unconventional gas resources to be commercialized. The extent to which a part of the huge amounts of NGH can be recovered economically will determine the extent of commerciality of the resource. If the cost of exploration and gas recovery can be made competitive with other gas resources, then NGH has the potential to become one of the largest gas sources on the planet. In addition to cost factors, the environmental risk and pollution potential of oceanic NGH is lower than either coalbed methane or shale gas and substantially lower than that of conventional gas (Chap. 10).

The special physical circumstances and properties of NGH allow a general system for dealing most efficiently with the unique resource to be outlined. These include the highly reversible chemical reaction through which it forms and converts back to its constituent gas and water, its unique presence in thermodynamic rather than geological traps, and its likely ubiquitous reservoir character in semi-consolidated reservoir hosts. We believe that because of the very similar drilling conditions within less than 1 km of the seafloor worldwide, significant opportunities exist for development and implementation of NGH-specific exploration, drilling, and production technology. In addition, the relative chemical purity and low temperature of the converted gas, and the controllable gas pressures within the reservoir and collector systems, also permits different materials that will be very different from applying simpler versions of existing technology and practices.

The cost of producing natural gas from NGH is commonly calculated as a sum of the conventional industry drilling and production costs plus the cost of NGH conversion from its stable form in its reservoir, and any additional costs that may arise, such as flow assurance, and handling and separation equipment. By implementing an NGH-specific set of technologies and methodologies, to which the

NGH specific costs are added, a much lower overall cost may be achieved. This will have a strong impact upon the perception of commerciality of the NGH resource. It may be possible to keep capital expenditure and operating costs well below those currently envisaged by the current conventional model of pricing of NGH commerciality. Implementing new methodologies and technologies for maintaining high production rates, safety, and reliability, NGH may prove to have a cost profile that will render it highly competitive, even taking into account conversion costs.

The first successful technical drill stem oceanic NGH conversion and gas production test was carried out on the 40 Tcf (gas-in-place) Nankai NGH deposit during March 2013 by JOGMEC, as part of their MH21 plan for production by 2018 (JOGMEC 2013). This followed earlier tests of NGH conversion in permafrost, but confirmed modeling predictions for the resource in situ, thus leaving production analogs behind. The low energy cost and performance of the depressurization process also appears to have additional benefits suggesting that it will remain the gas conversion process of choice. In addition, the newly designed and implemented gas-water separation technique used in the test appears to have worked satisfactorily. This proof of principal can now be scaled up into a long-term production system, and potentially one that can join other separation and processing systems being installed on the seafloor.

Moving to the Seafloor Existing exploration and production techniques and methodologies have now been brought to a level of adaptation and performance sufficient to begin planning for semi-autonomous or remote production from NGH deposits. An example of a major step toward remote production is the Total-operated Laggan-Tormore project to the east of Shetland in the NW North Sea that began operation in May, 2016 (Newman and Hainsworth 2016). The fields are in 600 m water depths with water temperatures almost steady at –0.1 °C and do not have a platform or vessel overhead during production. The field is controlled from, powered by, and produces through pipelines a distance of 143 km to a processing facility in Shetland. Pipelines and umbilicals are covered with stone to protect them. The wellhead processing facility has been designed for a 20-year life and will be serviced by ROVs from light support vessels overhead. A recirculating system of monoethylene glycol (MEG) provides for flow assurance. Seafloor structures comprising an industrial site can be assembled using right-sized, purpose-built seafloor construction vessels (SCV) that can also be used for decommissioning (Offshore 9/6/17).

The semi-autonomous operation of a remote seafloor wellhead and production systems being installed now and being planned, provides a framework for our vision for low-cost production from NGH. The wellhead installation consists of the usual wellhead apparatus, pipes and valves, pipeline chokes, power control and distribution system, multiphase flow meters, chemical injection apparatus, and control systems, with some of the controls responding to local sensors semi-autonomously. The production system is installed on two 30 × 40 m templates with suction-assisted fixing to the seafloor. Apparatus stands up to 21 m from the seafloor and each template weighs of 900 ton.

By focusing on the unique physical innovation opportunities inherent in NGH reservoirs (Chaps. 2 and 6), this understanding can be improved and NGH-specific innovative technologies can be devised. There presently exists a NGH technology development time window in which major innovation in the NGH field will be made. Technologies will generally be in the fields of: (1) Establishment of all equipment including exploration, drilling, and production equipment to the seafloor, (2) Specialized lighter duty vessels and equipment for supporting seafloor operations, (3) The optimization of an automated, robotic approach to seafloor E&P to produce a new range of autonomous and semi-autonomous service vehicles and equipment, (4) 24/7 operations without consideration of weather or surface conditions, (5) A merging of light drilling and tunneling practices, and (6) Innovative approaches for both rotary and a new type of coiled tube drilling including the merger of AUV technology down-hole. Implementation of new technology has the potential to result in lower materials and personnel requirements.

The core vision behind our NGH-specific technology is that everything involved with exploration and production, especially drilling or the NGH equivalent, can be moved to the seafloor. In proposing a complete move to the seafloor, we note that the move has already begun, with processing equipment, gas-water separators, pumps, and well control equipment already being migrated to the deepwater seafloor. Some exploration and sensor implanting drilling can be carried out from relatively small vessels as this would allow direct control of logging and pressure core sampling that could be logistically easier than drilling from the seafloor. As what could be seen as an intermediate step, Equinor has launched a new subsea concept combining existing and new technology on smaller seafloor templates so that operations can be guided using machinery in place from a vessel rather than a larger rig carrying much of the equipment (Offshore 4/19/16). The actual requirements of the NGH resource allow this to be done even for drilling, as the temperatures, pressures, corrosion potential and nature of the produced material are much more benign than the conventional deepwater case.

Although conventional drilling and processing equipment can be used initially for the NGH resource, we regard the development of a new suite of tools and practices as being of critical importance. The NGH resource appears to be huge and thus exploitation has the potential to provide a substantial part of natural gas demand. Exploitation could thus continue indefinitely into the future. Because of the low environmental risk of the NGH resource (Chap. 10), and its suitability of natural gas to provide the base load and spiker backstop for a renewable energy future (Chap. 1), we suggest that when environmental concerns increase in importance as part of a response to climate change mitigation, NGH may prove be the primary or sole natural gas resource of the future.

8.2 Exploration

Exploration methods for the identification and valuation of NGH deposits have been proven, at least to the point that good estimates of NGH in place can be made. Improvements in the seismic analyses techniques are currently under development, as are the background seismic characterization of host and related sediments and the geotechnical properties necessary to more exactly parameterize acoustic character for seismic processing.

Seismic survey and computer workstation analysis of digital data developed for conventional hydrocarbon exploration, ocean bottom seismometers, and secondary remote sensing exploration techniques such as electromagnetic survey (Chap. 3), combined with drilling and well logging and sampling can be applied to NGH exploration without expensive and time-consuming development of new technology. Following the successful Joint Industry Project organized and funded by the U.S. Department of Energy, during which geotechnical properties of NGH were successfully applied to NGH recognition and valuation, a dedicated body of knowledge for NGH seismic exploration now exists (Max and Johnson 2013).

We anticipate that considerable improvement in seismic analysis technique will lead to quicker identification (Chap. 3) and accurate valuation (Chap. 6) of deposits. Greater geotechnical detail, the use of phased frequency analysis, and the use of more acoustic properties should allow for greater discrimination of the 3D distribution of the NGH as well as more accurate valuation (Chap. 5). For instance, non-industry standard 4-D seismic surveys can be carried out in the kHz range in order to achieve highly detailed analysis cubes (OL 2013).

A variety of analysis techniques both for identification and classification of NGH deposits as well as for determining numerical estimates for gas-in-place are underway but are not discussed here in detail because we regard this as a field of specialist technique that is in development but has not yet settled into the ‘cookbook’ workstation approach that has been derived over decades for conventional reservoir analysis. The main point is that the basic exploration technique is already good enough to apply widely for locating and valuing a NGH deposit (Boswell et al. 2012; Frye et al. 2012). Our focus here is on new technology or new applications of existing technology in innovative ways that could provide for substantial improvements in the efficiency and cost of NGH development.

8.3 Drilling

Even if full capability, costly conventional drillships were to be used, drilling and preparation for NGH production would involve a much shorter drilling program because all NGH prospects will be within the relatively shallow NGH prospect zone in the GHSZ and well bores only will have to penetrate relatively unconsolidated marine sediments. The drilling requirements for NGH are relatively simple, which means that even if only existing conventional drilling equipment is used, it can be carried out from much smaller less expensive vessels using much less powerful drilling capability.

Smaller drillships that are now becoming available, such as the 'MOU Island Constructor' operated by 'Island Offshore', that likely have all the capability necessary to drill any NGH prospect (Offshore 10/26/15). In addition, for placing seafloor drilling equipment and operating it from the surface, many vessels of the size required to carry out NGH drilling already have a variety of large cranes that could be utilized, including a number of major research vessels (Bourillet et al. 2007; Gutenberg 2015). Ships having the character of platform supply vessels (PSV) could also be fitted with smaller drilling rigs operated from the surface, or cranes to service a seafloor drill. Smaller ships of this type would have all the capabilities required for NGH drilling operations at a fraction of the cost of major conventional drilling platforms.

A full-size deepwater conventional drill rig is not needed because its specifications far exceed those required for NGH drilling. A workboat-based drilling system that includes horizontal drilling to establish greater surface area for NGH conversion and an integrated gas production capability (Kolle and Max 2000) could be operated from a relatively inexpensive workboat using coiled tube drilling. Small drillships carrying specially designed coiled tube-like drilling equipment would be usable for NGH drilling operations without any significant modifications. This class of lightweight drill ship is already providing platforms for drilling into a significant depth of GHSZs using coiled tube drilling from the deck of relatively small vessels (Offshore 10/05/15).

NGH-enriched strata are generally much more benign than deepwater conventional gas reservoirs. Modest overpressure drilling can be done through pressure lock blowout preventers (BOPs) that will be adequate to insure drilling and reservoir safety. Hannegan (2005) identified some of the main issues, such as low-pressure differential between reservoir and seafloor, instability and subsidence, and temperature management, while suggesting that managing pressure in the drill hole can be accomplished using existing equipment. Although high cost conventional offshore drilling systems could be used to open up NGH until more NGH-specific drilling capability and support could be brought into play to lower costs, even they might be used economically if a drilling program were carefully planned. New low-cost NGH-specific drilling methods would ensure commercial viability by significantly increasing the percentage of the technically recoverable gas resource that could be produced economically (Chap. 11).

It is our contention that NGH drilling can be done entirely from the seafloor because of the particular character of NGH (Chap. 6). This means that site preparation requirements for both the seafloor drilling footprint as well as the production seafloor factory will have to be much larger than simple wellheads in the past. Increased expenses required for the scale of NGH site preparation, as envisaged by us, will be offset by the much less robust and expensive requirements of the NGH seafloor site. For instance, the huge and expensive BOPs required for conventional deepwater are not necessary for NGH. There is nothing particularly new about this, however, because more advanced and larger site preparation is already being practiced as production for conventional gas is being moved to the seafloor.

Carrying out exploration and production activities from the seafloor means that weather will no longer be a hazard to activity, especially where communications allow real-time monitoring and control from remote locations (Chap. 11). The advantages of seafloor drilling and production activities are particularly true for Arctic regions where sea ice may persist and disrupt, if not prevent, activity. This is because the sea ice will drift with wind and tide while a vessel drilling or servicing a well or receiving gas from the well has to maintain position. The sea does not have to be completely ice covered in order to disrupt drilling and well support from surface vessels.

The NGH will remain stable if the temperatures can be maintained within the field of NGH stability, and local refrigeration is easily provided from heat exchange with the naturally cold seawater nearby. Most conventional deepwater drilling for the first 500 m or more has been carried out open-hole, with the wellbore essentially exposed to seawater used for well flushing of cuttings. The Nankai March 2013 technical production test was accomplished with the top of the well cased after drilling to secure the well to water separation equipment and riser, but the bottom 60 m of the well in the NGH deposit was not cased and the test was thus carried out using open-hole in the pay zone, although a gravel pack had been put in place (Kawamoto 2013). This demonstrates that the drilling requirements for NGH are much more benign than for conventional deepwater gas deposits.

Although automated rotary drilling equipment is currently being used on the seafloor, equipment has been developed and improved to support site investigations, engineering geology, and to introduce a coring capability for upper near-surface sediments in which open hole drilling has been practiced (Chap. 7). In part, this new capability is to provide data that is necessary to engineer seafloor industrial site preparation. Expansion of this seafloor drilling capability would allow it to be used for both exploration drilling and for preparation of NGH deposits for production. Although it is probable that the technique could service the shallow depth, long-reach drilling that we envisage necessary for development of the NGH resource, new methods based on coiled tube drilling (Beckman 2014), which are only just beginning to be used on the seafloor (Chap. 7) may hold more promise owing to a potentially simpler and smaller equipment on the seafloor.

8.3.1 NGH Drilling Issues and Objectives

Virtually all drilling methods will work for NGH deposits and in the sediments necessary to reach them. It may be that both rotary and coiled tube-based developments for NGH drilling will work almost equally well or that one may prove overwhelmingly superior. We explore the possibilities of both here to establish the framework for development of NGH-specific drilling technology.

Drilling from the seafloor: The first objective in NGH-specific drilling is that it should be done from the seafloor. By drilling from the seafloor, the heavy-duty drilling and vessel support can be replaced by less expensive equipment that would be much better matched to the drilling requirements. This requirement is already being met partially by existing developments in autonomous seafloor drilling (Chap. 11), but so far without matching specific engineering to the requirements of NGH development in more than a general sense. Drilling from the seafloor for NGH will greatly reduce material and power requirements and simplify operations in which the distance from the sea surface to the seafloor may be longer than the length of a single NGH well bore. Drilling from the seafloor brings the drill much closer to the drilling target. Drilling from the seafloor (Eldridge 2013) would also open up the potential NGH resource in the Arctic (Max et al. 2013) as sea ice and foul weather could be largely ignored.

Rotary drilling: This is the traditional, well proven, widely practiced method in which drill string is composited during the drilling by adding drilling pipe sections to the top of the pipe as the drill extends into the ground behind the drill bit or tool (s) that cut their way through sediments and rocks of all types. Pipe sections are brought to the wellhead and added to extend the drill string as necessary. Over the years the drill tools at the bottom of the drill string have become more sophisticated and better methods of casing the well allow the wellbore to be protected from collapse and washouts during drilling and for the safe conduct of drilling fluids and subsequently produced oil and gas (+ water) to the wellhead from the reservoir. Rotary drilling is used exclusively for the main wells in ultra-deepwater.

Rotary drilling requires pulling and disassembling the entire drill string to replace tools and sensors that do the work at the drill face. This increases the likelihood for mechanical problems in the restacking and reassembly of drill stem sections. In contrast, coiled tube is simpler, with the coiled storage of what is in effect a single section of multiple pipes and cables, which has been sized for particular drilling jobs. A coiled tube approach thus is a strong de-risking element in choosing which method to use at a remote, unmanned site. Rotary steerable systems have been used in extended-reach drilling (ERD) since the 1990s (Fontenot et al. 2005). This technology allows for directional control through steering of the bit while continuously rotating the entire drill string.

Coiled tube drilling (CTD): This more compact method uses a continuous length of drill pipe referred to as a 'tube' that is stored rolled on a large drum from which it is unwound as it is forcefully inserted into the well during drilling. CTD is a newer drilling method, dating back to 1926—the Bannister concept for a flexible hose drill string in 1948 and the Priestman patent for a more conventional reeled rigid metal pipe (PW 2016), but the modern form of the rig only dates from the mid-1990s. The pressure on the drill bit comes from the insertion pressure at the wellhead. We do not discuss any details of the CTD method because although we propose using the basic element of CTD (a coiled tube connected to the drilling tools at the drill face), we do not be suggest the use of forced insertion to power the drilling.

Both rotary and coiled tube drilling could be used for developing NGH resources. In addition, each could be used to carry out different aspects of the drilling and production preparation work for which it is best suited. Although there is much greater experience with rotary drilling offshore, especially where down hole logging is required and in general in deepwater, we suggest that the scope for development of new and more efficient NGH-specific technology will be closer to CTD than it will be to a fully rotary approach. For an autonomous seafloor site, the combination of a CTD unit with a low-cost conventional rotary drilling rig should be able to meet all drilling requirements without need to change out drill rig infrastructure. In this application, the rotary rig is used to drill a quick and simple wellbore and sets casing just above the desired zone. CT is then used to drill a clean penetration into the pay zone for a completion.

Although it may still be preferable to begin the hole with rotary drilling and conventional setting of casing to secure the wellhead, the coiled tube approach will not require multiple magazines of drill stem and casing and with the provision of automated adding and removing of pipe sections and different drill heads and tools that need to be added to reach more conventional target depths.

Existing CTD has several advantages including: drilling from a pad that will only occupy about one third the space and one third the number of equipment loads required for rotary drilling; drilling smaller hole sizes reduces cuttings and drilling fluids and waste-disposal costs; drilling low-cost wells enables dedicated monitoring wells to be placed at optimum locations without disrupting production; and drilling sidetracks and laterals from existing well bores increases well bore contact with the reservoir and increases production.

Wellbore casing: Casing is the steel pipe that is inserted into a well in order to prevent the walls of a well from caving in or leaking into the well or, more important, leaking out of the well when conventional oil or gas is encountered and produced. Casing seals the well off into a single pressure vessel from the bottom of the well, where it will be inserted hopefully into a huge reservoir, to the blow out preventer and other controllable valves and throughput assemblies at the wellhead. A section of a well, such as the upper part, which is commonly drilled open hole, or a pay zone that may not be fully cased for some reason such as a technical test, may not be able to contain overpressures.

At present, the most efficient setting of casing is done using rotary drilling, especially in ultra-deepwater where CTD is not used to drill major wells. Once the upper section of casing is set to anchor the wellhead equipment, casing is set not far behind the drill tools, as the drilling tools penetrate and extend the well (Fontenot et al. 2005) often using variable hole diameter drilling tools. CTD is used for maintenance of existing deepwater wells but with a flexible ‘drill string’, setting casing at the same time as drilling.

Ability to configure complex wellbore geometry: In a more maneuverable and robotic downhole drilling capability it is possible to use new wellbore layouts so as to begin the gas-water separation process within the reservoir.

Sand flow issues: Undesired sand flow is the most important unresolved factor of production. The first NGH production test flowed very well and steadily for almost six days before developing sand flow issues, although the pay zone intersection was relatively short. This means that the velocity of produced gas and water passing through the sand pack into the well was at a relatively elevated velocity. Long term control of sand flow in the reservoir and in the immediate vicinity of the well interface is critical for long-term production of natural gas from NGH.

8.3.1.1 Seafloor Worksite for NGH Drilling

The relatively benign nature of the NGH resource, in which the natural gas is held as a stable solid in NGH at a relatively low formation pressure in its reservoir renders the resource amenable to autonomous drilling from the seafloor because a high degree of safety and very low risk can be maintained. Drilling from the seafloor allows equipment and infrastructure to be minimized, which will provide for a significant cost saving when contrasted to the requirements and costs for conventional drilling. Drilling from the seafloor is increasingly considered even for conventional hydrocarbons, despite the expensive work-arounds that have to be implemented (Chap. 7). Conventional drilling from the sea surface at ultra-deepwater sites can only be supported when the price of the energy resource can be maintained at a high enough level to justify development. Natural gas is essentially a low energy-density fuel. Thus, it is important to lower development costs. The NGH resource is highly amenable to low cost development.

A principal attribute of NGH unconventional resource that distinguishes it from conventional and the other unconventional gas resources is that pressurized free gas does not exist in the NGH reservoir within the GHSZ, although gas may exist below the GHSZ. Where a permeable stratum passes through the base of the GHSZ, free gas may be in direct contact with NGH, where it can bleed into remaining permeability. Natural gas is locked into the crystalline structure of the hydrate (NGH). Where the temperature of the drilling fluid is maintained lower than that in the GHSZ, in which the temperature rises from the seafloor to the base of the GHSZ, the NGH adjacent to the drill bore will be rendered even more stable. In a preferred embodiment of the drilling, care is taken to ensure that during exploration and development drilling only cooling water or other drilling fluid is used.

Like the other conventional and unconventional gas resources, the NGH technology regime is also constrained by a particular set of environmental and physical parameters, some of which offer opportunities for innovation (Chap. 6). Most important is the nature of NGH itself, regardless of the reservoir host sediment. Drilling can be carried out through strata that have effectively no free gas, although they can contain considerable gas within in the stable solid form, NGH. It is not desired to intersect a free-gas charged reservoir during exploration/development drilling. Gas that may occur below the GHSZ has not proven to be highly overpressured. Gas could leak into a well bore that reached below the GHSZ or into the metastable zone (essentially a phase boundary with a dimensional thickness) along the base of the GHSZ where free gas and NGH may coexist. Such leakage can be avoided by use of overbalanced drilling, in which the pressures within the well bore are higher than in the reservoir. The integrity of the wellbore can be maintained as pore fluids are directed away from the well bore. Pressure/temperature sensors can be distributed from near the tip of the drilling assembly back up the well bore along either casing or coiled tubing or contained in its own cable that may be affixed or free. The readouts from these are monitored by the electronic control system, which, in a preferred embodiment contains software for automatically recognizing pressurized gas influxes and implementing automatic counter pressure in the well bore.

To reiterate, most NGH reservoirs will consist of only semi-consolidated marine sediment that was not buried deeply or heated enough to be highly dewatered and lithified (as are conventional reservoirs or the shale reservoirs of oil/gas shales). This will be especially true in passive margins, such as the Atlantic and Arctic Oceans, where tectonic forces that could structurally compact the sediments have not affected sediments within the GHSZ. Reservoir deposits within sediments of constructive continental margins, however, may have undergone tectonic tightening and be well consolidated. Where NGH occurs in more consolidated, rock-like reservoirs, for instance in permafrost NGH, different drilling practices will be practiced from marine NGH reservoirs. The term ‘over-consolidated’ is used to refer to NGH reservoirs in which the sediment materials have to be largely broken up, and ‘under-consolidated’ is used for reservoirs in which the sedimentary materials can be essentially mechanically disaggregated.

In over-consolidated NGH reservoirs, or in sediments having a significant proportion of NGH, such as occur in the reservoirs of NGH deposits having commercial potential, drill bits of various types may be used; and the drilled materials are fragmented, ranging in size from powder to a larger fragment size that depends on the type of drill and drilling method used. The range of drilling tools and methodology is not discussed here. In generally under-consolidated materials, the object is to disaggregate the sediment, ideally into the original sediment grains. Disaggregation may be achieved without powerful cutting, possibly using light rotary, vibration, and water nozzle tools. The majority of the sediment associated with high-grade NGH turbidite deposits will usually be fine to medium-grained sands that are not cemented except by NGH.

Drilling in rock produces fragments that are often more than twice as dense as fresh water (1 kg/L), depending on the mineralogy of the rocks and their packing. To remove these fragments from the well and allow the drill tool to advance, drilling muds with artificially high densities (e.g., suspensions of higher density materials such as barite) are used to essentially float the fragments out of the well. Keeping the velocity of the drilling mud higher than the settling speed of the fragments in the drilling fluid (mud) also achieves the same result, even where the fragments may have a higher specific density than the drilling fluid. It is highly likely that drilling to a NGH deposit will largely disaggregate the sediment, reducing the requirement for moving large heavy cuttings.

Drilled fragments and sediment particles, even of over-consolidated sediments within the GHSZ, will have lower densities than the rock materials that have been buried more deeply and have been heated and compressed; especially where lithification has taken place. In under consolidated materials that can be thoroughly disaggregated, mineral grains, especially of phyllosilicates, can be carried in effective suspension even by relatively slow-moving drilling water.

8.3.1.2 Seawater as Drilling Fluid

Seawater taken from near the wellhead can be used as the drilling fluid because the special attributes of the NGH reservoir allow it. This is in strong contrast with the much more rigorous requirements for drilling fluids for high pressure and temperature conventional hydrocarbon deposits. This is one of the principle opportunities that allow for considerable cost savings in NGH resource development.

The formation pressure and pore water of NGH deposits will be similar because when the marine sediments were deposited, seawater filled the porosity. Fluid migrating from below is also generated largely from older marine sediments that once were deposited on the seafloor under similar conditions. Cold drilling fluid can be taken from a short distance above the seafloor to be used (with due regard to screening to prevent macrobiota from being pumped into the system). Use of seawater as a drilling fluid is facilitated by the tight range of the temperature, pressure (including gas partial pressures), and chemistry of seawater at depths above the NGH prospect zone (Chap. 3) worldwide.

It is intended that water from just above the seafloor be used during drilling operations. This is not just because it is the most convenient source of water to use but because of its low temperature worldwide. The sea surface and intermediate water in the open ocean characteristically becomes lower in temperature with increasing depth. Water temperatures worldwide are lower than about 4 °C (i.e. the inversion temperature of water) from a depth of about 5/8 of a mile water depth (1 km), which means that water near the seafloor above potential NGH deposits is denser than in warmer water from shallower depths. Use of seawater as the primary drilling fluid greatly simplifies requirements and costs for drilling fluid and also provides for refrigeration in the well during drilling so as to prevent the unwanted conversion of NGH. It may be injected into the hole without any treatment or addition of chemical additives because the nature of the NGH reservoir material does not require it.

Surfactants or other chemicals that would not cloud the water could be used to increase the density of the water used as drilling fluid (and only harmless or food-grade substances, or would increase the 'grip' of water on solid material should be used so as to insure minimum environmental impact on biosystems). All the production testing to date indicates that addition of chemicals to a depressurization-dissociation production system will probably be unnecessary. We think that increasing the velocity of the water flow rather than the weight of the drilling fluid would achieve the same result. As part of best drilling practice, however, all drilling should be done through a pressure seal and BOP established at the wellhead.

The temperature of the drilling fluid thus will be low enough to maintain conditions of NGH stability adjacent to the drilling, even where heat is produced by the friction of drilling. The cold seawater is being injected into warmer areas within the GHSZ, which has the effect of increasing NGH stability. There will be a general refrigeration effect, especially in the lower part of the GHSZ where the greater number of high-grade deposits is likely to occur. If some gas is encountered, especially as it may be in the lower part of the GHSZ where there may be a wide metastable zone in which NGH and gas could coexist, the refrigeration effect will promote the formation of NGH from the gas. Further, because the near-seafloor drilling fluid water is naturally undersaturated with respect to dissolved gases, at least some of the free gas could be taken into solution. The undersaturated water could dissolve NGH during drilling, so provision is made to expose as little of the drilling fluid to the drill face as possible, and for as short of a time as possible, so that it can carry out its primary function of removing drill cuttings.

The water throughput is maintained at a velocity that is sufficient to carry all sediment grains and particles, including tectosilicates commonly found in sands as well as the phyllosilicates commonly found in muddier sediments, along with sediment fragments containing mixed mineral grains. Deep-water turbidite sediments that are the likely host for deepwater NGH deposits likely will have few and only small, distal heavy mineral grains.

Using seawater as the drilling fluid renders the drilling operations highly environmentally friendly. Only naturally occurring materials will be used and removed during drilling. The sediment host at the depth of any NGH deposits is not likely to be associated with significant liquid hydrocarbons or potentially harmful chemicals produced under conditions outside of the normal range of deep microbiota, except where there is a very high thermogenic petroleum system. Even then, NGH formation is a process of rejection of any substance that is not a hydrate-former. This will also render drilling operations inherently much less expensive than conventional ultra-deepwater drilling for conventional gas deposits. These are not necessary for at least the more unconsolidated NGH reservoirs and associated sediments. Water as a drilling fluid has been used before for a variety of reasons but is not used in drilling for conventional hydrocarbon deposits because heavier drilling muds are required so as to be able to counter the higher pressures in the wellbore and prevent blowouts. Water is used here in preference for drilling oceanic NGH because of the previously explained characteristics of NGH in its reservoir that do not require many of the costly conventional drilling procedures. Water, or transparent water fortified with higher density additives is used in preference to enable the down-hole visual and microdrilling-assisted technology and to allow the maximum degrees of freedom from surface vessels.

The produced fragments from the pay zone of a NGH deposit are unique with respect to density of the fragments. Pure methane NGH has a density in the vicinity of 0.91 kg/L. Where higher density hydrocarbons are present in mixed or compound hydrate the density will be higher but still positive buoyancy. Because 99+ % of NGH drilled to date appears to be largely methane, it is consistent to use it for modeling purposes. Thus, the NGH fragments and a small amount of included sediment fragments will still have a density lower than the density of deep, cold seawater (~ 1.04 kg/L) calculated for anticipated drilling conditions in a pay zone, for instance, at 2500 m total depth below sea level, with a salinity of 34,000 ppm total dissolved solids (TDS) and 2 °C. These would float out of the well bore and could be deposited on the seabed along with other drill cuttings, where they would dissolve without forming a gas phase. NGH fragments that are unencumbered with sediment grains will be light enough to rise into the seawater, but below the local dissociation depth they will dissolve as they rise. Above it they will slowly dissociate and the methane will almost certainly dissolve in the seawater rather than breach the sea surface and pass into the atmosphere. Methane is chemically unreactive but provides nutrition for bacterial life at the base of the food chain. It is likely that if there is any environmental effect at all from NGH release to the seafloor, it is more likely to be beneficial and to have no detrimental effect on the atmosphere-ocean gas balance because all the gas will be dissolved at depth.

NGH drilling will be in poorly consolidated or semi-consolidated sediments, so overbalanced drilling may be utilized, with the pressure in the well bore maintained at pressures slightly above the formation pore pressures. The objective is to maintain wellbore pressures higher than formation pressures to prevent pressure-driven sediment sloughing into the wellbore before casing or its equivalent is inserted to stabilize the hole. Drilling muds are not being used so the permeability and porosity of the wellbore walls will not be diminished by mud caking adjacent to the wellbore. Provision is made for lining the wellbore and provision for working back a short distance up the hole to clear sloughs are discussed later. In order to monitor the wellbore both where it is lined and where it is not yet lined, sensor arrays are placed on the tube as well as on the bottom hole apparatus (BHA, Sect. 8.3.3).

A principle advantage of using seawater drawn from near the seafloor as a primary drilling fluid is that the material excavated during the drilling process can be disposed directly upon the seafloor for the same reason that the drilling water can be effused directly back into the sea. This is an environmentally more acceptable solution than that currently envisaged for treating and releasing produced water from conventional wells directly on the seafloor (Daigle et al. 2012). The produced NGH and formation water have not come into contact with liquid hydrocarbons and thus no liquid hydrocarbons will be expelled onto the seafloor from an NGH production activity. Where small amounts of liquid hydrocarbons may be present, existing processing equipment can be used for separation. The materials consist of marine sediments that have not gone through the elevated temperatures that deep burial would achieve. Cuttings will consist of marine sediments that have probably never been out of the pressure and temperature range of microbiota that occupy the seafloor. There are no harmful substances in it unless it has been mineralized by a very strong thermogenic system. Provision is made for cuttings distribution using equipment such as a walking (wheeled) extensible cuttings distribution pipe and/or conveyor, similar to those used in open pit mining operations. Such equipment will distribute the cuttings evenly over a predetermined area automatically. The drilling water can be effused from the same equipment to promote better mixing.

8.3.2 Active Tethered Drilling

We use the term ‘active tethered drilling’ (ATD) to indicate drilling equipment that consists of a tube or tubes rolled on a drum at or near the wellhead as part of a seafloor drill site, along with power and communications cables and all the other elements necessary to carry out the drilling. We envisage that the wellhead and upper part of the vertical wellbore will be put in place by a light duty rotary drilling capability, which will probably be part of the drilling pad equipment. It will be designed so that it can also provide for rotation of a reaming tube assembly or for wellbore lining. In particular, we envisage establishment of seafloor drilling capability based on the concept of Carriere and Gipson (2003) in which both CTD and rotary are combined in a single, small, easily transportable drilling unit. Coiled drilling tube insertion from the seafloor dramatically lowers the requirement of tubing length, which lowers the risk for jamming by cuttings and allows for better regulation of drilling fluid flow both to and from the drilling face. Tube assemblies can be specified for the drilling required at individual sites.

The main drilling arrangement has a superficial similarity to CTD but different in a number of ways. First, the impetus for putting forward pressure on the drill tools does not come from pushing of a CTD tube and its bottom hole assembly (BHA) into the hole, although some pressure can be placed on the pipe to assist forward penetration and so that the ABHA does not have to pull the entire tube assembly into the well. Second, the ABHA is much more active in that it communicates directly with the drilling controller and has within it equipment for controlling drilling direction and rate so that geometrically complex wellbores can be achieved.

ATD utilizes technology approaches from both rotary drilling and CTD to allow the maximum flexibility that will allow wellbores to have complex geometries that will allow better exploitation of the NGH resource. Redesign of existing hybrid drilling rigs (Brillon et al. 2007) for ocean bottom deployment will be easier, faster, and less expensive than designing a concept from scratch. The tube assembly will be a much more of a passive mechanical element than in conventional CTD. Issues concerning maintenance of well pressure during drilling, with CTD type BOP and lockout chambers are part of a more complete discussion (Preface) that we only summarize here. Although it should be well adapted to the relatively low power requirements of NGH drilling, more robust iterations of this technology could be applicable to conventional oil and gas deposits.

We envisage that a robotic approach will be optimized throughout, from the drilling infrastructure at the wellhead seafloor industrial site to the tip of the drilling tools. The main objectives of robotization are to improve the consistency of underwater operations, lower costs, reduce the number of people required, and reduce risk to humans and operations and maintenance. Although robotics is already being applied to virtually all technology and business fields, the manner in which seafloor equipment, down-hole equipment, and the exploration and preparation drilling communicate and cooperate will involve new applications to the field. Government is supporting robotic developments (Offshore 2017).

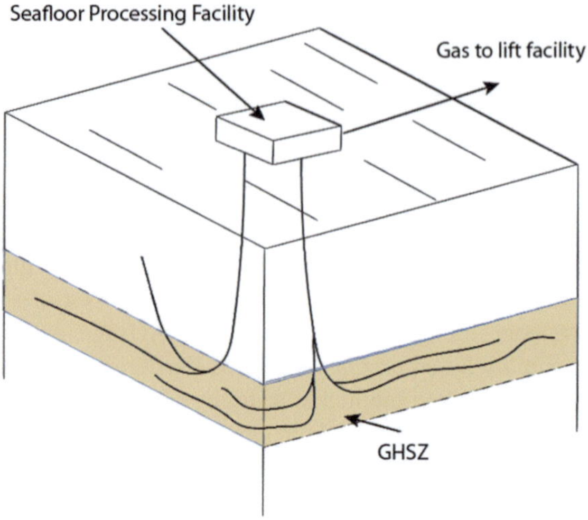

Fig. 8.1 Diagram of main elements of NGH-specific active tethered drilling. Note not all elements required to be fully operational for improved drilling practices are shown

A new range of semi-autonomous service vehicles/equipment operating in a manual override autonomous manner will service the seafloor operation. We apply the term 'active agents' or 'active robots' to this new class of technology that will merge AUV and ROV technology in a more intelligent form, giving the apparatus extensive decision-making capabilities. A form of this miniaturized active technology is already under development (Hydro 2016). These general developments are already taking place for supporting conventional drilling operations carried out from sea surface platforms, but because of the greater degree of autonomy and flexibility of downhole equipment, there is considerable potential for innovation.

ATD consists of three basic elements (Fig. 8.1). First is the seafloor infrastructure at the wellhead consisting of tube assemblies mounted on reels, which can be exchanged on the drill-mount controller. CTD tube assemblies can be retrieved relatively quickly. Tubes and tools can be interchanged in an automatic process in the same basic manner in which individual drilling pipe and casing sections can be held in magazines from which they can be added to drill stem used in seafloor autonomous drilling. This drilling infrastructure unit also contains power and communications systems that allow each tube assembly to connect to the local autonomous controller and remote oversight stations or control rooms. Second is the tube assembly in the well. Third is the ABHA at the wellbore face, which is where the main technology innovation is to take place.

In NGH reservoirs, better well control almost certainly will be obtained with long well sections in pay zone, multiple shallow angle wellbores can be driven from each seafloor well pad, in much the same way as multiple wells are now standard in the shale resource. For simplicity, Fig. 8.1 shows two wells that had been drilled from a single well pad but more wells, each with multiple wellbores may be used, especially on larger deposits (Chap. 11). In addition, more geometrically complex wells may be used. Processing, especially for gas-water-sand separation will take place at each well industrial site or at shared sites between two or more wells.

8.3.3 Active Bottom Hole Assemblies

The drilling is carried out with autonomous and semi-autonomous bottom hole assemblies, which we refer to in general as 'active bottom hole assemblies' (ABHA). These carry out the actual drilling and wellbore shaping, although other autonomous or semi-autonomous equipment will have their own functions in, for instance, well lining, sensing, and servicing. The ABHA is inserted into a wellbore on the tube assembly but is not driven forcibly into the well. The ABHA has the ability to stabilize itself in the well to provide the anchoring for drilling that extends the length of the wellbore. The ABHA contains locating equipment that is monitored from the surface so that the attitude of the ABHA can be changed to track the wellbore plan in a fully 3D sense. The extreme mobility of the ABHA is desired because NGH deposits may have considerable shale partings and other irregularities, such as patchy NGH development, and constructing a production wellbore is critical to intersecting as many of the otherwise isolated occurrences as possible so as to maximize recovery of the technically recoverable gas in the deposit.

ABHAs are a blending of existing CTD BHA capability, particularly its concept of tool insertion at the head of a single long pipe or tube, with the addition of drilling tool and well maintenance technology that introduces autonomous and ROV-like capabilities for working within a wellbore. The ABHA is much more important to active tethered drilling (ATD) than the BHA is to CTD. In ATD the tube has a more passive role, acting primarily as a carrier of drilling fluid and communications and power for the ABHA. As we envisage it, the ABHA is a completely new device designed to work in a very different way from CTD.

This design concept follows the development path in both rotary and CTD of more capable and directionally active and complex drilling tools at the business end of the tube drill stem. For instance, a development of conventional CTD BHAs for rotary drilling was the positive displacement drilling motor (PDM) located at the head of the drill string (MD 2016). These were introduced in the 1960s and allowed drilling without rotating the full-string pipe. These systems utilized hydraulic power. Better directional control was achieved using steerable motors and allowed construction of high-angle wellbores and supported extended-reach wells in the 1990s. Further developments in steerable drilling tool technology are underway for both rotary and CTD technology. Although an ABHA could be considered to be a logical progression in drilling technology, we regard it as a step increase in flexibility and capability that introduces a completely new level of capability.

The mode of operation of an ABHA is much more like that of an ROV deployed to carry out work at an underwater site, which even now may include some drilling capability. As is the case with new autonomous and semi-autonomous technology working underwater in conjunction with a remote control room, the ABHA carries out its instructions, with limited monitoring. It is remote from its controller at the wellhead or further away, where humans in a control room may monitor and override activity (Chap. 7). Although an ABHA remains linked to the lower end of a reeled tube assembly, it can consist of multiple units that can carry out individual functions or take part in joint activity. We envisage that these advanced ABHAs will have AUV capabilities which will allow for multiple drilling and wellbore maintenance functions without direct offsite control and the potential to operate mission specific ROVs within the well as part of the required set of functions.

ABHAs for drilling in oceanic GHSZ have a number of operational and functional features that distinguish them significantly from conventional BHAs. In addition, an ABHA contains equipment that no existing BHA currently has in order to engage in new wellbore activities. The new equipment adds functions that give it considerably more flexibility than a conventional BHA. For all subsequent discussion of equipment, the word 'powered' with respect to an ABHA includes all means of powering, which will predominantly be electro-mechanical or electrical-hydraulic.

The forward-most unit of an ABHA contains the drill motor and drill bits, plus outlets for drilling fluid and other devices. The larger ABHA contain location and orientation finding and sensors and sensor control units and communications equipment. Further following units contain suites of logging tools, other sensors, and possibly deployable tools including steerable cameras, grasping and jaw tools, and other tools that may be familiar to ROV operators. The ABHA may be at least partially segmented such that each of the component units may be oriented and operated semi-independently from adjacent units.

8.3.3.1 Positioning Drilling Units

Each ABHA and sections of the tube assembly will have high performance microelectromechanical (MEMS) accelerometer and gyroscopes developed for AUV guidance similar to those described by Whitley and Clarke (2015). Downhole designs will have to take vibration and mechanical stresses into account that are rare on AUVs. Existing equipment provides a springboard for further modification for use in down-hole tools to assist controlled directional drilling. The control system will drive the drill assembly according to a drilling plan, but sensor input will be used to modify the drilling plan if data suggests that deviation would be beneficial.

A conventional BHA is essentially unsupported by the tube assembly within a wellbore except by the weak compressive pressure transmitted through the tube. Maintaining wellbore integrity may be more difficult in conventional CTD in semi-consolidated sediments of the GHSZ than in conventional rock CTD. Where the walls of a wellbore have sloughed and the wellbore integrity has diminished, it may be difficult to position the BHA. If the orientation of the BHA is out of line it may be difficult to maintain a desired wellbore direction and maintain a smooth bore. A CTD BHA has essentially no means to independently orient itself in a wellbore because it is responding to mechanical control originating at the wellbore.

ATD overcomes this limitation by making the ABHA an independently maneuverable device. This can be done in a number of ways. First, propulsion can be used in the wellbore water in a direct analog of AUV propulsion, but in semi-consolidated sediments this could lead to wellbore washout. It is also possible to fix extendable pads on ABHA sections that allow the ABHA to essentially hold itself in place within the wellbore using gentle outward opposed pressure on the pads. It is also possible to have multiple extendible pad units that will allow the ABHA to essentially 'walk' either forward downhole or backward up the hole. A number of engineering solutions are possible but will not be discussed in detail here.

Being able to maintain satisfactory drilling rates depends to a large extent on the pressure that can be maintained on the face of drill tools. Exerting pressure on the drill tool in rotational drilling is accomplished mainly by gravity and mechanical 'pushing' of the wellbore. Exerting pressure on the drill tool in CTD operations is traditionally accomplished by pushing on the coiled tubing at or just before its point of insertion to the wellbore at the wellhead. For most purposes, this pushing on the tube is sufficient to accomplish the purpose of maintaining adequate drill rates. In ATD, force can be provided by the ABHA assembly itself. By a combination of walking the legs and extending and retracting a drilling tool, drilling can be advanced with the tube assembly, whose main job is to service the ABHA, only being used to provide minimal push rather than doing the pushing, as is done in conventional CTD.

It is anticipated that manufacture of ABHAs will focus on inexpensive designs rather than long life, as they will be interchangeable within the wellhead infrastructure. Ideally, the design life of an ABHA will be the time it takes to complete the exploration and development drilling for a field, with recovery and parts stripping followed by scrapping of worn parts during refurbishment of the units.

8.3.3.2 Maneuvering for Super-Directional Drilling

Although we are unsure as to whether drilling around corners will be necessary for NGH development, we envisage that the ABHA can have the capability to make relatively short corner turns that are dramatically tighter than are presently being done. In a conventional directionally drilled wellbore its course must be relatively open and switchbacks (more than one turn in one direction) may place the BHA into a position from which it cannot be extracted simply by pulling on the tube.

Although 'reverse pressure' can be exerted by mud injection, in conventional CTD there is no mechanical means for extracting the BHA up the wellbore except by pulling from the wellhead. In contrast, a segmented, 'walking' ABHA can move both backward and forward as well as being capable of reorienting itself while being able to firmly hold its positions within the wellbore while the tube is being eased out or retracted.

Tube push-pull capability: Conventional CTD BHAs can become stuck in a hole such that maneuvering away from the wellhead can be very difficult when there may be only brute force tensional pull used to extract a BHA. Also, at some combination of distance or hole complexity, further advance of conventional CTD BHA becomes impossible. While fishing for a broken drilling tool or any part that has become separated with the drill stem or tube/BHA assembly is often difficult even in a rotary drilling operation, the difficulty is much greater in CTD owing to its inability to rotate, lack of bending stiffness, and poor depth control (Byrom 1999). The most common fishing target in open-hole CTD is a BHA that has become stuck.

To overcome these limitations, where feeding or retraction of the tube from the wellhead is a limiting factor, a mechanical ram system and/or an extensible leg and walking capability of an ABHA can be used to move the tube by a feeding or retracting the tube at the wellhead at a rate and direction consistent with compatible driving motion by the ABHA at the drilling face end of the tube assembly. This will allow the tube to be snaked through wellbore configurations that would be impossible with driving force on the tube from the wellhead alone. Intermediate walking sections may also be affixed on the tube, from which they can receive their control communications using high-speed digital instruction sets using direct acoustic coupling of touching waterproof transducers with no water path, and with no need for hardwire connections.

8.3.3.3 Drilling Tools, Wellbore Width Control, and Sidetracks

The drilling requirements for NGH deposits are much less rigorous than for conventional drilling, which allows for a high degree of flexibility and inclusion of capabilities in ATD. For instance, when an ABHA can be held firmly in place in an orientation that is off line from the existing wellbore, sidetracks can be initiated without resorting to a whipstock device (Weatherford 2001), which is normally used to re-direct the orientation of the drill string to be placed in the wellbore. Also, it will not be necessary to mill through the steel wellbore pipe. Wider section of the wellbore may be created by multiple reaming that can provide rooms from which multiple sidetracks can be established. Because no whipstocks needs to be placed in the wellbore to initiate lateral drill holes (and these may often be permanent features requiring by-pass drilling to take the hole deeper), a number of lateral holes can be initiated at the same level or deeper, with no special provision for bypass drilling. In addition, variable width wellbores can be constructed, including the possibility of sump sections.

Three main different types of drilling methods can be practiced on an ABHA. The first is rotary cutting, using a rotating drill bit mount and the best drill bit for the GHSZ conditions. The second is jetting, which is now practiced not only in softer materials but also in rock materials. Jet drilling may be particularly applicable in a GHSZ because of the low mechanical strength of the sedimentary host materials. The third method is vibro-drilling in which a vibrating drill head is used to disaggregate the materials. For NGH exploration and production drilling, a combination of jet and vibro-drilling may provide the best drilling mode, at last in sediments without significant NGH in them. Light percussion and rotary drilling may also be a useful combination when drilling reaches a NGH-enriched part of a formation.

In the pay zone, where NGH exists in high enough concentrations to effectively cement the deposit, the materials that must be drilled will have the properties of solid rock. NGH is both hard and tough (Durham et al. 2003). In the most highly NGH-saturated reservoir host sediments, the combined material may be essentially rigid, with a low porosity and permeability characteristic of many rocks. However, NGH is different from any rock materials encountered in a marine environment in that it can be laser cut easily. The pay zone wellbore may be drilled without rotary drilling by the application of heat from a number of point sources including, but not limited to, low power, narrow beam laser cutting. The cutting will cause the NGH to dissociate in very narrow channels forming small chips so that the wellbore can be progressed by a disaggregation process. This chip cutting method is specific to NGH; especially it is desired not to produce significant free gas in the drill holes. Some gas production, however, could assist cuttings removal. Where NGH occurs, it will almost certainly be expedient to use angled laser cutting to create cuttings that can be floated up the cuttings course. Dissociation using a bulk thermal conversion process is unnecessary.

Down-hole drilling tools can be mixed. The drill motor can either be driven by injected drilling fluid or it can be driven by a geared electric motor, for which a back-up motor may also be installed on the same or a linked shaft. In addition, provision can be made for both driving forces to be used selectively. Second, the bit can be eccentric, that is, the shaft on which it rotates may be angled so that on a complete rotation, it sweeps a wider diameter than if it were in a fixed location. The shape of the drill bit in this instance should be rounder than normal, similar to some of the mills used for opening a sidetrack hole through casing. This allows wider wellbore sections to be created simply by drilling off axis.

Advanced drilling tools can be used in the benign pressure, temperature, and chemical conditions of NGH wells. If, for instance, the drill contains a seismic source (Theimer and Kolle 2007), a network of positioned receivers on the seabed can be used to control the position of the drill head in 3D space in a manner similar to that described by Spaziani and Lucos-Head (2012). Acoustic analysis while drilling would also allow more detailed NGH valuation. The same receiver array also could be used to monitor production across the entire deposit as one of the sensor inputs to guide NGH conversion and gas production. In conventional deposits, both polycrystalline diamond compact (PDC) and tri-cone drill bits are used to cut their way into rock. The semi-consolidated drilling targets in which NGH bodies will reside do not require the same robust cutting capability.

Tube assemblies and drilling fluid flow: A dual-tube assembly can be used to carry both a high velocity stream of drilling fluid down to the drilling face, and a high velocity stream of drilling fluid encumbered with drill cuttings back up the wellbore to the surface where the cuttings can be deposited. A dual tube assembly would allow for higher velocities to be maintained in both the downstream (supply) and upstream (cuttings) flows. This would be particularly advantageous in the upstream flow where higher velocities would insure better suspension of disaggregated cuttings. Two basic forms of dual-tube assembly each have different advantages. First, two tubes could be joined or wound, along with a third tube that would carry power, communications, and anything else not directly related to drilling fluid supply and removal. Coiling would preserve a generally round outside dimension, although the greatest roundness could only be achieved with equal-sized tubes.

An alternate design could consist of a concentric tube assembly that would maintain a generally circular tube cross-section. This could facilitate its movements within the tube and be less likely to jam or turn. In this case, the central tube will be the return or cuttings tube. It will have a larger unimpeded cross-sectional diameter than the encompassing tube around it. This would allow the largest particle size fragments and disaggregated sediment to be carried through the tube. The drilling fluid supply tube will be filled with at least crudely filtered seawater carrying only the smallest particulate matter that got through the intake screens. Thus, the width of the channel could be kept relatively small and still deliver the equivalent of a high cross-sectional volume. The electrical and communications cables could be carried within the outer or supply tube or on the outside of the drilling fluid tube assembly in a protected manner. As no cuttings will be passing up the outside of the tube assembly, there will be no potential erosion to sidewalls, which would make it a safer environment for the power and communications cables.

Drilling fluid throughput in a dual-tube assembly can have an injector pump and a separate extractor pump. Both pumps have variable pumping rate capability and can be automatically controlled according to a variety of normal drilling models, or manually overridden and controlled. By controlling the pumping rates of both pumps, it is possible to induce a suction component to the cuttings return as part of maintaining high fluid passage velocities. Two flexible rate and volume drilling fluid pumps allow the greatest control. They can be mounted at the wellhead and along the tube assembly in such a way that they can be changed out using ROV/AUV assistance at the wellhead. New applications for combined ROV/AUV capabilities, including locomotion using non-swimming mobility, may springboard off existing developments (Kruusmaa and Salumäe 2016). Additionally, overall flow can be reversed to assist in clearing possible blockages owning to cuttings particle jams.

The water between the outside of a tube and the wellbore walls will be essentially quiescent if a closed dual-tube assembly is used. The water in this region is maintained at pressures higher than formation pressures to limit pore fluid-driven sloughing from the wellbore walls. Provision for injecting or extracting water from this region may also be made or a cleaning extraction flow can be made by directing return water flow to this region.

Special drilling mud will almost certainly not be required for NGH and no oil is anticipated in association with NGH deposits. Thus, by slowly introducing fresh drilling water or ceasing its flow into the head of the wellbore should cause the water to clarify. Optical inspection can be carried out using a series of steerable mini-video cameras and digital output, video camera strings built into the side of the tube and protected with a transparent cover. These are solid state so they only need to be waterproofed and do not need any special pressure housings. These can be used in conjunction with other sensors to make detailed examination of sediments in the wellbore head.

Specialized tools may be lowered into a well using special ABHAs. These specialized service tools and narrow diameter wells with dedicated tubing, coiled storage handling equipment may constitute a new class of bottom-hole assembly. These will have been specially designed for miniaturized work to complement the tools carried by larger primary drilling ABHA. They also have a suite of communications and control systems so that they can be operated remotely, in a similar manner to an ROV on the end of a flexible tube deployed into the well from an assembly fitted to the tube without withdrawing the main ABHA. Special narrow tube equipment with specialized tools may be inserted, for instance for freeing jammed ABHA or associated equipment, dissociating unwanted NGH that has formed, or for visual or other inspection.

The ABHAs concept can also be used as the drilling system in rotary drilling where they may be affixed to the base of a rotary drill stem. In that role they can be operated in the same way as BHAs in CTD operations but with better stability and directional drilling capability, and with the possibility of a close-set casing. Operating in this mode, the drilling operation can be characterized as a hard-cable or stiff-wire ROV system. In this instance, the ABHA is attached to the end of the rotary drill stem with a clutch that can be activated at will by the controller. The ABHA would be tethered to a standard drill pipe system, which may include internal hydraulics, power and communications.

8.3.3.4 Reservoir and Environs Stability

Maintaining stability of the reservoir near well bores, and of the producing horizon and the sedimentary strata that support it, needs to be based on a good understanding of NGH reservoir geomechanical attributes. The semi-compacted nature of the sediments surrounding and within the NGH reservoir will almost certainly tend to compact more than more geologically mature conventional reservoirs. Stress induced by compaction associated with removing the solid NGH will weaken the reservoir as a whole. But compaction can be mitigated. The objective is to prevent as much compaction and to prevent sedimentary mass movement and deformation. In the first instance, keeping the gas/water removal ratio as low as possible will cause the reservoir to weaken the least. Second, the well itself should be designed for compaction and minor sediment movement.

It must be assumed that the entire NGH host strata will compact when the NGH is removed during production. Because one of the primary objectives is to remove as much gas as possible, well design has to take into account not only stability of the well within the pay zone, but the effect on surrounding sediments, which will also be relatively weak. A primary strengthening can be obtained from a geometrically complex wellbore system that can introduce vertical strengthening elements. Not all of the cased wellbore need be part of the producing system if its purpose is to primarily add strength to the reservoir as a whole. This remedy may be appropriate to NGH development because of potentially very inexpensive drilling and the benign character of the deposits.

In general, if a well can be engineered to be rigid where support is required and flexible where volumetric adaptation is unavoidable, a well will survive compaction in the host sediment. Because some compaction can be anticipated, it will be important to build some flexibility into the well liner so that if groundmass movement occurs, the integrity of the well can be maintained. Thus, flexible strength wellbore lining may also be an important aspect of NGH drilling and production.

8.3.4 NGH Well Conventional Casing Options

We use the term 'wellbore' to describe the course of a well, regardless of whether it is cased with steel or plastic pipe or any other kind of lining, including gravel packs and sand screens or liners that are not currently in use for conventional hydrocarbons. Also, the wellbore is understood to apply to any part of the well bore, whether it is in a producing or non-producing (with respect to extracting gas from the reservoir) situation or whether it extends from the surface to depth, is inclined at an angle or is of any geometric shape.

In conventional drilling, open wellbores are those where the walls of the well, in which some rock situations may prove sufficient because of their mechanical strength, support the wellbore without the need for casing or liner. Cased wellbores are those that have strong steel pipe forming the exterior of the well. These casings are usually cemented in place, yielding a very strong well capable of withstanding high reservoir overpressures.

Securing the physical integrity of a wellbore associated with hydrocarbon exploration and production is a concern of primary importance in order to keep the wellbore open during drilling operations, and allow for long-term maintaining of well control during assessment and production. This is especially true for NGH deposits where drilling is carried out in relatively soft sediments to access the NGH deposit. Setting casing, gravel packs, and sand screens, as well as other existing technology used in conventional and shale wells are perfectly adequate to prepare NGH wells for production. The requirements for NGH, however, are not as stringent and less robust casing and other well lining systems can be used that may prove to be easier to set, be less costly, and in the more innovative embodiments, allow for types of wellbores that are not possible to accomplish with conventional equipment.

NGH-specific casing: Because wellbore NGH walls may have a greater likelihood of sloughing into a wellbore, it is important that placing wellbore liners as near forward to the NGH drilling face as possible is accomplished. Casing can be inserted when rotary drilling or when combined rotary and CTD is used by a similar process of feeding it from the wellhead into the wellbore. In this instance, because of the lighter differential pressure loads with which the casing will have to endure than in conventional and other unconventional hydrocarbon drilling, an opportunity exists to use thinner steel casing. Because of the less demanding pressure and temperature differentials and minimum chemical hazard, NGH wells can be engineered for these lighter requirements. Casing can be thinner, lighter weight, and less expensive. In addition, it is possible to use other metal alloys with cost advantages, such as extruded anodized aluminum. A conventional screw-thread bayonet connection can be used. In addition, O-ring seals can be used that cannot be used in the more demanding conditions found in conventional wells.

It is envisaged that rotary drilling and insertion of casing will be made downward from the seafloor to some depth at which ATD is initiated. High-strength cased wellbore that will anchor the wellhead may be used for some further distance in a directional wellbore according to conventional drilling practice. Further down the vertical wellbore, however, lighter metal casing may be used. Given the unique production conditions of a producing NGH well in which pressure will be lower in the well than in the formation, polymer or plastic sections may be strong enough for production. NGH production will involve low pressure differentials between the reservoir and the well system, with pressure focused on the outside of the pipe, which will then be in a compressional stress field in the reservoir and possibly all the way up to the wellhead. Material strengths sized to requirements may allow significant cost savings.

Conventional drilling, which sets casing in decreasing diameters with depth and then selectively perforates producing sections as part of a second production step may not be necessary in 'lightweight' NGH development. Because some of the directional drilling and wellbore geometries for NGH conversion and production systems may be more complex than for conventional systems, and because of the soft nature of the semi-consolidated sediment, it may be advantageous to establish a new class of wellbore lining system.

Lower pressures in the NGH reservoir and lower relative pressures between reservoir and within the wellbores both during drilling and production allow mechanical wellbore stabilization to be engineered specifically for the NGH unconventional gas resource. The liner system does not have to withstand the high temperatures and pressures of conventional production wellbores, so a much less mechanically and materially strong wellbore liner system can be implemented for production of natural gas from NGH.

Conventional setting of casing involves sequentially inserting from the wellhead and forced in the well and then cemented in place. For deep wells, many sections of decreasing diameter casing are used because of increasing mechanical difficulty of inserting wide bore casing with depth. But wellbore lining in the softer sediments encasing NGH deposits requires rapid casing just behind the drilling tools to preserve the well before wall collapse can infill the well behind the drilling tools. Casing while coring isolates the formation while drilling as each foot of borehole is drilled and cased off simultaneously. This eliminates the need for separate casing or liner runs. With casing while drilling, casing is being incrementally forced downwards near the tip of the wellbore at about the same rate that the drilling face is advanced (Schlumberger 2016). This capability exists for conventional wells and would probably be the best method for setting NGH wells having a conventional geometry.

8.3.5 Active Wellbore Lining

Active wellbore lining is fundamentally different from conventional casing of wells from the wellhead. Active wellbore lining combines certain techniques of tunneling and drilling. In this embodiment, the wellbore is lined from the front or just behind the wellbore without adding casing from the wellhead. Each is effectively cemented into place just behind the ABHA drilling unit in much the same way that tunneling is carried out. Regardless of materials used or the manner of setting them described here, these are all referred to as 'active wellbore liners', which we will simply refer to as 'liners', although they are distinguished as a group from conventional liners/casing and their method of insertion into a well. In subsequent publications, we may refer to these as 'A-liners' where there might otherwise be confused with conventional liners. This allows for a number of new attributes including but not limited to:

(1) Variable diameter wellbores, with wide bore sidetrack 'rooms' at various places along the main wellbore being one of the uses for wide bore sections.
(2) Fluid flow engineering, for instance at sidetrack junctions to smooth flow and lower turbulence.
(3) Allowing the installation of equipment such as submersible pumps in dual bypass mode.
(4) Materials separation sections.
(5) Pay zone liners and solid liners can be set at the precise well locations where they should be based on real-time sensor data at the drill head.

Regardless of materials used or the manner of setting, these are all referred to as 'wellbore liners', or more simply, 'liners', to distinguish them as a group from conventional liners/casing and its method of insertion into a well. Unexpanded liner sections are designed to be carried down the wellbore in a compressed form and opened in such a way that they have as tight a fit with the wellbore as possible. The equivalent of cementing can take place as each liner is set. The wellbore can be lined from the front as the well is drilled. Thus, both impermeable and permeable liners can be inserted during a single phase of drilling operations, which means that exploration and development drilling and preparation for production can be accomplished in a single phase of drilling. Impermeable liners are utilized where no fluid exchange with the surrounding materials is desired, and permeable liners can be inserted where fluid exchange for gas production purposes is desired.

8.3.5.1 Examples of Liner Systems

A wide variety of liner section methods have been modeled that could be used to emplace sections. Five types are described here, although combinations, variations, and other types can be envisaged.

(1) **Folded pipe sections**: These can be inserted and then opened to a near circular section, snapping outward when finally locking into place. These are possible to design because the liner strength requirements are much lower than for conventional wells. In addition, materials can be polymers because the temperature and chemical environment is not as extreme as most deepwater gas wells. It is envisaged that autonomous assistance provided from within the ABHA will insure correct pipe section overlap and locking with the upstream section that had been set previously. There are a number of ways of segmenting liner sections that are delivered in a compressed or unextended form and then opened to encompass up to a 360° arc (even to provide overlap if desired). This can be a series of nested sections pre-curved to open to the desired hole diameter, twisted sections that open to the desired hole diameter, or compressed coiled systems. All of these may open using compressed spring drive or tension from the material itself that can be triggered when the liner is in the correct position. When the opening mechanism is activated, the segments slide over one another or 'untwist' so that a near-circular section can be achieved. Slide joins, over-center clip joins and other types of fasteners can be used to lock the sections into a more rigid assembly. O-ring seals or soft compressible seal material may be used to assist sealing at all joints, with regard for excess pressure residing on the outside of the well. Prepositioned pipe section-joining chemicals could include those that react chemically to form very strong cement that will result in essentially an unleakable solid joint. Although the strength of a joined casing may not be achieved and some leakage may be expected, this condition is not vital to subsequent successful production operations because of the different physical situation of NGH and the processes used to convert the NGH and produce the natural gas.
   Liner sections may be fabricated from metal or polymers (plastics) selected for their strength and resilience, although other materials may be used. Because of the relatively low temperatures that will be encountered throughout drilling and production activities, the materials do not have to have the high temperature resistance and pressure performance characteristics of casing used in deepwater conventional hydrocarbon deposits.
(2) **Soft liners**: These may also be used. For instance, very strong porous polymer material, such as but not limited to 'Kevlar', fabricated in a circular section and emplaced such that a strong metal spring is wound into a narrower diameter and held in that tensioned position by a number of mechanisms are activated when the liner section is placed in the appropriate situation. Pre-shaped overlap with the previous liner section will give a tight seal. An advantage of using fabric sections is that where water or gas transport through the material is desired, for

instance in a collector zone of a NGH production well section, it can be fabricated essentially as a permeable membrane, even to the extent of having successive spaced layers of decreasing diameter 'holes', each having a very high permeability. Metal spring reinforced fabric liners will be strong enough to be used to maintain open well courses. The pressure differences across the material will be much less than in a conventional well and the strength of the liner materials and specifications can be sized for requirements.

(3) **Foamed liners**. These consist of two layers of fabric that will form a double wall fabric when inflated reactive chemicals that form a polymer that can be foamed into place. Foaming processes are commonly filling a space between walls, as in the case of a foamed Styrofoam insulation in a building or into a preformed flexible mold. Foaming will fill any space, and by completely filling the space between two flexible retainers, such as the inside and outside of a wellbore liner section, form a solid and strong polymer or plastic pipe section at the point where it is inflated. This use was suggested by foaming requirements for an inflatable non-collapsible life jacket, in which the process of foaming and the options available are outlined in general (Max and Pellenbarg 2000).
Foaming agents can be selected for their product characteristics. For instance, the polymer can be dense, solid, not very flexible, and mechanically strong and resilient, or soft and flexible. Injectable liners can also be fabricated so that they are either essentially impermeable or permeable, although their permeability may not be as high as a fabric membrane type liner. A polymer for a well liner section could be of whatever mechanical character engineering required.
Wellbore liners would likely have few internal voids, higher density, higher strengths, but also having a highly elastic character so that if it were to be temporarily distorted by NGH formation, for instance during a well shut down, it would tend to regain its original shape following NGH conversion and re-start. The elastic character of well bore liner 'casings' allow production operators some flexibility in conversion strategy that using steel pipe might not.
Provision for water-cooling of the interior of an injection liner may be necessary in a NGH-enriched section as the foaming polymer process is exothermic.
Fortunately, the reaction rate can often be tailored and the foaming is taking place in a wellbore in which cold seawater can be circulated so unwanted dissociation of NGH can be avoided if care is taken. Where NGH occurs in the sediment outside liner, some thermal conversion can be anticipated, but gas will again form NGH when cooling takes place. The key to minimizing the conversion is to aid heat transfer from the outside of the liner to the inner walls, where heat transfer will cool the liner as a whole and allow ambient conditions outside the liner to be as little disturbed as possible. In sediment adjacent to liners in which little NGH is anticipated, no special provision has to be made for heat transfer because the amount of gas that could be produced from dispersed NGH before ambient conditions can be reestablished is minimal, with the regrowth of NGH absorbing free gas that may have been converted.

There are a number of advantages to using an injectable liner. First, the circular section can be over-sized with internal fabric ties between the inner and outer fabric wall, so that even where the wellbore is wider than planned, the liner will force itself outward and tend to fill irregular spaces while the inner liner will maintain a more circular section with fewer irregularities. Preformed thicker courses along the length, around the circumference, and spirally disposed, can act as reinforcement members, if desired. Where these protrude outward, keying with the sediments will be enhanced. Not having them protrude into the central well course will allow a smoother wellbore. In contrast to conventional casing, liners can have different internal and external shapes and enhanced overall stability and strength.

Second, where injection is initiated and there is an overlap between sections, seals will be very tight, even if no other sealants are used. Provision is made for the two liner sections to lap, with thinner sections that match so that a lap joint is achieved that will prevent section separation later, even with sediment compaction in the host sediment as NGH is converted and the gas is removed.

Third, the liner section, because of the tight fit with the country sediments, can be set with the same security of a cemented conventional casing section, with the exception that each injectable liner is essentially cemented, rather than in a conventional well where a cementing process usually affects a relatively short section of casing, often at the casing shoe.

Placing the liners may be accomplished either entirely automatically using trigger systems activated by the position of the previous liner, semi-automatic using mini-ROV visual situation rendering to the control room, or by mini-ROVs operating under remote optical control in an essentially quiescent water environment, especially when a double hose drilling fluid system is used. A combination of these methods may also be used. Inserting new liners that are suitable for the particular location in the well bore to be lined can be done from a magazine near the wellhead that has liner selection capability and automatic loading after following selection.

(4) **Composite material liner sections**: Where a sand screen is desired, it can be fabricated entirely as a permeable fabric section or by using a perforated metal or polymer section. Perforations can be large, and equivalent to holes that are burned through casing in production pipe or by the placing of sand screen sections. The permeable sections can be 360° of a section or only part of the section, with perforations set directionally, if desired.

(5) **Special materials**: Some special materials are available in industry that may be evaluated for active liner use. For instance, Shape Memory Alloy Subsea Fitting, which was pioneered by Shell and FMC, takes the place of welding tubing (Offshore 10/29/15). The technology is based on a particular nickel–titanium material that has the ability to return to its original shape after it has been distorted and then placed for joining with an adjoining section prior to being subjected to environmental changes of temperature or pressure.

8.3.5.2 Special Section Liners

In addition to lining a wellbore with either porous or impermeable section, it is possible to foam sections to achieve an in situ equivalent of cementing casing to lock it firmly into the strata in which the well has been drilled and to prevent leakage as well as forming filtering equipment such as well screens and gravel packs.

In conventional drilling circumstances, cementing the casing to geological formations is used to isolate the pay zone so that pressurized gas, oil or water does not migrate along the outside of the casing/liner into another formation or blow out at the wellhead or elsewhere on the seafloor. Isolating pay zones or producing horizons is one of the critical factors of efficient and safe production. In the case of NGH production, gas migrating up the outside of the well liners constitutes a safety issue. However, because wellbore liners can be sealed tightly against the bounding strata, the danger of leakage utilizing the wellbore can be minimized.

In pay zones, the formation may develop a tendency to compact as NGH is converted and the gas produced. It is planned that the degree of tightness elsewhere along the well insures that exterior gas migration along the wellbore penetration does not take place. Thus, the compaction that may open space in the vicinity of a well is not directly critical to sealing off the pay zone. Special provision can be made for special wellbore liners that will be able not just to stabilize the well, but to accomplish other objectives.

One of the major issues for production is to reduce sand flow into the well and minimize the amount of sand that is produced along with the pay gas. By configuring the external skin of a wellbore liner in a complexly shaped 'rubbly' form in which complex permeability paths are established for some distance outward from the inner wall of the wellbore liner, the equivalent to a gravel pack can be foamed in place. In cross section this will involve a smooth interior wall but with an outer zone of any desired thickness comprised of permeable foamed section that can be designed to act in the same fashion as a gravel pack to prevent sand and fines from flowing from the formation into the production well. An advantage of foaming gravel pack sections will be the relative ease in time saving and materials required to correctly place a gravel pack, especially in a horizontal or inclined well (Schlumberger 2012). Using a foamed gravel pack wellbore liner means that much longer sections of production wellbore can be lined in this fashion, which enhances long-term sand production mitigation.

In a 'gravel pack‘ section, pseudomorphs of gravel and fibrous mats can be manufactured from fabric that can be pressure-foamed to fill the internal spaces. When fully expanded with low-density, flexible polymer, the effect of product gas and water passing through it will be to filter particles from incoming flow. The permeability can be engineered to for size and depth, with designed porosity being coarser on the outside. Inflatable sections such as this can be engineered to be symmetrical or asymmetrical and sized for the sedimentary material in each deposit.

8.3.6 *Wellbore Geometry*

We envision that much more complex well bores may be used in NGH deposit reservoirs than have been used in conventional or unconventional gas resources because of the unique nature of the resource. Special geometry production wells allow for better control of NGH dissociation conversion utilizing the spatial reach of de-pressurization conversion envelopes to provide for very large-volume synchronous dissociation. In addition, it may be possible to initiate gas and water separation in the reservoir. The separation and controlled unmixing of produced natural gas and relatively pure water is unique to the NGH resource and is another potential cost-cutting area in which efficiencies can be gained. This topic has been the subject of initial trial designs, which we will not discuss further here. Nonetheless, we regard this option as a potential cost-reduction area requiring further modeling and testing.

8.4 Production Issues

This is not anticipated to be a difficult or costly issue, even though NGH conversion is considered to be a significant cost factor when using a conventional approach. Subsea processing is maturing to become one of the most innovative technologies in the offshore industry at present. Implementation of seafloor processing, including power distribution across the sites, is presently underway in ultra-deepwater. The distance between seafloor and sea surface has simply become so great that the difficulties of carrying out processing from the surface is no longer viable if operations are to remain reliable and further de-risked, and costs are to be reduced. The temperature, pressure, and chemical attributes of NGH production (which will consist mainly of fresh water mixed with formation water mixed with ambient pore water in the deposit prior to NGH conversion) will result in requirements for much less robust processing equipment.

Managing production should be facilitated by the character of the dissociation product of NGH and it effect on physical properties. For instance, the fresh water and gas will each lower the viscosity of the pore fluids. In addition, the dissociation product will be relatively pure water and gas with few of the minor 'pollutants' such as liquid or solid particulate material such as either paraffin or naphthenic based materials that may be associated with thermogenic conventional deepwater and ultra-deepwater deposits.

We suggest that design of seafloor processing equipment will focus on much less capable and expensive variants that will be effectively downsized from existing conventional processing equipment already in service. We envisage this as a relatively simple specification of materials and operational requirements. Once a set of NGH seafloor processing equipment is brought to design and fabrication, it is likely that virtually all NGH wells and fields will be serviceable for most, if not all, locations because of the similarity of the water depth range and the other requirements.

8.4.1 Sand and Sediment Fines Production

Although the end of the JOGMEC 2013 test was coincident with sand production issues, no data has been released to show whether this was a response to an unusual test condition or a time-related development that overwhelmed the gravel pack. The Japanese technical production test at their Nankai deposit produced a relatively steady gas flow for the better part of 6 days (JOGMEC 2013). At very nearly the end of the test period the well was closed coincident with sand production. In any case, however, this was a technical production test concerned with issues other than production in its own right (Fujii et al. 2015). Other experience with NGH production testing suggests that sand production can be well-managed through standard engineering controls as produced sand at the beginning of the Iġnick Sikumi CO_2–CH_4 gas hydrate exchange field trial stopped flowing after four days (Schoderbek et al. 2013; Anderson et al. 2014).

We suggest that controlling gas and water flow velocity by designing an optimal layout for wellbore geometry may be used to substantially retard the movement of fines toward production well bores.

Another aspect of NGH production is that with respect to sand flow, an NGH well may behave more like an oil well. This is because both gas and water will be flowing and the density of the water is close to that of oil, with relatively close (with respect to gas) viscosity and sediment particle carrying ability. In contrast, many deepwater gas wells are dry gas without oil or water. These can be allowed to flow faster into production wells because they are much less capable of carrying fines to the production well.

Therefore, we model the optimum production conditions for NGH as having:

(1) Relatively slow dissociation conversion over as large a volume as possible.
(2) As large a surface area on the production well bore as possible.

By producing gas in a large volume and having a multitude of migration paths, the velocity of the gas and water produced within the reservoir, and more importantly near and crossing into the production well bore, can be kept as low as possible. This will insure that the converted gas-water mass has as little potential as possible for carrying fines that would degrade production over time.

Modeling solutions will assist operators by better defining the natural characteristics of the natural resource to increase certainty in their operations, reduce costs, and improve productivity. In particular, geological modeling using a rock physics approach for the NGH reservoir and supporting strata (Moyano et al. 2011) is important in exploration and valuation. Geotechnical modeling is particularly important for designing drilling and wellbore-lining strategies with the important objective of controlling sand migration during conversion and gas and water drainage being possibly the most important issue. In addition, longer production wellbores with special geometries could both lower flow rates within the reservoir but activate dissociation and gas movement to collector well sections over quite large areas in order to minimize sand issues while dramatically scaling up gas production.

8.4.2 Produced Water

Only very rare conventional gas wells produce dry gas with little or no produced water.

Daigle et al. (2015) calculate that sands in stratigraphic sequences have 4–5 times the order of magnitude of permeability than fin-grained shales. This results in a dewatering of the shales and the development of focused flow within the sands. However, they note that even small amounts of clay in sand can have a major impact on permeability. This may be especially true for NGH-enriched sands in the lower NGH mineralized sequence in the Japanese Nankai deposit, which are highly interlaminated with discontinuous clay-rich shale partings and lamellae (Fujii et al. 2015). A potential issue that needs to be carefully addressed with respect to gas production from NGH deposits is that some of the clay minerals (smectites) commonly will expand on contact with fresh water. Because NGH dissociation produces fresh water along with the gas, the pore water geochemistry of host reservoir sediments may change enough to initiate swelling of clays and reduction in permeability.

Small amounts of clay in a sand may also act to prevent NGH formation as advective fluids will not be able to enter the sand in the volumes required, although the existing dataset is too small to parametrize the effect of clays of different types under NGH conversion/production scenarios. The energy industry has considerable experience in the effects of fresh water flood or acid secondary recovery operations. This experience has resulted in industry generally avoiding fresh water in reservoirs having swelling clays. And during the entire time of production, the water in the reservoir will be freshening. Clay problems in reservoirs are a major issue in reservoir geology and engineering and there is considerable experience in the field, much of which might be directly leveraged. There is considerable industry experience with conventional reservoirs having been shut down when acidizing jobs strongly drove alteration of clays that impeded permeability instead of opening tight porosity as it was supposed to do.

We are on very new ground with the production of fresh water from NGH dissociation. Although a swelling clay issue may turn out to be an insignificant problem or one that can be overcome inexpensively, we point out that understanding the clay mineralogy within NGH concentrations from cores taken on any of the drilling programs should be a necessary step to planning any NGH production. Although a number of production tests of NGH have taken place, no test has gone on long enough to date for produced water effect to be measured. Time is a particularly important factor for NGH production because it is anticipated that relatively steady production for many years to decades may be characteristic. In any case, systems for treatment of conventional well produced water for release near the ultra-deepwater seabed are under development (Daigle et al. 2012) and a relatively small part of that capability should be sufficient to treat and polish produced NGH water.

Production from NGH deposits does pose special challenges for commercialization. For example, hydrate dissociation frees significant volumes of free water, and the study by Walsh et al. (2009) shows nearly 2500 Bbl of water produced for every million cubic feet of gas produced from deepwater methane hydrate early in the development of a hypothetical deepwater reservoir. This figure drops to ~100–200 Bbl water produced per million cubic feet of gas produced during much of the life of the well before climbing rapidly after more than a decade into the hypothetical production scenario. We envisage, however, seafloor water separation so that sand lift from the seafloor will not be an issue for NGH production.

A further issue is that produced water is almost certainly going to be lower salinity than ambient at the seafloor and probably saturated with natural gas. Consideration of a full production scenario is that a solution has to be found for mixing produced water so that it does not form buoyant effervescent plumes from which free gas will form if it is released close to the seafloor.

8.4.3 Gas/Water Separation

A successful gas-water separation system was used for the JOGMEC 2013 test in conjunction with depressurization-dissociation. Even though water may be separated from gas on the seafloor, the water will contain considerable dissolved gas. Prior to dissociation, the pore water in an accreting NGH deposit will have at least a concentration of dissolved gas equivalent to the active driving force of crystallization (Chap. 1). Within the lower part of a NGH deposit that is in contact with free gas, pore water for some distance into the deposit may be close to supersaturation.

We envisage that very long pay sections of gas removal wells may be paired with crossflow wells to manage primary gas and water management and separation in the reservoir itself.

8.4.4 Reservoir Management

Naturally occurring NGH is stable in its reservoir regardless of how much gas is present. In order to produce the gas, the NGH must first be converted to its constituent gas and water. Although depressurization (Max and Johnson 2011) is proving to be the conversion process that achieves a number of technical objectives most efficiently, the principles of reservoir management remain the same. Because the location and rate of dissociation, as well as the extraction of gas in the reservoir can be controlled, optimal management of an NGH reservoir will be very different from a conventional gas well in which the inherent risk is much higher (Chap. 10).

NGH reservoir management will consist of balancing the rate and location of dissociation and pressure within the reservoir while simultaneously controlling the location and rate of gas extraction. This will allow pressure within the reservoir and immediately outside it that are in hydraulic continuity through pore water and gas to be maintained within design limits to minimize compaction of the NGH-depleted host sediments. By minimizing and localizing differential stress within the reservoir, controlling compaction has the potential to prevent differential movement of sediment that could lead to distortion or breaking of wellbores that could result if faulting and sediment mass movement occurs. In addition, local control of pressure within a reservoir has the potential to manage the movement of gas and water separately within the reservoir. It is possible to control different regions of the reservoir individually so as to achieve different pressure regimes using different wells or new designs for collector wells in which the rate of extraction can be controlled. NGH reservoir management will allow for maximum reservoir integrity as a function of gas extraction.

8.4.5 Flow Assurance

In general, when a NGH deposit is undergoing depressurization-dissociation, unwanted reformation of NGH is not an issue in those places in the reservoir-well system that are outside of NGH stability. Producing well decompression will be carried out using submersible pumps in the well (as was done in the 2013 Nankai test). In addition, the use of more submersible pumps and pumps from the base of the well floor to the wellhead, gas production will be accelerated as well as maintaining NGH conditions of instability within the well. Heat from gas compression for lift to the surface following liquid water separation can lower the concentration of dissolved water in the gas to below the driving force necessary to promote NGH nucleation and growth (Chap. 10).

Thus, by management of pressure within the well and the lift-transport system, flow assurance can be maintained without the need to apply inhibitor chemicals or make other intervention. This is another cost-savings area whose benefits remain to be quantified.

8.4.6 Production Risers/Pipelines

In ultra-deepwater, it is not feasible to use risers that have a roughly vertical profile to carry production from a wellhead to a production platform or vessel. Moored platforms have operational cost and operational limits that are being tested now. If virtually all of the production processing is carried out on the seafloor and drilling (which is more feasible for NGH than it is for conventional hydrocarbons), also moves to the seafloor, the logic that follows suggests that it is unnecessary to have expensive moored floating platforms. Turret-mounted floating production, storage, and offloading vessels can replace fixed floating platforms. This technology already

Table 8.1 Ultra-deepwater risers. After Milne (2012)

Coupled risers
Steel Catenary Risers (SCRs)
Weight-Distributed SCRs
Steel Lazy-Wave Risers (SLWRs)
Flexible riser systems
Un-coupled risers
Single Hybrid Riser (SHR)
Grouped SLOR
Hybrid Riser Tower
Buoyancy-Supported Risers (BSR)
Tethered Catenary Riser (TCR)
Catenary Offset Buoyant Riser Assembly (COBRA)

exists and can support NGH exploitation with little modification. The link between the sea surface and the seafloor, the risers, power and communications systems must all function for extended periods of time without disruptions but the requirement is not different for NGH than it is for conventional gas, especially if the gas has been fully processed on the seafloor.

Risers in deep water can be affected by currents moving in different directions at different depths and by other oceanographic forces. As water depths increase, the need for more flexible risers becomes critical to de-risk both production and environmental concerns. Even with semi-submersible systems, large swells and waves impose vertical strains. There are two basic types of flexible riser systems (Table 8.1), each having a number of variants (Milne 2012). These are not discussed in detail, as it would require an extensive summary. The main point is that risers that are perfectly adequate to service an NGH seafloor industrial site through exploration and production are already available. The fundamental purpose of all of these production risers is to bring gas and oil and possibly water (if it has not been separated and expelled on the seafloor) to the sea surface where the hydrocarbons can be treated, separated, and in the case of natural gas, compressed and transported. These complex risers are indispensable production equipment in water depths in which we anticipate that most NGH deposits will be found.

NGH gas and oil will have low temperatures and pressures that will not require costly steel casing. Composite, essentially non-metallic materials appear to have certain benefits over steel, especially in an upper section of a tethered connection to a vertical riser leg (Hopkins et al. 2015). Natural gas and water produced from hydrate will be of high quality and pressure differentials will be relatively low across pipe walls. Thus, it is likely that flexible composite pipes can be used for risers. This should lower the cost of risers in comparison with conventional hydrocarbon risers, and probably also lower maintenance costs while extending useful life. It is also possible that composite riser pipe may be used for flowlines within the industrial plant at the wellhead or field and possibly also used for long distance flowlines whose cost would justify long distance transmission to a production platform in shallow water or to land.

8.4.7 Communications, Monitoring, and Active Reservoir Control

Initiation of a depressurization dissociation-driven NGH conversion for natural gas production begins with pumping pore water from an NGH deposit to reduce the formation pressure to below that in which NGH is stable. Artificial conversion of NGH results in gas and water replacing NGH and filling pore space, which increases as conversion proceeds. In an ideal case, once depressurization-dissociation begins, removal of only gas would achieve the twin objectives of gas production and leaving as much water in the reservoir as possible because the water produced is only 80% of the volume occupied by the NGH. Unless complete gas-water separation can be made prior to gas lift to the surface or into a pipeline for transport, some water will also be removed from the deposit, although at some point, water flooding may be necessary to reduce reservoir compaction. In addition, NGH conversion with production of free gas has to be balanced against the rate and volume of gas that is removed to achieve a relatively small volume of gas in the formation while maintaining desired gas production.

In contrast to a conventional well that is essentially passive in that gas is being withdrawn from a large, high pressure gas reservoir in which gas occupies the pore volume throughout, with the exception of bound water, NGH wells have to be actively controlled by balancing NGH-water-gas-porosity issues with respect to an energy efficient process for recovering as much gas as possible with as little cost as possible.

It will be necessary to have an integrated, interactive and automated control system for each well. Where more than one will service a field, they must have the capacity for a higher level of integration when the dissociation envelopes of the wells begin to affect each other. Control systems will have to take into account thermodynamic effects of controlled artificial dissociation and provide a template for balancing gas production with natural thermal rebalancing of the reservoir to accommodate the heat loss during depressurization-dissociation so that the minimum artificial heating will be necessary. The control system controls the depressurization and the volume of gas in the reservoir and elsewhere in the recovery system to allow for long-term balance and the maintenance of a steady gas flow having a desired, pre-planned production rate.

Integrated control systems for NGH production will use computer programs that will be comparable in complexity to conventional production but very different. But algorithms embedded in the thermodynamic production models (Chap. 11) can be used as a starting point. In addition to controlling the rate of gas and produced water, production will be from mechanically weaker geological situations so a close integration with the stimulation is needed to provide for controlled conversion of the NGH into its component water and natural gas. Equipment must be programmed to operate independently, at least to carry out certain basic maneuvers and actions, or to stop in a rest position when the scope of its instructions has been exceeded and control can be implemented.

Autonomous control at well sites will increase as well as real time high-speed communications (Watson 2016). For application to autonomous seafloor industrial sites, Artificial Intelligence (AI) involves the concept of intelligent machines controlled by a computer program that is capable of carrying out a series of actions for which it may not have been completely programmed. The goal is to create automated solutions that can be used to control equipment, in particular autonomous underwater vehicles that carry out maintenance, repair, and operations in the absence of humans.

AI computer program will have enough of an understanding of the technology so it that can be used to resolve or accomplish a set of discrete tasks. These are under development today (OS 2017). Sensor input, analysis, 'learning', and fully autonomous self-control is not only possible, but in the case of self-drive cars, being implemented. Most current progress in AI predominantly utilize computer vision of one sort or another, for instance, facial recognition, facial expressions, automatic visual image analysis, and in the sound range, language processing in order to respond verbally to verbal interrogation; technology that is already implemented in smartphones. It is therefore consistent with existing practice that seafloor industrial sites will use extensive AI.

Sensor networks in the wellbores system will monitor the reservoir continuously. Gas fluid sampling will be key to efficient reservoir and production management (Nagarajan et al. 2007), but systems developed for multi-phase oil and gas wells will only require a small portion of the control programming for NGH production. Adequate data-rate telemetry built into the wellbore can be inserted along with casing or well liners. The Baker Hughes "TeleCoil" system for monitoring drilling activities includes some sensors and communications for accessing real-time down-hole data (BH 2016) and could provide a basis for an NGH well monitoring system along with other existing well communication systems that could be NGH-upgraded (Chap. 7).

In addition to sensor nets in the well, seafloor mounted sensors that can acoustically monitor the changing character of gas, water, and NGH in the reservoir provide input to the control system. Data links with remote control rooms will require considerable new communications capability but can be developed from existing systems (Chap. 7). In addition to electronic systems, software or algorithms for non-disruptive overriding of automatic and semi-automatic performance of various tools in the well will provide a level of backup security. Communications with remote control rooms will involve some lag times, even where there are no slow-speed links in the communication system. Thus, all deepwater technology will have to be autonomous to some degree.

8.5 Well Abandonment

Like all good field development and production plans, NGH wells will eventually have to be abandoned. Even though a strong petroleum system may continue to exist in an NGH well, and more NGH will form in the vicinity of a depleted deposit, a sizable deposit will probably not form quickly enough for NGH to be regarded as a renewable resource. It might be possible, however, to revisit NGH production on a multi-decadal scale. Thus, some NGH fields may require a different abandonment procedure than conventional wells, in which re-entry is not a consideration. When production is finished and artificial dissociation ceases, the natural formation pressure of a deposit will naturally reestablish and any natural gas in the depressurization envelope will reform NGH and stabilize the well and reservoir. If the basic permeability of the host sediment has been preserved by careful attention to geotechnical considerations (Small 2016), restarting production based on new NGH may be possible.

As more NGH forms from continued gas flux, the sediment of the depleted deposit will begin to strengthen. Like conventional wells, a NGH well will be sealed below the mudline and all surface equipment removed. Instead of cement, methane or compound hydrate may be used. Any gas coming into the well bore beneath the seal will form solid hydrate and become immobile, so the likelihood of any leakage beyond what would naturally take place from a natural active groundwater system is very slight. Abandoned NGH wells will not have to be monitored for pollution potential.

8.6 NGH as a Geotechnical Material

NGH traditionally has been regarded as a drilling hazard in conventional drilling as heat produced from drilling or upward passage of hot hydrocarbons in the well could cause dissociation of hydrate near the well. Even small amounts of NGH, on the order of 3–5% scattered grains, could cause mechanical weakening of the sediment that could initiate mass movement if they converted rapidly. Unintended conversion of even small amounts of dispersed NGH through heating associated with drilling and production can cause mechanical weakening of sediments. Weakening of sediment strength may result in sediment subsidence or flow, with resulting displacement of seafloor equipment and bore hole disruption. Industry has learned to refrigerate to counter heat production and this issue is no longer a serious problem.

It is possible to envisage an alternative case wherein increasing the strength or bulk modulus of seafloor to make it more load-bearing can be carried out by artificially causing NGH to form so as to mechanically strengthen the sediment and enhance foundation stability of seafloor industrial sites. We suggest that inducing

NGH to form in the sediment could open the field of geotechnical engineering, especially at the seafloor or in the upper part of a GHSZ where it is most stable (Chap. 2). This concept could be particularly useful in situations where expensive and difficult complete decommissioning of subsea structures is required (Offshore 3/29/16). In situations where the artificial formation of NGH has been used to strengthen sediments and increase the bulk modulus of seafloor sediments to provide a more stable base for seafloor equipment, simply ceasing any gas flow that stabilizes the artificial NGH allows the sediment to restore to its original condition at no cost other than removal of gas injection equipment,

8.7 Role of Intellectual Property

In general, when new technology is developed, it can be patented. Patents (intellectual property rights or 'IPR') may have considerable value. Patents can be the primary source of value to an innovative company in any field. A patent identifies a new apparatus or process as something new that is officially recognized by a government or more than one government through international treaties. Patents convey the right of ownership to a method or apparatus, although the only real right conveyed is the right to sue for infringement. Although patents no longer have to show working models or actual proofs of concept, the notional value of the patent is increased if the invention has not been anticipated by prior art, does not infringe upon an earlier patent, and if the patent can show that is has been tested in a laboratory that provides a proof of process at any scale.

Usually patents allow the holder to set license fees and royalties for the use of the technology by others. Gaining clear title, in which the patent is no longer subject to serious challenge, is often a matter of fighting many court cases, especially against 'trolls', companies that only file suits with the aim of achieving a settlement in order that a long and expensive court case should be avoided. Alexander Graham Bell, for instance had to fight over 300 court challenges before receiving essentially clear title. Any innovative company should have excellent patent attorneys and the funding to defend its patents.

Writing and filing of patents should be a primary consideration for any innovative company. Although the value of even giant companies may once have rested upon a single patent, such as the Hughs Tool Company's 'tri-cone' or walking percussion drill bit, without a dedicated and planned effort to memorialize inventions as patents, the future of a company cannot be secured. Therefore, achievement of as much IPR as possible should be a primary aim of the research program.

8.8 Technology Readiness Levels (TRL)

To establish a new paradigm, such as moving all exploration and production activity for NGH to a seafloor industrial site, a number of different technologies must be perfected and made to work together. Some of the technologies can be leveraged from existing technology, with different levels of alteration to suit the NGH circumstances. A number of new technologies must also be brought into practice. Not all of the individual technologies will have the same level of complexity or will require the same amount of development. In order to timeline planned development it is first necessary to identify the estimated amount of work required in each of the component technologies.

Establishing the timing along a development path for a particular technology provides metrics for estimating costs, time, and investment potential, amongst other issues. Substantial new technology may be developed and brought into practice for NGH exploration and production activities in response to understand the particular opportunities offered by the unique material in its reservoir. Table 8.2 shows the brief descriptions used by the European Union's current major research program (H2020 2014) a fuller description of the technology readiness (TR) level can be found in DOE (2011). Scientific and technical project management technique is only mentioned here in passing in order to make the general methodology known to less experienced persons. Technology developments, from ideas that have only been tested with respect to not violating any law of thermodynamics, often have a strongly non-linear development path. Two or more new technologies may often have to be mated in order to develop a reliable working technology.

A technology readiness assessment (TRA) scale is used to evaluate the level of technology maturity of any technology or practice using the Technology Readiness

Table 8.2 Technology readiness levels (TRL), European Union usage (H2020 2014)

TRL 1—Basic principles observed
TRL 2—Technology concept formulated
TRL 3—Experimental proof of concept
TRL 4—Technology validated in lab
TRL 5—Technology validated in relevant environment (industrially relevant environment in the case of key enabling technologies)
TRL 6—Technology demonstrated in relevant environment (industrially relevant environment in the case of key enabling technologies)
TRL 7—System prototype demonstration in operational environment
TRL 8—System complete and qualified
TRL 9—Actual system proven in operational environment (competitive manufacturing in the case of key enabling technologies; or in space)

Level (TRL) scale, which was pioneered by NASA (U.S.) in the 1980s when mission critical new technologies had to be developed and mated together for the U.S. space program. Although development of NGH-enabling technology will not be as complicated, nor possibly will the technology be human-critical, integration and applying or modifying existing technology will involve a degree of complexity that may be greater than one individual can manage. Thus, the TRL level assessment system (Table 8.2) provides a framework for the core of a development activity. One of the best examples of the use of TRL is in evaluating renewable energy production and storage capabilities to steer investment and adoption (NAS 2016)

The TRL scale ranges from 1 (basic principles observed) through 9 (DOE 2011). The basic scale has been adopted worldwide for application in offshore technology development, especially where government contracts are involved.

The fastest and most efficient technology development progress can be made through assessing existing technology in order to identify unused potential or new combinations of existing technology to be used in a new manner. It is common to find that some existing technology can be used in new ways. Also, combining underlying technology and practices may allow identification of new technology development paths that will enable lower cost operations. The new technology options discussed in this chapter should be ranked for their TRL level to determine development time factors. A strong development plan is a primary factor in technology risk reduction and safety.

Drilling costs have escalated as exploration has moved into deeper water, increasing the cost of finding and developing energy resources. Then there is the challenge of transporting the product to the market from remote locations, specifically if it is gas. A patent identifies a new apparatus or process as something new that is officially recognized by a government or more than one government through international treaties. Patents convey the right of ownership to a method or apparatus, although the only real right conveyed is the right to sue for infringement. Adding to this is the new regulatory and permitting regime resulting from the 2010 Deepwater Horizon incident. Once again, connecting Resource-Develop-Market-External factor dots in innovative ways will realize new possibilities (Khurana 2015).

The first effort is in finding new plays and larger resources. This means investing in 3D seismic, 4D seismic, subsalt imaging, and new inversion techniques. It also means engaging the regulatory authorities and non-government organizations early to ensure environmental aspects are assimilated from the onset. As for challenges of developing high-pressure and high-temperature reservoirs, new industry efforts such as the Project 20K for high-pressure wells are springing up. In development, the industry must prepare for increases in the complexity and size of the floating production systems by finding innovative ways of integrating large topsides with the hull, and to improve topsides efficiency.

8.9 Optimizing Leveraged and Innovative Technology for NGH Development

Leveraging existing technology (Chap. 7) is sufficient for short term development of the NGH resource. Cost reduction can be dramatic if technology and available vessels are applied and drilling and development plans take advantage of the many opportunities offered by the particular physical character and chemistry of NGH. Further cost savings could be realized if new technology made development more flexible and allowed accomplishment of objectives that are not possible in conventional gas deposits. Although we have not made a detailed analysis of the cost factors and potential gains set against development costs, when the number of areas in which cost savings could be realized is examined, the potential for dramatic cost cutting of development of the NGH resource appears to be very substantial.

References

Anderson, B., Boswell, R., Collett, T. S., Farrell, H., Ohtsuki, S., White, M. et al. (2014). Review of the findings of the Iġnick Sikumi CO_2-CH_4 gas hydrate exchange field trial. In *Proceedings of the 8th International Conference on Gas Hydrates (ICGH8-2014)*, July 28–August 1, Beijing, China, (17 pp.).

Beckman, J. (2014). ONS highlights next-phase deepwater drilling, subsea recovery challenges. *Offshore, 7*(10), 44–45.

BH. (2016). *TeleCoil intelligent coiled tube services*. Baker Hughes. http://www.bakerhughes.com/products-and-services/pressure-pumping/coiled-tubing-services/telecoil-downhole-communications-system. Accessed January 23, 2016.

Boswell, R., Collett, T. S., Frye, M., Shedd, W., McConnell, D. R., & Shelander, D. (2012). Subsurface gas hydrates in the northern Gulf of Mexico. *Marine and Petroleum Geology*, 21 pp. https://doi.org/10.1016/j.marpetgeo.2011.10.003.

Bourillet, J. F., Augris, C., Cirac, P., Mazé, J. P., Normand, A., Loubrieu B., et al. (2007). European ocean research fleets. Towards a common strategy and enhanced use position paper 10, European Science Foundation Marine Board (64 pp.).

Brillon, C. L., Shafer, R. S., & Bello, A. A. (2007). Pushing the envelope with coil tubing drilling. American Association of Drilling Engineers. AADE-07-NTCE-31 (16 pp.). http://www.slb.com/~/media/files/technical_papers/2007/07ntce31.pdf. Accessed March 7, 2016.

Byrom, T. G. (1999, June). Coiled-tubing drilling in perspective. *Journal of Petroleum Technology*, 57–61.

Carriere, G. J., & Gipson, T. C. (2003). Coiled tube drilling rig. U.S. Patent US 6,502,641 B1, Jan. 7, 2003 (12 pp.).

Daigle, H., Cook, A., & Malinverno, A. (2015, December). Permeability and porosity of hydrate-bearing sediments in the northern Gulf of Mexico. *Marine and Petroleum Geology, 68* (A), 551–564.

Daigle, T. P., Hantz, S. H., Phillips, B., & Janjua, R. (2012, November). Treating and releasing produced water at the ultra-deepwater seabed. *Offshore*, 74, 76–77.

DOE. (2011). *Technology Readiness assessment Guide*. U.S. Department of Energy, particularly Table 7. Hardware TRL Definitions, Descriptions and Supporting Information, (73 pp.).

Durham, W. B., Kirby, S. H., & Stern, L. A. (2003). The strength and rheology of methane clathrate hydrate. *Journal of Geophysical Research 108*, 11 pp. https://doi.org/10.1029/2002jb001872.

Eldridge, A. (2013). *Subsea industry: Drilling on the floor of the Arctic Ocean*. http://www.albertaoilmagazine.com/2013/10/offshore-Arctic-drilling/. Accessed 22 June 2015.

Fontenot, K. R., Lesso, B., Strickler, R. D., & Warren, T. M. (2005). Using casing to drill directional wells. *Oilfield Review, 2005,* 44–61. summer.

Frye, M., Shedd, W., & Boswell, R. (2012). Gas hydrate resource potential in the Terrebonne Basin, Northern Gulf of Mexico. *Marine and Petroleum Geology, 34,* 150–168.

Fujii, T., Suzuki, K., Takayama, T., Tamaki, M., Komatsu, Y., Konno, Y., et al. (2015). Geological setting and characterization of a methane hydrate reservoir distributed at the first offshore production test site on the Daini-Atsumi Knoll in the eastern Nankai Trough, Japan. *Marine and Petroleum Geology, 66*(1), 310–322. https://doi.org/10.1016/j.marpetgeo.2015.02.037.

Gutenberg. (2015). *List of research vessels by country*. Guttenberg Press. http://www.self.gutenberg.org/articles/list_of_research_vessels_by_country. Accessed November 2, 2015.

H2020. (2014). *Work programme, Extract from Part 19—Commission Decision C(2014)4995*. http://ncp-space.net/are-you-familiar-with-the-technology-readiness-levels/. Accessed August 22, 2015.

Hannegan, D. M. (2005). Methane hydrate drilling technology, OTC Paper 17448. In *Proceedings of the Offshore Technology Conference*, 2–5 May 2005, Houston, TX (3 pp.).

Hopkins, P., Saleh, H., & Jewell, G. (2015, September). Composite riser study confirms weight, fatigue benefits compared with steel. *Offshore,* 70, 72, 74.

Hydro. (2016). *Eelume the snake robots, Pros and Cons. Hydrographic Catelogue*. admin@hydrographic-catalogue.com. Accessed May 15, 2016.

JOGMEC. (2013). *News Release: Gas production from methane hydrate layers confirmed*. Japan Oil, Gas and Metals National Corporation, Tokyo (3 pp.). http://www.jogmec.go.jp/english/news/release/release0110.html. Accessed April 26, 2016.

Johnson, A. H. (2012). Global resource potential of gas hydrate—A new calculation. *U.S. DOE-NETL Fire in the Ice Newsletter Fire in the Ice, 11*(2), 1–4.

Kawamoto, T. (2013). *The first offshore MH production test. PDF presentation*. http://energy.gov/sites/prod/files/2013/06/f1/Takami%20Kawamoto%20-%20The%20First%20Offshore%20Production%20Test.pdf. Accessed December 22, 2015.

Khurana, S. (2015, January 29). Online exclusive: New strategy needed for deepwater development, unconventional plays. *Offshore Digital Magazine*. http://www.offshore-mag.com/articles/2015/01/online-exclusive-new-strategy-needed-for-deepwater-development-unconventional-plays.html. Accessed May 29, 2015.

Kolle, J. J., & Max, M. D. (2000). *Seafloor drilling of the hydrate economic zone for exploration and production of methane*. Tempress Technologies. http://ttinc10.qwestoffice.net/papers/hydratedrill.pdf. Accessed April 26, 2016.

Kruusmaa, M., & Salumäe, T. (2016). U-CAT for underwater archeology. *Sea Technology, 57*(4), 37–38, 40.

Max, M. D., & Johnson, A. H. (2011). Methane hydrate/clathrate conversion. In Khan, M. R. (Ed.), *Advances in clean hydrocarbon fuel processing: Science and technology* (pp. 413–434). Woodhead Publishing Series in Energy No. 19. Woodhead Publishing, Cambridge.

Max, M. D., & Johnson, A. H. (2013). Natural gas hydrate (NGH) Arctic Ocean potential prospects and resource base. OTC Paper 23798. In *(Digital) Proceedings Arctic Technology Conference*, Houston, Texas, USA, 3–5 December 2012 (11 pp.).

Max, M. D., Johnson, A. H., & Dillon, W. P. (2013). *SpringerBriefs in Energy: Natural gas hydrate arctic ocean deepwater resource potential* (113 pp.).

Max, M. D., & Pellenbarg, R. E. (2000). *Inflatable, non-collapsible, personal flotation device*. United States Patent (No. 6,066,017), May 23, 2000.

MD. (2016). *Positive displacement motors (PDM)*. Micron Downhole Tools. http://micon-drilling.de/Download/Catalog_PDM_EN.pdf. Accessed August 21, 2015.

Milne, G. (2012). *Deep water installation of steel catenary Risers (Subsea Asia).* Presentation 3 October 2012 Kuala Lumpur, Malaysia (3 pp.). http://www.subseauk.com/documents/presentations/installation%20of%20scrs%20subsea%20asia%203rd%20oct%202012.pdf. Accessed November 2, 2015.

Moyano, B., Jensen, E. H., & Johansen, T. A. (2011). Improved quantitative calibration of rock physics models. *Petroleum Geoscience, 17,* 345–354.

Nagarajan, N. R., Honarpour, M. M., & Sampath, K. (2007, August). Reservoir-fluid sampling and characterization—Key to efficient reservoir management. *Journal of Petroleum Technology (SPE)*, 80–92.

NAS. (2016). *The power of change: Innovation for development and deployment of increasingly clean electric power technologies (Especially Appendix D).* National Academy of Sciences (320 pp.). http://nap.edu/21712. Accessed January 25, 2018.

Newman, N., & Hainsworth, D. (2016). *In deep water: The UK's first subsea-to-shore gas plant.* Ingenia Online http://www.inenia.org.uk/ingenia/Articles/1035. Accessed March 7, 2017.

Offshore 10/5/15. (2015). Centrica gets go-ahead for North Sea coiled tubing shallow borehole. *Offshore Magazine*. http://www.offshore-mag.com/articles/2015/10/centrica-gets-go-ahead-for-north-sea-coiled-tubing-shallow-borehole.html. Accessed October 6, 2015.

Offshore 10/26/15. (2015). *Island vessel completes Butch shallow gas investigations*. http://www.offshore-mag.com/articles/2015/10/island-vessel-completes-butch-shallow-gas-investigations.html. Accessed November 23, 2015.

Offshore 10/29/15. (2016, March 30). FMC, Shell introduce new hydraulic tubing connector system. *Offshore Magazine.* http://www.offshore-mag.com/articles/2015/10/fmc-shell-introduce-new-hydraulic-tubing-connector-system.html.

Offshore, 3/29/16. (2016). *Offshore Magazine*. http://www.offshore-mag.com/articles/2016/03/dundee-team-examines-impact-of-subsea-structure-removal.html. Accessed March 20, 2016.

Offshore 4/19/16. (2016). Statoil launches new subsea concept. *Offshore Magazine*. http://www.offshore-mag.com/articles/2016/04/statoil-launches-new-subsea-concept.html. Accessed May 12, 2016.

Offshore. (2017). OGTC offers funds for North Sea robotics, subsea ideas. *Offshore* (Staff). http://www.offshore-mag.com/articles/2017/06/oil-gas-technology-centre-offers-funds-for-north-sea-robotics-subsea-ideas.html?cmpid=enl_offshore_offshoredailynewsletter_2017-06-09&email_address=michaelmax1@mac.com&eid=288179069&bid=1780145. Accessed June 9, 2017.

Offshore 9/6/17. (2017). Subsea construction vessel launched in Baku. *Offshore Magazine*. http://www.offshore-mag.com/articles/2017/09/subsea-construction-vessel-launched-in-baku.html?cmpid=enl_offshore_offshore_daily_newsletter_2017-09-07&email_address=michaelmax1@mac.com&eid=288179069&bid=1858522. Accessed September 7, 2017.

OS (2017). *Ventures invests $20 million in artificial intelligence development program.* http://www.offshore-mag.com/articles/2017/06/bp-ventures-invests-twenty-million-in-artificial-intelligence-development-program.html?cmpid=enl_offshore_offshoredailynewsletter_2017-06-09&email_address=michaelmax1@mac.com&eid=288179069&bid=1780145. Accessed June 9, 2017.

OL. (2013). *Methane Hydrate Community Workshop Report.* Consortium for Ocean Leadership report on meeting June 4–6, Washington, DC (40 pp.).

PW. (2016). *SME PetroWiki. Coiled tube drilling.* http://petrowiki.org/Coiled_tubing_drilling#When_CTD_is_the_best_tool_for_the_job.

Schlumberger. (2012). *Horizontal drilling*. http://www.glossary.oilfield.slb.com/search.cfm.

Schlumberger. (2016). T*DDirect casing-drilling and liner-drilling technology*. Accessed February 29, 2016.

Schoderbek, D., Farrell, H., Hester, K., Howard, J., Raterman, K., Silpngarmlert, S., et al. (2013). *ConocoPhillips Gas Hydrate Production Test Final Technical Report for DOE Award No.: DE-NT0006553* (204 pp.).

Small, A. (2016). Decommissioning subsea structures, moorings requires geotechnical considerations. *Offshore Magazine, 76*(3), 3536. http://www.offshore-mag.com/articles/print/volume-

76/issue-3/engineering-construction-installation/decommissioning-subsea-structures-moorings-requires-geotechnical-considerations.html. Accessed March 25, 2016.

Spaziani, A. L., & Lucas-Head, J. (2012). Effective communications and geosteering in thin, horizontal targets. Seafloor geohazards related to subsurface fluid migration. Abstracts volume, *Operations Geology Workshop* 5 Oct 2012, Aberdeen, Scotland, Geological Society of London, 60.

Theimer, K., & Kolle, J. (2007). *Microhole high-pressure jet drill for coiled tubing*. Final Report TR-117 for US Department of Energy (51 pp.).

Walsh, M. R., Hancock, S. H., Wilson, S. J., Patil, S. L., Moridis, G. J., Boswell, R., et al. (2009). Preliminary report on the commercial viability of gas production from natural gas hydrates. *Energy Economics, 31,* 815–823.

Watson, J. (2016). Shell Nigeria saves $1 M in digital oilfield infrastructure costs with IoT solutions powered by Ingenu and KONČAR. PennEnergy 5/10/16. http://www.pennenergy.com/articles/pennenergy/2016/05/shell-nigeria-saves-1m-in-digital-oilfield-infrastructure-costs-with-iot-solutions-powered-by-ingenu-and-kon-ar.html?cmpid=EnlDailyPetroMay102016&eid=288179069&bid=1399855. Accessed May 14, 2016.

Weatherford, 2001. *Whipstocks and whipstock mills*. Weatherford Drilling & Intervention Services. http://www.weatherford.com/en/products-services/well-construction/re-entry-services/cased-hole-systems. Accessed April 26, 2016.

Whitley, E., & Clarke, S. (2015). High-performance inertial microelectromechanical system. *Sea Technology, 56*(12), 21–25.

Chapter 9
Offshore Operations and Logistics

Abstract Carrying out energy-related offshore exploration and production depends first on having the equipment to be able to work in remote and appalling environmental and sea conditions. The history of the offshore energy industry is rising to the task of devising new technology, new types of vessels, communications, control systems, skills, and all that is necessary to be able reach, supply, carry-out work expeditiously, and be able to bring product to market. Working at sea is governed not only by conditions but also the laws of coastal states and by regional and international agencies concerned with safe and environmentally defensible practices. We see no barriers to supply and logistics for NGH exploration, including drilling, in the tropical and temperate world ocean, no matter how remote. But operations and logistics in the Arctic are severe and much more difficult for all exploration and production activities. This has driven the design and building of a worldwide network of ports and special vessels to support conventional oil and gas exploration that can be used to support NGH development activities.

9.1 NGH Exploration and Production Operations

The most critical aspect of being able to work and sustain offshore, energy-related activity in a sea area of choice is physical capability. As the energy industry has had to transition from relatively shallow targets on land to deeper and more difficult drilling opportunities, innovation and adaptation were always able to provide the tools for increasingly more difficult jobs. Working offshore followed the same technology development path. Drilling from barges in very shallow water gave way to gravity platforms resting on the seafloor. Industry is now fully capable for 'over-the-edge' (of the continental shelf break) drilling in deepwater and the ultra-deepwater (Chaps. 3 and 7). To better support the new technology, new classes of relatively less expensive research vessels are being designed and produced for both open ocean and Polar Research.

All exploration and production (E&P) activities at sea are constrained by the same basic set of operational and logistical (O&L) factors. Some sea areas may

M. D. Max and A. H. Johnson, *Exploration and Production of Oceanic Natural Gas Hydrate*, https://doi.org/10.1007/978-3-030-00401-9_9

have more benign sea-keeping conditions than others, but the effects of storms, currents and sea condition on ships, floating work platforms of all types, and infrastructure are predictable. Water depths, weather conditions, the nature of the seafloors, distance from operational ports, and the width of continental shelves are all first order considerations. But the energy industry has shown over the years that it can innovate, adapt, and continue to work increasingly far from shore and in remote and hostile areas, supporting operations many hundreds or thousands of miles from bases. Emplacement, maintenance, and protection of communications, power, and pipelines to shore will be required, but the technology for conventional hydrocarbons servicing these requirements is available and can be used without costly physical modification. For instance, IRM vessels, which are specifically designed to support inspection, repair, and maintenance operations are already joining commercial fleets. These relatively small vessels may also prove to be suitable to support a wider range of NGH seafloor activities. At present, much of the deep-water capability developed for conventional hydrocarbon E&P is fortuitously poised to support NGH development. It is only in locations with NGH potential and no conventional oil and gas operations that new support bases may have to be established.

In contrast to conventional offshore operations that require a wide range of equipment to meet requirements from shallow to deepwater, and in drilling capability that must reach many kilometers depth in a wide variety of different rock types having different drilling requirements, NGH E&P activity will be able to use the same narrow range of NGH-optimized technology world-wide. This is because NGH is only found in the geomorphologically and bathymetrically similar NGH prospect zone in deep and ultra-deepwater (Chap. 3). Virtually all offshore NGH has the potential to be serviced by a much more limited range of E&P, and this should have the effect of lowering overall CAPEX and O&M costs in the same way that maximum production and handling of a limited range of products has the effect of lowering any industrial or product costs.

Leveraging and slight modification of existing technology and the development of simple new technology for the Arctic are eminently possible in the short term. Many developments are already in advance of regulations; for example, polymer bearings could replace the need for petroleum-based lubricants that otherwise might leak into the seawater. Many weather-restricting issues can also be avoided by moving all industrial activities to the seafloor (Chap. 8). Even if no specialized NGH technology and practices are used, however, NGH development has already commenced using available tools. It is also likely that production from NGH will begin with the deep importing countries to which straight commercial economics may be less important than energy security and the effect of indigenous natural gas supply on the country's balance of payments.

9.2 Access

There are two principal aspects to sustain work in a sea area, physical capabilities and compliance with legal authorities of States and/or international bodies to achieve permitted access. Virtually all sea areas in the world are now regulated under the general authority of the United Nations, with which National law of States is generally compliant.

Coastal Nation-states and federations of states have legal rights to natural resources including minerals and energy, and marine resources such as fishing rights in the waters adjacent to their land areas. National laws governing permitted rights of access and conditions, such as those relating to safety and environmental concerns, and the agencies that regulate these activities for governments, can be very different from one legal authority to another. In addition, the manner in which regulations are established and enforced vary greatly. (We do not discuss individual regulatory environments but note that achieving permitted access can be expensive and lengthy.) This makes gaining and maintaining legal access for both exploration and production activities a CAPEX issue that must be taken into account. Specific issues related to NGH leasing are discussed in Chap. 10.

United Nations Convention on the Law of the Sea (UNCLOS) International agreement for the economic rights of states at sea is of paramount importance. Not only are the seaward boundaries between states at issue, but also the prolongation of national 'sovereignty' where the extended distance of national claims overlap the seaward claims of other nations.

UNCLOS is the governing legal instrument outlining the rights of States seaward from their traditional coastal waters. Taft (2006) outlines the basis for the 200-mile Exclusive Economic Zones (EEZ) and their relevance to NGH because a considerable number of NGH prospect zones (Chap. 3) lie outside of existing EEZs. Submissions for continental shelf extensions (Schoolmeester and Baker 2009) that will extend the EEZs of maritime states further from shore are now closed. Applications are being reviewed for approval (ECS 2016). This round of assigning the rights of States further from their shores is really about extending sovereignty into much deeper water, which is regarded as the last maritime zone. For instance, on 9 February 2016 the Rule of Law Committee for the Oceans, empaneled by the United Nations, began review of the second Russian submission for extension of their existing economic zones (LOS 2016). All parties have agreed that they will abide by the rulings.

There is some question of the standing of the United States application for an extended continental shelf initiative (Kimball and Hutchinson 2016) under the LOS in the current round of EEZ extensions that are currently being sought. The United States is the only major maritime nation state that has not signed the UNCLOS treaty. Nonetheless, the U.S. is seeking international recognition of its extended continental shelf claims using the terms of UNCLOS. The rights and obligations for maritime affairs highly consistent with UNCLOS have been published as Naval Instructions for the U.S. Navy and U.S. Coast Guard (NWP 2017).

Non-economic National Interests Local and indigenous communities and non-governmental organizations (NGOs) are increasingly vocal and important to political processes to which the writing of laws and regulations must increasingly respond. Much of this opposition is directed against the proliferation of fossil fuels. In general, they strongly support renewable energy as an alternative because of the probable climate change impact of large-scale combustion and effusion of CO_2 exhaust to the atmosphere.

Environmental NGOs (ENGOs) in particular, have been and are increasingly active in opposing the expansion of the hydrocarbon development into previously unexplored regions of the environmentally fragile Arctic. They have been particularly concerned about the effectiveness of cleanup technologies in the event of an oil spill in a pack-ice region (Mullin 2015). At certain times of year, a leaking well that is envisaged as being impossible to plug or remediate could cause immense damages. Because of the slow rate at which biological degradation of oil would take place in a cold, ice-covered sea, the environment could be damaged for very long periods of time.

Economic National Interests National regulations are primarily driven by States as license fees, royalties, and other income-generating instruments can become primary sources of income. Regulation of exploration and production activity is primarily an economic concern, both to energy companies and to States leasing 'blocks' offshore in which commercial terms between the State and the company have been agreed. In some states, the process is open and transparent, and provides a basis for legal redress. In other states the process may be more arbitrary and less open and with weaker legal frameworks.

9.3 Open Oceanic Regions

In general, not only is the world ocean interconnected oceanographically, but in the last 20 years or so, offshore energy E&P has spread to many continental margins that were once considered to be very remote. For instance, although the Nigerian offshore was subject to early E&P, the extent of exploration has now spread to literally along the whole Atlantic and Indian Ocean coasts of Africa. All of these coasts are passive margins having extensive and high-quality reservoirs underlying marine successions, which themselves have turbidite sand reservoir potential (Chap. 4). In support of energy E&P around the world, ports capable of supporting the large offshore platforms, supply vessels, some manufacturing and repair facilities now are strategically located in all the major offshore energy exploration regions (although they are not all capable of the same level of performance). Especially in regions having narrow continental shelves, servicing of the more seaward NGH deposits (should they be found) will be only a matter of a few more days of sea time than existing servicing and supporting of the more conventional fields for which the ports were originally developed.

In more developed regions, increasingly, infrastructure may be leveraged to enable NGH exploration (including drilling). Leveraging such development seems practical for locations such as the Gulf of Mexico (GoM), offshore Nigeria, offshore Somalia (Chap. 4)—where gas pipelines ashore are servicing the new gas production—and in regions of the eastern North Atlantic, as well as increasingly in SE Asia and the Eastern Mediterranean Basin, where new pipelines are also servicing new gas production. Where practical, NGH development can take advantage of existing conventional hydrocarbon exploration, equipment development and deep-water technology and techniques.

Wee foresee no barrier to supply and general logistics for NGH development virtually anywhere in open ocean regions and enclosed seas having nearby conventional offshore energy activity. Where there is no available infrastructure for production, the conventional solutions for stranded gas may be equally applied to NGH (Chaps. 7 and 11).

9.4 Arctic Ocean

The North Polar Region, including adjacent regions in the North Pacific and Atlantic, is a special case because of its harsh conditions and because it is an enclosed sea. The Arctic Ocean is almost entirely enclosed by the Eurasian and North American continents (GEBCO 2014; Marshall 2015). Because of ice cover and inhospitable conditions for bathymetric and geophysical survey, charts of the Arctic Ocean may have large areas in which the amount of survey detail is very small, unlike the surfaces of Mars and the Moon that are better known (Jakobsson et al. 2015). At this time only about 11% of the Arctic Ocean has been mapped with multibeam sonar; the rest of its sea area is represented through mathematical interpolation using a very sparse depth-sounding database. GEBCO is an informal international bathymetric map forum that regularly updates bathymetric maps and geophysical databases of the Arctic.

The Arctic Ocean is the first ocean to have agreed marine boundaries between countries (EEZs) and an agreed procedure for deciding extensions of these zones as well as a mechanism agreed by all Arctic States to settle disputes by binding arbitration. And this was before widespread development of regular marine transport. This was because the Arctic was not an ocean through which commerce could be practiced because of the sea ice, except for a few short summer periods mostly along the Eurasian coast. Thus, the stated objective of the United States of freedom of the seas and the new interest in an Arctic international sealane between the northern Pacific and the Atlantic Ocean (in which China and other nations are showing increasing interest), should cause no clashes, legal or otherwise. Some international agreements concerning search and rescue activities, anti-pollution response and mitigation requirements, and vessel engineering requirements to deal with winter sea ice have already been outlined.

Because of the dramatic warming of the Arctic, there is a well-established trend for more of the sea area to be open for longer periods of time in summers. The Arctic is an extremely important frontier area for oil and gas E&P in the immediate future and may contain up to 22% of the world's remaining conventional hydrocarbon resources (USGS 2008; WC 2014). Up to 85% or an estimated 412 billion barrels of oil equivalent (BOE) of resource potential is projected to be found the Arctic offshore. The Arctic appears to have a very strong petroleum system in virtually all its sedimentary basins, with temperature and pressure conditions along with thick continental slope sediments providing potentially excellent NGH host strata in GHSZ (Max et al. 2013).

The Arctic Council (AC 2016) is an informal intergovernmental forum for promoting cooperation, coordination, and interaction among member and observer nation-States in the Arctic. It was established in 1996 by the 'Arctic five' Nordic States consisting of the Kingdom of Denmark, Finland, Iceland, Norway, and Sweden, and the 'Arctic three' consisting of the Russian Federation, Canada, and the United States. The Arctic Council has no permanent bureaucracy, offices, or budget. All projects or initiatives are sponsored by one or more Arctic States. The Arctic Council does not implement laws or enforce its guidelines, assessments, or recommendations, but often provides support for negotiations between Council members, such as the 2010 agreement determining the boundary line between Russian Federation and Norwegian EEZs in the Barents Sea (WC 2014). Although the structure of the Arctic Council is regarded as being outmoded by some (Conley and Melino 2016), and despite its heterogeneous membership, the Council has functioned very well and we feel it is unlikely to change its structure or procedures significantly.

Countries with no Arctic coastlines have also associated with the Arctic council. In particular, China is likely to parlay its business experience and wealth to buy into joint ventures with energy companies in the Arctic (Byers 2013). Since China is one of the signatories of the (Spitsbergen) Svalbard Convention (1925), its citizens have equal rights of access and residence to this Arctic Archipelago. Since Spitsbergen is a high Arctic Island, technically it can be argued that this makes China an Arctic nation even though the island is primarily accorded protection by Norway. The Russian Federation dominates the Arctic Ocean geographically, with 40% of the Arctic land area and three quarters of the population. Russia and Norway are the only countries with excellent deepwater ports capable of supporting energy exploration activities, but only Russian Arctic ports are directly connected to the European and Eurasian railway system.

Though somewhat further south than Russian and Norwegian support facilities, Alaska, Canada and Greenland can support Arctic operations through ports at Dutch Harbor and Halifax, NS and St. Johns, NF. with limited port logistics support also available from Qaanaaq (Thule), Greenland. As the number of vessels employed increases in this multi-national ocean area, coordination of support resources, including the positioning of most vessels and activities should ideally be managed or serviced by a single entity, probably under the auspices of the Arctic Council. A number of other countries, for instance South Korea, Singapore, and

Japan, have indicated that they have an interest in the Arctic and are recognized as observers to council affairs. Interests include minerals resources to transit rights, amongst other issues.

There are essentially no territorial disputes on land in the Arctic where boundaries between countries appear to be as secure as anywhere on earth, aided by fortuitously placed bodies of water in a number of cases. Sovereignty, is defined for practical purposes, mainly with respect to national control of potential energy and mineral resources and fishing rights related to ocean boundaries. Since the Norwegian and Russian Federation agreed an EEZ boundary in 2012, only the United States and Canada have yet to agree their Arctic Ocean maritime boundary. Claims for projecting national EEZs northward is presently under consideration under the UNCLOS for all coastal states, in particular Europe and the Arctic (de Vivero 2007). The position of the United States, which has filed a claim for continental shelf extension along with the other Arctic Coastline States (Chap. 10), is uncertain because the United States Senate has not ratified the UNCLOS treaty, which is the internationally agreed mechanism and the authority to decide claims according to an agreed procedure. However, the US President has put UNCLOS into practice through Executive Order. As in North America, offshore Arctic energy exploration activity began in the 1970s in Norway and the 1980s in Russia. Both countries are adept at working in the Arctic (Fig. 9.1) because they have had cities and commercial activity in the Arctic for many centuries. Off northern Canada and Alaska, in contrast, there is largely wilderness that is uninhabited except for a few native peoples. Seismic surveys of the Norwegian western and southern sectors of Barents Sea began in the 1970s, followed by exploratory drilling in the 1980s. In 1984, Equinor made its first significant discovery in the Snøhvit field along the western margin of the Barents Sea continental shelf off NW Norway. Many more wells have been drilled and an LNG facility near Hammerfest.

Operations are difficult because of strong currents, severe storms, multi-year mobile pack ice and the 'tyranny of distance' from suitable ports that compound operational and logistical activities. Specially tailored seismic exploration technique development is necessary to overcome operational conditions, which become virtually impossible once pack ice gets becomes too profuse after summer melt seasons.

Sea Routes There are basically three potential sea routes between the Northern Atlantic countries and Asia through the Arctic Ocean (Fig. 9.2). The best known traditionally in the United States and Canada is the Northwest Passage, which involves a number of different passages through the Canadian Arctic Islands. This passes into the Labrador Sea in the Atlantic region and is most suitable for American rather than European ports. There is an on-going dispute between Canada and the United States about whether these passages are international or Canadian waters. The second, which is better known to Europeans, is the Northeastern or

Fig. 9.1 Semi-submersible in Tromsø Harbor, 30 May 2011 Image by M. D. Max

Eurasian Passage, which lies along the Arctic coast of Russia. This is in use today, virtually year-round, although ice-breakers are requited in winter. Ice-capable tankers and other vessels such as operated by Russian companies and government are suitable for summer and low ice conditions. The third 'shipping lane' is 'over-the-top', essentially through the center of the Arctic ocean and almost entirely within international waters, apart for the Bering sea passage between Russia and the United States. Although ideal in terms of distance and central North Atlantic access, the ice conditions are the worst. There are a number of speculative articles in the press about Chinese interests in the over-the-top passage linking with a major transshipping port in Iceland, but as yet there are no definite published planning documents available.

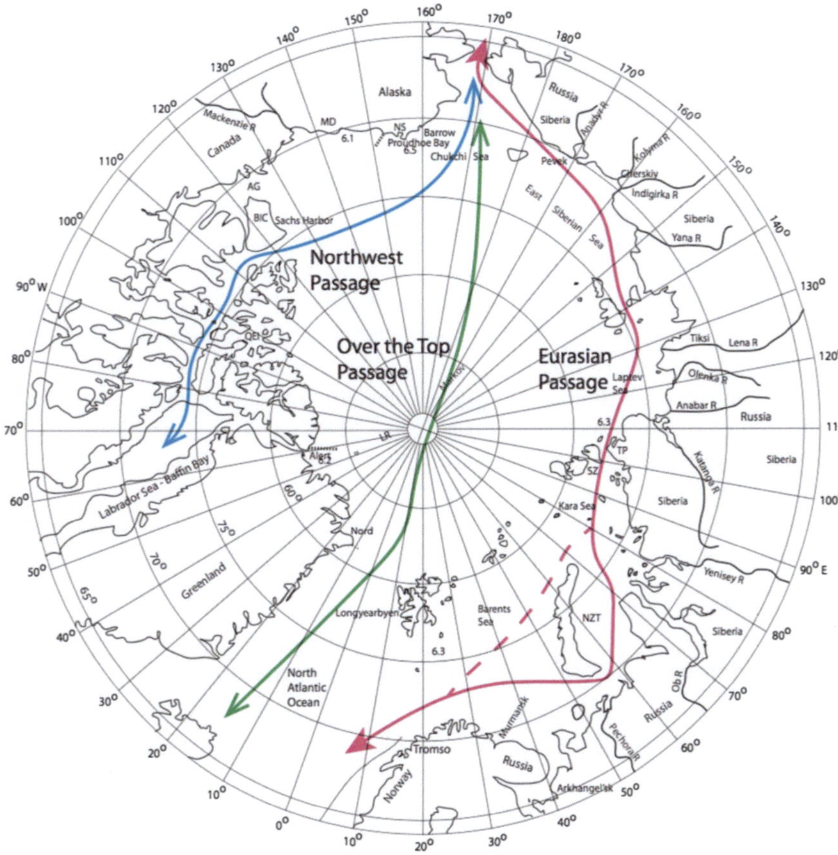

Fig. 9.2 Generalized position of the three main (possible) arctic shipping routes. Blue—Northwest Passage, Red—Eurasian Passage with alternate shorter track (dashed), Green—Over the Top Passage. Base map from Max et al. (2013). Other passages between Canadian Islands also possible

9.4.1 E&P Operations

The primary means of transport in the Arctic Ocean and coastal region is by ships with ice-strengthened hulls and icebreakers specifically designed for making navigable paths through ice, generally for other ships. Lloyds classification of icebreakers identifies nine different types (Bridges et al. 2012), many of which are restricted to rivers, harbors and inland waterways, such as the Great Lakes of the United States. The Canadian Coastguard identifies five types and also defines seven polar classes, which are based on type and thickness of ice and time of year limits (CCG 2012).

In general, Polar oceanic icebreakers can be divided into two types, heavy for very thick ice and medium for thinner ice. In addition, a variety of ships with ice-strengthened hulls and double hulled tankers exist, with more building as the Arctic Ocean becomes more navigable owing to less and thinner ice cover.

There is now a question, however, as to the advisability of building more than a certain number of expensive icebreakers. Although ore and LNG ships have transited from Norway along the Eurasian passage to northern Asia (China, Japan), they have done so accompanied by icebreakers. In the late summer of 2017, a Russian tanker travelled through the northern sea route on the Eurasian continental shelf without an icebreaker escort for the first time (Barkham 2017). The Christophe de Margerie, the first of 15 Arctic specification LNG carriers traveled from Hammerfest in Norway to Boryeong in South Korea in 19 days, about 30% quicker than the conventional southern shipping route through the Suez Canal and also avoiding choke points such as the Straits of Malacca. The tanker was built to transport natural gas from a new LNG facility on the Yamal Peninsula. Using the Eurasian northern sea route, for instance between Rotterdam the Netherlands and Yokohama Japan would save about 5000 sea miles and about 13 days instead of the conventional Suez Canal—Straits of Malacca route (Infographics 2017).

9.4.2 Factors Determining Icebreaker Requirements

Because icebreakers are very expensive vessels, only nations that had requirements to service Arctic or Antarctic bases, keep ports open or support other national needs considered building and maintaining them. Russia is the only Arctic country that has the most ice bound ports and a requirement for keeping the ports open during sea and fast ice periods. Many Norwegian Arctic ports commonly are ice-free year-round. The small number of indigenous peoples along the northern rim of North America, where there are few ports of any significance are commonly supplied for their winter needs during the summer, although sea ice has been an issue in the past.

One of the traditional ways in which the 'right number' of icebreakers were needed, at least on a relative level, was to compare the miles of Arctic coastline in Russia, Canada, and the US that are threatened by ice. The relative percentage of ice threatened coastline can then be calculated and compared. Assuming that Russia's 40 icebreakers are about right to protect their coasts using the 'Noah rule', the US should have four (4) or more for Arctic missions and another two for Antarctic missions.[1] Unfortunately, the Coast Guard finances their icebreakers on a pay as you go basis and there has been little coordination with other federal and state research agencies for cost-sharing. On their own merits, icebreakers lack a mission program for funding their operations. (Coast Guard charges user fees from

[1]Thanks to discussions with Captain Dennis (Mike) Egan, U.S. Coast Guard (Ret.).

NSF, and benefits from annual subsidies from the Navy (DOD) in addition to basic budget support from CG Operations and Logistics. Polar icebreakers are considered a national security asset and conduct a variety of logistics and scientific research missions.)

The 'right number' of icebreakers, however, is now much more complicated and the requirement for icebreakers has now extended beyond the traditional States bordering the Arctic Ocean. Countries with no sea ice along their coast have icebreakers, China, which was granted observer status with the Arctic Council, has 5 icebreakers (one elegantly named the 'Snow Dragon') and more building, is interested in establishing a new Arctic sea route for container transport that may be based in Iceland (Guschin 2016). Even in the case in which summer ice may virtually disappear from the Arctic Ocean, ice breaking will remain necessary to keep open year-round sea lanes.

The continuing requirement to service research bases and activities in Antarctica is one factor driving the national acquisition of icebreakers. Australia has one icebreaker and is building another, which gives it the ability to supply its Antarctic base. France has one icebreaker and another building. Germany has the major research vessel '*Polarstern*' in which Polar research can be carried out very comfortably. Major companies and national governments are becoming increasingly aware that the natural resources of Polar Regions are becoming increasingly accessible. Minerals, hydrocarbons, and food all appear to be abundant in Polar Regions and only icebreakers will guarantee access. Even if sea ice cover falls to very low levels, there will always be some sea ice in winter.

9.4.3 Eurasian Icebreaker Fleet

Russian and Norway have excellent port facilities on the Arctic Ocean and there are many excellent deepwater ports in Norway, such as Stavanger and Bergen, that have been used for years to support hydrocarbon exploration and production activities that have moved northward from the North Sea into the Arctic over time. Russia continues to pursue exploration activities in its western Arctic waters in the Kara, Barents, and Pechora Seas. In May 2012, Rosneft and Equinor signed an agreement to jointly develop shallow water areas in the Barents Sea and the Sea of Okhotsk (Max et al. 2013). This gives Russia access to Equinor vast offshore experience and its good safety and environmental record. The agreement also stimulates Russia's shipbuilding industry as they plan to build their own ice-class vessels and drilling platforms (S-R 2013). Russia has the most experience building and operating icebreakers and ice-capable vessels, and has more of these vessels in their naval list than virtually the rest of the world combined. In addition, Russia continues to commission building of ice-capable, multi-purpose vessels elsewhere to take advantage of the experience of other yards, for instance in Finland (Offshore 1/12/16).

Shell expended about $7 billion dollars on Arctic exploration ending after a short drilling expedition in 2015 ended with poor drilling results, a number of significant operational problems, including grounding of a floating drill rig along the Alaska coast (Dlouhy 2015), along with highly visible opposition from environmental groups. Shell has since abandoned its leases and has no further plans to carry out exploratory drilling in the Arctic Ocean. Apart from vessels that Shell could rent on the open market, there was little support available from either the Unites States or Canada.

The United States has let its heavy icebreaker fleet deteriorate from 6 to 2 since 1960, and it has failed to sign the U.N. Law of the Sea Treaty under which the EEZ Arctic extension designations have been made and will be extended. The US has two 40-year-old 399 ft. Polar Class icebreakers based in Seattle, but only one is fully operational. The heavy icebreaker Polar Star (capable of breaking 21 ft. of ice by ramming and crushing) remains seaworthy as a tribute to the ingenuity of Coast Guard engineers, although there is no provision for a solution to any catastrophic breakdown (Hunter 2017). The Polar Sea, a sister ship, is permanently berthed in Seattle and is being cannibalized for parts. The only other icebreaker is the newer 400 ft. medium icebreaker Healy that was commissioned in 1999, that is configured as a polar research vessel (capable of breaking 6 ft. of ice at 3 knots). There are no other US backup icebreakers or heavy ice-capable vessels. Without icebreakers and supporting infrastructure such as all-weather roads, deepwater Arctic access ports, and pan-Arctic high-speed communications including internet, any Arctic nation will remain essentially a bystander to developments (Aii 2017).

The U.S. Coast Guard is presently planning to design and build new heavy and medium icebreakers. Initial funding was made available in the 2018 budget. Each heavy icebreaker is estimated to cost about one billion dollars (Lamothe 2017). The new icebreaker(s) are being constructed through a joint US Navy—USCG project management office and the first request for proposals has been issued early 2018. The first ship is expected to be deployed about 2024. The Council on Foreign Relations supports the immediate funding and operation of six icebreakers and the signing of UNCLOS to enhance extension of the U.S. EEZ (Allen et al. 2017). Canada has 8 lights to medium icebreakers; the newest was built in 1986. Inexplicably for a major country with Arctic ambitions, Canada has not funded much development of its Arctic maritime capability either (Byers 2014).

9.4.4 North American Arctic Access

Canada has no port facilities in the Arctic. Canada's population is clustered along its southern border and there is only a scattered population in its western plains. Canada has virtually no roads or energy-related infrastructure such as pipelines to its Arctic coast. Canada has just one deep-water Arctic port, located in Hudson Bay far south of the Arctic Circle. The Canadian hinterland and Arctic is not well served logistically, especially during the winter. St. John's Newfoundland and Halifax, NS are the primary eastern Canadian ports providing harbor and logistic facilities for the Arctic. Halifax is directly connected with the Canadian and U.S. road and railway systems, whereas Newfoundland only has sea and air logistical links with the mainland.

The U.S. has no port facilities on the Arctic Ocean coast. The U.S. has only a small shallow water Arctic port at Nome and several ports with reasonable logistics and repair capabilities in Alaska at Seward, Ketchikan, Dutch Harbor, Adak, Prudhoe Bay and Anchorage/Cook Inlet which could service Arctic Sea E&P activity. The ports of Seattle and Portland can provide competitive services. The distance from Anchorage into the Arctic through the Bering Strait into the Chukchi Sea is longer than the distance from Anchorage to Seattle or Portland in the NW United States. This makes either of the U.S. ports only a few days sailing longer and their logistic advantage being located directly at major industrial support activities and logistics may render them superior ports to support U.S. Arctic activity than alternative Alaskan ports. In addition, bringing roads and other infrastructure to the Arctic coast of North America (in additional to the transportation infrastructure and ice roads of the Trans-Alaskan pipeline) will entail massive costs to repair damage being caused to existing roads, pipelines and buildings by melting permafrost, with little existing capability to supply from the sea during winter. Russia, in contrast, is adding specialist ice-capable supply vessels to its icebreaker fleet (Offshore 10/16/17).

Arctic conditions are uniquely different from the range of open-ocean conditions in that they feature deep cold during winter months that can result in the upper part of any floating structure becoming encrusted in ice and made unstable. There is also sea ice of all sorts, which impedes operations and, when driven by tides and wind, can inflict impact damage. The harsh Arctic environment results in working conditions only being available for part of the year. Onshore, energy related work is carried out mainly in the winter when the ground is frozen. Offshore, work is possible only during summer months, at least until the climate again returns to ice hazard conditions. A considerable range of technologies for conventional hydrocarbon development in harsh shallow water Arctic conditions will be unnecessary for NGH E&P. They include armored wellhead protective structures, artificial islands or caisson retained islands, and all other bottom founded structures capable of resisting ice impact and loading (DeGeer et al. 2014).

Exploration in the Barents Sea has proven the region to be a rich hydrocarbon province, and exploration north onto the continental shelf from the western Siberian gas and oil fields indicates a high hydrocarbon potential. Arctic conditions and floating ice also persist in the Bering Strait and in the Barents Sea, although the Barents Sea has not had annual ice cover lately in the winter. E&P activity is accelerating because of the perceived abundance of hydrocarbon resources in the Arctic. In addition, atmospheric and sea warming is making more of the region available for exploration and production activities.

Operational and Logistics Requirements for NGH in the Arctic Unless there is some environmental regulation that could restrict E&P for conventional oil and gas but allows E&P for NGH because of its extremely low environmental risk (Max et al. 2013), exploration will almost certainly spread north across the broad Eurasian continental shelf from the Barents and Laptev Seas and then west along the East Siberian Sea before deep-water targets are explored. Because conventional oil and gas prospects in the broad Eurasian and Alaskan continental shelves are so promising, infrastructure may not reach NGH prospect zones (Max et al. 2013) until after continental shelf hydrocarbon development progresses slowly pole-ward off the broad Eurasian continental shelf into deeper water.

When Max and Lowrie (1993) discussed the probable location, petrogenesis, and scale of the NGH resource in the Arctic Ocean, estimates were not tied to a petroleum system analysis and no NGH exploration targets were picked. NGH targets can now be picked and constitute part of the long-term hydrocarbon resource of the Arctic. Only in 2015 did the first gas pipeline extend Norwegian gas infrastructure north of the Arctic Circle (Equinor 2015). Establishment of infrastructure follows exploration and production; no one builds speculative long pipelines offshore. A continental margin zone in which the main region for NGH exploration currently resides (Max et al. 2013) is presently (April 2018) over 800 km to the north of gas infrastructure. NGH will probably remain low on the E&P list unless non-commercial factors intervene (Chap. 10) and a solution for stranded gas is found (Chap. 11).

The entire range of offshore technologies for Arctic conditions has to be very robust in order to resist the force of floating ice and seafloor erosion that could affect seafloor installations in the Arctic (DeGeer et al. 2014). Because the water depths of the NGH prospect zone (Chaps. 2 and 3) are below any ice effect, it is possible that exploration will jump across the shallow shelf to deepwater for both NGH and conventional hydrocarbons. Deepwater seafloor is essentially safe for long-term E&P activity. One of the realities of longer seasons of open water in the Arctic Ocean with a later return of seasonal sea ice is the increased frequency of sea storms with ice topped waves causing a severe threat to marine structures and undermining of coastal infrastructure.

9.4.5 Search and Rescue (SAR)

In addition to there being only scattered ports and airfields along the Eurasian mainland in Norway, northern Sweden, Finland and principally in Russia, flying weather is often very bad, which can strongly affect search and rescue operations. Forward basing of SAR sea and air equipment and personnel will be necessary along with oil spill mitigation equipment when exploration begins to move north. Ideally a multi-nation, Arctic SAR initiative could leverage the opening and reopening of Russian bases, such as those on Franz Joseph Land (FS 2017), from which air SAR would be based almost in the central Arctic. The Arctic has a tradition of cooperation because of the harsh conditions and the need to share resources. There is every reason to believe that an international SAR program could be perfected that used forward bases and equipment of many nations in a single, organized operational institution under the auspices of the Arctic Council or the United Nations. This would be in accordance with the International Search and Rescue Plan under the UN Convention on the Safety of Life at Sea (SOLAS).

Survivability in the Arctic is a matter of good planning, adequate provisioning of the right equipment, and of course, luck. Newly designed equipment would aid organization although existing equipment would have to be employed at first. Ice-strengthened survival craft and ice-capable special SAR vessels with large heated personnel spaces would have the potential to significantly extend rescue operations. Personnel transfer capsules or vehicle for helicopter or crane transfer between vessels and helicopters would provide limited floatation and insulation. Should a large vessel founder in the poorly surveyed Arctic, especially along the complex of seaways and islands of the Canadian Arctic, there had best be a prepared international response.

As part of an Arctic international SAR initiative, multipurpose rapidly deployed, air-droppable vessels such as light but high personnel capacity hovercraft or pre-positioned vessels capable of extended station keeping for long range SAR could be Hovercraft provide a very suitable platform for rapid travel on ice or water and have been used very safely on ice, broken ice, and sea in the Arctic for scientific work (Hall and Kristoffersen 2009). The advantage of a hovercraft, although it is more fuel-intensive than a normal vessel, is they are significantly less expensive that a small fully-ice capable ship and can travel much faster, even over rough ice (up to a design point). In addition, when they are 'parked', they are safe base camps even if the ice should break up, which is an increasing issue because of thinning and more broken ice cover (NAS 2012). Specialized hovercraft could be dropped from low-flying aircraft to shorten first response time.

9.4.6 Arctic Spill Response

The Arctic Ocean is an environmentally fragile region in which both the minimization of oil spills and optimizing cleanup has to be considered as urgent. Because of the remoteness of the Arctic region and the presence of deep-water capable operational bases only along the European coast, pooling of resources under a single response organization makes considerable sense. To this end, an agreement was signed under the auspices of the Arctic Council in 2012. The International Association of Oil and Gas Producers (IAOGP 2013) is coordinating the Arctic Oil Spill Response Technology joint industrial project (JIP). In 2013 an oil spill response agreement was signed by member states under the auspices of the Arctic Council (RAI 2013). As the Arctic warms and shipping and petroleum and gas exploration and production increases, the U.S. Coast Guard has huge responsibilities and very little dedicated Arctic capability (Tingstad et al. 2018).

Industry research, particularly by Norwegian institutions and companies, has already produced many reports and studies. Considerable attention has been placed on early recognition of an oil spill in a sea with mobile pack ice. A joint international project (JIP) committee was formed in 2012 to pool resources from nine international energy companies, BP, Chevron, ConocoPhillips, ENI, ExxonMobil, North Caspian Operating Co., Shell, Equinor and Total. The JIP has defined six core research themes cover dispersants, environmental effects, trajectory modeling, remote sensing, mechanical recovery, and in situ burning. Although in situ burning was perhaps the first method for dealing with spilled oil, it has proven to be effective for oil spills in icy conditions. Some of the combustion products contain potentially harmful chemicals, so burning generally increases the density of the residual product as the volatiles are driven off, making them more likely to sink to the bottom where they may do less harm than in the biotic-rich surface zone.

There is an extremely delicate balance between indigenous Arctic peoples and the Arctic environment. A major spill causing environmental damage that affects the human food chain would be extremely serious, possibly capable of breaking the link between indigenous occupation of the region that has lasted millennia. A breakaway oil spill could turn a large part of the Arctic into an icy wasteland.

The two current major points of international interest in the high Arctic are Fisheries and the potential for pollution of the fragile environment from ships and hydrocarbon exploration and production activities. Future points of interest will be seafloor mining and deepwater Arctic hydrocarbon E&P. In order to meet safety and life saving requirements in the little traveled Arctic region, location finding is vital. Polar orbiting satellites provide imaging, but this is not fully real-time. Communications and navigation are provided by GPS, but this system is not as dependable at high altitudes as it is in the more heavily populated lower latitudes. The Russian GLONASS system, which is similar to GPS, has more satellite visibility in the Arctic, and a number of navigations companies produce satellite navigation systems having both GPS and GLONASS. In general, LORAN positioning is used in the Arctic to back up satellite navigation. Although LORAN is not as accurate as satellite navigation and requires daily calibration, it is extremely dependable at sea.

9.5 Other Frontier Areas

There are many remote areas on earth but few that are not nestled against continental margins have either the thickness of sediment or suitable host sediment to enable thick accumulations of NGH to form. Even if rare NGH deposits are found related to unique sources of methane, for instance in thin sediments near plate margins (Johnson et al. 2015), water depths are liable to be extreme and sediment host impractical for economic recovery of natural gas.

The Antarctic or South Polar Region is the final frontier for hydrocarbon energy resources on Earth. The continental margin of Antarctica, however, is a long way from any significant market, and even those nearest to Antarctica, southern South America, southern Africa and Australia either have their own energy sources or have access to nearer sources of oil and gas. There are few nearby ports from which to base Antarctic operations and logistics. The South Polar Region around Antarctica has widespread, open ocean conditions with monster waves and with often extreme weather including hurricane-like storms and sea icebergs that can be worse than the Arctic. Support and logistics for the South Polar region are more limited than in the Arctic because of the limited habitable landmasses' proximity to the Antarctic continent. Logistics are currently provided from Christchurch or Wellington, New Zealand; Hobart, Tasmania (Australia); Cape Town, South Africa; and Ushuaia, Argentina and Punta Arenas, Chile at the southern tip of South America.

Most of East Antarctica is underlain by passive margin sediments related to the breakup of Pangaea (Chap. 4) that have strong potential for high-quality reservoirs (Veevers 1987). This province offers good prospects for subsided basin, clastic and carbonate reservoir, and marine turbidite reservoir hosts for conventional hydrocarbons.

In addition, there is evidence of widespread natural gas and the presence of GHSZ (Tinivella et al. 1998; Gambetta and Maldonado 1990; Camerlenghi and Lodolo 1994).

Exploration and production activity is currently prohibited by the Antarctic Treaty by the twelve countries whose scientists had been active in and around Antarctica during the International Geophysical Year (IGY) of 1957–58. In addition, "*No acts or activities taking place while the present Treaty is in force shall constitute a basis for asserting, supporting or denying a claim to territorial sovereignty in Antarctica or create any rights of sovereignty in Antarctica. No new claim, or enlargement of an existing claim to territorial sovereignty in Antarctica shall be asserted while the present Treaty is in force.*" (SAT 1961). The Treaty, which is an informal agreement under a stand-alone agency promoted by the United States, came into force in 1961 and, up to 2015, has since been acceded to by 53 nations.

If the post-2014 low oil and gas price structure persists while more resources are found and the fossil fuel demand-supply balance is restored, the transition to a post-carbon, renewable energy future (Chap. 1) will probably have taken place. It is entirely possible and likely that the South Polar Region never will become an active hydrocarbon exploration and production region.

References

AC. (2016). *The Arctic Council*. http://www.arctic-council.org/index.php/en/about-us. Accessed February 7, 2016.

Aii. (2017). Arctic Promise: Challenges and opportunities in realizing the next generation of U.S. Arctic Infrastructure. *Alliance for innovation and infrastructure*, 30. http://www.eenews.net/assets/2017/02/13/document_ew_04.pdf. Accessed February 26, 2017.

Allen, T. W. M., Whitman, C. T., & Brimmer, E. (2017). *Arctic imperitives*. Reinforcing U.S. Strategy on America's Fourth Coast. Council on Foreign Relations Independent Task Force Report No. 75, 84 pp.

Barkham, P. (2017). Russian tanker sails through Arctic without icebreaker for first time. *The Guardian* https://www.theguardian.com/environment/2017/aug/24/russian-tanker-sails-arctic-without-icebreaker-first-time. Accessed August 24, 2017.

Bridges, R., Crocker, J., DesRochers, C., & Upcraft, D. (2012). *Classification of Icebeakers—An update to requirements*. Lloyds Register Paper No. ICETECH12-123-RF, 6 pp.

Byers, M. (2013). *China could be the future of Arctic oil*. Aljazeera. Aug 22. http://www.aljazeera.com/indepth/opinion/2013/08/2013821135829162420.html. Accessed August 22, 2013.

Byers, M. (2014). Why Canada's search for an icebreaker is an Arctic embarrassment. *The Globe and Mail*. http://www.theglobeandmail.com/news/national/the-north/why-canadas-search-for-an-icebreaker-is-an-arctic-embarrassment/article16425755/. Accessed February 7, 2016.

Camerlenghi, A., & Lodolo, E. (1994). Bottom simulating reflector on the south Shetland margin (Antarctic Peninsula) and implications for the presence of gas hydrates. *Terra Antarctica, 1*, 154–157.

CCG. (2012). Canadian Coast Guard. Chapter 2 regulations and guidelines. In *Ice Navigation in Canadian Waters* (153 pp). http://www.ccg-gcc.gc.ca/folios/00913/docs/icenav-ch2-eng.pdf. Accessed February 6, 2017.

Conley, H. A., & Melino, M. (2016). A*n Arctic redesign: Recommendations to rejuvenate the Arctic council*. Center for Strategic & International Studies (28 pp, Washington, DC, U.S.A.). A report of the CSIS Europe Program, February.

de Vivero, J. L. S. (2007). *Atlas of the European Seas and Oceans: Marine jurisdictions, sea users, and governance* (146 pp). Ediciones del Serbal. Spanish Ministry of Education. ISBN: 978-84-7628-513-8.

DeGeer, D., Paulin, M., Cowin, T., Cocker, J., Ernst, W., McCullough, J., et al. (2014). 2014 Survey of Arctic & cold region technology for offshore field development: Challenges & solutions. IntecSea, Inc. & Offshore Magazine. Poster 110, One sheet.

Dlouhy, J. A. (2015). Shell abandons Arctic oil quest after $7 billion bid yields 'disappointing' results. *Fuelfix*, September 28, 2015. http://fuelfix.com/blog/2015/09/28/shells-arctic-oil-well-comes-up-dry/#34370101=0. Accessed October 15, 2015.

ECS. (2016). *Continental Shelf, the last maritime zone* (36 pp). UNEP/GRID-Arendal (originally published 2009). ISBN: 978-82-7701-059-5. Accessed February 7, 2016.

FS. (2017). Vladimir Putin visits Arctic archipelago and stakes claim to the oil-rich region. The Telegraph (newspaper, GB) foreign staff http://www.telegraph.co.uk/news/2017/03/30/vladimir-putin-visits-arctic-archipelago-staking-claim-oil-rich/. Accessed March 30, 2017.

Gambeta, L. A., & Maldonado, A. P. R. (1990). Geophysical investigations in the Bransfield strait and in the Bellinghsausen Sea-Antarctica. In B. St. John (Ed.), *Antarctica as an exploration frontier—Hydrocarbon potential, geology and hazards* (Vol. 31, pp. 127–141). American Association Petroleum Geologists Studies in Geology, Tulsa.

GEBCO. (2014). General Bathymetric Chart of the Oceans. Image reproduced from the GEBCO world map 2014, www.gebco.net. 1 sheet, 1 digital file. http://www.gebco.net/data_and_products/printable_maps/gebco_world_map/.

Guschin, A. (2016). China, Iceland and the Arctic. *The Diplomat*, 4. http://thediplomat.com/2015/05/china-iceland-and-the-arctic/. Accessed February, 5 2017.

Hall, J. K., & Kristoffersen, Y. (2009). The R/H Sabvabaa—A research hovercraft for marine geophysical work in the most inaccessible area of the Arctic Ocean. *The leading Edge. Society of Exploration Geophysicists, 28*(8), 4.

Hunter, D. (2017). Ensuring US Leadership in the Arctic. *Sea Technology, 58*(1), 27–28.

IAOGP. (2013). *OGP Environmental management in Arctic oil and gas operations: Good practice guide* (120 pp). Report 449, May 2013.

Infographics. (2017). *Russia's alternative to Suez Canal: Northern sea route*. Sputnik International Search http://onlinelibrary.wiley.com/doi/10.2903/j.efsa.2016.4501/full. Accessed September 7, 2017.

Jakobsson, M., Mayer, L., & Monahan, D. (2015). Arctic ocean bathymetry: A necessary geospatial framework. *ARCTIC. 68*, 7. 10.14430/arctic4451.

Johnson, J. E., Mienert, J., Plaza-Faverola, A., Vadakkepuliyambatta, S., Knies, J., Bünz, S., et al. (2015). Abiotic methane from ultraslow-spreading ridges can charge Arctic gas hydrates. *Geology, 43*(5), 371–374. https://doi.org/10.1130/G36440.

Kimball, S. M., & Hutchinson, D. R. (2016). USGS leverages extended continental shelf research to address deep-se science issues. *Sea Technology, 57*(1), 20, 22–23.

Lamothe, D. (2017). *In a changing Arctic, a lone Coast Guard icebreaker maneuvers through ice and politics*. Washington Post Link to: Freedbert, S. J. Breaking Defense (Raytheon). http://breakingdefense.com/2017/08/us-yards-can-build-icebreaker-for-under-1b-zukunft/. Accessed September 4, 2017.

LOS. (2016). *CLCS begins review of Russia's Arctic submission* (3 pp). http://us6.campaign-archive2.com/?u=f8f44615fa&id=30b96c8d33&e=8a752f6104. Accessed February 9, 2016.

Marshall, T. (2015). *Prisoners of geography* (p. 290). New York: Scribner.

Max, M. D., Johnson, A. H., & Dillon, W. P. (2013). *Natural gas hydrate Arctic Ocean deepwater resource potential* (113 pp). Springer Briefs in Energy.

Max, M. D., & Lowrie, A. (1993). Natural gas hydrates: Arctic and Nordic Sea potential. In T. O. Vorren, E. Bergsager, Ø. A. Dahl-Stamnes, E. Holter, B. Johansen, E. Lie, & T. B. Lund (Eds.), *Arctic geology and petroleum potential, proceedings of the norwegian petroleum society conference* (pp. 27–53), August, 15–17 1990. Tromsø, Norway. Norwegian Petroleum Society (NPF), Special Publication 2 Elsevier, Amsterdam.

Mullen, J. (2015). JIP examines Arctic oil spill response system. *Offshore Magazine*. http://www.offshore-mag.com/articles/print/volume-75/issue-11/arctic-update/jip-examines-arctic-oil-spill-response-systems.html. Accessed January 27, 2016.

NAS. (2012). *Seasonal-to-decadal predictions of Arctic Sea ice: Challenges and strategies* (93 pp). National Academies Press.

NWP. (2017). *The Commander's Handbook on the Law of Naval Operations* (182 pp), Edition August 2017. Department of the Navy. NWP 1-14M/MCTP 11-10B/COMDTPUB P5800.7A.

Offshore 1/12/16. (2016). Icebreaker vessels designed for harsh offshore Sakhalin operations. *Offshore* http://www.offshore-mag.com/articles/2016/02/icebreaker-vessels-designed-for-harsh-offshore-sakhalin-operations.html. Accessed February 1, 2016.

Offshore 10/16/17. (2017). Third icebreaker close to departure for Sakhalin-2. http://www.offshore-mag.com/articles/2017/10/third-icebreaker-close-to-departure-for-sakhalin-2.html?cmpid=enl_offshore_offshore_daily_newsletter_2017-10-18&email_address=michaelmax1@mac.com&eid=288179069&bid=1898186. Accessed 10/18?17.

RAI. (2013). *Arctic Council signs oil spill response deal*. Rai Novosti http://en.rian.ru/russia/20130515/181163818/Arctic-Council-Signs-Oil-Spill-Response-Deal.html. Accessed May 4, 2014.

SAT. (1961). *The Antarctic treaty*. Secretariat of the Antarctic treaty. http://www.ats.aq/e/ats.htm. Accessed April 26, 2016.

Schoolmeester, T., & Baker, E. (2009). *Continental Shelf, the last maritime zone* (p. 36). Arendal: UNEP/GRID-Arendal.

S-R. (2013). *Statoil and Rosneft move forward with exploration cooperation*. Statoil announcement. http://www.statoil.com/en/NewsAndMedia/News/2013/Pages/21Jun_Rosneft.aspx. Accessed February 9, 2016.

Statoil. (2015). *The first gas pipeline crosses the Arctic circle*. Statoil online. http://www.statoil.com/en/NewsAndMedia/News/2015/Pages/21Aug_polarled.aspx. Accessed February 7, 2016.

Taft, G. (2006). Regulatory and permitting environment for gas hydrate. In M. D. Max, A. Johnson, & W. P. Dillon (Eds.), *Economic geology of natural gas Hydrate* (pp. 267–288). Berlin: Springer.

Tingstad, A., Savitz, S., van Abel, K., Woods, D., Anania, K., Ziegler, M.D., et al. (2018). *Identifying potential gaps in U.S. Coast Guard Arctic capabilities* (117 pp). Homeland Security Operational Analysis Center (The Rand Corporation). https://www.rand.org/pubs/research_reports/RR2310.html. Accessed May 14, 2018.

Tinivella, U., Lodolo, E., Camerlenghi, A., & Boehm, G. (1998). Seismic tomography study of a bottom simulating reflector off the South Shetland Islands (Antarctica). In J.-P. Henriet, & J. Mienert (Eds.), *Gas hydrates: Relevance to world margin stability and climate change* (pp. 141–151). Geological Society, London Special Publication 137.

USGS. (2008). *Circum-Arctic resource appraisal: Estimates of undiscovered oil and gas north of the Arctic Circle* (4 pp). USGS Fact sheet 2008-3049.

Veevers, J. J. (1987). The conjugate continental margins of Antarctica and Australia: In S. L. Eittreim, & M. A. Hampton (Eds.), *The Antarctic continental margin: geology and geophysics of offshore Wilkes Land* (pp. 45–73). Circum-Pac. Council for Energy and Natural Research. Earth Science Series 5A, Houston, Texas.

WC. (2014). *Opportunities and challenges for Arctic oil and gas development* (29 pp). Eurasia Group Report for The Wilson Center. https://www.wilsoncenter.org/sites/default/files/Artic%20Report_F2.pdf. Accessed January 24, 2014.

Chapter 10
Energy Resource Risk Factors

Abstract Naturally pressurized, conventional gas deposits and unconventional gas deposits consisting of coalbed methane, tight gas, shale gas, and natural gas hydrate (NGH), are found in a wide range of geological physical systems. Each of these resources has particular petroleum systems, reservoir characteristics, and different potential hazards and opportunities for exploration and production that can be expressed as risk. Natural gas resource types have some risks in common but each type has some unique risks, while other types may not have some of the risks of others at all. We propose metrics and compare and contrast natural gas deposits for a full range of relative conventional and special risks. We also compare risk factors for different energy resources, with particular attention to traditional fossil fuels and other sources of energy. NGH appears to have the lowest overall technical risk factors, with particularly low pollution and environmental risk, which suggests that NGH resources could safely supply natural gas even in environmentally fragile regions such as the Arctic Ocean.

10.1 Factoring Risk into Energy Resource Development

Evaluation of the six most prominent technical risk stages for offshore energy prospects is roughly the same across the range of hydrocarbon resources (Fig. 10.1). For companies with no experience with evaluating NGH prospects, working up a risk profile should involve only minor alteration of their current practices for gas and oil deposits, although evaluators will need to have a thorough understanding of the differences between conventional and NGH risks. Overall, we regard the risks associated with NGH to be fewer and less significant than with conventional deepwater hydrocarbons (Chap. 7). For instance, Max and Johnson (2014) drew attention to the fact that crystalline NGH is stable in its host sediment, and can only produce free gas if artificially stimulated, a process that can be stopped very quickly.

Access Energy companies work years ahead on new prospects and trends. Even when a new play emerges because of new technology or discoveries in a new area,

M. D. Max and A. H. Johnson, *Exploration and Production of Oceanic Natural Gas Hydrate*, https://doi.org/10.1007/978-3-030-00401-9_10

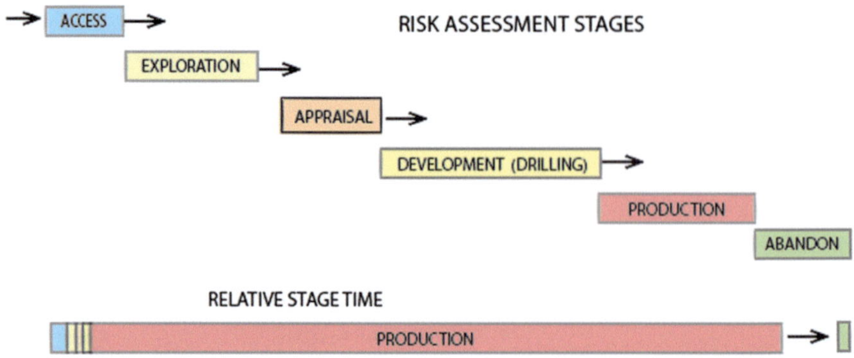

Fig. 10.1 Framework for technical risk assessment of marine energy resources. See also the production characteristics of marine gas resources in Chap. 11. Based on Wilson (2015)

companies will leverage local consultants and other companies with experience that will allow for baseline evaluation of the new opportunities. At this stage, the regulatory, political, and legal aspects, along with the steps involved in bidding or otherwise acquiring a lease block, are taken into account. A key component of the assessment process is the acquisition of exploration data of various types that will assist in deciding to declare an area prospective. There are many areas where drilling is restricted or prohibited, often because of environmental or geohazard risk. Examples include the U.S. Atlantic and West Coasts, where leasing and drilling have been prohibited, and the Arctic where special regulations have been imposed. One of the principal issues for NGH leasing, however, is whether NGH is included in an oil and gas lease or whether there will be less attractive situations such as where leases will be NGH only.

Exploration The physical aspects of seismic surveys, which are the principal initial means of NGH exploration, are essentially the same as for conventional exploration. Because NGH is only found in the seafloor-hugging GHSZ (Chap. 2), however, and virtually no data is required from below about 1.5 km depth into the sediment, NGH surveys may be analyzed in different frequency ranges or using setups and software designed to better reveal shallow exploration targets. Also, because of the shallower sub-seafloor depth at which NGH deposits will be found, a higher degree of angular separation is possible from an array of the same length used for conventional seismic data. This may allow for higher resolution of Vp in the exploration target region (Chap. 6). It should be possible to resolve the NGH exploration target at least as in as much detail as a conventional prospect, if not appreciably better. In basins where 3D seismic surveys have been acquired, a high-resolution shallow "data cube" may be extracted that will allow detailed analysis of locations having the highest potential for viable NGH deposits.

Although some drilling for conventional hydrocarbons will usually have been carried out during the exploration phase (Fig. 10.1), exploration for NGH will probably be confined to remote assessment of the NGH resource, up to and including valuation of gas-in-place (Chap. 6). Frontier areas with no stratigraphic data from nearby drilling are rare but they still exist. Exploration for conventional hydrocarbons may require the drilling of stratigraphic wells from which the geothermal history of a basin can be determined. The geothermal history is important for conventional exploration because critical moments can be defined when gas and oil were produced in source beds by thermal processes and then mobilized to migrate into traps. The first NGH drilling will then take place during the appraisal phase, when ground-truthing of remote conventional valuation has already been carried out.

Appraisal The main risk in NGH appraisal is in the degree of statistical degree of error on estimating gas-in-place. Errors may be inherently less in conventional deposits where the entire reservoir porosity can be filled about equally by gas. In NGH deposits pore filling can be very irregular, which makes determination of gas in place inherently less precise (Chap. 5). Production constraints are also high because there is no long-term production data on which to base estimates of technically recoverable gas. Geohazard and geotechnical risk will also be reassessed in the light of increased in situ data. This is probably the risk stage where the methodology required for valuing NGH concentrations will vary the most from conventional risk estimate practices. Because NGH is a solid crystalline material (Chap. 2), rather than a gas or liquid that spreads about equally through available porosity, it may be best assessed on a cell basis, much like a classic mineral deposit (such as a lead-zinc deposit), but using seismic, magnetic, and other remote sensing method data as an indirect rather than a direct process for first order valuation (Chap. 3).

Development Development involves preparing a deposit for production. We suggest that although conventional drilling methods can be used, the more technologically advanced and potentially less costly methods that we propose (Chaps. 7 and 8) can be safely used for development of NGH reservoirs. Risk evaluation of conventional drilling methods are well understood and the risks can be estimated in a relatively narrow range that is not much different from evaluations of conventional hydrocarbon deposits, taking into account the less mechanically strong NGH drilling target and surrounding marine sediments. Risk evaluation of new drilling methods and wellbore layout, however, will have higher and less predictable risks, at least until a number of developments have taken place that will allow risk ranges to narrow. This, at least in part, reflects a desire for most companies to be early adaptors rather than innovators.

Production Production risk is largely concerned with predictability of production characteristics such as initial production rates and decline rates. Ideally, shut-in time is minimized as a factor of reservoir and NGH conversion performance. Geotechnical concerns, such as movement of sediments and faulting as an unwanted response to gas extraction, should be fully taken into account in planning so that any mitigation can be rapid and production resumed quickly. Again, with a new resource in a geomechanically weak reservoir, risks will be higher than for conventional hydrocarbon deposits until enough production data can be gained. Nonetheless, over time production data should increasingly serve to lower risk. Obtaining a fuller understanding of production risks may take years or even decades, but we suggest that some in situ experimentation with varying reservoir pressures and other production attributes will yield datasets that will narrow risk estimations relatively quickly.

Abandonment Risk factors of abandoning NGH production wells will be substantially lower than for conventional hydrocarbon wells. As conventional fields mature, abandonment decisions are based on decreasing production (even with secondary recovery), operating expense, and product prices (Chap. 11). All standard issues for conventional hydrocarbons apply for NGH. These include timing of cessation, retained decommissioning liabilities from previous contracts, and the options of batch decommissioning. Environmental and procedural requirements also apply. But once artificial methods of NGH conversion are stopped, gas remaining in the reservoir that lies in the GHSZ will convert quickly, on a timescale of weeks to months, to the stable solid crystalline NGH as natural conditions of pressure and temperature are restored. Any NGH on the surface or on any wellhead equipment left on the seafloor will dissolve in seawater and the well will naturally reestablish formation pressures. Because we anticipate that most NGH deposits will have little oil associated with it (Max et al. 2013), there will also be a very low risk of pollution.

One of the risks factors that is often ignored is incidental impact, or what a military refers to as 'collateral damage'. These impacts are incidental to the activity under scrutiny and are often not foreseen or recognized except in hindsight. Impacts that are environmental in nature may only develop indirectly, often over a period of time during which the relationship between cause and effect emerges only slowly. For instance, the effect on global climate of burning fossil fuels was not clear (Revelle and Suess 1957; Revelle 1983) for almost 200 years after the initiation of the industrial revolution and the beginning of significant artificial effusion of CO_2 into the atmosphere in the 20th Century. The scientific case, however, is now virtually beyond challenge (USGCRP 2017).

Other risks can be postulated more directly, for instance those associated with shale resource development such as fracking, increased natural gas leakage, and seismicity induced from injection wells. Economic risks are usually offsetting in some way where cause and effect are directly related. For instance, the drop in hydrocarbon prices beginning in mid-2014 led in less than a year to consumer benefit due to reduced fuel prices, but also the loss of jobs in the energy industry worldwide. Exploration and production activities have been devastated by the lower oil and natural gas prices. Energy risks may be more complicated than they might at first appear.

A common risk factor to all fossil fuels is that combustion produces CO_2 that is almost always exhausted into the atmosphere. The preponderance of evidence suggests that the colossal amounts of CO_2 being added to the atmosphere is at least partly responsible for global warming and climate change (IPCC 2014). From the standpoint of energy-related, man-made CO_2 in the atmosphere, natural gas is the lowest risk of all fossil fuels. It produces less CO_2 per Btu of energy than either coal or oil (Fig. 1.3), and its exhaust tends to have almost no other pollutants, especially where high temperature combustion can be maintained as in a combined cycle turbine generator. In addition, natural gas can be produced from a number of different geological sources, which substantially mitigates at least the risk of availability through diversity of supply. If any fossil fuel has to be combusted for energy, natural gas is superior to all other industrial fossil fuels (Chap. 1).

In addition, there are five major non-technical risk phases (Rudloff and Schultz 2016) that are not discussed in detail here.

(1) Regulatory changes and government or environmental scrutiny that affect operations and business plans and income structure.
(2) General economic conditions in markets or between countries and trading blocks that may restrict growth or profit opportunities.
(3) Lack of ability to deal with cyberthreats that could disrupt operations or damage the brand. In addition, physical disruption from terrorism is possible due to the difficulty of protecting infrastructure (pipelines and pumping stations) over very large areas and concentrated highly combustible plant such as refineries and other processing and transshipping locations.
(4) Resistance to change within the organization that are necessary to meet changing conditions.
(5) Acquisition and retention of talented employees necessary to achieve industrial and business targets.

10.2 Risk Factors of Major Natural Gas Resources

Natural gas produced from any of the major resources (Table 10.1) is about the same, once it is separated from water, oil, and solid particulate material that may be produced with it. Methane is usually the dominant gas fraction and sometime the only gas present apart from traces of other gases. Higher energy density hydrocarbon gases and natural gas liquids are sometimes part of the mix along with small amounts of other gases, especially in regions having a strong thermogenic petroleum system, such as the northern Gulf of Mexico.

10.2.1 Gas Purity

The longer a natural gas can coexist with oil and coal, and to a lesser extent with any carbonaceous sediments, the greater will be the tendency for the gas to include dissolved acids, non-hydrocarbon volatiles and other gases. In addition, where

Table 10.1 Comparison of production parameters and characteristics of conventional and unconventional gas deposits

		Conventional gas (CG)	Unconventional gas resources			
Issue		All including deepwater and ultra-deep	Coalbed methane (CBM)	Tight gas sands (TGS)	Shale (SH)	Oceanic natural gas hydrate (NGH)
(1) Gas purity[a]		Often multiphase with water, oil and particulates	High	High	Often multiphase with water, oil and particulates	High
(2) Sediment host		Geological traps	Coal measures	Sandstones	Shales	Partly consolidated, course silt-sand
(3) Flows under own pressure		Yes	No	No	No	No
(4) Recovery technique	Stimulation	All	No	No	No	No
		All	Decompress-ion	Fracking	Fracking	Decompression
(5) Injection of materials and water required		Any, often as part of secondary recovery	No	Water chemicals Proppants	Water chemicals Proppants	No
(6) Temperature		Often extremely high	Low	Low–mod	Low–mod	Low
(7) Pressure		Often extremely high	Low	Low–mod	Low–mod	Formation pressure Low
(8) Impact on water available resources		Low	Low, mainly drilling	High, fracking	High, fracking	None. Seawater circulated during drilling, fresh water produced
(9) Water and air environmental risk		Some produced water late in production	Large volume of produced water requiring treatment	Treatment or disposal of flowback and produced water	Treatment or disposal of flowback and produced water	Almost pure water produced, disposal on seafloor

(continued)

Table 10.1 (continued)

	Conventional gas (CG)	Unconventional gas resources			
(10) Blowout risk and atmospheric greenhouse feedback potential	Naturally high pressure-risk blowout (e.g., Macondo 2016)	Very low	Naturally high pressure-risk exists	Naturally high pressure-risk exists	Very low
(11) Reservoir and production performance	Well understood by industry	Well understood by industry	Becoming better understood	Becoming better understood	Production issues unproven

[a]Purity includes chemicals, dissolved matter including metals, and dissolved or liquid oil and natural gas liquids
[b]Actual or method preferred by us
[c]Primary recovery
[d]Secondary recovery. Sand hosts for NGH can include high quality reservoirs (Chap. 4)
Note We use the term 'Macondo' instead of the essentially synonymous 'Deepwater Horizon', which has been made into a movie (cinema)

multi-phase hydrocarbons and water are produced, some of the pollution associated with non-gas materials follows the gas, even after some processing.

CG The purest conventional gas is biogenic methane that has migrated clear of its source bed. Gas of thermogenic origin, in which high temperatures have reduced otherwise possible liquid and solid contaminants to natural gas and solid or very viscous residue from which only the gas will be recovered, can also be associated with very few contaminants. Thermogenic gas can be very pure methane, although it may also contain substantial percentages of higher density hydrocarbon gases that have also been 'cooked up'. The presence of the higher density gases ethane, propane, and butane will increase the Btu value of the mixed gas and commercial value of the natural gas, which may involve separation of natural gas liquids.

CBM, SH These gas resources are intimately associated with carbonaceous and other matter, some of which may be produced along with volatiles and both dissolved and finely particulate contaminants when produced. These gases are still associated with their source beds.

TGS Almost pure dry methane that has migrated into sands after leaving its subjacent source bed(s) using a fault and microfracture system for migration instead of the normal petroleum system that usually features permeable strata for at least some of the migration (Fall et al. 2012). Tight sands currently produce about 6 Tcf of gas per year in the U.S. yearly.

NGH NGH is different from all other gas resources because it has been purified from pollutants as part of its crystallization process, during which all materials other than hydrate-forming gas are expelled and can be dispersed in pore water. As a result, natural gas derived from it will tend to be purer than the other gas sources. In addition, NGH has been and will likely continue to be found in relatively clean sandy reservoir beds without associated oil or source-rock related non-gas materials. There is a particularly strong contrast to the geological settings of coalbed methane and oil-gas shale. Whereas NGH is sequestered in a high state of purity and effectively chemically isolated from potential pollutants in the reservoir until converted and free gas is produced, the gas in CBM and shale has resided in contact with solid and liquid carbonaceous materials for long periods of time. Conventional gas condensates, for instance, often require purification. The presence of mercury, amongst other pollutants in gas reservoirs, which is particularly prevalent in the Asia-Pacific region requires considerable care and produces concentrated hazardous materials (Intersea 2017). In contrast, when reservoir conditions are altered to produce NGH, the free gas will be in contact with other materials for only very short periods before extraction.

In cases where the hydrate-forming gas is the result of thermogenic processes, complex gas mixtures can occur and the level of non-hydrocarbon gases can be high (Hachikubo et al. 2015). In addition to affecting the chemistry and crystal structure of compound or non-methane hydrate, this may have the beneficial effect of increasing the energy density of produced gas, which can significantly affect the overall commercial value of the gas.

Liquefaction of natural gas requires a purified feedstock $N_2 < 4\%$, $CO_2 < 50$ ppm (by volume), water < 1 ppm, $H_2S < 4$ ppmv, Hg < 10.0 ng/Sm^3, and all C5+ components that would freeze during refrigeration must be removed. Because natural gas from a NGH resource can be expected to be much purer at the wellhead than conventional gas, CBM, or shale gas, processing costs prior to liquefaction may prove to be substantially lower.

10.2.2 Sediment Host

CG Conventional deepwater gas is found in relatively pure sands with permeability suitable to allow flow, although there can substantial fines. These can either be deeply subsided pre- or syn-rift clastic, sub-areal, or shallow water sands that formed as part of the establishment of the continental slope adjacent to a marine basin area (Chap. 4); or marine turbidites that form as part of the marine sedimentary succession and related to sequence stratigraphic sea level lowstands (Chap. 3).

CBM Coalbed methane is often found in unrecoverable coal measures that usually have many thin coal and shale layers, along with some sand partings and often with much dispersed coal-like carbonaceous material scattered through the sediments. Major coalfields, such as the Powder River Basin coals, however, may produce coalbed methane also. Coalbed methane deposits are found in continental basins, for instance in the Rocky Mountain region of the U.S. The methane is dissolved in the organic matter and pore water. Fracture permeability is generally good, even on the fine scale and producible horizons are much thicker than individual beds. The gas is produced by depressurization effected by pumping out some formation water, which lowers pressure and causes the gas to exsolve.

TG Tight gas is generally found in relatively pure sands of continental subaerial or shallow water origin that have been substantially lithified. Tight gas sand deposits are found in continental basins, for instance in the Rocky Mountain region of the U. S. Although porosity may still be substantial, permeability is low, typically less than 0.1 millidarcy (mD). Gas production has to be induced by fracking.

SH These are fine-grained shale rocks that are typically very rich in organic matter (1–20% total organic carbon) (Chopra et al. 2012). Many may have been source rocks that generated mobile hydrocarbons that migrated before their permeability was reduced by compaction and lithification. Other shales may have had their permeability decreased to the point where both gas and oil would be trapped in place. Most of the U.S. shale gas beds were deposited in restricted euxinic shallow water or marine basins in which organic matter was buried faster than it could be consumed by bacterial activity, however it is possible that some were deposited in deeper water (Max and Johnson 2012).

NGH The most common sedimentary host for oceanic NGH will be turbidite sandy beds deposited under marine conditions (Max and Johnson 2014). The Nankai deposit to the SW of Tokyo, offshore Japan (Noguchi et al. 2011) and the Walker Ridge NGH deposit in the northern Gulf of Mexico (Boswell and Collett 2011) are both continental slope turbidites whose position and disposition was in part determined by the geomorphology of the deep continental margin and continental slope on which they formed. These beds are the same as older beds in the marine sedimentary succession that are more deeply buried and for which the offshore energy industry has perfected exploration procedures that can be applied. The shallower NGH-related turbidite sand hosts almost certainly have less consolidated sediments, whose mechanical properties will vary more than in the deeper conventional gas deposits.

The possibility that NGH deposits may be found in high quality reservoirs (Chap. 4) means that greater attention needs to be paid to the geodynamic and geological history of deepwater prospect areas, especially those in which clastic-dominated sedimentary basins were transitioned into spreading centers whose margins on the consequent continental shelves underlay marine sediments deposited when modern oceanic conditions were established.

10.2.3 Flows Under Own Pressure

Gas in conventional deposits may flow under its own pressure, although in deep and ultra-deepwater artificial lift is commonly utilized to increase production rates. Artificial lift enhances natural drive energy of the reservoir. According to Schlumberger (2015), less than 5% of wells flow naturally at satisfactory rates and nearly all of the world's oil and gas production is reliant on artificial lift to maintain production rates, especially on older wells. But lift only accelerates or improves on a tendency to flow naturally (Sect. 2.3). In dramatic contrast, all of the unconventional gas resources require invasive artificial stimulation of some sort to free the gas so that it can be produced.

10.2.4 Recovery Techniques

Conventional gas deposits will flow spontaneously but a variety of stimulation techniques are commonly applied in mature fields to prolong production. These include acid treatment, injection and production well pairs, and fracking, along with some other minor techniques. All the unconventional gas resources require stimulation. Both shale and tight gas require fracking while both coalbed methane and NGH can be produced using depressurization. NGH can be converted in a number of other ways (Max and Johnson 2011), however conversion appears to be best achieved using depressurization.

10.2.5 Injection of Materials and Water Required

For primary production, conventional hydrocarbons require no injection to initiate primary production. Both coalbed methane and NGH do not require any material to be injected to stimulate gas production. Both tight gas and shale gas require fracking, with substantial injection of water and chemicals, along with proppants such as sand or artificial sand-like particles to hold open fractures in order to maintain permeability.

10.2.6 Temperature and Pressure

Deep conventional hydrocarbons usually have elevated temperature and pressure. Temperatures are commonly in excess of the boiling point of water and have been found at over 300 °C, although this varies depending on the geothermal gradient and reservoir depth. Formation pressures are a combination of lithostatic and hydrostatic pressures and can be very elevated. Hydrostatic pressure is a combination of seafloor depth and the depth below the seafloor. The combination of high temperature/high-pressure natural gas and oil (and water) introduces costly and extensive engineering and safety requirements.

Tight and shale gas (Table 1.1) may also have elevated temperatures, but almost certainly not as high as deep conventional hydrocarbons. Both CBM and NGH will have lower pressures and temperatures, particularly. Coalbed methane may once have passed through a higher temperature phase that converted much of the organic matter to gas, a condition known to produce high quality anthracite coal in which the gas has been driven off and the grade of the coal will be a good indicator of the thermal history. In typical CBM deposits, however, coals are dispersed to thin strata and the formation as a whole contains dissolved and absorbed gas that will be released upon depressurization. NGH, in contrast to all other gas resources, is temperature limited. The base of the GHSZ is the lower limit of NGH stability. NGH deposits of commercial interest will have temperatures below 40 °C, and probably below 30 °C while like CBM, pressures will be hydrostatic in NGH concentration to the base of the GHSZ.

10.2.7 Impact on Water Resources

CG, CBM, TGS Conventional and coalbed methane drilling are carried out using little to moderate amounts of water during drilling. Like shale, tight gas sands can only be produced by fracking.

SH Water stress and scarcity already will be exacerbated by fracking, which requires large amounts of water. The availability of water with which to carry out the fracking is an important issue because much of the fracking may be taking place in regions that are already water stressed, intensifying competition for available water resources. Water scarcity is an increasingly serious problem both of human consumption and for agriculture and industry. In many parts of the world, reduced water supplies are largely the result of increasing demand, depleting surface resources, pollution, falling water tables, seawater intrusion, and drying of large areas. Warming of the atmosphere and changes in weather patterns are apparently accelerating water stress and enhancing these risks in many regions. Water demand is increasing at the same time that rainfall and natural water supplies are expected to decrease dramatically. Because of its weight and the volumes necessary, distribution infrastructure, transport and their related costs can add substantially to development of resources requiring large amounts of water.

In more water stressed areas, such as the Tarim basin in NW China, water supply could become a limiting factor in shale gas development. The cost of providing water over long distances to highly water-stressed regions in which shale resources occur will probably be prohibitive. In the Sichuan and Chongqing regions where most of the initial 200 Chinese shale wells are located, water supplies are adequate although drought forecasts are not promising (Sandalow et al. 2014). Also noted is the poor Chinese record for environmental management.

Produced frack water is composed of both flowback of injected water and chemicals along with the water produced from the formation. The produced water typically contains gas, oil, and a substantial list of contaminants (Beckman 2015). Both may require extensive underground injection disposal and other often expensive treatment and transport (Dunne 2017).

A number of water treatment options are under development but no industry standard commercial water treatment method currently exists, and thus a demand remains for disposal of frack water in injection wells (EPA 2015a). Injection of water drawn from natural sources remove water from the hydrological budget, but extraction is highly variable depending on use (Gallegos et al. 2015). This extraction of water has the potential to intensify water shortages, especially in areas of high water stress. An undesirable environmental impact may be created if too much water is withdrawn from surface and groundwater resources. Each well also consumes valuable resources (Fig. 10.2).

Although the environmental emphasis is on the polluted water that is produced along with oil and natural gas, the process of fracking consumes large amounts of water. For instance, over 80% of the water used in hydraulic fracturing in West Virginia is surface water taken from streams and rivers and about 92% of the water does not return to the surface; about 1.6–2.2 gallons of water are consumed for each million cubic feet of gas produced from the Marcellus Shale. In Pennsylvania, about 94% of water used does not return to the surface and 3.2–4.2 gallons of water are used to produce each million cubic feet of natural gas, figures that are consistent with water consumption by the shale energy industry over the entire U.S. (Hansen et al. 2013). The availability of water is also important, and if there is no water abundant to frack, water reuse and water treatment are also very important.

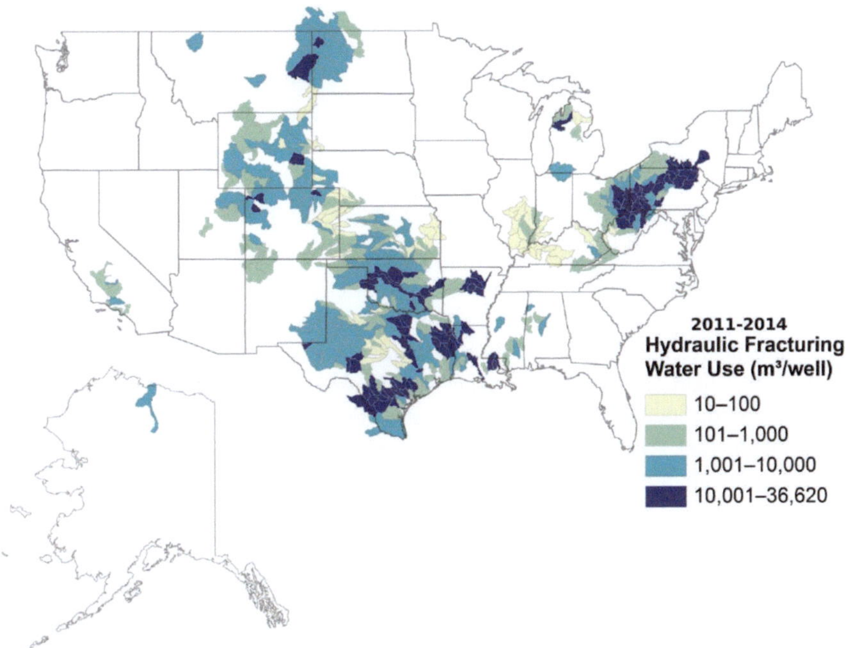

Fig. 10.2 Hydraulic fracturing water use in the United States. From USGS (2015a)

NGH In contrast, NGH is projected to use relatively small amounts of local seawater during drilling and to produce virtually pure, fresh water as a byproduct of conversion and gas production. The water can be dispersed at the seafloor or mixed with seawater along a gas riser or from its own buoyant dispersal pipe.

10.2.8 Water and Air Quality Risk

Air Quality All of the conventional and unconventional natural gas resources have in common the potential for gas leaks to the atmosphere from gas storage and transport infrastructure. Current leakage of methane and ethane from industrial systems appears to be higher than previously thought (Höglund-Isalsson 2017). There are two main levels of risk for air quality degradation due to natural gas leaks.

(1) Small leaks: Individually, these are small but add up as they all usually vent directly to the atmosphere. They are usually due to poor engineering or maintenance, or industrial accidents within plants or transportation systems. For instance, hundreds or perhaps thousands of small leaks coalesced into a huge natural gas plume from leaks from nearly 40,000 gas wells around the town of Cuba, in energy-rich northern New Mexico has been imaged from space by

NASA (Warrick 2014). In 2009, total U.S. methane emissions to the atmosphere were about 700 million metric tons (MM) tonnes or about 11% of U.S. greenhouse gas emissions (EIA 2011). Energy-related methane emissions were the greatest single source, at about 300 MM with agriculture and waste management responsible about equally for the remainder. Energy-related emissions are leaks from pipes, pipelines, industrial plant, incompletely combusted methane, all of which are amenable to regulation and its enforcement.

It has been estimated that from one to nine percent of all natural gas produced leaks into the atmosphere. According to the Union of Concerned Scientists (UCS 2017), at the high end this is the positive greenhouse equivalent of 600 MW produced by coal-fired power stations. Regulations already exist (in most countries) and stringent enforcement of natural gas leakage regulations are necessary to reduce the risk of leakage and increase both the economic and climate value of natural gas as a replacement fuel. The main problem with leaking natural gas to the atmosphere is the effect on the atmosphere greenhouse character.

(2) Large leaks. There are examples of major leaks that may directly affect human health. For example, the major natural gas leak from the SoCal Gas Aliso Canyon gas underground natural gas storage facility in the Porter Ranch neighborhood of Los Angeles, California continued for weeks from October 23, 2015 before the location of the leak was identified in late December 2015 (AP 12/29/15). This is probably the largest single emission of methane to the atmosphere documented in the United States. The storage facility is using the vacant porosity in an old oil field from which most of the oil has been removed. This blowout doubled CH_4 emission rate of the entire Los Angeles Basin, and released 97,100 a total of metric tonnes of CH_4 to the atmosphere (Conley et al. 2016). The gas leak is large enough for a no-fly zone to have been declared and resulted in the evacuation and temporary housing of over 2,300 residents.

At its peak in November 2015, the leak was emitting about 58,000 kg/h or about 25% of all emissions in California (Wilson 2016). 13 research aircraft flights between 7 Nov 2015 and 13 Feb 2016 showed leak rates directly to the atmosphere of up to 60 metric tonne/h of CH_4 and 4.5 metric tonne of C_2H_6. New California State regulations require all gas flow from Aliso Canyon wells to pass through newly installed and pressure-tested steel tubing inserted in the old well bores to prevent any further leaks (SocalCA 2017). Nonetheless, the full use of the Aliso Canyon storage facility is subject to court adjudication (Melly 2017a) because of the possibility of venting following an earthquake that ruptures the reservoir itself or damages pipeline or well infrastructure. SoCalGas, the operator, resumed pumping gas into the Aliso Canyon Field 31 July 2017 but on 11 September 2017 notified the California State regulators that 13 of the 39 injection wells were shut down after detecting an unanticipated pressure buildup (Melly 2017b). The long-term functioning of the Aliso Canyon facility will probably remain uncertain for years.

Although environmental discussions often focus on gas leaks from pipelines and industrial infrastructure, and emphasis is placed on how much more gas could leak, it should be remembered that natural gas systems comprise only about a quarter of the present methane effusion to the atmosphere. Enforcement of regulations could make industrial leakage a smaller volume even if more industrial infrastructure was established, whereas cutting down on other methane leakage sources such as landfill, biodegradation, manure, animals, etc. do not appear to get much attention, probably because they are difficult and expensive to deal with and do not so much attract the attention of environmentalists concerned with fossil fuels. There are currently (October 2017) no federal regulations in the United States on gas storage. These regulations are left to States and local jurisdictions (Daly 2016), and the near-term future of new federal or other regulations are in doubt following the 2016 elections. States regulate this aspect of natural gas. **(CG, CBM, TGS)**: Conventional deepwater is too far away from human populations to offer any risk to aquifers that might be used for human water supply. Seafloor and marine pollution, however, remains an issue that can be overcome with regulation and proper practices. Coalbed methane and tight gas produce water at the surface that may need treatment owing to an often high salinity and the presence of dissolved metals and noxious substances such as sulfur.

SH One of the main issues targeted by environmentalists is that existing potable water resources may become polluted, either with the natural gas or with other pollutants associated with fracking and production. There has not been much data presented on this emotional issue until recently. In June 2015, the EPA released a report (EPA 2015b), which clearly stated that pollution of existing potable groundwater resources was not a significant problem; although Morrison (2015) observed that they had overlooked chemical tracer analysis. The report concluded that they did not find evidence that fracking or injection have led to widespread, systemic impacts on drinking water resources in the United States, and that the number of identified cases where drinking water resources were affected is small relative to the number of hydraulically fractured wells. This does not rule out individual cases of pollution existing from place to place, and it is likely that lawsuits over allegedly contaminated water will continue.

Fracking Hydraulic fracturing releases water, gas, and oil from an otherwise impermeable shale or tight reservoir. It is the enabling technology for development of the shale resource. Although the terms 'fraccing', fracing, and 'fracking' have all been used for 'hydraulic fracturing', we use the term 'fracking' throughout. Fracking was originally developed as a secondary recovery technique for conventional hydrocarbon deposits in order to increase recovery from existing fields under production in the mid-20th century. In modified form since the mid-2000s, fracking has proven to be the enabling technology for opening the shale oil and gas play to commercial exploitation. Fracking is a means of breaking up impermeable deposits holding considerable trapped oil and natural gas. Whereas in conventional fracking the existing permeability of a reservoir is enhanced, in shale and tight gas plays, hydrocarbon-rich (often >10% total organic carbon) strata that is essentially impermeable is fractured so that an artificial permeability is created, allowing hydrocarbons present in isolated porosity to flow.

Both the flowback water (water that was originally injected carrying ‘frack’ chemicals) and the produced water (water released from the tight strata by the fracking) may carry organic and other dissolved and finely particulate dissolved metals and other pollutants derived from the reservoir (Harkness et al. 2015). An often overlooked contaminant is radioactive waste produced along with water and oil. North Dakota, for instance, generates up to 75 ton of radioactive waste daily, largely from oil filter socks, tubular nets and other filters that strain liquids during processing. Under rules promulgated under the Obama administration, companies had to keep ‘cradle-to’ grave records on oilfield wastes, including dump sites that now have regulations for the maximum of 5 picocurie threshold for radionuclides (MacPherson 2015). Under the Trump administration, however, regulations are being reviewed and the situation is expected to be in a state of flux for some time. TENORM, the technologically enhanced radioactive material that does not meet the industry-supported rules is to be transported for disposal out of state. A downstream consequent hazard is radioactive waste transported by truck or train.

It has been found that there is no inexpensive solution to treating frack water. Either water treatment (Gunderson 2015) or deep injection in disposal wells is commonly required, although water reuse is possible if multiple wells are drilled and completed in sequence. Pad drilling, in which a number of wells are drilled from a single surface installation, may facilitate reuse of water and reduce the initial water requirement, but water treatment or disposal remains an issue. This is not a small matter; over 280 billion gallons of frack water was produced in 2014. Frack water volumes will generally track the number of wells fracked, although reuse of water could diminish overall volumes somewhat. On the positive side is risk reduction that may be achieved by using new expandable hydraulic fluids to cut water and chemical use.

About 8 million gallons of water and to 40,000 gallons of chemicals are injected with the water for each frack (and multiple fracks may take place over time). Over 500 chemicals may be present in the fracking fluid, locally including some known carcinogens and toxins. Many of these are not disclosed. There are about 1.1 million active fracked gas wells in the U.S. that have used about 72 trillion gallons of water and 360 billion gallons of chemicals (Loki 2015).

Earthquake hazard In addition to water being used by the shale energy industry in fracking, the waste water is often contaminated beyond the point it can be treated and the contaminants removed from the water. Highly saline water can be evaporated, but when the water also contains volatile chemicals, these will also evaporate and will affect air quality. The solution for these highly contaminated water residues is to inject them into deep permeable rock formations that are significantly below drinking water aquifers. Injection of waste water from conventional and fracked wells has been practiced for many years although not all states license waste water injection.

The EPA has determined that out of the 9,000 currently active injection wells, nearly all of the induced seismic is coming from only very few of them. The likelihood of associated earthquakes appears to be a function of injection rate and local geology rather than being related primarily to the particular fluid being injected (Weingarten et al. 2015; EPA 2015b). In addition, the injected water is derived from many sources other than fracking operations. Again, as with the finding on the practice of fracking, problems from injection would seem to be isolated. Hicks et al. (2018) found that Wastewater injection-induced earthquakes in Oklahoma. Although operational and geologic parameters in triggering such earthquakes is unclear, they determined that the injection depth above the crystalline basement was the most important parameter when considering the seismic potential. This modeling strategy may provide a way to improve forecasts of the impact of proposed regulatory changes on induced seismicity.

Geologic reappraisal can lead to withdrawal of licenses to operate disposal wells, although the State of Oklahoma did not react with tough measures following a spike in earthquakes including one that knocked out electric power to part of Tulsa in December, 2015 (Borenstein et al. 2016). New regulations governing rate and/or volume of injection are the alternative to seeking other measures of fluid disposal.

Induced seismicity or earthquakes directly related to fracking are rare but where they occur they appear to be very special cases that are dependent on particular geological situations. Schultz et al. (2018) describe a fracked volume that was associated with induced earthquake activity in the Duvernay play of the Western Canada Sedimentary Basin of Saskatchewan, and their evidence is convincing.

Water ponds art typically used for flowback and produced water at every shale drill site. Expansion of drilling activities has brought oil and gas development literally to backyards of homes and communities where air contamination is increased. In particular, the major danger is not from leaking gases, as is the case with chemically inert methane. The threat is from the large frack water ponds that are open to the atmosphere. Indeed, evaporation is encouraged to reduce the amount of water that would otherwise have to be treated or disposed of. Most of the potential dangerous material is volatile and will evaporate where air will carry the contaminants for considerable distances. Thus, there is increased threat for human exposure to new contaminants such as BTEX: benzene, toluene, ethylbenzene, and xylene; along with methane, carbon monoxide, carbon dioxide, and a large number of unrevealed chemicals returned in the flowback water (Srebotnjak and Rotkin-Ellman 2014).

Waste disposal is usually expensive and is a driver for industry engaging in illegal or dangerous practices such as large-scale evaporation, unauthorized and probably illegal dumping on the surface or shallow injection, which may find its way into water aquifers. But this is fundamentally a regulation and enforcement issue, not a systematic problem with injection wells. California environmental groups, Earthjustice on behalf of the Center for Biological Diversity and the Sierra Club, for instance, have begun to file lawsuits against state oil and gas regulators seeking to halt illegal oil industry dumping of waste water into drinking water aquifers.

The topic of fracking has crystallized into an antagonistic relationship between environmentalists and industry. In areas where adequate water resources occur and remain unpolluted, the main issue is the quality of 'frack water', which may require extensive treatment or disposal, either of which adds to the cost of producing tight oil and gas, which affects profits and commerciality, and is a persistent risk,

NGH Environmental water risk associated with conversion of NGH to its constituent gas and water is extremely low. Not only is pure water produced when NGH is converted (Chap. 2) but any water brought to the seafloor will first dilute the pore water into which it will mix and then into the seafloor water. As NGH is not usually associated with liquid hydrocarbons or other potential pollutants, and it does not have the potential to contaminate potential aquifers or surface water resources on land, its water risk is negligible.

Life cycle water usage Natural gas power generation has the lowest life cycle water requirements of all fossil fuels when it is used in combined cycle (CC) turbine generation plants (Chang and Lammi 2016, Fig. 9). The efficiency of gas turbines is improving. Both General Electric and Misubishi have new gas turbines that operate at up to 64% efficiency (Proctor 2018). Although oil can also be used in the efficient combined cycle turbine generators, it is not as effective or as clean burning, and also does not produce as clean an exhaust. In contrast, coal and nuclear, as well as open cycle geothermal, have substantially higher water requirements. CCs produce direct power from combustion in a gas turbine electric generating unit that is similar to a jet engine auxiliary power unit used on many aircraft, but dramatically scaled up. Additional power is produced indirectly from the gas combustion by heating water to produce steam that is also put through a turbine (GE 2016). Where water is recycled the life cycle water requirement is lower than where it is evaporated in cooling towers to cool the water before release back to the environment.

10.2.9 Blowout Risk and Atmospheric Greenhouse Feedback Potential

CC, CBM, TGS, SH Because conventional gas and oil reservoirs are typically highly pressurized with respect to surface ambient conditions, they have the potential to blow out from the moment a reservoir is reached. Tight and shale gas have the potential to blow out or leak gas from improperly secured casing. Coalbed methane pressures can be controlled in the reservoir and thus CBM does not have a high potential for blowout risk.

The process of producing gas from a conventional deepwater reservoirs having very high pressure and temperature has to be very concerned with safety because of the constant threat of blowout should anything go wrong. Controlling the gas flow and more important maintaining the capability of stopping gas flow at the wellhead is vital for production security. Stopping gas flow is done mechanically by using chokes, valves, or powerful rams on the Blow Out Preventer (BOP). Well closing systems are designed to be fail safe but the risk of disastrous blowout remains, no matter how well intentioned and operated any well may be.

NGH Because NGH is very responsive to environmental changes that result in pressure and temperature variation, especially at the base of a thick GHSZ, it was thought possible that the sequestration and release that could be expected during glacial - interglacial cycles could be responsible for varying the amount of natural gas in the atmosphere (Max et al. 1997). Even though NGH dissociation is endothermic and extracts heat from its immediate environment when warmed, it does not dissociate rapidly, especially if warming is slow, as would ocean be warming in response to climate change. This factor provides a measure of reaction buffering or naturally "self-limiting or "self-preserving" (Stern et al. 2003) that acts to prevent run-away dissociation. Natural mechanisms for blowout of gas trapped beneath a NGH seal near the base of a GHSZ have been proposed by Dillon et al. (2001), based on seismic interpretations of apparent blowout features on the Blake Ridge, off the U.S. SE coast. But this would have involved trapping of large volumes of gas produced by dissociation over a long period of time, with some trigger such as earthquake to breach the free gas concentration and cause it to vent suddenly.

An outgrowth and end member of this general model was the 'clathrate gun hypothesis' (Kennett et al. 2003), which proposed that climate-driven dissociation of NGH that results in sudden injections of large amounts of methane into the atmosphere could be derived on a short-term basis through global warming. This has led to a model for catastrophic positive feedback effect of global warming (AN 2014). The occurrence of large amounts of NGH in deep ocean sediments, especially where they overlie free natural gas in the sediments below the base of the GHSZ, offers a physical model for gas blowout. In the case of seafloor warming, the thermal signal will penetrate into the seafloor and warm the upper sediments. When the geothermal gradient equilibrium is reestablished, the base of the GHSZ will move up the phase boundary (Chap. 2) and a certain thickness of the preexisting GHSZ will become unstable, causing the NGH to destabilize. Under certain circumstances, the increased buoyancy of the gas trapped beneath the GHSZ might have the potential to cause fracturing of the overlying sediment and escape over a short period of time.

This would constitute a natural blowout. Bock et al. (2010) in a study of the hydrogen isotope ratios from North Greenland ice-cores from the 22,700–41,000 ybp interval, conclude that the deuterium ratios could best be explained by production of the methane in Polar wetlands in which an increase in rotting of vegetation increased methane production about six-fold to about 32 Tg CH_4/y. This increase, combined with and increased production from tropical wetlands of about 118 Tg CH_4/y could have provided the bulk of the methane 'boost' associated with Dansgaard-Oeschger sudden warming events (with a recurrence time some multiple of 1,470 years. But there is no evidence of NGH participation (Bock et al. 2010). Möller et al. (2013) note that relatively stable $\delta^{13}C$ values that persisted during large changes in the atmospheric loading of methane responded simultaneously to climate changes across these periods rather than being sourced from NGH. Whiticar and Schaefer (2007) note that the evidence for no input from NGH refers specifically to biogenic methane as the $\delta^{13}C$ values relate only to biogenic methane.

Although it remains possible that thermogenic NGH gas releases are possible during the Late Pleistocene and particularly the Younger Dryas warming period, but it is also possible that thermogenic methane is related to conventional natural gas seepages, which are well known. Sowers (2006) concludes Late Quaternary atmospheric CH_4 isotope records suggest that marine NGH was essentially stable through atmospheric warming events. Methane produced from rotting vegetation on land will almost all find its way directly into the atmosphere. Maslin et al. (2010) concur that there is no evidence for NGH being involved in the initial methane spike in the atmosphere.

Bock et al. (2010) also require for their numerical modeling an additional $\sim$25 Tg CH_4/y, which they ascribe to 'a constant production rate of $\sim$25 Tg CH_4/y'. We regard it more likely that the source of this additional methane is the now widely recognized venting of methane in shallow water, which, because it only has to traverse a relatively narrow column of seawater in which methane NGH will not form, will largely pass into the atmosphere. Ruppel (2011) notes that temperature changes even in relatively shallow water will take multiple thousands of years to affect the stability of NGH in permafrost terrane (or the now inundated one-time permafrost terrane in shallow Arctic continental shelves).

Biastoch et al. (2011) have modeled the ocean warming effect on NGH and conclude that "Compared to the present day annual emissions of anthropogenic methane, the global amount of methane release from melting hydrates in the next 100 years is small and will not have a major impact on climate". In deep- and ultra-deepwater where over 95% of NGH occurs, the penetration of heat into marine sediments could at most produce negligible amounts of natural gas from NGH conversion (Reagan and Moridis 2008; Garg et al. 2008; Beaudoin et al. 2014, Fig. 3.6) and could take thousands of years between warming the seafloor and penetration of the heat signal to the base of the GHSZ (SBC 2015).

In shallow water in the North Sea, Vielstädte et al. (2015) observed that the natural gas leaks associated with abandoned wellheads commonly developed bubble-driven methane fluxes from the seafloor. However, more than 98% of the gas released at the seabed was dissolved in the 81–93 m deep water column and did not reach the atmosphere. They further note that diffusive sea-air gas exchange could provide methane release to the atmosphere, but a very small part of less than 2% of shallow water vented gas is almost certainly insignificant with respect to other atmospheric methane sources.

Although natural gas venting from the seafloor is a worldwide factor in distribution of materials in the biosphere, there is to our knowledge no definite isotopic evidence that the methane venting from shelves is largely derived from NGH. It is therefore difficult for us to subscribe to the dramatic proposition that the global warming is presently underway is primarily responsible for the methane production in shallow water continental shelves through climate-driven dissociation of NGH. The methane venting seen is the Arctic shelves is, however, evidence for the Arctic being a very active natural gas petroleum system, which is a prerequisite for the formation of large concentrations of NGH in deepwater continental slopes (Max and Johnson 2014).

Barnard et al. (2015) describe the activity of the NGH system within a deep water column in which NGH is stable can contribute to increasing dissolution of vent gas such that even large emissions from the seafloor are unlikely to reach the sea surface and add to the global greenhouse effect of the atmosphere. First, ascending gas is highly turbulent and rapidly breaks up into smaller bubbles. Second, NGH rims will form on bubble surfaces, and will break into very small pieces with high surface areas. Third, as natural gas—rich vent water ascends, natural processes substantially alter the mass to surface area ratio of the gas. Because dissolution is a diffusional surface area phenomenon, the rate of dissolution of the natural gas—NGH mix is naturally accelerated, especially after the mass rises above the pressure required for NGH stability.

Probably the best documented deepwater blowout that mapped the oil and gas from unconstrained venting took place in the study of the Macondo well. The well was located about 150 km to the SE of the present delta of the Mississippi River, with its wellhead at a subsea depth of about 1.6 km, the arbitrary boundary between deepwater and ultra-deepwater zones (Fig. 3.1). At this wellhead depth, if 1 m^3 of NGH were to dissociate and the water could equilibrate (flow away from the space previously occupied by the hydrate)—that is an open hydraulic system—the dissociated gas would be about at the pressure of the water and would not have a measurable overpressure. Oldenburg et al. (2012) estimated flow rates for oil were between 50,000 and 70,000 BPD and natural gas was thought to average about 170 MMSCF/d using numerical simulations. Pressure in the gas-oil reservoir was 16,800 psi or about 1142 bar (PFL 2017). The pressure in the Macondo well following run-away may well have been higher than would be expected from normally dissociating NGH in the much shallower and lower pressure GHSZ, especially in an open flow situation. Although dissociation of NGH has been held to be responsible for or partly responsible for the blowout, (GLS 2017) following earlier recognition that NGH could constitute a drilling hazard (Folger 2010), there appears to be no evidence for well blowing through casing of casing joints. It is possible that NGH formed on the BOP and jammed it; solid NGH can be mechanically quite strong (Durham et al. 2003).

The impact of the blowout on the environment was a primary concern both during and after the spill. Measurements and sample analyses obtained during the course of the blowout provides a body of information that establishes a standard by which other blowouts can be compared. From quantification of the amount of oil at different locations (i.e., wellhead, ocean surface, and atmosphere) (Ryerson et al. 2012), McNutt et al. (2012) reached the conclusion, reiterated by Joye (2015) with a different perspective and data set, that although a smaller amount of higher gravity, less water-soluble volatiles associated with the oil release reached the atmosphere, virtually all of the released methane that was not dissolved in oil and about 50% of the oil, appears to have remained deep in the sea. Although emissions to the atmosphere in the early stages of the blowout before dispersants could be introduce at depth, it is possible that longer term blowout could have resulted in some ventilation of natural gas associated Benzene and Toluene the atmosphere without dispersants (Kaufman 2017). In an Arctic region to along the Svalbard margin of the Barents Sea, Graves et al. (2017) have sampled methane in shallow subsurface sediments at the up-slope limit of NGH stability at about 400 m water depth, in deeper water and in shallower water sediments in the Barents Sea. They found no evidence that that positively indicated that dissociating/converting NGH was an immediate precursor to the venting. As the Arctic is the region undergoing the most dramatic warming in the current phase of global climate change, if the heat signal from atmospheric warming has not contributed to NGH dissociation, it is unlikely to have done so in other ocean regions where the thermal stress is not as great.

One of the concerns of global warming is to what extent destabilization of NGH at the base of GHSZ could lead to large free gas ponds that could blow out because of overpressure and upward buoyancy. Li et al. (2017) use detailed multi-layer seismic interpretation or near-3D to show that gas produced as a result of inter-glacial warming at the base of a GHSZ offshore Mauritania created vent structures but that much of the gas appears to have reformed NGH when it rose into the GHSZ. This is evidence for the suggestion of a NGH gas conservation cycle in which buildup of NGH through multiple cycles of formation and dissociation can be accomplished. This agrees with the vent model for developing NGH concentrations higher in a GHSZ based on a lateral gas supply from a persistent vent (Max et al. 2006).

NGH is stable in its deep-sea environment and dissociation is naturally buffered. Mestdagh et al. (2017) have reviewed the issue of NGH being a significant positive feedback mechanism in a global warming situation and conclude that the largest fraction of NGH, which is found in deepwater oceanic and permafrost regions convert on timescales of thousands to hundreds of thousands of years. They also add that even shallower water NGH may convert in sub-1000 year timescales but that their significance does not appear to be great.

Shallow water methane venting risk from NGH formed during the glacial episode remains an open question, but all the references we have seen show vents with roots well below the base of GHSZs. For instance, Serov et al. (2017) discuss in detail adjustment of the thickness of the GHSZ in response to ocean warming of the seafloor in troughs in the SW Barents Sea, which infers that NGH conversion would have produced methane that could have reached the sea. But vent roots identified on a reflection seismic section clearly shows vent roots well below the colder climate paleo-GHSZ. In any case, most of the Eurasian Arctic continental shelf is too shallow to support a GHSZ from which NGH could be converted by further warming and the area of troughs with GHSZs is proportionally relatively small, although identified as having potential for holding NGH concentrations in coarse grained sediments (Max et al. 2013). We agree with Ruppel and Kessler (2017) that there is no conclusive evidence that NGH-derived methane is reaching the atmosphere now or in the future on human timescales as a result of global warming

It is important to remember that the potential for atmospheric warming due to methane venting from the seafloor cannot be considered in isolation. Methane is only one factor in a complex system of methane production and consumption. Although methane is chemically unreactive without life system metabolism of some sort, it is important as a nutrient at the base of the food chain in which different microorganisms consume it to produce oxygen, CO_2, and other byproducts while themselves providing food for higher organisms. Pohlman et al. (2017) for instance, note that while warming of the Arctic Ocean is expected to cause dissociation of NGH that in shallow water will likely vent to the atmosphere, biological up-take of CO_2 has the potential to more than offset the positive warming potential of vented methane. They note that the negative radiative forcing expected from this CO_2 uptake is over 200 times greater than the positive radiative forcing would be if the methane reached the atmosphere.

The immediate environmental risk of blowout leading to a strong positive feedback for heating in the atmosphere caused by changing natural conditions that can be perceived on a human timescale on the Earth's surface from deepwater thick GHSZs is thus regarded by us as very low. By analogy with the history of the Macondo methane, it is unlikely that natural gas from even a massive blowout of gas beneath a NGH-rich GHSZ in deep or ultra-deepwater, would reach the atmosphere in any significant quantities. Thus, we regard the 'clathrate-gun hypothesis', especially in its short-term climate feedback response form that has excited environmentalists, to be unlikely in practice.

10.2.10 Reservoir and Production Performance

NGH is the only resource of natural gas whose reservoir and production performance are not yet known well enough for high confidence in predicting reservoir stability, production decline rates, effective mitigation of sand production, and other parameters that will all affect commercial competitiveness with the other natural gas resources. Quantification of NGH reservoir and production uncertainties is difficult because there is no body of data and experience upon which to draw. Thus, analysis should be conducted to better define of uncertainty and risk. Accordingly, NGH development is still at the stage where we need to know what needs to be known (Ma 2011). Nonetheless, the first technical production test on oceanic NGH has been carried out successfully on the 40 Tcf Nankai NGH deposit during March 2013 (JOGMEC 2013) and has demonstrated both a depressurization system for NGH conversion and for gas-water separation, as well as being able to sustain acceptable levels of natural gas production from a vertical wellbore having only a small intersection with the NGH-rich strata. Optimized well configurations could provide better control of the conversion and gas extraction process (Chap. 8).

Longer-term production testing, on the order of two or three months, is necessary to develop a production profile and creaming curve comparable with conventional gas before committing to full commercialization. Ideally, one of the criteria for these tests will be that the gas can be recovered and brought to market so that large amounts of gas do not need to be needlessly flared, as is common on shorter tests. Longer term testing is part of the Japanese MH21 plan, and has been recommended to the U.S. Secretary of Energy by the DOE Methane Hydrate Advisory Committee for proven NGH deposits in the northern Gulf of Mexico (DOE-MHAC 2015). Ideally, gas produced from this production testing can be captured and brought to market to both provide some income to defray test costs and to prevent flaring.

10.3 Risk of Overdependence on Single Fuels

Putting all one's eggs into one hydrocarbon resource box introduces some risk as no resource is of indefinite size and many things could happen that could render any particular resource to be less than required. There are three primary factors to consider. The first is the conceptual risk in restricting fossil fuel supply to natural gas (from any resource) as a bridge fuel to a renewable energy paradigm (Chap. 1). The second risk concerns the long-term dependability of natural gas supply for individual countries such as the U.S., which companies are currently making plans based on the promise of a long-term natural gas supply from the newly developed shale resource. A third risk area, of course, is the sum of all those things of which we are not aware; the 'unknown unknowns'. Additionally, and most important for many environmentalists and concerning the availability of resources for renewable energy, greater use of natural gas could increase the difficulty of achieving long-term global warming emissions reduction goals (Deyette et al. 2015).

The Union of Concerned Scientists considers that the dramatic expansion of the use of natural gas creates numerous risks for the U.S. economy and health, and on a wider scale, of climate. They (UNC 2015) have analyzed the potential for risk owing to overdependence on natural gas on a state-by-state basis using a range of metrics. Although a state's electricity generation may not be representative of the fuel mix, consumers may be exposed to varying risk based on states' individual regulations and practices. Although some states have more or less risk to overdependence on natural gas, so long as natural gas is widely available at low or acceptable prices to consumers, overall risk will remain low.

Over-Restricted fuel choice Diversification is usually the prescription for overall success. This is as sure in evolution and economics as it should be for energy supply, and it is why the development of a multi-energy source market strengthens security and makes it less likely that serious energy supply issues will emerge (Chap. 1), even from hostile activity. Nonetheless, if there is only one supply for base load energy when renewable energy delivery is abnormally low, energy supply gaps could develop.

Natural gas has replaced coal in electricity generation. Coal traditionally fired over 50% of electricity production whereas now it provides about 30%. The replacement of coal by oil will almost certainly continue. This has the potential to create numerous risks. In the first instance, an electrical power system that becomes dominated by any fuel can introduce market distortions affecting price. There have already been examples of monopoly practices wherein companies managed supply to control energy prices, for instance in California. Single fuel use might introduce price volatility for consumers, but natural gas is sourced from conventional deposits, coalbed methane, shale, and likely soon from NGH. The variety of sources from which natural gas is recovered introduces a redundancy in gas supply, which should act to limit price volatility. If the security of that base falters, however, then restricted fuel choice has the potential for having a major economic impact at some future time.

Dependability of natural gas supply This discussion is limited to the United States for brevity, but similar analyses could be done for every country or economic block in the world. For instance, although Russia is regarded as being the long-term supplier of natural gas to Europe, offshore locations may require substantially costlier infrastructure to land the gas and the resource on land may not have sufficient reserves without a significant price rise (Huber 2016).

Indigenous energy sources are very important to all countries, even if they have the foreign currency to import fuel. Without the development of coalbed methane and shale oil and gas, the security of the U.S. could be held to ransom by energy exporting states, as it was in 1973 following the U.S. support for Israel in that year's war. Energy independence may be a useful political objective but different hydrocarbons are used for different things and it makes more sense to trade commercially rather than to try to throw up walls around a county's energy resources. Countries such as the United States, which has abundant energy resources and an economy is in the best position to trade. Countries such as Russia, which has abundant energy that for long retarded the development of a broad-based economy because of its purchasing power, are only recent, and possibly with the prodding assistance of sanctions, broadening their economy. The Gulf States (Persian Gulf) are still mainly energy producers without broad economies. At least abundant energy allows countries to buy in almost everything else but countries without energy economies may only remain near subsistence level (Chap. 1). No resource lasts forever. For instance, Likvern (2016) shows excellent graphs of Norwegian oil production showing that their oil production is decreasing and despite new discoveries, such as the Johan Sverdrup field, their high production rates achieved from 1990 to 2008 are very unlikely to be achieved again. All resource extraction peaks and decreases subsequently, which is why multiple energy resources including the transition to renewable energy is critical on a broad front.

Even small countries with secure energy supplies may take additional steps for energy independence. An example is the Republic of Ireland, which initiated production of gas from the Corrib Field to the NW of Ireland in January 2016 even though it has indigenous gas in the Irish Sea. Corrib Field is in water depths of around 350 m and has the potential to supply over 60% of the country's current natural gas requirements (PGN 1/6/16) through a 20-inch pipeline to shore in NW Co. Mayo.

Since 2005, the shale resource has developed at a rate that may never have been seen before and appears to offer the possibility of near-term U.S. energy independence. The key drivers for shale gas development are the gas demand in the northeast U.S., the switch to natural gas from coal for electricity generation, the change by the petrochemical industry from imported to domestic gas, and the huge natural gas production numbers combined with optimistic estimates for the size of the shale resource. A vibrant services industry, including pipelines or other gas transport capability for bringing the gas to market, is also required. Local companies have started producing fracking sand to support development, and other niche services providers have also emerged. The U.S. natural gas production has risen to over 2.4 Tcf (December 2017, with large growth potential, although the cost of shale gas development and its price are not yet aligned because of natural gas oversupply. This is in contrast to 1978, for instance, when monthly production was between 1.5 and 1.7T $ft.^3$/month.

Although this topic is worthy of a much longer discussion, we only mention it briefly because it is controversial and still not well understood. However, it is important to note that the shale resource, which only began significant production in the 1980s with the initiation of the "gel frack" combined with horizontal drilling, may not have as much upside as its promoters assert. A principle attribute of tight reservoir resource extraction is that a great number of wells are required because the life of an average well has only been 14 months to 2 years (Chap. 11). Thus, the rate at which drilling has to be carried out has to be constantly high, if not accelerating in order to increase overall production.

Because the average life of a shale well is commonly on the order of 18-24 months on the first frack and shorter periods for subsequent fracks, there is a significant departure from a major field production profile (Chap. 11). Refracking can be carried out for about $1.8 million each or about a quarter of a primary frack, although most refracking, like initial fracking concerns oil shale or oil/gas shale rather than for natural gas alone (Collins 2015). Bush (2012) cites a cost range of from $900,000 to $2,800,000 for several operators in the Haynesville Shale. He further estimates that refracks yield a 39% increase in production with a break even natural gas price about $2.30/Mcf, which is less expensive than the exploration, drilling, and preparation costs of a first frack and is economically viable at $1.75 million per refrack.

In addition to water availability issues, flowback and product water issues, cost of sand, proppants, pumps, ponds, electricity, chemicals, potential for increased methane leakage, contamination of aquifers, injection wells, earthquakes, etc., the main issue is exactly how big is the shale play at different price levels for oil and natural gas?

The shale gas and oil industry appears to have financed itself by popularizing the growth of production and the promise of virtually endless supplies of natural gas from the shale resource; however, those resources may be much more limited. Hughs (2015a) agreed with Berman (2013) that shale gas and shale oil wells deplete in the extremely short time of a year to 18 months and except for a few wells in western Pennsylvania became a commercial failure for dry natural gas as prices dropped. Lower prices, however, has introduced greater caution in lending. Berman (2017a, b) notes that, "shale gas companies spent an average of \$1.42 for every dollar they earned in the first quarter of 2017", with the exception of Gulfport and Chesapeake. "Including those two operators, companies spent \$2.12 for every dollar they earned." Gas priced below \$3 MMft3 means that the cost of borrowing that is increasing from 2016 may considerably squeeze the whole shale gas industry (Azar 2017). Because of reselling of leases, however, a number of companies have made their money by flipping acreage—getting into a play early when leases were cheap, having some discoveries, and then selling the leases to other companies and this is probably not reflected in Berman's calculations. Major companies getting into shale late took a proportionally bigger loss than earlier entries as they bought acreage or whole companies when the prices were near peak.

From late 2014, the number of drilling rigs in the U.S. dropped by 68% although by spring 2017 drilling was again increasing. Over 100 companies went out of business, defaulting on at least \$70 billion of debt, but the resurgence of drilling shale in 2016–2017 is again built on debt (Schumpter 2017).

Berman and Pittinger (2011) suggest that the true structural cost of shale gas production was then in the range of \$8.00–\$9.00/Mcf to break even on full-cycle prices, and \$5.00–\$6.00/Mcf on point-forward prices. Full-cycle break-even is far less today as the cost structure has decreased markedly. Nonetheless, the shale gas industry may be much less self-financing at low gas price points. 21 of the major shale operators had negative cash flows from 2010 through 2013 (Rogers 2014). With even lower prices since 2014, the financial positions of many shale gas producer companies have not improved markedly, which was reflected by the increasing bankruptcies of shale gas and small service companies beginning in the second half of 2015. The market appears to be relatively stable in the early part of 2018.

Rapier (2015), reported that a representative sample of large cap exploration and production companies saw cash flow swing from +\$14.7 billion in 2005 to −\$6.4 billion in 2009 and to an estimated −\$32.7 billion for 2016. The underlying problem with shale gas and oil companies has been that most of them were only marginally profitable to 2008 and have been fueled by debt and negative cash flow. Expenditures on shale exploration and development have contributed to negative cash flows reported by major integrated companies and refining and marketing companies. In a way, shale E&P companies appear to have become sub-prime investment vehicles. The shale gas resource gives every indication that it had become financialized in part because of the availability of cheap money. This will probably be different for the NGH resource, which because it is in deepwater will require a higher CAPEX and a longer development and production time.

There is considerable variation in production among the 27 shale resource basins. Some shale gas fields may already be in decline. Major new field discoveries in the US are needed to keep production still increasing. The Bakken, Niobrara, Permian Basin, Eagle Ford, Haynesville, Utica and Marcellus shales accounted for 92% of domestic oil production growth and all the domestic natural gas production growth during 2011–2014. Production in the Permian Basin and Utica was still growing in early 2016 (EIA-DPR 2016). Average per-well productivity is declining, so ever-more wells are required simply to maintain production.

This shale resource “exploration treadmill” imposes a limit on the long-term potential of the shale resource unless new technology for recovery of more hydrocarbons are developed. Over the short term, U.S. production of both shale gas and tight oil is projected to be robust, even at the post mid-2014 low cost period, but production data indicates that this will not be sustainable for multiple decades. But the underlying amount of the shale resource remains huge especially on a world basis, if expensive to produce. For instance, the U.S. Geological Survey estimates that there are 53 trillion cubic feet of natural gas, 172 million barrels of oil, and 176 million barrels of natural gas liquids in the Barnett Shale of the Bend Arch—Fort Worth Basin Province of Texas (USGS 2015b), but all these estimates are revised. The negative view of sustainability and potential of the shale resource is based on the apparent fact that four of the top seven U.S. shale gas plays (stratigraphic horizons in basins) may be in decline now. The life of shale fields may be much shorter than originally thought, but this may be a factor of more efficient takeoff of gas and oil, which yields sharper production peaks (Chap. 11).

It is estimated that production from the top seven plays will underperform EIA’s reference case forecast by 39% from 2014 to 2040 period (Berman 2013). A preview of the shorter life of the shale resource may already be prefigured by an oil and gas production decline in the Bakken shale of the Williston Basin (Southern Canada–South Dakota) that began in January 2015 is probably not reversible (Berman 2017a, b). In 2016 both gas and oil production from the Eagle Ford in Texas also began to drop precipitously (Peters 2016). SRSROCCO (2018) proposes that if the production trends of shale oil (which drives a great deal of shale gas production) follows that of conventional oil, a similar symmetrical slope trend of increasing and decreasing production from the shale resource may be a suitable model. The slope of the trend of the increase of conventional oil production from about 1925 to 1973 is about the inverse of that for its depletion from 1973 to about 2007, when significant shale oil began to hit the market. In other words, there is a time-production volume symmetry. If this proposition is correct, the dramatic increase of production of both gas and oil from the shale resource may be followed by an equally dramatic collapse in their production once the peak is passed. If true, shale oil and gas production may not be far away from its peak in the U.S. But this the critical assumption of symmetry of the growth and decline rates may be completely in error.

Actual future production may depart from EIA estimates even before resource shortages because of long-term risk of front-loading of production and cash flow. Nelder (2012) points out that only an 11-year supply of natural gas from the shale resource presently can be classified as ‘reserve’. Hughs (2015b) shows that five of the shale (basin) plays collectively peaked in August 2012 and by mid-2014 were producing 21% less gas than at peak, which appeared to answer his early question about the durability of the shale resource (Hughs 2013). Whether the decline is dominated by technical rather than economic issues, however, is uncertain. Inman (2014) suggests that banking on multiple decades of abundant natural gas from the shale resource may be wishful thinking. More recently, with more production data and experience of the shale resource available, Polson and Loh (2017), in an MIT study, suggest that the EIA overstates oil shale production forecasts by at least 10%.

Because productivity of shale wells declines rapidly, a very large number of new wells must be drilled just to maintain current production levels. Approximately 130,000 additional shale gas wells will need to be drilled by 2040 to meet gas production projections, on top of the 50,000 wells drilled in these plays through 2013. Of the top five shale gas plays, the Marcellus, Eagle Ford, and Utica are likely to maintain and grow production.

In contrast to this view of the future of shale gas and oil, BP (2016) estimates that US shale gas is expected to grow by around 4% p.a. between 2014 and 2035, by which time it will account for around three-quarters of total US gas production or, almost 20% of global output. In addition, BP estimates that the share of shale gas in global gas production will more than double from 11% in 2014 to about 24% by 2035. In addition to the amount of natural gas available in the resource, the cost of producing the gas and its price determines the extent of the resource where no oil is produced jointly. Some of the sweet spots are still commercially attractive at $60 oil but the economics deteriorate away from the sweet spots, and some areas within an identified shale play have never met economic hurdles, especially in the post-2013 period of low oil prices. Fawcett (2015) for instance, notes that much of the shale gas produced is incidental gas produced during shale oil production, which has been the primary economic driver of the shale hydrocarbon industry since 2008. Thus, continued production of shale oil in the 2014 + low price environment, will have an impact upon gas production and availability and price.

Analysts suggest that if WTI (benchmark West Texas Intermediate oil) stays below $60 a barrel for any extended duration, it could push as many as 30 per cent of the shale industry's B- and CCC-rated borrowers into default, and zero their gas production along with their oil. Increased gas demand, however, could come to the rescue of shale gas and possibly feedback to partly support high cost oil production. The Cove Point, MD LNG import facility has been approved for LNG exports and construction of the liquefaction facility is in progress. The Chenier, Louisiana LNG plant shipped its first load of U.S. gas as LNG in February, 2016. Other LNG liquefaction facilities are under construction at Sabine, LA, Hackberry, LA, Freeport, TX, and Corpus Christi, TX. In addition, low natural gas prices are not sustainable over the long term. Apart from commenting on the economics, we leave this discussion to others.

The shale gas potential outside of the US appears to be lower than is usually reported by gas energy enthusiasts. Argentina, for instance, was reported to have up to 400 years natural gas supply (Costa 2016), but early work suggests that these shales cannot be as effectively fracked as the majority of U.S. shales because they are not sufficiently brittle. But both Mexico and China are confident enough in their natural gas supply to invest in new petrochemical industries (Tullo 2015). In China, a large shale gas resource has also been modeled, but the availability of water may prove to be a critical deficiency in its development, in addition to geological issues that will not be discussed here.

A critical factor in the development of U.S. shale gas, however, is largely absent outside of the U.S. In the U.S. local landowners who also own subsurface mineral rights would benefit immediately from development of the resource beneath their property, and the tax and regulatory structure was generally amenable to developing new technology that would enable the resource. In other countries, the ownership disconnect between surface rights and subsurface rights, along with almost vindictive government tax, leasing, and regulatory issues (See Sect. 10.8), are sufficient to stifle industry even from basins that could be productive. The tax structure for extractive non-renewable natural resource revenue administrations differs not only strongly from country to country, but from most other types of taxes (Calder 2014). The degree to which tax policy is understood by any authority that can establish a tax or fee is directly related to investment risk.

Profit motive and the market are not the only inhibitors of new shale gas fields. Certain critical deficiencies must be overcome to elevate any shale oil and gas development to high levels of production and minimize decline rates. Operators have continued to bring down drilling costs, with longer horizontal laterals and introduce slim-hole drilling, walking rigs and proppants that are most suitable for the shales there. Improved completion technology has increased initial production rates and slowed production decline rates. There also is a need to maintain a well-trained workforce with the required houses, hotels, school and hospitals. This involves the recruitment and training of engineers and field workers. And in conformance with new development anywhere, new infrastructure, pipelines, roads and logistic trains are required as production ramps up.

Regional Natural Gas Shortage California has been switching to natural gas in preference to oil and coal for decades. The State is now very dependent on natural gas as a fossil fuel as a backup to periods of low power delivery from renewables, particularly solar. Natural gas-fired power plants generate about three-fifths of California's electricity in 2015 (EIA 2016). Because California produces only about 1/10th of the natural gas it consumes, most of the gas has to be imported through a relatively small number of pipelines (EIA 2016) and local storage of gas is important. This has reduced the fuel options available for backing up base load and spiker demand (Chap. 1). The natural gas blowout at Porter Ranch in the Los Angeles Basin (NPR 2016) was scheduled to be closed permanently but safety modifications have allowed it to reopen (10.2.8). Maintaining natural gas fired power stations is very important to load shifting, for instance when solar energy that at peak can deliver over 5% of the State generating capacity, goes off line at the end of each day (Sect. 1.3).

Building gas storage is very expensive. For this reason, underground storage of natural gas is used. The three main types of underground storage are in depleted gas and oil fields (Aliso Canyon-Porter Ranch, 10.2.8), in caverns in salt formations (Otto 2017) and in depleted aquifers. Evaluating the geology and risk factors of natural underground storage is becoming more important as more marginal storage opportunities are pressed into service. Whereas the Aliso Canyon blowout was due to corrosion of a well pipe (10.2.8), blowout or leaks from geological faults or other structural weaknesses must be well considered. In the United States, there are over 400 underground storage facilities that hold about 110 Bcm of gas, with depleted oil and gas fields accounting for about 85% of capacity, with the remainder split almost evenly between salt and aquifer (Newman 2017). Local gas storage facilities can accept gas slowly over long periods of time but can release it fast enough to power local power generation. Relatively inexpensive local gas storage allows for distributed power generation.

Unforeseen risks An example of one of the lessor challenges to the emerging dominance of natural gas as the major fuel choice is new technology that could resolve a number of issues at the same time. Probably the main inhibitor of the continued use of fossil fuel is that colossal quantities of CO_2 are produced that are widely regarded as accelerating global warming. Also, the energy density and easy storage and handling of liquid hydrocarbon fuels provides high-quality energy on demand and will favor liquid fuel over energy stored in other ways. R&D on the unlikely objective of recycling CO_2 to manufacture liquid fuels using new catalysts and processes (Service 2015) has yet to succeed. If a commercial process were developed that would compete with fossil fuels (possibly encouraged by government regulation favoring climate change mitigation), every fossil fuel power station could provide the CO_2 industrial feedstock for producing liquid fuel. This would result in a recycled fuel paradigm that would at least arrest the increase in CO_2 effusion into the atmosphere.

10.4 Environmental Risk to Energy Projects and Production

Although most media attention given to climate change focuses on the direct effects to large scale economic effects and human occupancy of coastal regions (NAS 2016), there is also direct impact on energy production in general. The global warming that is currently taking place is generally referred to as climate change, as changes are not expressed as uniform warming. In fact, some areas are cooling. Climate change is expected to alter the balance of temperature, water vapor in the air, available water supplies, sea levels, vegetation, and other environmental factors and this presents measurable risks (Beatty 2016). The current focus on energy and climate primarily concerns climate change factors as a function of energy source. That is, renewable energy sources (Chap. 1) do not emit CO_2 into the atmosphere whereas non-renewable, or combustion fuels do. The interaction and effect of climate on renewable energy sources is more obvious, but climate may also affect both the acquisition of non-renewable energy and its use in power plants.

Feedback effects from changes in climate have the potential to affect different types of energy source (Jonassen 2015). Renewable energy sources are very environmentally responsive. For instance, high winds can damage wind generators while little or no wind will produce little power. Reduced precipitation and stream flows can reduce the level of hydrokinetic energy generation on land while at sea climate change may affect tidal flows. Reduced sunlight caused by clouds, haze, or rain will affect solar electricity output at any location and the latitude of increasingly poleward locations will have lower potential solar energy.

Environmental feedback effects are much less likely to affect non-renewable energy generation. Once fuel has been delivered to power plants, they will function about the same no matter what the weather is—within bounds, and output about the same power per unit of fuel. Although catastrophic events such as earthquakes and tsunami can damage any type of power plant, the greatest danger is with a nuclear power plant from which uncontrolled radioactivity can escape from the reactor containment vessel. A disrupted fossil fuel power plant can be shut down simply by stopping fuel supply, in contrast to a sudden disruption of a nuclear plant, such as happened at Fukushima-Daiichi in eastern Japan, which is still leaking high levels of radioactivity years later. Environmental risk to geothermal is low, consisting mainly of water usage, disposal of mineral-laden produced water, and air emissions. Operational risks, however, can be high because the best geothermal sites are often in the most geologically active areas where faulting, magmatic intrusion, and high geomechanical stress can deform wells.

High sea state and foul weather conditions can have a serious impact upon drilling and production of gas and oil. Normal operations may have to be suspended for weeks at a time, especially if infrastructure such as ships, platforms, pipelines and land installations are damaged or shut down as a damage prevention measure. At power stations, however, all combustible fuel types face about the same level of risk from climate in that severe weather, especially lightning storms and high winds (including tornados and hurricanes), is about the same. Variations in air temperature and humidity will affect the cooling potential of conventional combustion power plants, and heated water that is not used in some way, for instance in heat pump heating or cooling systems, may cause thermal pollution of waterways. Downstream activities are also at hazard from weather. For instance, chemical plants can be disrupted, which can lead to hazmat incidents (PGN 11/17/15). Climate modeling of anticipated changes suggest that some sea areas will have increasing surge and higher waves, which can place increased demands on engineering and costs of working at sea (McGrath and Lynch 2008).

Probably the greatest risk to offshore energy infrastructure of all types are major earthquakes along tectonically active continental margins, such as around the Pacific ‘ring of fire’ including the SE Asian subduction zones abutting the Indian Ocean (Offshore 16-1 2016). These natural earthquakes are related to crustal collision zones between continental plates. The amount of energy in these zones is gigantic and earthquakes have the potential to be much larger than an earthquake induced in injection wells, unless in a rare case an older deep crustal structural line adjustment is triggered. The risk is great enough, as E&P moves out into ultra-deepwater for U.S. Department of Energy sensor system for early detection of earthquake potential to be deployed, at least in the U.S. EEZ.

When investment is being considered in developing different power options, it will be important to consider the effects of climate upon the factors governing the stability and predictability of power generation. Although it may not be said with confidence that any particular storm was caused by climate change, it can also be said that there will be changes including the likelihood of more and stronger storms locally because of the greater amount of energy in the atmosphere generally.

10.5 NGH Environmental Risk

A fundamental question is whether NGH exploration and production operations can meet the environmental and commercial expectations of the public. We think that the answer is definitely "YES", especially in environmentally sensitive regions such as the Arctic where the factors that distinguish the environmental impact of NGH from conventional hydrocarbons may be most important (Max et al. 2013). Although the U.S. has stopped exploration drilling in the Arctic recently, Arctic drilling will begin again (Page 2015). A major oil spill in the Arctic could be devastating whereas a natural gas vent from NGH production alone will have little impact because there should be relatively little free gas in the reservoir at any time and it will almost certainly be at a lower pressure than the seafloor. Although small leaks may take place if the reservoir is ruptured by faults, it will all be dissolved in the deepwater where oceanic NGH deposits will occur (Chap. 3). Because of the low pressure in a producing reservoir that will drive NGH conversion, blowout is physically impossible.

NGH is stable and effectively inert in its reservoir host sediment. This distinguishes it dramatically from conventional resources that can blow out if proper safeguards are not rigorously maintained. If NGH deeper in the GHSZ were to be exposed to overlying seafloor water, such as during drilling, it would only become more stable as its pressure-temperature point would move further into the field of stability and away from the phase boundary (Chap. 2). NGH can thus be regarded as an environmentally secure resource, although in a different manner from the other unconventional gas resources. In order for gas to be produced, the NGH must be artificially stimulated so that it may be converted to its component gas and water or dissolved as the first step of a gas production methodology. NGH will not naturally dissociate to its component gas and water so long as stability conditions are maintained, even in well bores that can be easily refrigerated using seafloor water.

During depressurization, which is the most likely method of conversion (Max and Johnson 2011), pressure in the reservoir is lower than normal formation pressures, which strongly reduces the likelihood of leakage, much less blowout. The reservoir chills during dissociation because it is an endothermic reaction that removes heat from its environment. Thus, in producing NGH deposits, a heat deficit will exist. In the case of leakage, sediment movement, or faulting that could produce pathways for free gas to escape, simply ceasing depressurization and allowing pressure to equilibrate will almost immediately cause free gas within the reservoir to crystallize to NGH within no more than one or two days (Moridis and Reagan 2014).

The thermal deficit will provide a heat sink into which the heat produced by the exothermic hydrate formation reaction can be absorbed. Production models include the maintenance of low gas mass in the reservoir such that NGH-forming potential will consume all the free gas. As such, a built-in mechanism for preventing even small blowouts of free gas can be maintained. There is no need for any provision for expensive intervention through drilling or other methods aimed at sealing a wellbore, such as may be necessary for conventional, high pressure natural gas deposits.

A further risk to the environment is pollution potential. Because NGH concentrations are generally not spatially and directly associated with oil deposits, any leakage will involve only pure water and colorless, chemically unreactive natural gas, although there may be very minor amounts of associated H_2S hydrate. Whereas oil spills are instantly identified as pollution, a deepwater natural gas leak would not have significant environmental impacts, especially as leakages in deep water are unlikely to reach the surface. Conventional coal waste and fly ash (Chapter 10) and oil leaks may constitute a general health risk that can affect humans as well as a wide variety of plants and other animals, including ocean life. Pollution from burning fossil fuel varies considerable depending on the type of fuel and natural gas from all sources is less polluting that burning oil and coal (Chap. 1, Table 1.2). For every 4–8 ton of coal burned, 1 ton of fly ash is produced (Ritter 2016), which has every characteristic of a hazardous substance. In contrast, natural gas derived from NGH has none of the toxic materials commonly found in conventional oil and coal and will have little to none of the CO_2, N_2, and SOx often found in conventional gas. Methane leaks can now be imaged easily (Gålfalk et al. 2015) so that both natural and man-made leaks can be identified and remediated.

Oceanic NGH tends to be relatively pure because it is dominated by biogenic methane produced at temperatures too low to allow higher density hydrocarbon gases or complex hydrocarbons (i.e., oil) to form. The natural gas produced from NGH appears to have about the same purity worldwide. In addition to the basic NGH-forming component, any chemical or dissolved ionic material migrating with the dissolved gas that is not a NGH-former is rejected from the crystallizing NGH into the pore water where it will dissolve and be dispersed.

Biogenic NGH is not normally associated with oil or gas condensates that are derived from thermogenic sources. In addition, NGH usually has little nitrogen, H_2S, CO_2, and other contaminants that are often found in conventional deposits; and almost pure water is produced when NGH dissociates. Even if some converted gas and water were to leak from a NGH deposit, there would be no pollution, in the common sense of the word. Venting or leaks of gas from NGH conversion would have virtually no biological impact (except possibly to stimulate the base of the food chain) or visual effect.

Thus, NGH deposits have an inherent geosafety both during exploration drilling, in which simple circulation of seafloor water will prevent NGH conversion and during production. Not only is NGH the most environmentally friendly of fossil fuels from the point of view of minimizing CO_2 effusion into the atmosphere, but it is also the most environmentally friendly of the resources because of its extremely low potential to cause pollution that might be harmful to animals and plants. NGH may prove to be a preferable gas resource because of its low environmental risk, even in environmentally fragile regions such as the Arctic Ocean (Max et al. 2013).

10.5.1 Tracking of Ocean Environmental Impact

Drilling on land in the United States has been accelerated since 2005 by advanced technologies and energy demand that has resulted in the drilling of about 50,000 new oil and gas (O&G) wells per year through central North America since 2000. Multiple-hole pad drilling to shale horizons has transformed millions of hectares of the Great Plains into quasi-industrial regions. The scale of these developments has transformed large regions in a way that may not be obvious when focused on a single region or environmental parameter. Allred et al. (2015) have empirically quantified the relationship between drilling and production activity for central North America by defining a net primary production unit (NPP), a fundamental measure of a region's ability to provide ecosystem services that can be measured using satellites and historical data.

The concept of NPP can be extended to the seafloor and to any facilities directly connected to seafloor from an environmental point of view, such as pipeline and cable landings. Although a similar measure of the impact of subsea activities at all depths, and at least far enough into the seafloor to include the bulk of the marine ecosystem, the manner in which the NPP has been determined on land offers a useful example of enumerating some measure of an environmental impact unit (EIU).

NPP factors in the obvious alterations to a landscape such as drilling pads, roads, settling ponds, and the like, replacing agricultural land. This is in addition to longer term changes introduced by the damming effects of runoff and channeling runoff water different from the previously existing natural surface. NPP is meant to identify the direct loss of vegetation resulting from oil and gas activity. Although it is not clear that a compensation factor for climate change in their study has been made, the impact on the environment they note has been made over too short a time span for a background factor to significantly alter their conclusions.

Oil and gas development has resulted in a loss of about 4.5 Tg of carbon or 10 Tg of dry biomass in central North America. The 'feeding' ability loss is equivalent to about 120 million bushels of wheat or about 6% of the wheat produced in 2013 within the region. In addition, Allred et al. (2015) note that high-volume hydraulic fracturing uses 8,000–50,000 m^3 of water per well for the initial fracturing. This water was removed from the natural hydrologic cycle in their study area in a 12 year period. They have made no quantitative estimate of impact upon the effect on agriculture but, as over half of their area was already water stressed, it is likely that there was some effect.

Loss of NPP is almost certainly a long-lasting, semi-permanent condition in semi-arid grassland regions because reclamation commonly takes place following cessation of oil and gas activities, and then because reclamation is expensive, land ownership in a region is often complex. As only certain aspects of the degradation are identified for remediation, it may be incomplete. It is likely that the loss of NPP will continue to increase with aggregate drilling. Safeguarding the water-energy-food nexus should reach beyond traditional laws and regulations that often deal with single issues. Ideally, land use policy should take into account the long term and environmentally complex impact of energy-related activities on NPP but regulations tend to not be integrated under the auspices of a single agency, making integrated assessments unlikely in the short term.

10.6 Geohazards

We are primarily concerned with the general case for NGH exploration and production in open ocean areas because that will contain the greatest acreage and in which methane is dominant hydrate-forming gas. In the open oceans seafloor water temperature in deepwater and ultra-deepwater is in a narrow cold range just above freezing. Methane gas hydrate gas hydrate stability zone thickness can be calculated using seafloor temperature and heat flow information (Wood and Jung 2008). Except in a few areas where thermogenic, high density natural gas is a significant part of the gas flux, the open ocean GHSZ (Chap. 2) will be relatively close to the methane-model. In restricted bodies of water that are much shallower than open ocean and which may have very high thermogenic gas flux, such as the Caspian and Black Seas (Hill et al. 2015), the GHSZ may occur in much shallower and warmer water and marine sediments (Max et al. 2006). These areas require careful seismic and drilling assessment to assess the NGH geohazard.

NGH geohazards are primarily conceived of as a geotechnical consideration when conventional drilling is carried out through hydrate-bearing sediment. The focus is at or near the seafloor where inadvertent conversion of dispersed NGH can take place when equipment is placed and operated on the seafloor or where the exothermic reaction of setting cement provides sufficient heat to cause the NGH to dissociate. The dispersed gas thus produced has the potential to weaken the sediment, allowing mass flow that can disrupt well bores and cause infrastructure such as processing equipment and pipelines on the seafloor to be displaced. Industry is now well aware of the near-surface NGH issues and can use seafloor water for refrigeration to keep any NGH within its stability field. This issue is generally considered to be resolved.

Downhole, the same physical constraints on NGH stability pertain. The presence of NGH may be a drilling hazard downhole if dissociation results in gas entering a wellbore or preventing an adequate cement bond for casing. Hazards may exist not only in deepwater NGH but on continental shelves and permafrost regions where NGH deposits may occur (Boswell et al. 2012) outside of geological traps. The likelihood of unwanted NGH conversion is most significant at the base of the hydrate stability zone where hydrate is very close to the phase boundary (Chap. 2) and small increases in temperature can result in dissociation. Maintaining drilling fluids at low temperature should be sufficient to keep hydrate from dissociating, and the presence of gas hydrate in sediment has been observed to enhance wellbore stability. Beneath the hydrate/free gas phase boundary, sediment may contain free gas at slightly elevated pressures (possibly up to 4–5% above normal formation pressure). Thus, the likely position of the base of the GHSZ should be determined prior to drilling, and the likelihood for gas or NGH to be present taken into consideration in the drilling plan for drilling safety. Casing may be set within hydrate-bearing sediment as long as the cement refrigeration counteracts heat generation.

A longer-standing geohazard is high pressure 'shallow' gas zones, which may cause local over pressures during drilling. These can occur within the gas hydrate stability zone (GHSZ), even though there is abundant pore water and these gas-charged sands are often within the GHSZ. This demonstrates that gas can reach high levels within the GHSZ without completely forming NGH and its possible presence is not to be ignored (Gayyum 2015). A standard shallow geohazard analysis is usually carried out using bright spot analysis of seismic data for any potential drill site or seafloor location on which processing facilities are liable to be installed. Drilling safety assessment for NGH deposits will follow almost exactly the same methodology as for conventional hydrocarbon drilling but extra care will be necessary in the vicinity of a NGH concentration and especially at the base of the GHSZ NGH/free gas transition.

Because NGH deposits are not overlain by thick overburden and the reservoirs and the sediments that enclose them may not have been well compacted, reservoir stability will be a concern common to all production. As NGH is converted to its constituent natural gas and water, the reservoir weakens. The geotechnical character of a producing deposit will have to be monitored (Chap. 8) continuously with production fail-safes for rapid shut off of conversion to be in place. In addition to the possibility of faulting and local (but almost certainly inconsequential) gas release to the sea, there is also a possibility that production will induce weaknesses that could result in mass flow that could affect sediments all the way to the seafloor.

10.7 Risks of Non-NGH Energy Sources

Combustible fuels All energy sources (Table 1.1) have a wide variety of associated risks. We highlight a few here but do not carry out a full risk analysis. In the first instance, the conventional and shale gas sources have blowout risk because of their high reservoir pressures. Reservoir pressures can only be controlled in the reservoirs of NGH and coalbed methane. Anderson et al. (2014) from their production testing, consider that NGH destabilization is self-limiting, which should dispel any fear of the potential for uncontrolled destabilization. Other combustible energy resources risks are mainly in the area of pollution. Coal, is especially contaminated with pollutants (Table 1.2), and some oil may also have considerable pollutants.

Fig. 10.3 Generalized bathymetry of lower Lake Powell above the Glen Canyon dam (LP 2015). Cook et al. (2009) discuss the deep incising of the river. Red is shallower and blue, which traces the river course, is deeper. Reservoir profiles and bathymetric data are available from: Reservoir Surveys, US Department of the Interior—Bureau of Reclamation

Hydrokinetic Power is concerned with extracting energy from water moving downhill under the influence of gravity. Moving water caries substantial energy, as anyone who has tried to wade a fast-flowing stream or has been in a strong beach surge knows empirically. The force of moving water was initially harnessed by water wheels in antiquity, with water-driven flour grinding mills and reciprocating saws being common across the Roman Empire. Combining dams that are used to conserve and control river water so that a more constant supply can be provided and the release of water through huge turbine electric generators is the most efficient and almost pollution-free energy resource.

Reservoirs can have huge aerial extents but a large part of any reservoir can be very shallow (Fig. 10.3). When reservoirs are full, and the water floods over shallow water, it can appear that there is much more water available than is actually present. When large expanses of dried reservoir bottom are exposed, well less than half of the available water may remain. The morphology of reservoirs is very important to water management and predicting the energy production potential necessary for planning when more than one energy source is involved (Chap. 1). For instance, Lake Powell, which has been created by the Glen Canyon Dam (Fig. 10.3), has a V-shaped main channel and only a relatively narrow volume above the main channel where there is a full section of water (Randle et al. 2015). Lake Mead, which has been created by Hoover Dam, is more bowl-shaped with a relatively flat bottom, and has a much greater area of full water section (Ferrari 2003). Lakes Powell and Mead are the largest man-made reservoirs in the United States. The effect on river flow has been most noticeable since the beginning of the current drought affecting the western U.S. For instance, the Colorado River, whose flow is impeded by these dams, has only flowed to the sea twice since the 1990s.

Although reduction of water that can be used for human consumption and agriculture is important in its own right, there is an energy risk as lower water head reduces the force through the turbines. For instance, in the current drought conditions of the western U.S., peak hydrokinetic electricity generation has fallen 10% since 2011 for the States of Oregon and Washington (Hoza 2015a), largely because of diminished precipitation. In California, hydroelectric power has averaged about 1/6th of the state's net generation over the last decade; its share varies with annual precipitation. With adequate snowpack, hydrokinetic power can account for more than one-fourth of California's total net generation, but, in 2014 in the fourth year of drought, it supplied less than 1/10th (EIA 2016).

Drought's effect on water supply is one of the principle issues in the western U.S. as well as other places. Natural climate fluctuation is the primary driver of precipitation variability but anthropogenic warming is estimated to have accounted for 8–27% of the observed drought anomaly in 2012–2014 and 5–18% in 2014 (Williams et al. 2014). Although natural variability dominates, anthropogenic warming has substantially increased the overall likelihood of extreme California droughts. The first indications of what climate change may mean to hydrokinetic power are beginning to emerge. For instance, in Tanzania, which is an East African coastal state whose western border is marked by a river—lake system that feeds northward into the Nile River, all of its hydropower plants are being switched off because of a lack of rain and low water levels behind dams. Hydrokinetic energy had produced up to 35% of Tanzania's electricity, but hydrokinetic reservoirs had fallen in autumn, 2015 to less than 20% of capacity. Water for human consumption had to be reserved. Tanzania is energy poor, with only 24% of the population connected with electric supply. All but 105 MW of Tanzania's demand of 870 MW must be imported from Kenya, Uganda and Zambia (BBC 2015). Hydrokinetic energy generation may not be possible if precipitation at the beginning of the normal rainy season in December falls short. Drying may significantly affect hydrokinetic energy generation much more widely in the not-too-distant future.

Ocean Energy consists of harnessing both tidal and wave movement to mechanically generate electricity (Chap. 1). Wave energy is inherently riskier because it depends on not just the movement of water but the presence of waves, and in the ideal case, a certain type of ocean waves whose frequency and amplitude best fits the energy recovery device. Wave energy is generally concerned with extracting energy from an up and down motion. Part or all of these devices are floating, which creates mooring requirements and impede, or could actually be dangerous to, ships and small boats. Waves are generally caused by wind, and long period swells are the result of storm waves that have traveled long distances.

We would regard the risks inherent in wave energy as much greater than those of tidal energy because if there is no wind for a period, there are no waves.

Tidal energy is an ocean analog for hydrokinetic energy because it generally extracts energy from water flowing laterally in streams. Tidal movements are the result of the interaction of gravity attraction between the Earth and Moon, the angular momentum of water induced by the Earth's rotation, the Arctic conveyor of deep plunging cold saline water and its counter currents, and other oceanographic forces. The water flow associated with tides is predictable decades in advance (Chap. 1) and thus there is a very low risk in calculating the power generation profile of any generating installation, whether fixed or moored. A commercial consideration is the selection of sites where the tidal range is sufficiently strong to warrant the cost of installation.

Renewable Energy risk varies considerably with the type of energy (Chap. 1). Wind energy is inherently more difficult to predict than solar, for which fairly dependable calculations can be made. Thus, the risk factor for wind is higher than for solar, but both can vary widely within a certain range depending on weather. Geothermal energy potential varies considerably from place to place. In Iceland, the availability of geothermal energy is very high and extracting energy from the hot volcanic edifice of the island is relatively inexpensive and completely predictable. T Even in places where geothermal energy is not abundant, however, accurate calculations can be made on a long-term basis for its availability. Thus, the risk factor for geothermal energy is very low.

Nuclear Energy is often promoted as a dependable, non-CO_2-emitting source of base load power, which in some regional cases such as the SE United States could limit the demand for natural gas (Bennet 2015). Alabama, Arkansas, Florida, Georgia, Mississippi and Tennessee, currently have 19.3 GW of installed nuclear generating capacity, representing about 20% of the 98 GW of national nuclear generating capacity. Projected new nuclear generation is intended to rise to replace coal-fired generation (NRC 2016), but even if built, it would not come on line for decades. In this region, new demand to be filled by natural gas generation is regarded as non-base load power. There is a very low risk for energy production and predictability but unanticipated events such as took place at the Fukushima Daiichi power station located low along the shoreline in the tectonically active eastern geological margin of Japan.

The most important risk factor for nuclear energy is not the complete loss of power, which occurred, but the melt down of at least one of the reactors, which continues to poison the local land area and which still continues to leak dangerous radioactivity into the Pacific Ocean (Hunziker 2015; WNA 2015) despite tens to hundreds of millions of dollars spent to contain the radioactivity. While extremely rare, this risk is only associated with nuclear power stations (Greenpeace 2017). While nuclear accidents may be so rare as to be characterized statistically as low risk, nuclear disasters can have radiological health impacts that last from a few minutes, in the case of a short venting of radioactive steam, to those that involve core meltdowns whose effects will last many human lifetimes. It only takes one disaster like a Chernobyl or a Fukushima to be equivalent to thousands of complete plant failures or plant destruction of other energy sources.

If any other type of power plant suffers a catastrophic event, conventional physical and chemical cleanup is involved in the vicinity of the plant. In a nuclear accident involving reactor meltdown and uncontrolled leakage, pollution may continue to intensify for many decades and huge areas could be affected. No other type of power plant has the same combination of low likelihood and the extreme seriousness of a major catastrophe. In addition, economically, nuclear energy is a relatively poor risk. 'Conventional' large nuclear power plants cannot be constructed without taxpayer assistance through government loan guarantees and other financial instruments and because no private insurance is available for a catastrophic accident. For instance, the radiation decontamination of Fukushima Daiichi for 2016 alone is set at 522.3 billion yen (Japan Times 2015) and may not fully include the impact upon women and children, who are asymetrically affected (Ulrich 2017). The existence of this nuclear power risk should be given careful consideration anywhere a new station is proposed or an old station relicensed. With increasing low-pollution renewable energy being installed, it may be time for risk analysis to take a longer view of nuclear energy and to properly factor the rare but catastrophic risk.

10.8 Regulations, Leasing, Tax, Matters, and Law

The energy industry works worldwide because energy resources occur on land or in the national sea areas of a large number of countries. Oceanic NGH is in the provenance only of coastal states, including inland seas such as the Mediterranean and Black Seas. Both regulatory and tax issues have severely stalled the development of shale gas and oil, principally in Europe where water quality and environmental issues are relatively more important than in most other countries. Not only can regulatory and issues be complex and very different from country to country, but taxes can extend into the downstream for gas and oil products, rendering investment uncertain because of the predictability of regulations and taxes, for instance in Romania where new tax and energy-related royalties will give companies a better platform for investment (Ilie 2015). On the other hand, the Congress of the United States is bringing forward legislation to dramatically restrict regulation of the energy industry, in concert with the policy of the Trump administration (Showstack 2017). Although regulations have been referred to periodically in this chapter, it is important to understand that regulations are different in each country or political association such as the European Union, and regulations can be established, removed, replaced, or changed.

Legal and tax matters can be show stoppers, especially for small contractors working offshore for the first time, in new environments, or in the waters of a new country. This can be the case even when working with a larger partner or major that has experience in the new business setting. Mexico, for instance, is among a number of countries that operated a government controlled offshore energy company (Pemex) and has recently made the decision to allow more commercial activity in their water. Mexico offered a leasing round in July 2015 in order to turn around a decade long production decline. The offering failed for a number of reasons, with less than half the number of leases offered being awarded. This has caused Mexican authorities to alter their leasing rules (Stevenson 2015). The changes affect fiscal terms and cash guarantees along with clarifications on contract risks, as well as possibly revising the lease block system and rules for when future reservoirs pass off the lease block. In general, Mexico will be introducing far more flexibility with respect to the composition and alteration of consortia.

Wide area prohibition on exploration and production activities of conventional hydrocarbons can have a strong impact upon the energy industry and hydrocarbon availability. Exploration activity has been banned for large areas of the East Coast of the United States and the entire sea area of Florida for a considerable time. For instance, The Blake Ridge in the continental slope off the Carolinas coasts, in which NGH was evaluated as a science project in Leg 164 of the ODP program (which found NGH is muddy sediment unsuitable for production), is part of the entire offshore region that is under the drilling and leasing moratorium.

There is now some pressure to ban exploration and production activities off the entire West Coast of the United States (AP 2015), with full details at CR (2005). Although it would seem counterproductive or at least illogical to pursue a policy of energy independence while banning exploration and production from most of the U.S. offshore, under some governments, environmental concerns can trump the prospect of indigenous energy availability. In the United States under the Trump administration, however, environmental matters do not have the same importance.

The shale resource, however, is receiving particular attention from environmentalists and these concerns are finding their way into regulation. The amount of saline and polluted water that must be disposed of from the fracking of shale and injected into deep wells has received considerable attention. However, water from conventional hydrocarbon operations often also requires disposal. In addition to frack water, about 560 billion gallons of water are produced along with oil and natural gas (Harkness et al. 2015). Nebraska, for instance, is giving new scrutiny to oil and gas regulations for injection wells; regardless of the type of operation that has produced the water (Schulte 2015).

Thus, because of the increasing environmental concerns over frack water, regulatory risk appears to be blowing back into the energy industry as a whole. Similar environmental concerns are threatening a moratorium on shale gas exploration in the Northwest Territories of Australia despite the jobs and wealth that would be created from shale development (APPEA 2015). In the U.S., the Oklahoma Corporation Commission's Oil and Gas Conservation Division (OGCD) regulators took action in September 2015 to curtail injection well activity when it ordered two disposal wells near Cushing, OK to be shut and cut volumes at three more disposal wells after analyzing seismic activity data. The OGCD is implementing a new plan to change the operation of certain disposal wells, but the plan may be altered as more data becomes available. The OGDC has ordered American Energy-Woodford to reduce volumes at its Calyx 35-2 and Clay wells. American Energy was also ordered to shut and plug back its Louis 6-3 well and shut its Wilson 11-15 well.

And it has ordered FHA Investments to reduce volume at its FHA 1-15 well (PGN 2015). The OGDC issued a directive on 16 February 2016 to operators of 245 injection wells to reduce wastewater injection volumes by 40% or about 500,000 Bbl in a 5,281 square mile area of western and central Oklahoma (Heywood 2016). This move came three days after Oklahoma recorded its third largest earthquake. This directive will probably result in more wells shut-in than would have been shut-in on the basis of gas price and should accelerate research into industrial water treatment.

California ordered 33 (of which only 21 were active) permitted waste water injection wells in the vicinity of Los Angeles shut down in October 2015 that were either leaking into aquifers or injecting into aquifers (Torres 2015a). More than 150 other wells are also being studied for their impact on groundwater resources. Outside of the United States, governments have a greater say in energy resource production because property owners do not own the mineral rights and do not stand to benefit directly, as they do in the United States. Regulation and enforcement can also be self-defeating. Poland, for instance, apparently still under the influence of a planned economy system, enacted cumbersome regulations that have made exploration efforts more difficult. Poland mandated that companies must submit a five-year operational plan before breaking ground to set up the drilling unit. Each change to the plan has to be notified, considered, and acted upon, a process that can take regulators months, or even years. Reuters reports that the only companies that are likely to continue drilling are the state-run gas distributor PGNiG and the refining company, PKN Orlen (Reed 2015a).

In the European Union fracking is permitted only by Denmark, Britain, and Poland and operations in UK are even more heavily regulated than in the US. In addition, the Ukraine permits fracking but is a special case because of the intimate involvement of the US and Ukrainian governments. Early drilling on shale in the Ukraine has revealed 'poor' geology, which combined with an upward revised tax system from Kiev, has caused major western companies to pull out without proving commercial resources (Varfolomeyev 2015). France, Germany, Romania and Bulgaria, are among other European countries that have placed moratoriums on fracking. The reserve and economic potential for shale gas and oil in Europe is proving to be much lower than in the United States, apart from widespread environmental and political resistance to fracking. Conoco Phillips and Chevron ceased their shale exploration efforts in Hungary and Poland in 2014 and 2015, respectively, following Total, ExxonMobil, and Marathon Oil over the previous two years. Two French companies, GDF Suez and Total, as well as UK firm Cuadrilla Resources, were planning to drill in the UK (Reed 2015b), but this has now been cancelled because of environmental concerns. Local environmental resistance to fracking, no matter how justified, has the potential to significantly reduce or remove the shale option (Inman 2014). Roseacre Wood near Blackpool would have been one of the UK's first shale gas production locations, but the application to drill was refused in July 2015. The UK possibly has relatively large shale potential but environmental concerns and local political interests may strongly reduce or even stop drilling and potential production. So far, only one British shale location has been fracked and there is no great rush of applications. Proposals to initiate fracking have also been rejected in Denmark and New Zealand (Hill 2015)

Elsewhere in the U.S., environmental concerns may stop fracking both onshore and onshore California and actually stop fracking as a proven secondary recovery technique in conventional hydrocarbons where fracking has been carried out over 203 times over two decades in waters off Long Beach, Seal Beach and Huntington Beach (PennEnergy 6/1/15)

State legislators in both Ohio and Pennsylvania are considering an increase or application of a severance tax, which is a tax levied when resources are 'severed' from the ground. Currently, Ohio has severance taxes of $0.10/Bbl of oil and $.0.025/Mcf, while Pennsylvania has a flat fee 'impact tax' on every well drilled, regardless of production. Pennsylvania is considering joining other States that are considering or have impact taxes (Ohio, West Virginia, North Dakota, Colorado, Texas, Louisiana, and Oklahoma) in establishing or revising severance taxes on gas and NGL (Hoza 2015b).

In addition, at the national level, new regulation can be imposed on industry and implemented through using executive authority. An example is an action of the Environmental Protection Agency, which had proposed regulations aimed at cutting methane emissions from the oil and gas sector by 40–45% over the next decade from 2012 levels but these are being revised by the Trump administration. Methane leakage, which is has also come under scrutiny because of the widespread development of the shale resource is a general production issue that is largely resolvable through regulation and enforcement, and adherence to proper practices.

The regulations on methane were one part of the Obama administration's strategy to curb greenhouse gases and combat climate change. The targets in were in line with a January announcement by the Obama administration that it wanted to reduce oil and gas industry methane emissions by up to 45% from 2012 levels by 2025. The EPA reported that the oil and gas industry contributed 29% of US methane emissions (EPA 2015c-2). In contrast, the American Petroleum Institute has reported that methane emissions from hydraulically fractured gas wells were down 79% since 2005 and total methane emissions for all natural gas systems are down 11% since 2005 as a result of industry innovation and better engineering (Carroll 2015).

Fracking, the enabling technology for the shale play, does not have the same widespread support outside of the United States, and even within there are local areas that are strongly against fracking. Argentina, which is ranked as potentially the second largest shale gas producer in the world, took initial steps to encourage exploration investment of both conventional and non-conventional (shale) hydrocarbons (Costa 2016). Fiscal constraints in Argentina have kept oil companies on the sidelines for many years. With inflation soaring and limited access to foreign funds, the cost of doing business is higher than elsewhere. The next Argentinian government must restore confidence and return the country to the global financial markets to ensure investment continues to flow into the nascent shale industry.

In general, there is an on-going conflict of interest between the energy industries, if not industry in general, and governments wishing to secure greater safety and environmental security. Industry does not like to spend money to be safer or more secure than it judges necessary, and their lobbyists have conveyed this message to U.S. State governments and the Congress loudly, if not always effectively (Burdeau 2015). This conflict of interests between the regulators and regulated often leads to uncertainty that will affect business and investment when, for instance, legal challenges as to the efficacy of specific regulations have to progress through the court system (PGN 10/1/15).

It is common for new regulations to be issued following a significant accident, such as the 2013 Macondo Blowout in the northern Gulf of Mexico (Parish 2015). New regulations were put out for comment in August, 2012 and issued after comments in October 2015 by the U.S. Bureau of Safety and Environmental Enforcement (BSEE) have the potential to reshape regulations in U.S. waters (Brooks 2015). BSEE announced that the federal government would hold oilfield service companies responsible for failure to comply with rules and regulations, with BSEE as the arbiter. The impact of these regulations has yet to be tested and with the end of the Obama administration regulations may be suspended or changed.

It is also common for regulations to be issued following a political initiative in response to a non-energy issue. Regulations regarding methane leaks were in the advanced stages of preparation as part of the Obama Administration's climate agenda but are likely to not now be enacted while most of those already enacted are likely to be overturned under the post-Obama administrations. The proposed rule has been issued for comment (Reed 2016). "The proposed rule on venting, flaring and leaking will help curb waste of our nation's natural gas supplies, reduce harmful methane emissions and provide a fair return on public resources for federal taxpayers, Tribes, and States," Interior Secretary Sally Jewell announced. The Bureau of Land Management estimated it will cost the O&G industry $125 million to $161 million to enact the new rule. Virtually everything in the energy industry is thoroughly regulated, but these regulations cover much higher environmental and safety risk conventional hydrocarbons. The entire permitting process and safety oversight will have to be rewritten for NGH development if its development is to be facilitated. But because less robust regulations will be necessary; safety levels will still be high and conformance almost certainly much less expensive. We suggest that special regulations governing NGH can be written that will allow for a lower level of risk than for any conventional gas resource. This will have the effect of establishing lighter and less expensive to comply with regulations and would have an enabling effect on production from NGH through dramatically lowering the cost of exploration and production.

Law of the Sea The United Nations Convention on the Law of the Sea (UNCLOS) is important because most NGH prospect zones lie within (Chap. 3) the Exclusive Economic Zones or their soon-to-be extensions under which national ownership is conferred internationally (Taft 2006). The United States is the only major consumer that is also a major producer. There is some question of the standing of the United States application for an extended continental shelf initiative (Kimball and Hutchinson 2016) under the UNCLOS in the round of EEZ extensions that are currently being sought because the U.S. is the only major country that has not signed the UNCLOS (UN 1982). On 1 March 2016 Courtney et al. (2016) proposed House Resolution 631 "Calling upon the United States Senate to give its advice and consent to the ratification of the United Nations Convention on the Law of the Sea" in the latest attempt to remedy this anomaly.

10.9 Energy Prices

The sudden drop in oil prices caused by the maintaining of high production rates by OPEC, and particularly by Saudi Arabia, has caused prices to drop to less than one third of their over $100 per barrel price in the spring of 2014. This price drop was unexpected and had not been predicted by any authority in energy economics. We expect that oil price volatility will continue for some time as part of a complex market balancing process. With the slowing of economies, particularly in China where the very high increases in GDP (Chap. 1) that were once obtained have become more moderate, the rate of increase in world oil demand is slowing despite the gratuitous return to buying gas guzzling vehicles again, particularly in the United States.

With increased global oil and gas production, there is no predictable floor to oil prices (Shilling 2015). Shankleman and Warren (2017) identify increased efficiency and increased use of renewables and electric cars will driving a lower oil demand, depending on the rate vehicle numbers increase, the adoption of electric vehicle and other variables. By inference, the higher prices necessary to produce from increasingly remote and expensive resources such as deep- and ultra-deepwater have to be justified by demand. But even deepwater costs are diminishing Cheong, et al. (2017). Should the price of oil diminish to the point where shale oil is no longer commercial, then much of the co-produced (or 'incidental') gas will no longer be available and this should drive the price of gas up, at least in the U.S., counter to the oil price trend. Gas is fast replacing coal and should the price of oil drop and stay low for a period, a complex fluctuating oil market will develop.

Although the drop in energy prices is just another price downturn in a long cyclical history of pricing in the energy industry, this is the one that we are living through now. The one constant is that a historical approach to entirely unprecedented international conditions of supply, demand, and other market drivers cannot be used to predict anything other than gross generalities. As usual, companies with a longer exposure to the industry have been more careful than smaller and younger companies on the risk inherent of taking on debt and making exploration and development commitments. The price downturn appears to have been caused mainly by high production that may or may not have a political root rather than a technical or purely financial one. A low oil price environment increases pressure on all oil and gas projects; however, deepwater developments are calculated over the long term. Currently, industry consensus indicates mergers and acquisitions have increased. Deepwater oil and gas activity may not have been seriously set back yet

(Dupre 2015) because of its longer time scale, and in many cases much of the CAPEX may already be sunk. In addition, there is an expectation of oil price recovery in the mid- to long term. While the economic feasibility of deepwater fields varies, typically long-term oil prices of $80-85/Bbl ensure the viability of the majority of developments. (Adeosun 2015).

In recent years, record deepwater rig demand has resulted in unprecedented deepwater activity in Latin America, particularly offshore Brazil where most of the activity will be in the ultra-deepwater (Kliewer 2015). It is likely that deepwater exploration will slow, but with the very large deepwater developments, activity is likely to remain focused even during the post mid-2014 low price era (Beaubouef 2015). The Middle East and Western Europe that have had historically low levels of deepwater activity are likely to experience considerable growth over the next five years, primarily due to the installation and consequent expansion of major deepwater pipeline infrastructure.

The big unanswered question is how long the OPEC countries can continue to have a decline in revenue. Some countries such as Saudi Arabia can get by for quite a while smaller and more costly producers who depend on profits from energy sales are feeling the pain. Saudi Arabia may require a minimum oil price of $90/Bbl to meet its internal social subsidies that are oil-funded. For now, it is a consumer's market for oil importing nations. There are always winners (i.e., China and other energy importers) and losers (Russia, Venezuela, among others).

In the long run, the price of oil may affect the price of natural gas, especially in the US. This is because the price of gas (April 2016) was generally below the production price of shale gas. A considerable amount of the gas available in the US is 'associated gas', that is, gas produced along with the shale oil and conventional oil. If the low price regime for oil begins to affect production, the amount of gas produced will fall, which will cause indigenous gas prices to rise. But even in the current (first quarter 2018) low price environment oil production, largely from shale, is outstripping predictions.

10.10 Business Cycles

The normal business cycle for the energy industry consists on the upside of increased demand driving up prices followed by increased exploration and production to meet the increased demand. When supply from an overdeveloped production exceeds demand, a downside in the cycle initiates and prices drop. The price drop always knocks out the less economically careful and more deeply leveraged companies. A trough is reached when enough production is lost so that demand again becomes greater than supply, and another upside initiates. The entire cycle is strongly affected by politics as energy is a world business, and nothing is simple. In addition, energy prices may quickly respond to the broader economy, as witnessed in 2008 as the global banking crisis led to the "Great Recession", and a sharp decline in energy demand.

As this Chapter was first written in the Spring of 2016, layoffs continue to roll through the industry as a result of the steepest decline in oil prices since the 1980s, with prices being highly volatile on a generally downward to static trend in late 2016 or later. Under the circumstances, significant government funding for new technology is probably not to be expected until prices go back up and companies are again securely in a profit situation. But low prices have not stopped innovation in the private sector. In addition, many of the investments begun when oil and gas prices were high have had to be abandoned owing to the lower prices in the post mid-2014 energy world. In particular, monetization of the low cost gas available in the U.S. owing to the shale boom by converting it to LNG and selling it in Asia where prices were approaching $20 per Mcf, is essentially over so long as there is an abundance of gas and a buyer's market. Increased supply and transport has increased competition and lowered prices to the point where at least 85 gas projects that were planned and, in some cases, initiated are being terminated (Buurma, et al. 2015).

Because the development time for deepwater projects is long enough to ride out the normal troughs in the cyclical energy industry, these will largely proceed as planned although some may be postponed. Also, as they will be on production for decades after they start production, they will probably be producing increasingly into a rising price market that has taken out more costly production. The main cutback may come from evaluating expenses more closely, while keeping production up to generate cash flow. It is likely that industry was not being as careful as they could have been with project management when prices kept going up. This post mid-2014 downturn will probably last well into 2016 or even beyond and can be viewed as a normal, if deep possibly politically motivated correction to a business cycle downside caused by increased production and reduced demand.

10.11 Exploration Risk

Risk crystallizes for companies with the first drill hole into a prospective structure identified by indirect methods, such as seismic analysis (Chap. 3). No matter how promising a workup of exploration data is, no matter how much gas and oil is estimated, until the first exploratory drill hole into a structure or potential trap is completed, the means to reduce risk remain guesswork—although of a highly applied scientific nature. The first drill hole returns data on the reservoir lithology, presence of hydrocarbons, trap integrity and the nature of the potential source. This initial test may also recover data and samples from which the thermal history of the basin can be analyzed for its hydrocarbon-generating and timing nature. Initial formation testing may be undertaken to estimate eventual flow rates. In the early years of the oil industry, wells such as Spindletop near Beaumont, Texas in 1901 became a self-powered 'gusher', proving the abundant presence of pressurized oil. Wells on land are no longer allowed to blow out like this. Drilling provides evidence that has the potential to remake risk from interpretation to fact. Drilling provides measures which dramatically narrow statistical error of risk analyses.

Given the high cost involved (such as $250 million average cost for a deepwater well, multi-billions for LNG facilities with uncertain markets, etc.), the energy industry continually lives with risk. Even after a field has come on-stream and every factor is performing according to operational and economic modeling, events completely out of the control of the energy company can arise that could either have the benefit of increasing or destroying profits. Throughout the risk analysis process (Sect. 10.1) companies need to quantify risk before making decisions. If they can't run economics on a venture (of any type) they tend not to invest. Energy companies are not venture capitalists. Energy companies engage in expensive prospect workups before drilling not only to better define the prospects but also to define and reduce risk.

The recent drilling by Shell in the Arctic Ocean Alaskan continental shelf about 50 km to the NW of Barrow is a case in point. The Burger structure on which the 2000 + m deep exploration well was sunk is huge, and its upside potential was calculated as being a world-class potential of a major petroleum basin. The industry tends to support a technology that helps them to reduce (or at least understand risk elements). Shooting more seismic data, enhanced seismic processing, and various studies are all done to better grip on risk. Billions of dollars were invested on this risky play because the risk analysis was positive. Drilling showed that at least this structure in the shallow Alaskan waters (43 m) on the continental shelf in the Chukchi Sea was not worth following up in late September 2015 (Torres 2015b). Oil and gas is a risky business.

10.12 New Technology: Innovation and Risk

The offshore energy industry combines innovation along with an aversion to risk, especially where major deep- and ultra-deepwater projects are involved. This is due to the high cost associated with heavy drilling capabilities and massive production facilities required for development of the difficult conventional hydrocarbons. In the field of NGH development, it will be important not just to leverage developing deepwater technologies, but to make giant steps in the way exploration, and especially drilling technology, is applied to the field of NGH commercialization. Simply using existing deepwater technologies that were developed for conventional hydrocarbons will almost certainly be noncommercial. Therefore, new methods and practices must be found and applied if NGH is to be commercialized. This implies that a step function in risk is necessary.

A general theme in late 2015 is that operators are becoming much more risk-averse due to the low oil price and are relying more than ever on proven technology. Small service companies that are trying to bring out new tech are running into increased difficulties to get anyone to consider something that isn't already in use, and possibly by more than one customer. On the other hand, there is always the buzz of new technologies from the bigger service companies such as Schlumberger and Halliburton. So far as we understand, new technology development is focused on addressing existing issues of exploration (seismic imaging, Controlled-source electromagnetic (CSEM) surveying, drilling completions, reservoir management, etc. When oil and gas prices are low, budgets are often cut to meet immediate demands for cost savings, cuts are often too deep.

The trend into low prices is resulting in canceled work, job losses, and holdback of exploration in the U.S., but the squeeze has also generated innovation. The rest of the world that is dominated by national oil companies has underestimated the entrepreneurial impact of the small US operators. While the common wisdom is that with low oil prices US production would drop precipitously, onshore production continued to rise. The key has been better completion technology resulting in more production from fewer wells. An example is of Devon Energy (DVN). When Devon bought Mitchell (the company that initiated the shale revolution), they had 37 rigs working. In 2013 they had 7 rigs working and were able to increase daily production. In August of 2015 they had 1 rig working but their production was still growing. The U.S. independents are able to avoid bureaucracy and have demonstrated repeatedly that they can adapt very quickly as long as they are not over-leveraged financially.

Salazar and Rauniar (2015) have eloquently summarized the main issues surrounding the introduction of new technologies that are particularly applicable to the marine energy industry. Even though almost everyone at least gives nominal support to the concept that innovation is necessary, supporting research and development that can lead to real innovation is not easy to achieve. Fundamentally, conservatism rules, especially when oil and gas prices are low and when money is tight. This is because in a risky world, when money is tight, risk is greater. A wrong turn during an upside in the business cycle may be overcome because of the availability of in-house funding through higher cash flow and easier credit, whereas in a downside of the cycle, availability of funding can more easily have an impact upon survival. Innovations are happening all the time; even though a particular company may have an innovation, there is no guarantee that a better innovation would not become available before your innovation could have consolidated a part of the market. Nonetheless, especially where government directed funding in harsh operating areas such as offshore northern Norway, or where rewards in opening up new areas such as in ultra-deepwater off Brazil, risks where new areas have been pioneered.

Perceived risk is perhaps the greatest obstacle to development and introduction of new technology. Virtually no one wants to be first; an early adapter strategically leaves the steepest part of the learning curve that has the greatest risk to others. There is no formula for overcoming distrust of the new and innovative. A common innovation adoption model does not exist.

Disruptive technology Disruptive technologies can impact and even destroy existing businesses. The telephone, for instance, was a disruptive technology that essentially destroyed the 60-year-old telegraph industry. The personal computer and word processors destroyed typing pools and back office practices that had been perfected over 150 years. The internal combustion engine destroyed the horse drawn vehicle businesses, which stretched back into antiquity—except for those businesses that adapted their product to the new motive power. Long-range passenger airplanes severely damaged the passenger steamship and railway businesses, and the list could go on and on. The only certain thing is change, which usually rides on the coat tails of new technology.

In addition to making any particular technology dramatically more efficient, related businesses could suddenly be enabled. For instance, silicon photovoltaic technology could be replaced by more efficient and less expensive semiconductor systems. Backup battery technology could have the potential to provide spiker or backup power to a part of a region or to an entire region, replacing the need for natural gas fired power stations. In fact, it would appear that a mix of solar and batteries may be the backup and spiker technology of choice instead of natural gas at least locally in California (Martin and Chediak 2018). The ability to foresee the upside of technology disruption as an opportunity to guide investment, is one of the most difficult of all business issues because there is often no direct frame of reference or analog to guide decision making. It is difficult to detail a business model for businesses that does not yet exist.

Future fossil fuel use, which is ripe for new disruptive technology, is being affected by two converging issues in addition to potentially disruptive technologies that are now all around us. The first is that CO_2 emission regulations are increasingly coming into force and as sea levels rise as a response, more regulation can be anticipated. It is no longer a question of whether fossil fuel will be left in the ground; it is a question of how much, and of which of the fossil fuels will survive longest. The second is that markets are always changing. The trends are clear; an increasing amount of natural gas is being burned in electric power stations while displacing oil and more importantly, coal, and vehicles may be weaned off liquid hydrocarbon fuel. Energy companies are adapting to the possibility of a 'post-carbon' economy by increasing investment in renewable energy and in new vehicle technology. And the transport mix is changing—but to what?

10.13 Downstream Issues and Risk Factors

NGH (Chap. 2) is a relatively newly recognized material with poorly known and undemonstrated risks in the context of handling, storage, and transportation. NGH occurs in extremely large quantities in nature (Chap. 2) and is on the cusp of natural gas production (Chap. 11). This section outlines the qualitative risk of exploration and production in general terms of facilitates risk, accident likelihood, leak probability, and potential consequences to population and environmental receptors, in addition of potential impact on gas distribution as a whole. We also assess risk potential of artificially fabricated NGH for gas storage and transport applications.

10.13.1 Natural Gas Hydrate Resource Cycle

The cycle of NGH crystallization and dissociation (Fig. 10.4) has been studied in both nature and in laboratories concerned with producing it as a natural gas resource as well as to accomplish a number of industrial processes using its highly reversible chemical reaction. NGH offers a number of opportunities for production of natural gas that may considerably lower production and environmental risk.

The NGH resource cycle consists two parts. The first part is migration of concentrated dissolved natural gas or free gas upward into the gas hydrate stability zone (GHSZ) where NGH spontaneously forms under the right conditions of pressure, temperature dissolved concentration (Max and Johnson 2014) (Chap. 2). Formation of NGH concentrations may take place over a geologically short time of as little as thousands to hundreds of thousands of years, which allows the heat produced during the crystallization process to be dissipated naturally by pore water movements and thermal conduction.

The second part of the resource cycle is a forced dissociation that can be accomplished by one of 5 means (Max and Johnson 2011) for the production of natural gas from NGH. A technical production test by the Japanese company JOGMEC in March 2013 proved the technical feasibility and longer production tests are scheduled off Japan and India in the summer of 2017. This must be a much more rapid change of state process in order to produce the natural gas at an acceptable rate (Max and Johnson 2016). Models for the forced dissociation of NGH in gas production scenarios indicate that a thermodynamic equilibrium in which heat to compensate for the thermal demand provided by dissociation can be naturally established under certain conditions that will allow for continuous, long-term production of natural gas.

Derisking of other natural gas that contains too contamination or non-combustible gases (the price of natural gas is determined by its Btu content) can be achieved by mixing it with NGH-derived natural gas, which is purer. This means

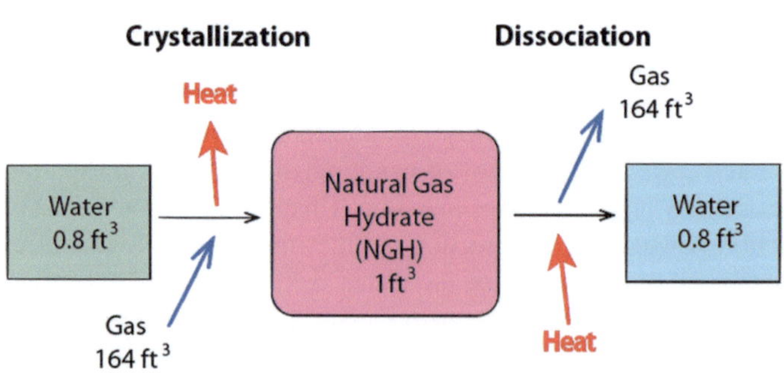

Fig. 10.4 Gas hydrate cycle. Under slow growth natural conditions, rejection may approach 100%, resulting in very pure aggregates of intergrown NGH crystals

that some natural gas derived from conventional resources with higher contamination to be commercially viable. This asymmetrically raises the value of the NGH-derived gas above that of the value of it alone.

10.13.2 Synthetic Implications of NGH Used as a Storage and Transport Media

In addition to the production characteristics of NGH, it has been considered as a storage and transport media, which would make it a competitor of compressed natural gas (CNG) and Liquefied Natural Gas (LNG). Because of its unique properties as a concentrated natural gas media, NGH offers potential for synthetic production of NGH under industrial conditions for any natural gas that may be inherently safer and lower overall risks.

NGH can also be easily synthesized, and has been proposed as a material that can be used as a chemical engineering means of separating water and gas (Max and Pellenbarg 1999) and as a transport media (Gudmundsson 1996; Kanda 2006; Nakai 2012). According to the Mitsui company of Japan (MES 2017), the use of artificial natural gas hydrate (NGH) for natural gas storage and transport has the advantage of being commercial economical, with CAPEX to be reduced 20–25% comparing to LNG system as total transportation chain cost, which makes it easy to handle in moderate pressure and temperature conditions of about -20 °C. The Mitsui company considers that the most important aspects of artificial NGH use is that there is no risk of explosion and no hazardous substance discharge when the NGH is converted to its constituent water and natural gas.

The production and environmental risk of NGH production are extremely low. Both the amount of gas and the pressure within the reservoir can be controlled, in contrast to conventional deepwater gas, which has very high reservoir pressures. Risks associated with NGH production are much lower than for conventional gas. Whereas conventional deepwater gas may reach the seafloor wellhead at temperatures as high as 300 °C and pressures up of around 9000 psi (e.g., a reservoir in 3+ km water depth and 3 km below the seafloor), gas derived from dissociated NGH will be below 40 °C and, with a hydraulic link to the seafloor, essentially at the pressure of the seafloor or only a few atmospheres pressure greater. In addition, NGH is usually not associated with liquid or solid hydrocarbons and has an extremely low level of contaminants. Because it is most likely to be produced by a process of depressurization, it cannot blow out and because the presence of gas depends on artificially dropping the pressure by pumping out water and gas, if a breach of the reservoir occurs, gas production can be stopped within seconds by stopping the conversion process, at which point free gas will form NGH and stabilize the reservoir.

In situations where NGH may be used as a storage or transport media (Max and Pellenbarg 1998), a rupture in the containment vessel will lead to a thermal demand produced by dissociation that will be much higher than that found in nature. Sudden exposure to pressure-temperature conditions that are dramatically out of its stability field will drive dissociation to a maximum not found in nature. Suddenly exposing NGH to conditions well outside of its field of stability will not result in run-away or explosive conversion. This is a result of two factors.

First, dissociation is governed by diffusion mechanisms only takes place on the boundary between NGH and surrounding gaseous (e.g., atmosphere) or water media. It is proposed to mechanically compress NGH fabricated for storage or transport into pellets about 1 cm diameter, which will very substantially reduce the surface area available. This results in a high mass to surface area ratio, which means dissociation of the mass will proceed relatively slowly.

Second, when dissociation operates at a high rate, the exothermic reaction can chill the dissociating surface to the point where water ice will form. The rim of ice on the edge of the NGH mass prevents and retards dissociation and can dramatically reduce the dissociation rate. This is a well-known buffering phenomenon termed 'self-preservation' that has been observed to increase dissociation time of a particular mass of NGH from what could be minutes to hours to days or even weeks under normal conditions of diffusion (Stern et al. 2003).

Reduction of surface area and the self-preservation phenomenon introduces a basic safety factor. If a storage or shipping container would be breached, for instance when a NGH container on a cargo vessel or railway car would be breached by an accident, the NGH exposed to normal atmospheric conditions would only vent gas very slowly. NGH can be lighted and it burns with a slow steady flame using methane that is produced slowly even in the presence of an open flame. Similar evidence for NGH persistence outside it stability field comes from simple observations that NGH that has been dredged from the seafloor and cores from NGH saturated sediments brought to the deck of the drill ship over a period of hours in an aqueous environment (which has the potential to remove heat faster than in air) are still slowly dissociating NGH hours after being brought into conditions of extreme instability.

In contrast to CNG, in which venting would be rapid enough to create electrostatic ignition, a first pulse of gas from dissociation would lower the temperature of the NGH within the container, which is in itself a fire-retardant, and the evolution of more gas would slow as the surface area of NGH not coated with ice would decrease significantly. In addition, the produced gas would be saturated with water vapor, which would also act to retard ignition and continuous combustion. Because gas would be vented more slowly, dispersal or controlled combustion could be used to reduce risk to immediate property and allow for more rapid repair of infrastructure.

LNG is the predominant form of natural gas transfer media owing to its high energy density and relative stability. LNG was first used to transport from the southern US to Europe but in the latter part of the 20th century new LNG production terminals were built in the Middle East and more recently in Australia for delivery to Asia, principally Japan, China, South Korea, Taiwan and SE Asian countries. More recently, LNG is again being exported from the southern United States. A world market in LNG is developing (Max and Johnson 2016). The most recent development in LNG is the construction of floating LNG production vessels, which are necessary to bring 'stranded' gas (location too far or remote to use pipelines) to market.

LNG also produces combustible gas from liquid at a slow determined almost entirely by the ambient temperature and the rate at which heat can be brought to a breach in a container where gas may be venting. But the gas produced is almost pure methane and highly combustible. Part of our risk analysis will establish the relative risk of LNG and NGH.

The impact on CAPEX of new fiberglass CNG containers, which should dramatically reduce the cost of CNG transport, has yet to be fully considered by industry.

NGH has been created in a number of laboratories for a number of purposes from basic research into physical properties to potential industrial applications, usually by bubbling natural gas through water. The main NGH transport to date has been between laboratories or in pressurized drill core from a ship to a laboratory using pressure vessels that meet the containment standards for containment of gas that might be produced if the cooling systems used for transport break down. The volume of each vessel is relatively small and inappropriate for commercial use in a natural gas transport scenario.

Current use of NGH for storage and transport are experimental, although pilot industrial apparatus, including large vessels, have been manufactured and brought into test. Most of the data is proprietary, with only summary conclusions available (MES 2017). NGH is not presently be used as a storage or transport media and the only port at which NGH has been produced in pilot industrial quantities is at the Chungoku Electric Power Company, Inc. located on the western coast of Japan, at which a NGH pellet production plant with a capacity of 5 tons per day has been constructed near the Yanai LNG-based power station (Nakai 2012). A NGJ bulk carrier has been designed ant patented by Mitsui Corporation of Japan.

Large areas which are do not have widespread pipeline infrastructure, in particular the continent of Africa, coastal maritime transport is a viable alternative to the cost of building and maintaining major pipeline systems.

Most comparative valuations of natural gas transport alternatives focus on the gas alone. But NGH transport can also deliver near-potable water. Rehder, et al. (2012), for instance, have carried out a basic economic evaluation and concluded that there was no inherent economic benefit from using NGH. The water required for fabricating NGH can be derived from seawater as part of a desalination process. Because the water in NGH constitutes 80% of the volume and a greater percentage of weight, for every m^3 of NGH 0.8 m^3 of water also arrives at the delivery location. In many places where the gas is to be delivered or near at another location on the return voyage where a shipload of fresh water would have considerable value, for instance China, India, and around the coast of Africa (PA 2017), the water will raise the overall value of the NGH transport process asymmetrically in contrast to CNG and LNG.

10.14 Safety Management

Compressed combustible gases constitute potential fire and explosion hazard and as such, regulations are instituted to provide for safety and to minimize risk. Three general areas of energy and industrial activity have been increasing in difficulty and risk since the beginning of the industrial revolution in the late 18th century. First, is acquisition of energy, second is transport, and the third is use, either as an industrial feedstock or directly for energy. Each has a different set of issues for which safety and good management are required.

(1) The acquisition of energy, on which all modern societies are built, has migrated from relatively easy of access, near-surface coal and petroleum deposits that were relatively close to markets and industrial activity to increasing remote ocean regions having difficult operating conditions.
(2) Transport requirements are increasing both in difficulty, which introduces systematic issues and in length, which increases statistical risk. As exploration and production of hydrocarbons extends further into deepwater and into increasingly remote parts of the earth, safety and environmental concerns become riskier.
(3) Processing and industrial use of fossil fuels, as well as end-user applications such as heating and cooking.

Federal regulation of industrial practices began in 1887 and concerned explosives at the inception of the Interstate Commerce Commission (ICC) (Table 10.2) that was created to improve safety across industry and the country. Regulations promulgated superseded State regulations, which varied greatly, where they existed at all. The ICC was dissolved and broader regulatory authority was transferred to the Department of Transportation (DOT) in 1966.

Because the nature of regulations is now so broad and the degree to which regulations are necessary or advisable in certain instances are the subject of intense discussion and government review , we do not discuss them in detail here.

10.15 Risk-Cost-Benefit Analysis

In general, evaluation of risk is always difficult and imprecise because this involves quantification of future events in which there is considerable unpredictability about the interaction of known systems and the high probability for misplaced confidence and surprises. Nonetheless, Fischhoff (2015), incorporating recognized practices (CSA 2015), suggests that methodology includes procedures for incorporating

[1]Based on a presentation made by Brigham A. McCowan, Alliance for Innovation and Infrastructure on 9 Jan 2017 at the 96th Annual Meeting of the Transportation Research Board.

Table 10.2 Milestones on the road to Hazmat Regulations in the United States[11]

Date	Action
1887	Establishment of Interstate Commerce Commission (ICC)
1905	Prepare regulations for safe transport of Explosives
1908	Explosives and Combustibles Act
1911	Bureau of Explosives
1921	Bureau of Explosives
1960	Extends ICC's ability to use outside organizations to carry out work
1966	ICC dissolved 1 January 1966
1966	Establishment of Department of Transportation (DOT)
1966	Granted authority to regulate transport of hazardous materials
1966	Hazardous Materials Regulations Board created
1975	Hazardous Materials Transportation Act and Subsequent revisions
2004	Pipeline and Hazardous Materials and Safety Administration (PMSA)
2004	Public Law 108-426—Norman Y. Mineta Research and Special Programs Improvement Act

All but the initiation of explosives transport regulations by train by the American Railway Association were enacted directly by the Congress of the United States

expert advice, observations, and assessment of residual uncertainty in a hierarchy. Analytical logic of risk involves identifying system components and their bilateral and group interaction and then attempting to enumerate levels of risk. Examples of probabilistic risk analysis are found in: "Risk Analysis, Reliability Engineering and System Safety, IEEE Transactions on Reliability", and other journals. Modeling risk assessment can be carried out in a series of logical stages (Fig. 10.5)

The general structure of risk analysis, assessment, and management can be applied to any issue, each of which will have its own set of components and different levels of confidence. Experience of evaluators is important because it introduces the concept of a time factor to projections that is difficult to understand for anyone who has not had the experience of making early decisions that had to be revised as events unfolded. As Napoleon is attributed to have said, "No plan survives the first contact with the enemy". Although an original plan may have to be altered, sometimes dramatically, a formal risk assessment provides a plan that can be used to guide reassessment and incorporation of new technology and practices.

10.16 The Case for NGH

(1) **The concept of risk** in the energy sector is very broad, ranging from unwanted physical outcomes, such as blowout, leakage, pollution, and the like to economic failure. Although this chapter mainly discusses prospect E&P risk issues,

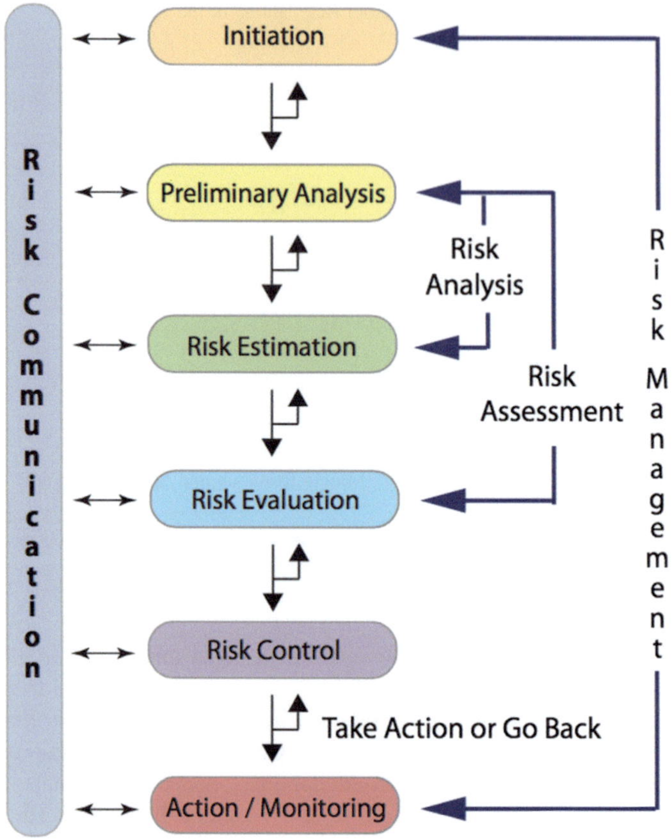

Fig. 10.5 Risk analysis process, after Fischhoff (2015)

there are larger issues that are more difficult to evaluate, especially in the long term. The principal factors lying behind the various risk factors of particular fuels and industrial practices and cost structures have a common denominator. That is, how efficiently can a particular traditional combustible fuel be converted to electricity as a factor of cost, environmental and societal impact, and other input and output parameters.

(2) **Size of the fuel resource**. All hydrocarbon fuels resources are very large. Coal, oil, and natural gas resources are now recognized as being very large and most types of their sub-resources are accessible and competitive economically except for NGH, for which long term production has not yet begun. We argue that NGH, which is potentially the largest natural gas resource with the lowest environmental risk, may be the resource of choice, especially if its impact upon

[1] Based on a presentation made by Brigham A. McCowan, Alliance for Innovation and Infrastructure on 9 Jan 2017 at the 96th Annual Meeting of the Transportation Research Board.

the environment and the thermal character of the atmosphere becomes a dominant factor.

(3) **The hydrocarbon-renewable energy mix**. The dramatic decrease in cost of renewable energy and a consequent upswing in its installed base, is definitely the wave of the future. The principle questions are how fast can renewables be merged into the energy sector and what impact this will have on the energy business. Increased renewable energy use will inevitably reduce pollution of all types, and also reduce the growing warming trend in the earth-atmosphere system, but factors such as energy distribution, different types of environmental impact, and changes to business models remain to be more fully evaluated. We argue (Chap. 11) that natural gas is the best fuel to back up very widespread renewable energy production, and that NGH gas has potentially the lowest environmental risk and greatest flexibility of any hydrocarbon fuel.

(4) **Demand**. What hydrocarbon fuel will emerge as the dominant fuel, and which resource of that fuel will emerge as dominant. It would appear that electricity use will dramatically increase, particularly in mobile platforms such as cars and trucks. Relatively small improvements in battery performance will result in vehicle ranges and recharging (fueling) that compete directly with combustion-driven motor vehicles. This will dramatically lower the demand for liquid petroleum fuels in the same way that the change from oil-fired to natural gas fired central heating in the United States has almost completely obliterated the home oil business. Natural gas provision and demand are rising much faster than liquid hydrocarbon use and coal demand is falling fast as market forces allow old coal fired power plants to close and to be replaced by natural gas fired plants (IEA 2015; 2013). We argue that natural gas demand will increase dramatically, in contrast to other fuels, and that the NGH resource will best match the requirements of a more environmentally sensitive future.

References

Adeosun, M. (2015). Falling oil prices—how will they impact the deepwater market? *Offshore from Douglass-Westwood*. http://www.offshore-mag.com/articles/print/volume-75/issue-5/international-field-development/falling-oil-prices-how-will-they-impact-the-deepwater-market.html. Accessed June 14, 2015.

Allred, B. W., Smith, W. K., Twidwell, D., Haggerty, J. H., Running, S. W., Naugle, D. E., & Fuhlendort, S. D. (2015). Ecosystem services lost to oil and gas in North America. *Science, 348*, 401–402. www.sciencemag.org/content/348/6233/401/suppl/DC1.

AN. (2014). Horrific methane Eruptions in the East Siberian Sea. *Arctic News*. http://arctic-news.blogspot.com/2014/08/horrific-methane-eruptions-in-east-siberian-sea.html. Accessed August 18, 2015.

AP. (2015). *Senators want drilling ban on outer continental shelf*. Associated Press October 8, 2015 reported in: PennEnergy 10/8/15. http://www.pennenergy.com/articles/pennenergy/2015/10/senators-want-drilling-ban-on-outer-continental-shelf.html. Accessed October 9, 2015.

Anderson, B., Boswell, R., Collett, T. S., Farrell, H., Ohtsuki, S. White, M., & Zyrianova, M. (2014). Review of the findings of the Iġnick Sikumi CO_2–CH_4 gas hydrate exchange field trial.

In *Proceedings of the 8th International Conference on Gas Hydrates (ICGH8-2014)*, July 28–August 1, Beijing, China (17 pp.).

AP 12/29/15. (2015). *Gas company finds underground location of leaking LA well*. The Associated Press. http://www.pennenergy.com/articles/pennenergy/2015/12/gas-company-finds-underground-location-of-leaking-la-well.html?cmpid=EnlDailyPetroDecember302015&eid=288179069&bid=1265179. Accessed December 30, 2015.

APPEA. (2015). Shale gas moratorium a risk to NT jobs, development. http://www.appea.com.au/wp-content/uploads/2015/09/150914-Gas-moratorium-a-risk-to-NT-jobs-development.pdf. Accessed September 15, 2015.

Azar, A. (2017). *Reserve base lending and the outlook for shale oil and gas finance* (33 pp.). Columbia/SIPA Center on Global Energy Policy. Columbia University in the City of New York.

Barnard, A., Sager, W. W., Snow, J. E., & Max, M. D. (2015). Subsea gas emissions from the barbados accretionary complex. *Marine and Petroleum Geology 64*, 31–42. https://doi.org/10.1016/j.marpetgeo.2015.02.008.

BBC. (2015). Tanzania closing hydropower plants. *BBC News* (October). http://www.bbc.com/news/world-africa-34491984. Accessed October 12, 2015.

Beatty, A. (Rapporteur). (2016). *Characterizing risk in climate change assessments: Proceedings of a workshop* (101 pp.). National Academies Press. ISBN 978-0-309-44551-1. https://doi.org/10.17226/23569.

Beaubouef, B. (2015). Deepwater Gulf remains active amidst low oil. *Offshore*. http://www.offshore-mag.com/articles/print/volume-75/issue-5/international-field-development/deepwater-gulf-remains-active-amidst-low-oil-prices.html. Accessed June 21, 2015.

Beaudoin, Y. C., Waite, W., Boswell, R., & Dallimore, S. R. (Eds.) (2014). *Frozen heat: A UNEP global outlook on methane gas hydrates* (Vol. 1, 80 pp.). United Nations Environment Programme, GRID-Arendal. © United Nations Environment Programme. ISBN: 978-92-807-3429-4. Job No: DEW/1866/NO.

Beckman, A., et al. (2015). Considerations for accepting fracking wastewater at water resource recovery facilities. *Water environment Federation*. http://wef.org/uploadedFiles/Access_Water_Knowledge/WastewaterTreatment/FrackingFactsheet%20Final(1).pdf. Accessed April 26, 2016.

Bennet, M. (2015). A Nat gas killer? New Nukes in the Southeast. *Oilpro*. http://oilpro.com/post/13590/nat-gas-killer-new-nukessoutheast?utm_source=DailyNewsletter&utm_medium=email&utm_campaign=newsletter&utm_term=2015-0527&utm_content=Article_10_txt. Accessed August 21, 2015.

Berman, A. E. (2013). Let's be honest about shale gas. *Houston Geological Society Bulletin*, September 2013 (Hardcopy handout), *33*, 35.

Berman, A. E. (2017a). Why investors should beware of the Bakken. *Zero Hedge*, 7 pp. http://oilprice.com/Energy/Crude-Oil/Why-Investors-Should-Beware-Of-The-Bakken.html. Accessed March 2, 2017.

Berman, A. E. (2017b). Shale gas not a revolution. *OilPro*. http://oilpro.com/post/31864/shale-gas-not-revolution?utm_source=DailyNewsletter&utm_medium=email&utm_campaign=newsletter&utm_term=2017-07-06&utm_content=Feature_1_txt. Accessed July 6, 2017.

Berman, A. E., & Pittinger, L. F. (2011). U.S. shale gas: Less abundance, higher cost. *The Oil Drum*. http://www.theoildrum.com/node/8212. Accessed January 23, 2016.

Biastoch, A., Treude, T., Rüpke, L. H., Riebesell, U., Roth, C., Burwicz, E. B., Park, W., Latif, M., Böning, C. W., Madec, G., & Wallmann, K. (2011). Rising Arctic Ocean temperatures cause gas hydrate destabilization and ocean acidification. *Geophysical Research Letters, 38*(8), 9 pp. L08602, https://doi.org/10.1029/2011gl047222.

Bock, M., Schmitt, J., Möller, L., Spahni, R., Blunier, T., & Fischer, H. (2010). Hydrogen isotopes preclude marine hydrate CH_4 emissions at the onset of Dansgaard-Oeschger events. *Science, 328,* 1686–1689.

Borenstein, S., Kissel, K. P., & Murphy, S. (2016). More quakes rattle Oklahoma but state avoids tough measures. *PennEnergy* 1/19/16 (Associated Press), 5 pp. http://www.pennenergy.com/

articles/pennenergy/2016/01/oil-gas-news-more-quakes-rattle-oklahoma-but-state-avoids-tough-measures.html?cmpid=EnlDailyPetroJanuary192016&eid=288179069&bid=1282667. Accessed January 22, 2016.
Boswell, R., & Collett, T. S. (2011). Current perspectives on gas hydrate resources. *Energy & Environmental Science, 4,* 1206–1215. https://doi.org/10.1039/c0ee00203h.
Boswell, R., Collett, T., Dallimore, S., & Frye, M. (2012). Geohazards associated with naturally-occurring gas hydrate. *Fire in the Ice*, *12*(1), 11–15.
BP. (2016). The shale revolution continues. *BP Global*. http://www.bp.com/en/global/corporate/energy-economics/energy-outlook-2035/shale-projections.html. Accessed February 17, 2016.
Brooks, A. (2015). New energy world means offshore company regulation. *Oilpro*. http://oilpro.com/post/19834/new-energy-world-means-offshore-co…mpaign = newsletter&utm_term=2015-11-04&utm_content=Feature_1_txt. Accessed November 4, 2015.
Burdeau, C. (2015). Republicans oppose new safety rules on offshore oil drilling. *PennEnergy*. http://www.pennenergy.com/articles/pennenergy/2015/09/republicans-oppose-new-safety-rules-on-offshore-oil-drilling.html?cmpid=EnlDailyPetroSeptember172015&eid=288179069&bid=1179270. Accessed May 16, 2015.
Bush, T. (2012). Refracs yield 39% production increase, with a breakeven natural gas price around $2.30/Mcf. *Oilpro*. Reprinted from http://www.roselandoilandgas.com/web/haynesville-shale-refracturing-shows-positive-returns/. Accessed October 12, 2015.
Buurma, C., Stapczynski, S., & Malik, N. S. (2015), 85 gas projects dying on the vine as LNG's promise falls short. Bloomberg. http://pubads.g.doubleclick.net/gampad/jump?iu=/5765/worldoil.em&sz=800x600&c=1444657097226. Accessed October 12, 2015.
Calder, J. (2014). *Administering fiscal regimes for extractive industries: A handbook* (105 pp.). Washington, DC: International Monetary Fund.
Carroll, C. (2015). API reports that methane emissions fall under existing regulations. *PennEnergy*. http://www.pennenergy.com/articles/pennenergy/2015/07/unconventional-oil-gas-api-reports-that-methane-emissions-fall-under-existing-regulations.html?cmpid=EnlDailyPetroJuly272015&eid=288183525&bid=1134570. Accessed August 22, 2015.
Chang, I., & Lammi, H. (2016). The great water grab. How the coal industry is deepening the global water crisis. *Greenpeace International*, 60 pp.
Cheong, S., Cho, S., & Murtaugh, D. (2017). Now there's another source of oil that's starting to get cheap. *Bloomberg*. https://www.bloomberg.com/news/articles/2017-05-30/trouble-brewing-for-opec-as-once-costly-deep-sea-oil-turns-cheap. Accessed May 31, 2017.
Chopra, S., Sharma, R. K., Keay, J., & Marfurt, K. J. (2012). Shale gas reservoir characterization workflows. *Society of Exploration Geophysicists Annual Meeting*. https://doi.org/10.1190/segam2012-1344.1.
Collins, G. (2015). Frack 2.0: Refracking could enable a second wave of production growth in the Bakken and Eagle Ford. *BakerHostetler North America Shale Blog*. http://www.northamericashaleblog.com/2015/05/21/frack-2-0-refracing-could-enable-a-second-wave-of-production-growth-in-the-bakken-and-eagle-ford/. Accessed June 6, 2015.
Conley, S., Franco, G. Faloona, I., Blake, D.R., Peischl, J., & Ryerson, T. B. (2016). Methane emissions from the 2015 Aliso Canyon blowout in Los Angeles, CA. *Science*, first release, 6 pp. https://doi.org/10.1126/science.aaf2348.
Cook, K. L., Whipple, K. X., Heimsath, A. M., & Hanks, T. C. (2009). Rapid incision of the Colorado River in Glen Canyon—Insights from channel profiles, local incision rates, and modeling of lithologic controls. *Earth Surface Processes and Landforms, 34*(7), 994–1010, 17 pp. https://doi.org/10.1002/esp.1790.
Courtney, Y., Davis, B., McDermott, K., Moulton, L., Garamendi, Johnson, Castro, & Smith. (2016). *4th Congress 2d Session H.Res. 631*, 9 pp.
Costa. (2016). Argentina holds gas reserves for the next 400 years. *Oilpro*. http://oilpro.com/post/14200/argentina-holds-gas-reserves-next-400-years?utm_source=DailyNewsletter&utm_medium=email&utm_campaign=newsletter&utm_term=2015-06-09&utm_content=Article_5_txt. Accessed April, 26 2016.

CR. (2005). *Congressional record for 15 May 1998* (Vol. 151, Part 8). https://books.google.com/books?id=IEoMEW8Xy5wC&pg=PA9265&lpg=PA9265&dq=Senators+want+drilling+ban+on+outer+continental+shelf&source=bl&ots=og08zXccFH&sig=AsiLYnBmwNTH29hf8hY7gh2qHmI&hl=en&sa=X&ved=0CEYQ6AEwCTgoahUKEwio5dDn7crIAhXG8x4KHToqCvE#v=onepage&q=Senators%20want%20drilling%20ban%20on%20outer%20continental%20shelf&f=false. Accessed April 26, 2016.

CSA. (2002). CAN/CSA-Q850-97 (Raffirmd2002) Risk management = guideline for decision-makers: A national standard of Canada. *The Standard Council of Canada*, 62 pp.

Daly, M. (2016). After Calif. leak, feds seek to curtail natural gas blowouts. *PennEnergy*. http://www.pennenergy.com/articles/pennenergy/2016/10/after-calif-leak-feds-seek-to-curtail-natural-gas-blowouts.html?cmpid=EnlDailyPetroOctober192016&eid=288179069&bid=1561635. Accessed October 20, 2016.

Deyette, J., Clemmer, S. Cleetus, R., Sattler, S. Bailie, A., & Rising, M. (2015). The natural gas gamble: A risky bet on America's future. *Report, Union of Concerned Scientists*, 56 pp. http://www.ucsusa.org/clean-energy/coal-and-other-fossil-fuels/natural-gas-gamble-risky-bet-on-clean-energy-future#.VfnRS7Sm3cY. Accessed August 18, 2015.

Dillon, W. P., Nealon, J. W., Taylor, M. H, Lee, M. W., Drury, R. M., & Anton, C. H. (2001). Seafloor collapse and methane venting associated with gas hydrate on the Blake Ridge—Causes and implications to seafloor stability and methane release. In C. K. Paull & W. P. Dillon (Eds.), *Natural gas hydrates occurrence, distribution, and detection. American Geophysical Union Geophysical Monograph, 124*, 211–233.

DOE-MHAC. (2015). Methane Hydrate Advisory Committee (MHAC) meeting. *Meeting summary*, May 7, 2015. http://energy.gov/sites/prod/files/2015/06/f22/MHAC%205-7-15%20Meeting%20Minutes%20Final%205-29-15.pdf. Accessed May 18, 2015.

Dunne, E. J. (Rapporteur). (2017). Flowback and produced waters: Opportunities and challenges for innovation. In *Proceedings of a Workshop* (100 pp.). National Academies Press. ISBN 978-0-309-45262-5. https://doi.org/10.17226/24620.

Dupre, R. (2015). Online exclusive: Oil and gas M&A activity on the uptick in 2015. *Offshore magazine* 5/21/15. http://www.offshore-mag.com/articles/2015/05/online-exclusive-oil-and-gas-m-a-activity-on-the-uptick-in-2015.htm. Accessed June 6, 2015.

Durham, W. B., Kirby, S.H., & Stern, L. A. (2003). The strength and rheology of methane clathrate hydrate. *Journal Geophysical Research, 108*(B4), 2182, https://doi.org/10.1029/2002jb001872.

EIA. 2011. *Emissions of Greenhouse Gases in the United States 2009* (86 pp.). U.S. Department of Energy.

EIA. 2016. *California, state profile and energy estimates*. U.S. Energy Information Administration. https://www.eia.gov/state/analysis.cfm?sid=CA. Accessed February 1, 2016.

EIA-DPR. (2016). *Drilling productivity report for key tight oil and shale gas regions*. U.S. Energy Information Administration. Year-over-year summary January 2016. Accessed January 23, 2016.

EPA. (2015a). Natural gas extraction—hydraulic fracturing. *Environmental Protection Agency*. http://www2.epa.gov/hydraulicfracturing. Accessed September 28, 2015.

EPA. (2015b). *Assessment of the potential impacts of hydraulic fracturing for oil and gas on drinking water resources, executive summary* (28 pp.). United States Environmental Protection Agency External Review Draft/EPA/600/R-15/047c/June 2015. www.epa.gov/hfstudy. Accessed September 28, 2015.

EPA. (2015c). Overview of greenhouse gases. http://www3.epa.gov/climatechange/ghgemissions/gases/ch4.html.

Fall, A., Eichhubl, P., Cumella, S. P., Bodnar, R. J., Laubach, S. E., & Becker, S. P. (2012). Testing the basin-centered gas accumulation model using fluid inclusion observations: Southern Piceance Basin. *Colorado, American Association of Petroleum Geologists Bulletin, 96*(12), 2297–2318. https://doi.org/10.1306/05171211149.

Fawcett, M. (2015). Is the shale revolution becoming a casualty of its own success? *Alberta Oil Magazine*. Canada's leading source for oil and gas news. http://www.albertaoilmagazine.com. Accessed September 18, 2015.

Ferrari, R. L. (2003). *2001 Lake Mead sedimentation survey* (195 pp.). U.S. Department of the Interior Bureau of Reclamation Technical Service Center Denver, Colorado (including appendices).

Fischhoff, B. (2015). The realities of risk-cost-benefit analysis. *Science 350*, 7 pp. https://doi.org/10.1126/science.aaa6516.

Folger, P. (2010). Gas hydrates: Resource and hazard. *Congressional Research Service*. 7-5700, RS22990, 9 pp.

Gålfalk, M., Olofsson, G., Crill, P., & Bastviken, D. (2015). Making methane visible (+Supplementary article, 14 pp). *Nature Climate Change*, 2 pp. https://doi.org/10.1038/nclimate2877.

Gallegos, T. J., Varela, B. A., Haines, S. A., & Engle, M. A. (2015). Hydraulic fracturing water use variability in the United States and potential environmental implications. *Water Resources Research (American Geophysical Union), 51*(7), 5839–5845. https://doi.org/10.1002/2015WR017278.

Garg, S. K., Pritchett, J. W., Katoh, A., Baba, K., & Fujii, T. (2008). A mathematical model for the formation and dissociation of methane hydrates in the marine environment. *Journal Geophysical Research, 113,* 1–32. https://doi.org/10.1029/2006/JB004768.

Gayyum, F. (2015). Shallow geohazard analysis a growing area of concern. *Offshore, 75*(5), 86.

GE (2016). Power generation https://powergen.gepower.com/resources/knowledge-base/combined-cycle-power-plant-how-it-works.html. Accessed April 25, 2016.

GLS. (2017). *The Gallowglaich. Melted shallow gas hydrates blowing through burst disk triggered Macondo.* https://thegallowglaich.wordpress.com/2013/08/15/melted-shallow-gas-hydrates-blowing-through-burst-disk-triggered-macondo/. Accessed August 7, 2017.

Graves, C. A., James, R. H., Sapart, C. J., Stott, A. W., Wright, I. C., Berndt, C., et al. (2017). Methane in shallow subsurface sediments at the landward limit of the gas hydrate stability zone offshore western Svalbard. *Geochemica et Cosmochemica Acta, 198,* 419–438.

Gudmundsson, J. S. (1996). Method for production of gas hydrates for transportation and storage. US Patent 5,536,893, July 16, 1996.

Gunderson, J. (2015). Shale and unconventional energy sources: Water-intensive requirements drive solutions in water management treatment. *Industrial Waterworld.* http://www.waterworld.com/articles/iww/print/volume-15/issue-2/features/shale-and-unconventional-energy-sources-water-intensive-requirements-drive-solutions-in-water-management-treatment.html. Accessed April 27, 2016.

Hachikubo, A., Yanagawa, K., Tomaru, H., Lu, H., & Matsumoto, R. (2015). Molecular and isotopic composition of volatiles in gas hydrates and in sediment from the Joetsu Basin, Eastern Margin of the Japan Sea. *Energies 8*, 4647–4666. https://doi.org/10.3390/en8064647.

Hansen, E., Mulvaney, D., & Betcher, M. (2013). Water resource reporting and water footprint from Marcellus Shale development in West Virginia and Pennsylvania. *Downstream strategies prepared for earthworks oil & gas accountability project* (88 pp.). CA: San Jose State University. http://www.downstreamstrategies.com/documents/reports_publication/marcellus_wv_pa.pdf. Accessed September 17, 2015.

Harkness, J. S., Dwyer, G. S., Warner, N. R., Perker, K. M., Mitch, W. A., & Vengosh, A. (2015). Iodide, bromide, and ammonium in hydraulic fracturing and oil and gas wastewaters: Environmental implications. *Environment Science Technology, 49*(3), 1955–1963. https://doi.org/10.1021/es504654n.

Heywood, M. (2016). *Oklahoma orders producers to cut wastewater injection by 40% Oil Pro.* http://oilpro.com/post/22393/oklahoma-orders-producers-to-cut-wastewater-injection-40. Accessed February 17, 2016.

Hicks, T., Aspinall, W., Cooke, R., & Germon, T. (2018). Oklahoma's induced seismicity strongly linked to wastewater injection depth. *Science, 359*(6381), 1251–1255. https://doi.org/10.1126/science.aap7911. Originally published online February 1, 2018.

Hill, D. (2015). What would have been the first fracing project in Denmark's history has been cancelled. *Oilpro*. http://oilpro.com/post/17666/would-have-been-first-fracing-project-denmark-history-has-been-ca?utm_source=DailyNewsletter&utm_medium=email&utm_campaign=newsletter&utm_term=2015-08-18&utm_content=Article_5_txt. Accessed August 18, 2015.

Hill, A. H., Hampson, K. M., Hill, A., Golightly, C., Wood, G. A., Sweeney, M., & Smith, M. M. (2015). ACG field geohazards management: unwinding the past, securing the future. In *Proceedings Offshore Technology Conference*, Houston TX, 407 May, 2015 (18 pp.).

Höglund-Isalsson. (2017). Bottom-up simulations of methane and ethane emissions from global oil and gas systems 1980 to 2012. *Environmental Research Letters, 12*, 11 pp. https://doi.org/10.1088/1748-9326/aa583e. http://iopscience.iop.org/article/10.1088/1748-9326/aa583e/pdf. Accessed February 9, 2017.

Hoza, M. (2015a). No NW hydro? No problem. We got gas. *Oilpro* (3 pp.). http://oilpro.com/post/12785/no-nw-hydro-no-problem-we-got-gas?u…paign=newsletter&utm_term=2015-05-07&utm_content=Article_15_txt. Accessed April 27, 2016.

Hoza, M. (2015b). The tax collector comes to the Marcellus. BTU. *Emerging Market Commentary*. http://oilpro.com/post/13983/tax-collector-comes-tomarcellus?utm_source=DailyNewsletter&utm_medium=email&utm_campaign=newsletter&utm_term=2015-06-05&utm_content=Article_16_txt. Accessed June 6, 2015.

Huber, R. (2016). Does Russia have the gas Europe needs. *Oilpro*. http://oilpro.com/post/27000/does-russia-have-gas-europe-needs?utm_source=DailyNewsletter&utm_medium=email&utm_campaign=newsletter&utm_term=2016-08-29&utm_content=Article_4_txt. Accessed August 29, 2016.

Hughs, D. (2013). *Drill, baby, drill: Can unconventional fuels usher in a new era of energy abundance?* (165 pp.). http://www.postcarbon.org/publications/drill-baby-drill/. Accessed August 22, 2015.

Hughs, D. (2015a). *The U.S. is counting on a long-term abundance of oil & natural gas. But what if the boom is just a bubble?* http://shalebubble.org/key-findings/.

Hughs, D. (2015b). *Drilling deeper: A reality check on U.S. government forecasts for a lasting tight oil & shale gas boom. Especially Part 3: Shale gas* (315 pp.). http://www.postcarbon.org/wp-content/uploads/2014/10/Drilling-Deeper_FULL.pdf. Accessed November 18, 2015. Accessed February 2, 2016.

Hunziker, R. (2015). The Fukushima fix. *Counterpunch*. http://www.counterpunch.org/2015/09/21/the-fukushima-fix/, September 21, 2015.

IEA. (2013). *Medium-term gas market report (MTGMR) 2015* (12 pp.). International Energy Agency.

IEA. (2015). *Medium-term market report 2015. Market analysis and forecasts to 2020*. International Energy Agency OECD/IEA, 2015. https://www.iea.org/Textbase/npsum/MTGMR2015SUM.pdf. Accessed July 16, 2016.

Ilie, L. 2015. Romania to send oil and gas royalty tax law to parliament in September. *Reuters News Service*, 25 May 2015. http://www.reuters.com/article/2015/05/25/romania-energy-tax-idUSL5N0YG15120150525. Accessed April 27, 2016.

Inman, M. (2014). Natural gas: The fracking fallacy. *Nature, News and Comment*. http://www.nature.com/news/natural-gas-the-fracking-fallacy-1.16430. Accessed December 4, 2015.

IPCC. (2014) Climate change 2014: Synthesis report. In R. K. Pachauri & L.A. Meyer (Eds., Core Writing Team), *Contribution of Working Groups I, II and III to the Fifth Assessment Report of the Intergovernmental Panel on Climate Change* (151 pp.). IPCC, Geneva, Switzerland.

Intersea. (2017). *Mercury removal from rich MEG—Key considerations* (6 pp.). Joint Industry Project (WorleyParsons Group). http://oilpro.com/media/lgc?id=17. Accessed February 7, 2017.

Japan Times. (2015). Japan's radiation decontamination budget to hit record 522.3 billion yen. *The Japan Times*, 17 December 2015. http://www.japantimes.co.jp/tag/fukushima-no-1/page/2/. Accessed February 22, 2016.

JOGMEC. (2013). *News release: Gas production from methane hydrate layers confirmed* (3 pp.). Japan Oil, Gas and Metals National Corporation, Tokyo. <http://www.jogmec.go.jp/english/news/release/release0110.html>. Accessed 26 Apr 2016.

Jonassen, R. (2015). *Energy investment risk and future climate—Part 2: Climate risk to energy projects*. http://www.energy-learning.com/index.php/78-what-s-new/139-energy-investment-risk-and-future-climate-part-2. Accessed May 26, 2015.

Joye, S. B. (2015). Deepwater horizon, 5 years on. *Science, 349,* 592–593. https://doi.org/10.1126/science.aab4133.

Kanda, H. (2006). Economic study on natural gas transportation with natural gas hydrate (NGH) pellets. In *Proceedings of 23rd World Gas Conference* (11 pp.). Amsterdam. http://members.igu.org/html/wgc2006/pdf/paper/add10399.pdf. Accessed February 9, 2017.

Kaufman, R. (2017). Deepwater horizon dispersant cleared the air, new model shows. *EOS, 98* (11), 8–9.

Kennett, J. P., Cannariato, K. G., Hendy, I. L., & Behl, R. J. (2003). *Methane hydrates in quaternary climate change: The clathrate gun hypothesis* (216 pp.). Washington DC: American Geophysical Union. ISBN 0-87590-296-0.

Kimball, S. M., Hutchinson, D. R. (2016). USGS leverages extended continental shelf research to address deep-sea science issues. *Sea Technology, 57*(1), 20, 22–23.

Kliewer, G. (2015). Ultra-deepwater to capture most of Brazil's offshore. http://www.offshore-mag.com/articles/print/volume-75/issue-5/international-field-development/ultra-deepwater-to-capture-most-of-brazil-s-offshore-capex.html. Accessed May 25, 2015.

Li, A., Davies, R. J., & Mathias, S. (2017). Methane hydrate recycling offshore of Mauritania probably after the last glacial maximum. *Marine and Petroleum Geology, 84,* 323–331.

Likvern, R. (2016). Norwegian crude oil reserves and extraction per 2015. *Oil Pro*. http://oilpro.com/post/23569/norwegian-crude-oil-reserves-and-extraction-per-2015?utm_source=DailyNewsletter&utm_medium=email&utm_campaign=newsletter&utm_term=2016-04-13&utm_content=Feature_1_img. Accessed April 27, 2016.

LP. (2015). *Lake Powell multibeam mapping 2005*. Ocean Mapping Group, University of New Brunswick, Canada Department of Earth Sciences, Duke University, NC, USA National Park Service, Page AZ, USA Bureau of Reclamation, Denver, CO, USA. US Government copyright free publication. http://www.omg.unb.ca/Projects/LakePowell/OMG_UNB_Lake_Powell_Overview.html. Accessed October 24, 2015.

Loki, R. (2015). 8 dangerous side effects of fracking that the industry doesn't want you to hear about. http://www.downstreamstrategies.com/documents/reports_publication/marcellus_wv_pa.pdf. Accessed April 30, 2015.

Ma, Y. Z. (2011). Uncertainty analysis in reservoir characterization and management: How much should we know about what we don't know? In Y.Z. Ma, P.R. La Pointe (Eds.), *Uncertainty analysis and reservoir modeling. American Association of Petroleum Geologists Memoir, 96,* 1–15.

Macondo. (2016). Investigation report executive summary. Drilling rig explosion and fire at the Macondo Well (24 pp.). U.S. Chemical Safety and Hazard Investigation Board Report No. 2010-10-1-OS.

MacPherson, J. (2015). New year brings changes in North Dakota oilfield waste rules. *PennEnergy*, 12/30/15. http://www.pennenergy.com/articles/pennenergy/2015/12/new-year-brings-changes-in-north-dakota-oilfield-waste-rules.html?cmpid=EnlDailyPetroDecember302015&eid=288179069&bid=1265179. Accessed December 30, 2015.

Martin, M., & Chediak, M. (2018). A powerful mix of solar and batteries is beating natural gas. *Bloomberg*. https://www.bloomberg.com/news/articles/2018-02-12/a-powerful-mix-of-solar-and-batteries-is-beating-natural-gas?utm_source=Sailthru&utm_medium=email&utm_campaign=Issue:%202018-02-13%20Utility%20Dive%20Storage%20%5Bissue:14033%5D&utm_term=Utility%20Dive:%20Storage. Accessed February 13, 2018.

Maslin, M., Owen, M., Betts, R., Day, S., Dinkley-Jones, T., & Ridgwell, A. (2010). Gas hydrates: Past and future geohazard? *Philosophical Transactions of the Royal Society A, 368,* 2369–2393.

Max, M. D., John, V. T., & Pellenbarg, R. E. (2000). Methane hydrate fuel storage for all-electric ships: An opportunity for technological innovation. In G. D. Holder, P. R. Bishnoi (Eds.), *Gas hydrates challenges for the future. Anals of the New York Academy of Sciences, 912*, 460–473. Special publication of papers from the Third International Conference on gas hydrates.

Max, M. D., Johnson, A. H. (2011). Methane hydrate/clathrate conversion. In M. R. Khan (Ed.), *Clean hydrocarbon fuel conversion technology* (pp. 413–434). Woodhead Publishing Series in Energy No. 19. Woodhead Publishing Ltd. Cambridge, U.K. ISBN 1 84569 727 8, ISBN-13: 978 1 84569 727 3.

Max, M. D., & Johnson, A. H. (2012). Could gas hydrate in fine-grained sediments be a precursor for some shale gas deposits? *Petroleum Geoscience, 18,* 231–238. https://doi.org/10.1144/1354-079311-035.

Max, M. D., Johnson, A., & Dillon, W. P. (2006). *Economic geology of natural gas hydrate* (p. 341). Berlin, Dordrecht: Springer.

Max, M. D., Johnson, A. H., & Dillon, W. P. (2013). Natural gas hydrate—Arctic Ocean deepwater resource potential. *SpringerBriefs in Energy*, 113 pp.

Max, M. D., & Johnson, A. H. (2014). Hydrate petroleum system approach to hydrate natural gas exploration. *Petroleum Geoscience, 20*(2), 187–199. Geological Society of London (Published in Online First 21 March 2014). https://doi.org/10.1144/petgeo2012-049.

Max, M. D., & Johnson, A. H. (2016). *Exploration and production of oceanic natural gas hydrate: Critical factors for commercialization* (405 pp.). Springer International Publishing AG.

Max, M. D., & Pellenbarg, R. E. (1998). Clathrate-based fuel storage and transport media: Potential impact. In *Proceedings Fuel Chemistry Division of the American Chemical Society*, San Francisco, April 13–15, 1997 (Vol. 24, pp. 463–466).

Max, M. D., & Pellenbarg, R. E. (1999). *Desalination through methane hydrate*. United States Patent (number 5,873,262), February 23, 1999.

Max, M. D., Pellanberg, R. E., & Hurdle, B. G. (1997). *Methane hydrate, a special clathrate: Its attributes and potential*. Naval Research Laboratory Report NRL/MR/6101–97-7926 (74 pp.).

McGrath, R., & Lynch, P. (Eds). (2008). Ireland in a warmer world. Scientific predictions of the Irish climate in the twenty-first century. *Community Climate Change Consortium for Ireland (C4I), National Development Plan of Ireland* (116 pp.).

McNutt, M. K., Camilli, R., Crone, T. J., Guthrie, G. D., Hsieh, P. A., Ryerson, T. B., Savas, O., & Shaffer, F. (2012). Review of flow rate estimates of the Deepwater Horizon oil spill. In *Proceedings of the National Academy of Sciences, 109*(50), 20260–20267. Also supporting information at www.pnas.org/lookup/suppl/doi:10.1073/pnas.1112139108/-/DCSupplemental. Accessed August 17, 2015.

Melly, B. (2017a). *Appeals courd blocks California gas facility from reopening*. Associated Press 28–29 July 2017. http://www.nbc12.com/story/35992844/appeals-court-blocks-california-gas-facility-from-reopening. Accessed July 29, 2017.

Melly, B. (2017b). A third of wells shut down at site of trobled gas facility. *PennEnergy* (Associated Press). http://www.pennenergy.com/articles/pennenergy/2017/09/oil-and-gas-a-third-of-wells-shut-down-at-site-of-troubled-gas-facility.html. Accessed September 13, 2017.

MES (2017). Mitsui Engineering & Shipbuilding co. Ltd. https://www.mes.co.jp/english/business/energy/energy7.html. Accessed February 9, 2017.

Mestdagh, M., Poort, J., & De Batist, M. (2017). The sensitivity of gas hydrate reservoirs to climate change: Perspectives from a new combined model for permafrost-related and marine settings. *Earth-Science Reviews*, 69 pp. https://doi.org/10.1016/j.earscirev.2017.04.013. Pre-publication draft.

Möller, L., Sowers, T., Bock. M., Spahn, R., Behrens1. M., Schmitt, J., Miller, H., & Hubertus Fischer, H. (2013). Independent variations of CH_4 emissions and isotopic composition over the past 160,000 years. *Nature Geoscience, 6*, 885–891.

Moridis, G., Reagan, M. (2014). Response of a gas-producing hydrate deposit following a well-shut in. *Fire in the Ice, 14*(1), 6–8. National Energy Technology Laboratory, U.S. Department of Energy.

Morrison, J. (2015). Fracking study gives mixed results. *Chemical & Engineering News,* June 15, 2015, 23.

Nakai, S., (2012). Development of natural gas hydrate (NGH) supply chain. In *Proceedings of the World Gas Conference*, Kuala Lumpur (10 pp.). http://www.ipt.ntnu.no/gudmundssonNGH/pressrelease.pdf. Accessed February 9, 2017.

NAS. (2016). *Attribution of extreme weather events in the context of climate change* (186 pp.). The National Academies Press. ISBN 978-0-309-38094-2. https://doi.org/10.17226/21852.

Nelder, C. (2012). *What the frack?* (Slate, New America & ASU. http://www.slate.com/articles/health_and_science/future_tense/2011/12/is_there_really_100_years_worth_of_natural_gas_beneath_the_united_states_.html. Accessed January 23, 2016.

Newman, N. (2017). Market drivers for an increase in gas storage. *Eniday*. https://www.eniday.com/en/sparks_en/increasing-gas-storage/. Accessed July 25, 2017.

Noguchi, S., Furukawa, T., Aung, T. T., & Oikawa, N. (2011). Reservoir architecture of methane hydrate bearing turbidite channels in the eastern Nankai Trough, Japan. In *Proceedings of the 7th International Conference on Gas Hydrates (ICGH 2011)*, Edinburgh, Scotland, United Kingdom, July 17–21, 2011 (9 pp.).

NRC. (2016). *Location of projected new nuclear power reactors*. United States Nuclear Regulatory Commission. http://www.nrc.gov/reactors/new-reactors/col/new-reactor-map.html. Accessed May 12, 2016.

NPR (2016). California regulators order company to permanently close leaking gas well. *National Public Radio*, 24 Jan. http://www.npr.org/sections/thetwo-way/2016/01/24/464195563/california-regulators-order-company-to-permanently-close-leaking-gas-well. Accessed January 27, 2016.

Offshore, 16-1. (2016). Research shows link between San Andreas and Cascadia faults. *Offshore Magazine, 76*(1), 44–45.

Oldenburg, C. M., Freifeld, B. M., Pruess, K., Pan, L., Finsterle, S., & Moridis, G. (2012). Numerical simulations of the Macondo well blowout reveal strong control of oil flow by reservoir permeability and exsolution of gas. *Proceedings National Academy of Sciences U.S. A, 109*(50), 20254–20259. 10.1073/pnas.1105165108. http://www.ncbi.nlm.nih.gov/pmc/articles/PMC3528597/. Accessed May 12, 2016.

Otto, B. (2017). The importance of detailed geological characterization for future expanded use of gas storage in the sustainable energy context. *Petroleum Geoscience, 23*(3), 327–338.

PA. (2017). *Understanding natural gas and LNG options* (243 pp.). Power Africa, a U.S. Government Partnership. U.S. Department of Energy.

Page, S. (2015). Don't let shell fool you. Arctic drilling is not a thing of the past. *Climate Progress* (12 pp.). http://thinkprogress.org/climate/2015/10/23/3714537/russia-norway-arctic-drilling/. Accessed December 23, 2015.

Parish, J. R. (2015). Proposed regulations clarify the meaning of a safe drilling margin in the OCS. *Offshore Magazine*. http://www.offshore-mag.com/articles/print/volume-75/issue-7/departments/regulatory-perspectives/proposed-regulations-clarify-the-meaning-of-a-safe-drilling-margin-in-the-ocs.html. Accessed August 11, 2015.

PennEnergy, 6/1/15. (2015). *Environmentalists want California to stop offshore hydraulic fracturing*. http://www.pennenergy.com/articles/pennenergy/2015/06/environmentalists-want-california-to-stop-offshore-hydraulic-fracturing.html?cmpid=EnlDailyPetroJuly22015&eid=288179069&bid=1110823. Accessed July 2, 2015.

Peters, E. (2016). Eagle ford oil and gas on steady decline. *Oilpro*. http://oilpro.com/post/29011/eagle-ford-oil-and-gas-steady-decline?utm_source=DailyNewsletter&utm_medium=email&utm_

campaign=newsletter&utm_term=2016-12-06&utm_content=Article_7_txt. Accessed December 12, 2016.
PFL. (2017). Drilling the Macondo prospect. DW Physics, planet for life (5 pp.). http://www.planetforlife.com/dwphysics/drillmacondo.html. Accessed April 12, 2017.
PGN. (2015). Earthquake concerns: Oklahoma shuts in two disposal wells, cuts volume at three. Petro Global News. http://petroglobalnews.com/2015/09/earthquake-concerns-oklahoma-shuts-in-two-disposal-wells-cuts-volume-at-three/. Accessed April 27, 2016.
PGN 10/1/15. (2015). Federal court halts Obama fracking rules. *Petro Global News*. http://petroglobalnews.com/2015/10/federal-court-halts-obama-fracking-rules/. Accessed October 1, 2015.
PGN 11/17/15. (2015). Texas tornado levels Halliburton chemical plant. *Petro Global News*. http://petroglobalnews.com/2015/11/texas-tornado-levels-halliburton-chemical-plant/. Accessed February 8, 2016.
PGN 1/6/16. (2016). Shell starts gas flows at offshore Ireland Corrib Field. http://petroglobalnews.com/2016/01/shell-starts-gas-flows-at-offshore-ireland-corrib-field/. Accessed April 27, 2016.
Pohlman, J. W., Greinert, J., Ruppel, C., Silyakova, A., Vielstädte, L., Casso, M., Mienert, J., & Bünz, S. (2017). Enhanced CO_2 uptake at a shallow Arctic Ocean seep field overwhelms the positive warming potential of emitted methane. *PNAS*, May 23, 2017, 6 pp. *114*(21), 5355–5360. Published ahead of print May 8, 2017. https://doi.org/10.1073/pnas.1618926114.
Polson, J., & Loh, T. (2017). U.S. vastly overstates oil output forecasts, MIT study suggests. *Bloomberg*. https://www.bloomberg.com/news/articles/2017-12-01/mit-study-suggests-u-s-vastly-overstates-oil-output-forecasts. Accessed February 13, 2018.
Proctor, D. (2018). Efficiency improvements mark advances in gas turbines. *Power Magazine, 162* (1), 26–29.
Randle, T. J., Lyons, J. K., Christensen, R. J., & Stephen, R. D. (2015, undated). Reclamation. Managing water in the west. *Colorado River Ecosystem Sediment Augmentation Appraisal Engineering Report* (71 pp.). Grand Canyon Monitoring and Research Center. Accessed May 23, 2015.
Rapier, R. (2015). The shale boom's red ink. OilPro. http://oilpro.com/post/19945/shale-boom-red-ink?utm_source=DailyNewsletter&utm_medium=email&utm_campaign=newsletter&utm_term=2015-11-09&utm_content=Feature_1_img. Accessed November 9, 2015.
Reed, J. (2015a). With ConocoPhillips' exit, Poland shale is now on life support. *Oilpro*. http://oilpro.com/post/14053/country-shale-hopes-all-but-dead?utm_source=DailyNewsletter&utm_medium=email&utm_campaign=newsletter&utm_term=2015-06-08&utm_content=Article_13_txt. Accessed June 8, 2015.
Reed, J. (2015b). Remember the talk about a global shale revolution? Don't hold your breath. *Oilpro*. http://oilpro.com/post/11504/remember-talk-global-shale-revolution-dont-hold-your-breath. Accessed June 8, 2015.
Reed, J. (2016). Obama issues new methane rules, adds to O&G headwinds. http://oilpro.com/post/21759/obama-issues-new-methane-rules-adds-to-og-headwinds?utm_source=DailyNewsletter&utm_medium=email&utm_campaign=newsletter&utm_term=2016-01-25&utm_content=Article_17_txt. Accessed April 27, 2016.
Reagan, M. T., & Moridis, G. J. (2008). Dynamic response of oceanic hydrate deposits to ocean temperature change. *Journal Geophysical Research-Oceans, 113*. https://doi.org/10.1029/2008jc004938.
Rehder, G., Eckl, R., Elfgen, M., Falenty, A., Hamann, R., Kahler, N., Kuhs, W. F., Osterkamp, H., & Windmeir, C. (2012). Methane hydrate pellet transport using the self-preservation effect: a techno-economic analysis. *Energies, 5*, 2499–2523. https://doi.org/10.3390/en5072499.
Revelle, R., & Suess, H. E. (1957). Carbon dioxide exchange between atmosphere and ocean and the question of an increase of atmospheric CO_2 during the past decades. *Tellus, 9,* 18–27.
Revelle, R. (1983). Probable future changes in sea level resulting from increased atmospheric carbon dioxide. In National Research Council Carbon Dioxide Assessment Committee (Ed.), *Changing climate* (pp. 433–448). Washington, DC: National Academy of Sciences.

Ritter, K. (2016). *A new life for coal ash* (pp. 10–14). American Chemical Society C&EN, February 16.
Rogers, D. (2014). The Aspen Institute, *Energy Policy Forum*. Statement by Deborah Rogers. Table: Major drillers' free cash flow is almost entirely negative (3 pp.). http://www.aspeninstitute.org/policy-work/energy-environment/our-policy-work/energy-policy-forum.
Rudloff, D., & Schlultz, M. (2016). Top risks in oil and gas. How oil and gas companies gauge the risks they face. *Oil & Gas Journal 9/11/2016,* 11 pages. https://www.ogj.com/articles/ogfj/print/volume-13/issue-9/features/top-risks-in-oil-and-gas.html. Accessed 25 September 2018.
Ruppel, C. D. (2011). Methane hydrates and contemporary climate change. *Nature Education Knowledge, 3*(10), 29.
Ruppel, C. D., & Kessler, J. D. (2017). The interaction of climate change and methane hydrates. *American Geophysical Union Reviews of Geophysics, 55*, 43 pp. https://doi.org/10.1002/2016rg000534.
Ryerson, T. B., et al. (2012). Chemical data quantify deepwater horizon hydrocarbon flow rate and environmental distribution. In *Proceedings of the National Academy of Sciences, 109*(50), 20246–20253. This article contains supporting information online at www.pnas.org/lookup/suppl/doi:10.1073/pnas.1110564109/-/DCSupplemental. Accessed August 17, 2015.
Salazar, R., & Rauniar, R. (2015). Diffusion of new O&G technologies: How beliefs & norms shape acceptance. *Oilpro*. http://oilpro.com/post/15221/diffusion-new-og-technologies-beliefs-norms-shape-acceptance?utm_source=DailyNewsletter&utm_medium=email&utm_campaign=newsletter&utm_term=2015-07-08&utm_content=Article_4_txt. Accessed October 21, 2015.
Sandalow, D., Wu, J., Yang, Q., Hove, A., & Lin, J. (2014). *Meeting China's shale gas goals*, September 2014 (83 pp.). Working draft for public release. Columbia University SIPA, Center on Global Energy Policy.
SBC. (2015). *Natural gas series FactBook*. Gas Hydrates SBC Energy Institute (SBC-EI) Schlumberger, 99 pp. https://www.sbc.slb.com/~/media/Files/SBC%20Energy%20Institute/SBC%20Energy%20Institute_Gas%20Hydrates_FactBook.pdf. Accessed December 22, 2015.
Schultz, R., Atkinson, G., Eaton, D. W., Gu, Y. J., & Kao, H. (2018). Hydraulic fracturing volume is associated with induced earthquake productivity in the Duvernay play. *Science, 359*(6373), 304–308.
Schlumberger. (2015). Integrated lifting solutions for enhanced well production. http://www.slb.com/services/production/artificial_lift.aspx. Accessed February 2, 2016.
Shankleman, J., & Warren, H. (2017). This is what the demise of oil looks like. *Bloomberg*. https://www.bloomberg.com/graphics/2017-oil-projections/. Accessed May 31, 2017.
Shilling, A. G. (2015). Get ready for $10 oil. *Bloomberg View* 8/5/15. http://www.bloombergview.com/articles/2015-02-16/oil-prices-likely-to-fall-as-supplies-rise-demand-falls. Accessed Aug ust 5, 2015.
Showstack, R. (2017). Proposed bill would loosen ocean drilling restrictions. *EOS*. https://eos.org/articles/proposed-bill-would-loosen-ocean-drilling-restrictions?utm_source=eos&utm_medium=email&utm_campaign=EosBuzz102017. Accessed 10/20/17.
Schulte, G. (2015). Oil and gas regulations to get new scrutiny in Nebraska. Associated Press 14 September 13. PennEnergy. http://www.pennenergy.com/articles/pennenergy/2015/09/oil-and-gas-regulations-to-get-new-scrutiny-in-nebraska.html. Accessed September 14, 2015.
Schumpter. (2017). America's shale firms don't give a frack about financial returns. *The Economist* 25 Mar 2017, 5 pp. http://www.economist.com/news/business-and-finance/21719436-exploration-and-production-companies-are-poised-go-another-investment-spree-americas. Accessed March 24, 2017.
Serov, P., Vadakkepuliyambatta, S., Mienert, J., Patton, H., Portnov, A., Silyakova, A., Paieri, G., Carroll, M. L., Andreassen, K., & Hubbard, A. (2017). Postglacial response of Arctic Ocean gas hydrates to climatic amelioration. In *Proceedings of the National Academy of Sciences (U. S.A), 114*(24), 6215–6220.
Service, R.F. (2015). Tailpipe to tank. *Science, 349*(6253), 1158–1160.

SoCalCA. (2017). *State inspections confirm safety of Aliso Canyon natural gas storage facility*. California Department of Conservation. http://www.conservation.ca.gov/dog/Documents/Aliso/News%20Release.pdf. Accessed July 20, 2017.

Sowers, T. (2006). Late quaternary atmospheric CH_4 isotope record suggests marine clathrates are stable. *Science, 311,* 838–840.

Srebotnjak, T., & Rotkin-Ellman, M. (2014). NRDC issue brief IP:14-10-A. Fracking fumes: Air pollution from hydraulic fracturing threatens public health and communities, 12 pp. http://www.nrdc.org/health/files/fracking-air-pollution-IB.pdf. Accessed September 17, 2015.

SRSROCCO. (2018). Future U.S. oil production will collapse just as quickly as it increased. https://srsroccoreport.com/future-u-s-production-will-collapse-just-quickly-increased/. Accessed February 11, 2018.

Stern, L. S., Kirby, S. H., Burham, W. H., Circone, S., & Waite, W. F. (2003). Laboratory synthesis of pure methane hydrate suitable for measurement of physical properties and decomposition behavior. In Max, M. D. (Ed.), *Natural gas hydrate: In Oceanic and permafrost environments* (2nd ed., pp. 323–348). London, Boston, Dordrecht: Kluwer Academic Publishers (now Springer).

Stevenson, R. (2015). Will Mexico get it right in round one part two? *Oilpro*. http://oilpro.com/post/18183/mexico-get-right-round-one-part-two?/campaign=newsletter&utm_term=2015-09-04&utm_content=Article_14_txt. Accessed September 4, 2015.

Taft, G. (2006). Regulatory and permitting environment for gas hydrate. In M. D. Max, A. Johnson, W. P. Dillon (Ed.), *Economic geology of natural gas hydrate* (pp. 267–288). Berlin: Springer.

Torres, N. (2015a). Oil and water: California shuts down 33 wastewater wells for injecting into aquifers. *Petro Global News*. http://petroglobalnews.com/2015/10/oil-and-water-california-shuts-down-33-wastewater-wells-for-injecting-into-aquifers/. Accessed April 27, 2016.

Torres, N. (2015b). Arctic bust: Shell abandons $3 billion Alaska well. *Petro Global News*. http://petroglobalnews.com/2015/09/arctic-bust-shell-abandons-3-billion-alaska-well/. Accessed April 27, 2016.

Tullo, A. H. (2015). Mexico's chemical rebirth followed by: Argentina's new gas source may lead to chemical developments. *Chemical & Engineering News*, 25–26. American Chemical Society, September 7, 2015.

UCS. (2017). *The climate risks of natural gas* (3 pp.). Union of Concerned Scientists. http://www.ucsusa.org/clean-energy/coal-and-other-fossil-fuels/infographic-climate-change-risks-natural-gas#.WM7D0xiZOL4. Accessed March 19, 2017.

Ulrich, K. (2017). Unequal impact, women's & children's human rights violations and the Fukashima Daiichi nuclear disater. *Greenpeace*, 56 pp. http://www.greenpeace.org/japan/Global/japan/pdf/20170307_UImpact_fin_report.pdf. Accessed March 14, 2017.

UN. (1982). *United Nations convention on the law of the sea* (202 pp.). United Nations.

UNC. (2015). *Rating the states on their risk of natural gas overreliance* (8 pp.). Union of Concerned Scientists Technical Document. www.ucsusa.org/naturalgasoverreliance.

USGS. (2015a). *Hydraulic fracturing use in the United States*. http://www.usgs.gov/newsroom/images/2015_06_30/water_use_for_fracking.jpg. Accessed September 24, 2015.

USGS. (2015b). Assessment of undiscovered shale gas and shale oil resources in the Mississippian Barnett Shale, Bend Arch–Fort Worth Basin Province, North-Central Texas. *National and Global Petroleum Assessment*. U.S. Geological Survey Information Sheet (2 pp.). Accessed December 18, 2015.

USGCRP. (2017). Climate science special report: A sustained assessment activity of the U.S. global change research program. In D. J. Wuebbles, D. W. Fahey, K. A. Hibbard, D. J. Dokken, B. C. Stewart, T. K. Maycock (Eds.), *U.S. Global Change Research Program*, Washington, DC, USA (669 pp.). https://assets.documentcloud.org/documents/3920195/Final-Draft-of-the-Climate-Science-Special-Report.pdf. Accessed August 13, 2017.

Vielstädte, L., Karstens, J., Haeckel, M., Schmidt, M., Linke, P., Reimann, S., et al. (2015). Quantification of methane emissions at abandoned gas wells in the Central North Sea. *Marine and Petroleum Geology, 59,* 1–13.

Varfolomeyev, O. (2015). Ukraine fails to make shale gas breakthrough. *Eurasia Daily Monitor*. http://www.jamestown.org/programs/edm/single/?tx_ttnews%5Btt_news%5D=43992&cHash=35aaa4010ef513ecd3d9146c257451f0#.Vgm4QrSm3cY. Accessed April 27, 2015.

Warrick, J. (2014). Delaware-size gas plume of West illustrates the cost of leaking methane. *Washington Post*. https://www.washingtonpost.com/national/health-science/delaware-sized-gas-plume-over-west-illustrates-the-cost-of-leaking-methane/2014/12/29/d34c3e6e-8d1f-11e4-a085-34e9b9f09a58_story.html. Accessed February 2, 2016.

Weingarten, M., Ge, S., Godt, J. W., Bekins, B. A., & Rubinstein, J. L. (2015). High-rate injection is associated with the increase in U.S. mid-continent seismicity. *Science, 348*(6241), 1336–1340.

Whiticar, M., & Schaefer, H. (2007). Constraining past global tropospheric methane budgets with carbon and hydrogen isotope ratios in ice. *Philosophical Transactions of the Royal Society A, 365,* 1793–1828. https://doi.org/10.1098/rsta.2007.2048.

Williams, A. P., Seager, R., Abatzogiou, J. T., Cook, B. I., Smerdon, J. E., & Cook, E. R. (2014). Contribution of anthropogenic warming to California drought during 2012–2014. *Geophysical Research Letters, 42*(16), 6819–6828. https://doi.org/10.1002/2015GL064924.

Wilson, A. (2015). Managing marine geohazard risks throughout the business cycle. *SPE*. www.spe.org/jpt. Accessed September 2, 2015.

Wilson, E. (2016). Epic methane leak eases up. *Chemical & Engineering News, 94*(4), 35.

WNA. (2015). *Safety of nuclear power reactors. Fukushima accident.* World Nuclear Association (Nuclear Industry Association, Non-Governmental). http://www.world-nuclear.org/info/Safety-and-Security/Safety-of-Plants/Fukushima-Accident/. Accessed September 18, 2015.

Wood, W. T., & Jung, W.-Y. (2008). Modeling the extent of Earth's marine methane hydrate cryosphere. In *Proceedings of the 6th International Conference on Gas Hydrates (ICGH 2008)*, Vancouver, British Columbia, Canada, July 6–10, 2008 (8 pp.).

Chapter 11
Commercial Potential of Natural Gas Hydrate

Abstract The entrepreneurial opening up of coalbed methane, shale, and tight gas was achieved mainly because small companies took initiative in wildcatting new methods. In addition, the U.S. Department of Energy provided material support over decades. NGH will probably also have to go through its wildcatting stage in which different exploration and production models are tested in the bridge from science to application. The current burst of oil and gas production from shale is based on mobile drilling rigs, relatively small fracking infrastructure, and innovative inclined or 'horizontal' drilling that keeps the well in the pay horizon for long distances. Because natural gas hydrate (NGH) is found in deepwater, the available drilling, logistic, and processing technology is inherently more expensive. Accordingly, less expensive exploration and production techniques are necessary for commercial production of natural gas from the NGH resource. The term '**commercial**' probably should never be used in relation to any extractive material until after the concentrations have been identified and valued and estimates of technical recoverability have been made. The topic of commerciality is complicated because there are a large number of financial and other variables that have nothing to do with geology or with simple supply and demand. The chemical and physical characteristics of NGH, as well as the location of the resource world-wide, provide the potential for innovative technology and methodology to make NGH competitive on a producer basis. With the emerging world market and world price in compressed natural gas utilizing the new build-out in LNG and possibly CNG transport, natural gas could soon be as tradable as oil is today. Depending on the emphasis that is given by governments to reduction in from fossil fuels, Natural gas could become the main fossil fuel backing up renewable energy possibly into the far future because of its low CO_2 emissions per energy produced. The very low environmental impact/risk of NGH could enable it to become the preferred source of natural gas.

M. D. Max and A. H. Johnson, *Exploration and Production of Oceanic Natural Gas Hydrate*, https://doi.org/10.1007/978-3-030-00401-9_11

11.1 State of the Industry

The publication of this book in 2016 took place at the confluence of major consolidations within the hydrocarbon energy industry, along with a dramatic expansion of the renewable energy industry. The two are interconnected but not in a direct fashion. The post-2014 crash in oil and gas prices is not due to the emergence of large scale renewable capability but rather to oversupply that may have been initially politically-driven. Concerns for global climate that are expressed as taking action that diminishes the effusion of CO_2 into the atmosphere are also becoming a very strong influence on fossil fuel selection. There has been considerable high level political action since the COP21 Conference held in Paris in late 2015, although the resulting state of mitigating CO_2 remains mainly aspirational rather than enforceable targets. Nonetheless, regulations in mainly in Europe are imposing costs on fossil fuel industries, particularly on coal, that will act over time to affect the mix of fossil fuels used and the mix of fossil and renewable energy production.

Like most other people, we have no definite way of predicting what the price of fuels will be and how this will impact their use from place to place, although we feel that it is safe to say that natural gas will almost certainly emerge as the dominant fossil fuel in its own right because of climate mitigation, demand, availability, and its flexibility. The only sure thing in the energy business is change and in the proven ability of the industry to meet a continuing succession of technical and operational difficulties and hazards. These can be especially severe in deepwater and ultra-deepwater offshore (Barton et al. 2015). At some point, demand will reemerge although more costly production almost certainly will be lost, and market forces will act to raise prices in a severely competitive situation in which the participants will be a combination of survivors from the pre-2014 energy business and innovators driving down costs and opening up new resources.

Triepke (2015) summarized the emergent situation as one from which everything in the industry will have bottomed, from an optimists' point of view, not much later than two years from the price collapse in late 2014. Service company margins, drilling day rates, and supply and logistics, as well as probably every other business concerned with the offshore, will all be working at close to break-even. Barring any major disruption to the international political structure, an upturn will develop as the post-2014 decline cannot continue indefinitely. As the bottom of everything resolves, uncertainty will begin to fade and the exploration and production field will have been cleared of all the least competitive equipment and practices. Rebalancing will provide a new starting point for the future energy industry, not only for fossil fuels but also for renewables and the whole business of bringing energy to customers. It is clear that demand for gas and oil is going to rise as population increases and countries such as India and China, amongst a number of African countries which are aggressively trying to bring more of their citizens in their energy economy. What is not clear is where the energy is going to come from if the long-term investment is not put in place.

In the new environment, risk taking by those having the strongest balance sheets or available credit will be in a position to advance first. Companies that prepared for the future during the post-2014 collapse will have an advantage in being able to pick up promising leases and initiate exploration. The likely destruction of non-innovative industry 'dinosaurs' may leave the ground more open to innovators and entrepreneurs than prior to the collapse when bad practices were being rewarded as well as good. In particular, new technology and new cost-cutting approaches to the broad range of exploration and production should be particularly attractive.

The management structure of surviving companies probably will also show profound changes from pre-mid-2014 with substantial simplification of approaches and reduction of requirements, with due regard for maintaining safety. Morrow and Wallace (2016) suggest that: (1) Non-safety related company standards should be dropped, (2) Reduce the number of specialists to the minimum and keep a sharp eye out for the point of diminishing returns beyond which effort is not rewarded by commensurate reward, (3) The focus of owner teams to review and design quality should be replaced by using modified technology where possible, and (4) Rethink design approaches to reflect loss of experienced personnel. In general, focus should remain on the profit centers. Where the development of new technology involves intellectual property (IP) this should also be considered as a profit center but one with a longer timeline and a measurably higher risk and reward.

On a purely cost-competitive basis, low prices for fossil fuels appear to have caused reduced investment in renewables. But alternative energy, particularly solar, continues to make strides in efficiency and output while maintaining durability and its extremely low maintenance character. New, more efficient wind turbines and the likely increase in tidal and geothermal energy will have improved lifetime cost profiles and are likely to compete directly on a cost basis with fossil fuel. In addition, regulation and subsidy for both fossil fuel and renewables will continue to distort competitive conditions in favor of renewable energy. Whatever oil and gas prices are, the trend of imposed or regulatory cost will favor renewable over fossil energy at the same time as the cost of renewable energy continues to fall. The scale of renewables manufacturing, improved materials cost and efficiency of energy delivery, fossil fuels will find a market value that have already made coal and some oil increasingly uncompetitive on a purely cost basis. As oil prices begin to rise, climate concerns focusing on CO_2 emissions may become the dominant factor. From an operational point of view in which continuity of dependable energy supply is required, fossil fuel, which will most likely be natural gas (Chap. 1), will provide underlying base load for the intermittent renewable energy sources, and also the spiker energy needed to meet short term or sudden increases in energy demand.

11.2 Shale Gas and Oil Are Game Changers

The structure of the oil and gas industry has a number of components that do not all work in synchronization. Historically, estimates of global oil, and to a lesser extent gas supply, have been based on maintaining a rough balance of resource supply and demand. Although supply and demand would rarely match perfectly, over and under production and demand were managed to try and maintain price stability. This 'business model' was exploded in 1973 with the political use of oil production by OPEC (Organization of Petroleum Exporting Countries, established 1960). Political factors and international competition for energy resources that has destabilized much of the Middle East, is very complex and is not a subject of this book except to point out that a combination of political instability that could suddenly alter production and the variation in demand that is no longer controllable in the same manner as before, has introduced new volatility into both production and oil and gas prices.

In the past, oil and gas prices were driven largely by giant and super-giant conventional fields. Energy companies were intimately concerned with maintaining a rough parity of E&P costs between the least and most expensive producers. Less expensive to produce oil and gas is of course more profitable, but when producer countries control E&P rather than commercial companies, the income from energy is commonly factored into their budgets, which has the effect of introducing a higher than purely commercial break-even price level. OPEC's low-cost producers are Western-dependent countries such as Kuwait and other Arabian Peninsula countries (Fahey 2015), but their budget break even costs are much higher. For instance, in 2013–2014, Libyan budget projections were for oil prices at $185 a barrel (in the last year before President Gadhafi was over thrown; there now is no national government and no budget), while Iran's budget would balance at 500,000 Bbl/d at $133, and Russia pre-2015 had been projected at $100/Bbl. Times have changed. The 2016 Russian budget is based on an estimated average price of $50/ Bbl for the year.

By the early 2000s, the relatively low-cost oil coming from the giant and supergiant fields was no longer sufficient to meet global demand. In addition, forecasting supply and price was complicated by the changing technology as E&P expanded into ultra-deepwater. Profound changes in a wide range of costs, technological improvements and changes in the market also made projecting production and supply more problematic. Because in an open market situation the price of oil is no longer always determined by chasing the lowest price but by the sale price of the most expensive oil needed to meet the top end of market demand (Bishop 2015). The development of the shale resource, which is has a relatively high oil break-even price of $40–$50/Bbl, was based on being able to sell at a 'top of the market' price.

There are many small players in any natural commodity resource extractive industry, and they have serious cash flow problems soon after the initiation of a period of price drop. Shale development is done in the same way as an on-time manufacturing process, with everything (financing, drilling, production, finance payoff and profit) planned with short time lines. As a result, when potential cash flow falls below a certain level, the entire process stalls. By the end of the second half of 2014, as the price of a barrel of oil had fallen from $105+ to below $70, the rig count for oil-gas plays stayed high based on oil and condensate prices. At the same time rig count in the Haynesville (where there is essentially no liquids production) fell from 180 in mid-2010 to 23 in the last week of November 2014. As the price for oil dropped below $30/Bbl in January 2016, however, the developing credit crisis on which shale resource E&P is based froze because few wells or fields in the shale resource are profitable below this level. Selling into a rising price market can be spectacularly profitable but selling into a diminishing price market can be disastrous. As continuous drilling is necessary to keep opening new production to meet post-peak production for wells and fields, some companies will fare better than others.

Although it would appear (in the absence of war or sanctions) oil prices appear to have stabilized at a level where most production is sufficiently profitable, it is likely that oil prices will not rise until the oversupply situation is resolved, and even then it is not likely that prices will recover to the $100/Bbl price level for a far longer time (Blas and Smith 2016). Industry has responded to market forces by cutting costs, reducing personnel, and using lower cost methods at lower breakeven levels. Huge staff reductions have taken place with many smaller companies going bankrupt. While some large companies have the financial resources to weather the current period of low oil prices and have not had large staff reductions (Reed 2016), smaller companies that are not deeply in debt have options. Finding lower cost conventional oil is entirely possible, especially for smaller, more flexible companies because of its longer production and higher long-term profit profiles. For instance, Det Norske, aims to pursue stand-alone offshore developments that can be economic below $40/Bbl (Offshore 1/18/16).

High liquids prices have kept the oil shale drilling high and kept the gas market glutted with the incidental gas that is produced along with the oil. While the amount of oil in a reservoir is generally depleted from a constant volume, the amount recovered is affected by oil price. In some cases, as much as two-thirds of the oil may remain in the reservoir. As production decreases, secondary and tertiary recovery programs may be undertaken if the oil price warrants the investment. Shale gas and oil are thus intimately entwined. If the oil price should drop, the amount of incidental gas produced will also drop, which could cause the gas price to rise. Only if gas gets to the $\sim$ $5.00/MMcf range for the shale resource dry gas sweet spots and $6.00/MMcf for larger areas would there be a widespread reopening of the shale gas resource. Most conventional gas would presumably also be profitable at these levels.

At the bottom of the production food chain for conventional oil are stripper wells, each of which may produce less than 10 barrels of oil per day or less than 60 thousand cubic feet of gas per day (NSWA 2014), although the IRS uses a cutoff of 75,000 cubic feet per day. About 80% of the roughly 500,000 producing oil wells in the United States are classified as stripper wells. The stripper oil wells in the United States produce nearly 1 million barrels per day of oil in aggregate. Production from this bottom of the producing oil and gas market is complicated because tax policy, which varies from place to place, may be the most important factor.

11.3 Underlying Economics of Natural Gas

We do not intend to go into much detail on the various financial factors that determine profitability of individual wells, fields, or segments of the fossil fuel industry. That would be a very large task for technically knowledgeable economists, but a general description of the financial elements of the upstream fossil fuel industry shows how much more complex this commodity is from general commerce business models. It is important to note that the economic fundamentals for natural gas hydrate (NGH) are exactly the same as for conventional oil and gas. Thus, full financial valuation of NGH only awaits adequate data, some of which will be on the cost side where leveraged and new technology could dramatically lower the E&P cost. The value of the produced natural gas will be determined largely by market supply and demand. We have made the case that natural gas is the uniquely best-qualified fossil fuel to backstop a greatly enlarged implementation of renewable energy (Chap. 1). Thus, it is likely that demand for natural gas will continue to accelerate while regulations governing CO_2 emissions and environmental risk (Chap. 10) will also favor increased use of natural gas.

Valuation: Present and future field-level prices determine overall income and estimated income upon which financing and planning are based. The most important aspect of future assessments, are the SEC (Security and Exchange Commission) averaged prices which defines parameters to be used by public reporting energy companies in quantifying, pricing and valuing oil and gas reserves. The SEC figures use an average of the prior four quarters oil prices in order to better estimate prices for the volatile commodities by averaging prices for the year. The valuation is extremely important because without sufficient income or estimated income, the collateral value will not be sufficient for any debt coverage and the property will essentially be worthless.

The amount of natural gas in a reservoir is referred to as 'gas-in-place'. The technically recoverable gas that can be extracted conventional wells under normal circumstances is about half, with possibly another 20% recoverable using pumped extraction and boosting (Chap. 8). The amount of gas that is economically recoverable being some fraction of the technically recoverable gas that can be sold at a justifiably high enough profit to carry the costs of development and return investment in a reasonable time (Boswell and Collett 2011). Income is everything in the energy business where a very large amount of money spent on all stages of earlier investments are extremely well known and can diminish risks. At least a couple of years of production data from NGH will be required before real risk assessments can be made.

Costs: All funding and development costs, lease operating expenses, company G&A (General and Administrative), and most important, financing costs, are the upfront CAPEX obligations that have to be made substantially in advance of any income. Should production performance or market conditions deviate too far from planned operations and income, debts can only be restructured with great difficulty without declaring some form of bankruptcy. Plugging and decommissioning costs are often not included or are understated and can be very costly, especially for companies at field end of life with no income (O&G 2016). Lease operating expenses (LOE) or the direct extraction cost is the cost of maintaining and operating property and equipment on a producing lease, and may be difficult to quantify accurately because the term of production is very important in determining these costs. For instance, in normal shale wells, production life from a single frack is on the order of 1.5–3 years with multiple fracks each extending the life of the well and aggregate gas production. The LOE on short term wells can be in the $20–$40 range (Harrington 2016). For a longer-life well, such as offshore conventional or, in our view, NGH production, the LOE will be much lower.

Return on Investment (ROI) is often not as easy to estimate or measure for an individual resource to consumer example. The amount of energy or value of the energy recovered from any resource can be measured in two ways. First, as a direct measure, EROEI is the amount of energy delivered divided by the amount of energy required to deliver that energy. At its simplest, this results in a number that can be expressed in energy (Joules or Btu) or volume (ft^3 or m^3), which are interchangeable. A more complex calculation involves including financial costs and income, which can be highly variable. EROI is a total cost parameter, in which cumulative income from the supply of power to consumers and self-use of power is divided by the cumulative energy costs (CAPEX for financing and construction, operational and maintenance costs over the lifetime of the installation, and decommissioning costs and any downstream costs such as pensions) are also expressed as a number. Ideally, the net energy should be strongly on the positive or production side of the equation where the larger the EROEI or EROI, the greater the efficiency, and presumably the profit.

EROI is strongly example-based and may vary in all components from exploration and production to processing and use. EROI tends to be highly specific and often makes difficult comparing any two resources to user example of any particular example. Rodgers (2017) believes that intermittent energy is more expensive in terms of its EROI than is commonly believed while Tverberg (2017) goes into the detail of intermittent energy assumptions and calculations. The financial aspects of EROI are also not fixed but may vary over time, which renders early calculations revisable. For instance, bonds may be struck that are State or Government insured and these may be rolled over into what amounts to a government grant rather than repaid. Any financial changes such as these alter the EROI.

The EROI from the broad array of energy sources can be ranked from hydrokinetic power, which is the most efficient, to biodiesel, which is commonly the least efficient. An interesting aspect of plotting both fossil fuels and renewable

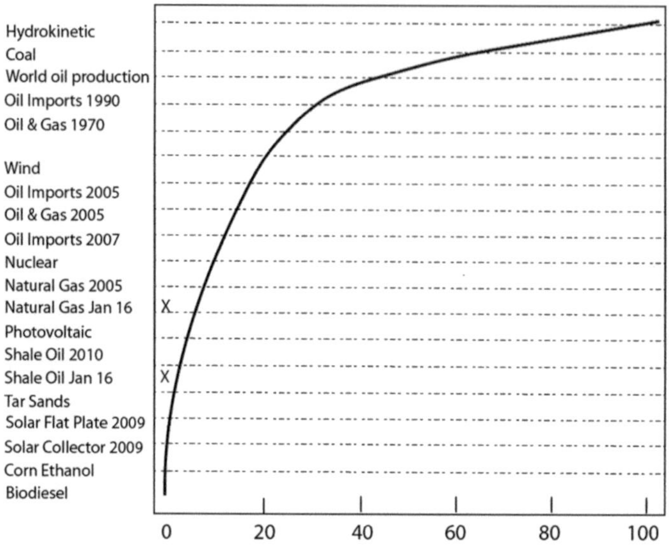

Fig. 11.1 EROI for US energy resources. Redrawn after Murphy and Hall (2010). X indicates low oil and gas price equivalent for January 2016, which reflects prices about 1/3 or normal

energy together from the point of view of EROI is that renewable energy is not uniformly more expensive, as many opponents of renewable energy appear to believe, but fits on a curve of EROI rankings (Fig. 11.1). The estimated positions for January 2016 EROI (using our estimates for sub- $2/MMcf/d natural gas and sub- $30/Bbl oil) show that a rough comparability is possible. A fitted curve shows the relative positions of EROI from Murphy and Hall (2010). As energy use intensifies, an increasing proportion of energy output and economic activity must be diverted to obtaining the energy to power economies, which may leave less funding for powering activities that drive growth (Hall et al. 2014).

The energy payback period, which is a convention measure for investors, is the time it takes an installation to produce an amount of energy equivalent to that on the cost side. The EROI for crude oil has declined over time, and may continue to do so as secondary recovery techniques need to be applied to extend recovery. Deepwater fields are more remote and require more energy not only to develop but also to maintain operations. A more straightforward measure of the economic outcome considers the two cost sides of the equation, which is less of an engineering measure than EROI. This measure is more qualitative because all processes are not reduced to an absolute factor such as heat (Btu) and costs for both energy and materials will vary over the lifetime of any installation and often do not vary in a linear or direct relation to one another.

Directly comparing renewable from non-renewable energy using EROI may not tell the whole story. EROI reflects energy in and energy out. It is less sensitive to CAPEX. The EROI remains high for wind power even if the location was poorly selected and the wind seldom blows or is weak. Sub-prediction performance over time is also a risk for photovoltaic energy and hydrokinetic power (Chap. 10). When the "fuel" is free, EROI is high, even when operating costs, servicing debt, or other non-energy costs are high. EROI is more often used by academics and government agencies, while calculations that ignore the energy balance, such as Return on Capital Employed (ROCE) are more often used by industry.

$$\text{Economic outcome} = \frac{\text{Cost of supplying energy}}{\text{Income from supplying energy}}$$

The longevity of any installation becomes important to the economic outcome or any segment of it because, over time, factors related to both sunk energy (investment plus CAPEX) and financial costs/income both improve on the income or produced energy sides of the equations.

11.3.1 Funding NGH E&P: Lessons from the Shale Patch

At the beginning of exploitation of any resource, acquisition of funding is always a problem because the risks are highest, with new and often untried technology being developed or employed before predictable production can be undertaken. In contrast, with more than 100 years of production in a mature resource such as conventional gas and petroleum, enough experience exists so that estimation of risk can be well understood.

With the development of a new resource such as shale, in which production went from essentially nothing to nationally significant, financing had to be largely speculative Reserves-Based Loans (RBL) could not be used as collateral because there was no history of production and there was little evidence to underlie production rates, estimates, or production curves that would guide financing. Nonetheless, because early shale production and the development of new technology could be carried out on land with relatively small rigs, small companies with limited resources could undertake exploratory production, but even with relatively small investment shale is a debt-fueled industry (Freeman, 2017).

NGH development is currently in progress and its pre-development situation is similar to that of the shale resource before significant production began in 2005. The state of the reservoir was well understood, as was the aim of creating suitable porosity to allow the hydrocarbons to flow and be recovered. Long lateral drilling, multipad drilling, and fracking provided the technology to access the resource. Although these were developed over a period of decades, the achievement of consistently workable technology appeared to be rather faster than it actually was. Accessing the shale resource was the result of hard work, trial and error, private company innovation, and research support from the U.S. Department of Energy extending over decades.

NGH is an emerging gas resource with little proven production but immense potential. At this writing (October 2017), only a short 5-day production test of the Nankai NGH deposit in 2013 (JOGMEC 2013) and one or two short production tests in the Nankai NGH deposit in the summer of 2017. The Chinese production test in the southern sector of the South China Sea may have lasted almost a month. Little actual data is available from all these production tests that would significantly derisk funding a what would amount to the first long-term NGH production. NGH is a deepwater or ultra-deepwater resource, and essentially more expensive than a development on land or shallow water—if existing deepwater equipment is used. Current progress with NGH production is currently being undertaken by governments because governments of countries, especially those that have an energy deficit such as, but not limited to, China, India, South Korea, and Japan will provide the funding to develop strategic or commercial production of NGH. Because important industrial states have NGH potential, it is likely that initial NGH production will be underwritten by governments, as is being done now. But it is also likely that the subsequent development of much more efficient technology and practices will be carried out increasingly by private industry.

One of the lessons from the shale patch is that development of new technology can be achieved by small companies with vision. It is possible, if not probable, that a low-cost development of new technology undertaken by private companies could be transformative with respect to strategic production and eventual commerciality of NGH (Max and Johnson 2016). Another of the lessons is that initial production results of as little as 6 months will provide a base of experience that will be suitable to generate funding that will not be completely speculative. As more production data becomes available, risk diminishes. However, because many coastal states with NGH potential are currently importing much of their energy, replacing imports with National production will have a strategic rather than a purely commercial (e.g., cost competitive) basis.

11.4 Supply, Demand, and Natural Gas Resources

Commercial potential does not mean commercial success. For instance, until 2004, an apparent shortage of natural gas supplies or indigenous energy resources in many countries promoted widespread interest in NGH there in addition to the deep importing countries, including the United States and Canada. Most of the existing national NGH programs were established during a period of gas scarcity and minimal transport capacity. Today, even with the surge in gas production from U.S. shale, the demand for natural gas is driving megaprojects for conventional gas using floating vessels to produce liquefied natural gas (LNG) for delivery to countries in the Asian Pacific region (Musarra 2015). The development of shale gas resources in China and elsewhere will probably introduce a rebalancing of gas transport. The main interest in 2015–2016 in NGH is only in deep energy importing countries, which will almost certainly develop their indigenous NGH resources (SBC 2015). Countries with a surplus of energy relative to demand have less impetus to pursue NGH. This is especially the case for the U.S., where natural gas prices are generally below production costs for dry gas owing to abundant production from the oil and gas shale resource (Berman 2013).

Hubbert (1956) appeared to successfully predict 'peak oil' but his predictions began to fail in 1995 when crude oil production began to deviate from those predictions (Hussein 2015). Hubbert, of course, essentially extrapolated existing technology, plays, and allowance for estimated future discoveries; and established a paradigm that lasted almost fifty years. Development of new plays, such as deep and ultra-deepwater turbidites, new technology for developing tight plays, and the emergence of new resources for natural gas and oil have changed the hydrocarbon energy resource picture considerably by adding new production. At the same time, increase in demand has been balanced to some extent by energy-saving inventions and efficiencies have reduced demand per capita. Environmental concerns, rather than pure economics are increasingly affecting the demand, availability, and cost of natural gas, as well as other fossil fuels (Chap. 1).

Following the spread of technology that has enabled production of oil and natural gas from tight and shale resources, and the desires of most low-cost producers, such as the Gulf States and Saudi Arabia and Russia, not to lose market share, oil prices had fallen by more than half but prices again are rising (as of March 2018), even though the energy market is and may remain highly volatile. As of early 2016, there is a glut of oil and natural gas, which is now increasingly transportable as liquefied natural gas (LNG). For instance, the LNG contract price for end of year 2015 was often substantially below \$10 and prices remained volatile and on a lower trend as more supply comes into the market in 2016. That is half of the price from the previous year, and will strongly affect the profit on monetization of U.S. natural gas exports. Indeed, because so much incidental gas (gas produced as a byproduct of shale oil production), most dry gas shale wells are 'under water' as the actual production price is higher than current market prices. Not only is this slowing exploration and development of new conventional resources, but new resources, such as NGH, will be stalled until demand increases again or it can be produced at competitive prices.

The unseasonably warm weather of summer 2015 cut natural gas demand and helped take the price down 30% over previous months. The abnormally warm early winter in the eastern U.S. and Europe also cut energy demand. In addition, the global delivered natural gas price has fallen to below \$12 Mft^3 in Japan and only slightly less in India and China, where disruptions of imported gas transport have caused some local shortages (Bloomberg 2015). Demand in East Asia is generally down but must recover if industrialization is to proceed. The Japanese economy is in recession. The Japanese government is pushing hard to restart Japanese nuclear plants, despite the continuing ocean pollution from Fukushima-Daiichi, and some plants were restarted in late 2015.

Falling oil prices, however, are likely to result in higher gas prices because much of the gas produced from the shale resource in North America is incidental gas produced along with oil. As of January 2016, the price of oil is still falling. Scott (2015) estimates that if the price of crude oil falls to \$30/Bbl or below, this will lead to a dramatic cutback of US shale production and a rise in the price of natural gas to about $6. At this price point the Marcellus Shale gas would become highly economic. If the gas price ranges to $6–8, a lot of the less profitable parts of the shale basins will start to look more promising. A lot will depend on what the price is for associated liquids (condensate). At $60–80/barrel some of the gas plays with decent liquid yields will allow for good economics. That will include the Eagle Ford, Marcellus, and Utica. The small liquid yield from dry shale gas is a commercial issue rather than an absolute benefit due to separation costs.

11.5 The Emerging World Gas Market

Natural gas trade has not shown signs of developing a world market until recently. The global gas market is expanding rapidly (Clemente 2017) as gas displaces both coal and oil, resulting in an increasing market for LNG of about 4–6% (2016 into 2017). Oil is a widely traded commodity with a mature global market, has been very different from natural gas or LNG markets in having fewer suppliers and customers.

- LNG prices and contracts have not been as transparent
- Prices have not been widely benchmarked even within regions
- Futures markets are more volatile
- The number of buyers and sellers has been small
- LNG cannot be pipelined long distances as with oil
- LNG tankers are much less common than oil tankers (Lee 2014).

But beginning in about 2015, disruptive changes have been occurring in the gas market. Ample LNG supply connecting to new pipeline infrastructure in consumer countries and the use of long-distance pipelines connecting gas sources in central Asia with customers in Europe are the tipping point of a world gas market. More LNG and more pipelines are anticipated in the immediate future. Abundant transportation increasingly will allow gas to be traded with a global price that is expected to become independent of oil (Addleshaw Goddard 2017).

Increasing natural gas transportation capability will allow the commodity to truly 'go global'. LNG investment in both manufacturing LNG by producer countries and in gasification in importing countries is growing dramatically worldwide. Asia is the current center of gravity of the global LNG trade and has the potential to develop the greatest demand. Gas supply in Asia has primarily been provided by Australia. Older fields off NW Australia are increasing production and new fields are scheduled to come on stream. Producers everywhere are gearing up to meet the Asian and European demand, although in 2016 supply considerably exceeds demand and LNG prices have been driven down. Nonetheless, producers are competing for market share.

In the European Union, presently composed of 28 countries, 26 LNG import terminals are in operation, with gasification capability of a little less than

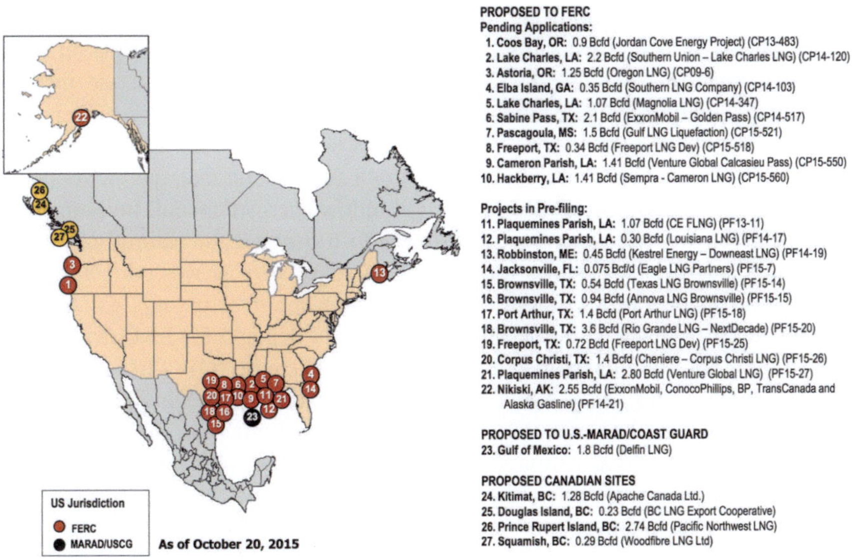

Fig. 11.2 Proposed North American LNG export terminals from FERC (U.S. Government), 2015

195 Bcm/year (6.9 Tcf/year), with an additional 23 Bcm/year (812 Bcf/year) under construction. Total LNG import capacity exceeds recent Russian exports to the region (DW 11/9/15). New LNG supply export from the U.S. Gulf Coast and possibly from Alaska will add to supply. There are many LNG plants in the United States and more are under construction (Fig. 11.2). The focus is to monetize the value of U.S. natural gas, which on a Btu basis is much more valuable in LNG import countries, particularly in Asia. Some chemical companies and other industrial natural gas consumers worry that if natural gas exports become too large, it will have the effect of raising domestic gas prices and inevitably cutting profit margins by raising the cost of their principal industrial feedstock, which would undercut their international competitiveness (Tullo and Johnson 2013). However, there is at present no indication of this paradigm.

The list in Fig. 10.1 does not include LNG plants like Cheniere LA, where export began in February 2016 (Otani 2/24/16) and Cove Point, where the permits already have been awarded. If all of these plants are permitted, built and operated at 100% of capacity, the proposed LNG production volumes will be very large. Some individual facilities will have large production volumes, including Exxon's Sabine Pass plant with 2.1 Bcf per day. Production from facilities approved and under construction (November 2015) is 9.22 Bcf while pending applications propose another 14.33 Bcf of production for export. The US will shortly have 5 LNG plants in production; all will be for export with more possible. Possibly the most important factor, however, is that LNG from the Gulf Coast can reach Japan via the Panama Canal in about 20 days while a cargo from the Persian Gulf region takes 30 days (Chia 2017) and transits a much less secure route.

The LNG industry worldwide has always faced an uphill battle because of its capital-intensive nature. Recently, however, the low price of natural gas in some producer countries, like the U.S., has driven investment to monetize the gas resource as a liquid with a high energy density. The projects now under construction are set to come on stream broadly as scheduled. This is because large upfront capital costs have already been incurred. But beyond that, new LNG plants will struggle to get off the ground in a period of low energy demand. Investment on LNG facilities has risen recently in response to a demand for natural gas. If the more optimistic of the forecasts for the U.S. shale resource proves correct, the US could be a major LNG exporter. The traditional approval process for onshore projects remains slow. Also, the downturn in world demand for LNG will inhibit expenditure, but LNG development is projected to increase (DW 2015).

The International Energy Agency (IEA 2016) projects major shifts in global gas trade over the next five years that will assist the development of a globalized gas market, even if increased nationalism increasingly permeates international affairs. Bloomberg (Malik 2017) notes that the LNG market has become more fragmented and now characterized by a spot market similar to the global oil market. The cause for this increased market flexibility is considered to be primarily due to the initiation of U.S. exports of shale gas, which is expected to rise rapidly through 2019.

In addition to LNG deliveries for power generation and industrial and domestic use, new markets for LNG are springing up with the broadening of the supply base. For instance, the second LNG-powered offshore supply vessel (OSV) chartered from Harvey Gulf International Marine, has begun routine work in the Gulf of Mexico (Offshore 3/12/16). TOTE (Bonney 2015; JOC 2015) launched the first LNG-powered container ship in April 2015. Construction of ferries and possibly cruise ships using LNG fuel instead of oil are planned or under discussion by at least three ship builders. Other ships are in construction and at least two diesel-powered ships are undergoing refit to use LNG. As refueling stations are installed at both LNG production and gasification facilities, the use of LNG for ships will become increasingly attractive.

Thus, the growth of both supply and demand, in addition to a broadening of the use base, are accelerating. The maritime use of LNG, which is essentially a pollution-free fuel with the except for CO_2 and water vapor emissions, is particularly attractive from an environmental point of view, particularly with those familiar with the normal black clouds of exhaust from ships powered by older diesel or bunker fuel engines.

Over the next two years at least, the LNG and the separated liquefied petroleum gas (LPG) market will need to cope with a flood of new supplies coming from new countries such as Vietnam (PEES 2016). GIIGNL (2015) projects global LNG export capacity to increase by more than 40% by 2020, with 90% of the additions coming from Australia and the US. Nearly half of the incremental supply is due to be on line in 2016 and 2017. As LNG supplies surge over the next five years, Europe will be an important market for U.S. LNG; and European LNG imports are projected to roughly double between 2014 and 2020. Many of the largest LNG plants are in Australia and the Middle East, far closer to Asian markets than are plants on the U.S. Gulf Coast. Despite the foreseen increase in LNG intakes, the report does not expect a meaningful reduction in European imports from Russia, which are projected to remain in the 150–160 Bcm (5.3–5.7 Tcf) range.

LNG from Alaska is particularly attractive to Asian countries, especially Japan (Itoh 2013). Whereas there is potential for disruption of LNG shipping from the Persian Gulf and SE Asian with choke points such as the Straits of Malacca, and potentially the South China Sea, there is virtually no potential for disruption of shipping between Alaska and Japan. Alaska, however, is a special case in that its LNG plant is in the "pre-permitting" stage because the cities along the Pacific coast do not now have a source of gas. Thus, it is strongly in contrast with LNG that will be shipped from Louisiana and Texas, whose plants have immediate access to a superabundance of natural gas.

The United States used 27.47 Tcf of natural gas in 2015, of which over 10 Tcf was industrial/commercial, over 8 Tcf was electrical power generation, and over 5 Tcf was residential (EIA 2016a). About 2 Tcf was for production, transmission, and processing. Use of natural gas in transportation (buses, cars, and trucks) was below 0.05 but based on the increased use of natural gas for internal combustion motors and as a feedstock for fuel cells, this is a market segment in which there could be considerable scope for increased use. New bus conversions will likely yield a higher US gas demand than shipping.

The United States has excess natural gas (as of January, 2016) and may continue to have for a considerable period of time. A period of 'American dominance' of the oil and gas market based on shale production is possible (Snow 2017), but the economics of competing gas suppliers limits the extent U.S. gas can penetrate markets in which less expensive gas can be supplied. Although the demand for U.S. and other export LNG will vary depending on the demand in importing countries, deep importing countries such as China, South Korea, India, and Japan will require LNG imports unless they can become more energy sustaining if not independent by exploiting their indigenous NGH deposits. Improvements to the size of ships that will be able to transit the enlarged Panama Canal will allow very large gas carriers (VLGC) to use the canal, dramatically opening up the potential from LNG export from the U.S. Gulf Coast to Asia. The Panama Canal will now be able to accommodate 88% of the global LNG carriers, compared with only 8.6% previously (Thomas 2016).

11.6 A World Price for Natural Gas

Oil has been priced as a world commodity for over 100 years. There are oil benchmarks that allow all oil to be classified and priced. The two main or benchmark oil types are West Texas Intermediate (WTI) or Texas light sweet and Brent crude from the North Sea. In addition, Nigerian Bonny Light, Dubai Crude, Oman Crude, Urals oil and the OPEC Reference Basket may be used. This is possible because the transport costs to bring oil from producing areas to markets were a relatively small part of the overall price, and generally were priced into an average delivered cost. As oil and gas prices decrease, the relative impact of transport costs increases, which in a low energy cost environment is a stimulus for countries to develop indigenous, nearby supplies.

The value of natural gas, whose energy density can be increased by compression for transport in a CNG or LNG tanker (Chap. 7), traditionally has a delivered price that can substantially increase its price in the producer country. In a producer market such as the United States, with an extensive pipeline network, there may be only a small markup between wholesale and retail prices, especially where there is competition between two or more providers. Long-range gas transport involves compressing the energy density of the gas to LNG, which can involve high CAPEX and shipping costs. For instance, prior to the onset of low prices in late 2014, natural gas sold in the continental US was priced at less than $4 per Mcf whereas the same amount of gas delivered as LNG from the Middle East to Europe was in the $10–$12 ranges and the gas delivered to Asia, particularly Japan, was priced at nearly $20. This led to a strong promotion for LNG exports from the U.S. to monetize the value of North American gas, which was being produced in abundance from the shale resource. On 24 February 2016 the first shale gas LNG was exported from the Sabine Pass terminal in Louisiana (Gronholt-Pedersen 2016).

A revolution is taking place in the gas industry. The producer price of gas tends to be low, particularly with respect to the tight and shale gas being produced in the United States. In addition to gas from major traditional exporters, such as those around the Persian Gulf and new exporters that are springing up in the Americas and around Africa. Some previously stranded gas will be liquefied on site and transported directly to market without first touching shore. Liquefaction is the primary means by which the gas can be monetized, that is, sold for a much higher price per Btu than in a market area closely associated with the area of production. In the United States, an extensive existing pipeline network exists to distribute natural gas, and this sets a very low base price against which delivered gas cannot easily compete. With the increase in LNG manufacture and transport, more gas can be shipped from all over the world to markets. As the quantity of delivered gas increases, a complex competitive market between different LNG suppliers and indigenous low-cost gas is inevitable. Over time, this should have the effect of merging the level of producer and delivered prices for natural gas. It is very likely that a world gas price will emerge as a result of the greater availability of delivered gas and lower price of many indigenous natural gas-rich countries, such as the United States. It is possible that an impact of natural gas export from the United States, will be to hasten a tightening in the availability of gas sooner than many might expect as it is being extracted from National supply.

For global oil supply, there has been a lot of additional oil coming on stream. China's is marginal at present as the new production may be barely replacing declines in existing fields. The big increases are from Brazil, West Africa, and the Gulf of Mexico (primarily deep water). The pre-salt plays offshore Brazil are now producing over 700,000 barrels per day and that trend is relatively new. Oil from Iraq, Libya, and a few other countries (while small individually) add up to a significant addition. In February 2016, Iranian oil again began to be exported following suspension of international (U.N.) sanctions. Iranian oil could add over 2 million Bbl/d to the international total. Oversupply of oil was about 2 million Bbl/d at the beginning of 2016, with global production at 94.1 million Bbl/d, but as of March 2016, some higher cost U.S. production is beginning to decline (Watts 2016). It would not take much of a production decline to eliminate the oversupply, and this would have the effect of driving prices up.

The major western oil companies are struggling through the current downturn but most have continued to pay a reasonable dividend. The investors in trouble are those who bought the small shale companies. Most of those companies have lost half or more of their share price and pay little or no dividend. As noted before, a lot depends on how low the oil and gas prices go, and how long they stay low. A key factor for the major producers is that cutting back on production is not as easy or trouble free as it might sound once a steady production has been achieved. Reservoirs can be damaged when high-capacity wells are shut in. An interesting aspect of the oil and gas business is that the large deepwater fields are on stream for many years and many more of them are being developed. By comparison, each oil shale well has a very limited duration, even with recompletions. For the deepwater Gulf of Mexico, it is likely that even at $40/Bbl, the majors can cover their operating expense (OPEX) on the big fields. The capital expense (CAPEX) is another matter, but it is probable that the majors are viewing CAPEX as a sunk cost, accounted for in prior years as platforms were built and wells drilled. This is true for a self-funded company but not where the money has to be borrowed (as is common in the shale industry).

11.7 NGH Production Factors

There are a number of factors that have to be taken into account in considering production from a NGH resource. Elaborate financial models exist for predicting commercial potential of conventional deepwater hydrocarbon deposits. These take into account the full range of factors governing commerciality, with the critical factor of the lowest price per barrel or energy equivalent (BOE). Mahabadi and Jang (2014), for instance, model water and gas flow in a NGH reservoir subject to depressurization dissociation. However, the sediment host has tended to be coarsely modeled, both with respect to sedimentology and detailed stratigraphy. The effect of course modeling on production is not well understood (Boswell et al. 2017). Although gas and water movement through permeability has been the focus, and sediment complexity would likely introduce negative factors, heat management within a depressurized dissociation producing reservoir might be improved if dissociation was not uniform. A P50 resource estimate is commonly used. This would include water depth, logistics, exploration, production, landing costs, CAPEX and OPEX, amongst other factors (Etherington et al. 2005) but must input the production modeling from a NGH deposit. We focus on the technical issues concerned with critical production factors that allow for comparison with other natural gas resources, such as those discussed by Anderson et al. (2006).

11.7.1 NGH Conversion Techniques

Stable NGH must be controllably converted to its constituent gas and water so that the gas can be collected and produced from the deposit. There are three methods for causing dissociation (Table 11.1), which is the endothermic chemical reaction for breaking down NGH crystal structure. Controlled dissociation methods have been discussed in detail by Max and Johnson (2011) and have been colorfully summarized by SBC (2015).

The essential nature of NGH occurrence in a reservoir has been ground-truthed by downhole measurements, capturing pressurized core on that can be used for experiments, and in carrying out technical production tests. Early production modeling up to the late 90s suggested strongly that NGH production would be all but impossible without heating because it was assumed at that time in most models that the NGH was a solid impermeable mass in the reservoir. A more open model for the habit of NGH crystalline aggregates was strongly suggested as a result of the production testing in the permafrost NGH deposit in the Mallik test tests from 2002 that model was disproved. The best method for long-term NGH conversion (Max and Johnson 2011) appears to have been resolved by in situ testing during the March 2013 tests at the Nankai deposit of SE Japan. These showed that the front of dissociation emanating outward from a depressurized well bore was on the order of metres per day. This suggests that substantial permeability in addition to that predicted in present production models existed even in a highly concentrated NGH deposit.

Depressurization dissociation: The weight of scientific and practical experience is agreed that depressurization is overwhelmingly the most energy and hydraulically efficient (Chap. 5) method for long-term production (Moridis and Reagan 2007a; Moridis et al. 2008). Depressurization has been used successfully in both Mallik permafrost hydrate (USGS 2014) and the marine environment for technical production testing by Japan in 2013 (JOGMEC 2013) and Japan and China in 2017 (Brown 2017).

Table 11.1 NGH conversion methods

Regime	Process
Dissociation	Thermal stimulation, (Heating)
	Depressurization
	Inhibitor injection
Molecular exchange	Diffusion replacement
Dissolution	Water circulation

Thermal Stimulation Dissociation: Heating alone would be much less efficient than depressurization (Hancock 2008; Walsh et al. 2009), largely due to the time required for a dissociation front to propagate through the NGH-bearing sediments and the essentially wasted sensible energy that would have to be expended in heating reservoir sediments that form the greater part of the reservoir. If dissociation-induced cooling begins to inhibit dissociation, some heating of collector wellbores might be necessary to ensure flow assurance. Zheng et al. (2011) modeled gas production using thermal stimulation for a NGH concentration that has been identified in the Shenhu Area, South China Sea. They concluded that the production cost and the rate of gas production did not meet the industrial criteria for production.

Thus, more energy is required either by heating, depressurization, or any other means, to alter conditions from stable to unstable. As all of the large concentrations recognized to date are immediately at the base of the gas hydrate stability zone (GHSZ), however, it would appear that close to the minimum energy for NGH conversion would be required for most oceanic NGH deposits as they occur near the base of the GHSZ, which is the NGH phase boundary.

Inhibitor injection dissociation: In this method a chemical that affects the position of the phase boundary would be injected into the reservoir to render some NGH unstable. This method has never been seriously considered because the cost of the chemicals, issues associated with its injection, and the risk of leakage of the chemicals into the environment. The combination of overall greater and more complicated infrastructure that would be required combine to give this method a very low chance of commerciality.

Molecular Exchange: This method consists of pumping a Class 1 hydrate forming gas with a preference for hydrate formation similar to that of methane into a NGH deposit. Because the injected gas would be at a very high vapor pressure compared with the methane in the NGH, the injected gas would replace the methane, which would be expelled, concentrated, and produced. The original driver for this method was to use it for geological sequestration of CO_2, with the incidental positive byproduct of methane production.

In 2011 and 2012, the Iġnik Sikumi Gas Hydrate Exchange Field Trial was conducted in northern Alaska to test the viability of gas injection as a hydrate production strategy and to assess the potential for using N_2–CO_2–CH_4 molecular exchange. This was originally conceived of as a straightforward CO_2–CH_4 exchange to produce methane from NGH reservoirs. This scientific field test was designed simply to test the concept. It was carried out by ConocoPhillips in partnership with the U.S. Department of Energy/NETL, JOGMEC, and the U.S. Geological Survey in a well within the Prudhoe Bay Unit of the Alaskan North Slope using a 'huff-and-puff' system that was known to have no commercial significance because it would save the cost of drilling other wells (Schoderbek et al. 2012).

The Iġnik Sikumi test was also the longest-duration field test to-date of a depressurization of a NGH deposit. Important insights were gained about: (1) the nature of NGH in the western part of the Prudhoe Bay Unit; (2) the requirements for gas injection into NGH deposits and the efficacy of permeability even within high NGH pore fill; (3) the response of NGH reservoirs to varying depressurization; (4) the interaction of water within NGH-bearing strata here; and (5) thermal, physical, and chemical property changes that took place in the NGH reservoirs during mixed gas injection at relatively high pressure and subsequent depressurization.

Dissolution: A recirculating water system was suggested by Max et al. (2006) in part because of the low energy required by dissociation and in part because it could possibly work in muddy sediments having vein, nodule, and dispersed NGH. Dissolution would be most applicable for a deposit confined to specific regions. Analogs of mining practices could be employed so that a large area of seafloor would not become geotechnically weak and reservoir stability could be maintained, even though there was little or no primary porosity in which produced gas and water would flow. We regard this as a method for possible use on second order NGH deposits in muddy sediments. It would not be competitive with depressurization for production in sands, even though less energy is consumed in dissolution (Max and Johnson 2011).

11.7.2 Production Rates

The degree to which a well can be considered as 'good' or not often comes down to where it is located. Onshore wells can be good producers at 2 or 3 MMcf/d, and a sustained 1 MMcf/d would be good for some operators. A considerable part of the 'good' or 'bad' equation depends on CAPEX and OPEX. Some of these wells are cheap to drill, and once on-stream production continues to flow for many years with little additional work. The metric often used is number of days to payout—recovery of all initial costs (Downing and Livingston 2015). Gas composition is also a factor. Gas with a lower production rate but an energy content of ~1,000 Btu/cf (and no H_2S) would be preferred to a higher rate and poorer quality.

The rate of production for NGH is appears to be competitive with conventional gas. The simple drill stem technical test carried out at the Nankai field SE of Tokyo in March 2013 sustained a relatively steady flow rate on the order of 706 Mcf/d. In a production test or in actual production, a much longer well intersection with the NGH reservoir would be used, which would almost certainly bring production up to acceptable standards. Models for NGH production from better known permafrost hydrate estimating cumulative production and gas recovery for 50 years from large concentrations (Fekete 2010).

A production rate that would be considered "good" depends on where the well is relative to pipelines, gas composition, price, floating natural gas (FLNG) vessels and support, and other factors. We propose that commercial production levels in shallow water and onshore U.S. of about 3 MMcf/d would be a good well and 5 MMcf/d or more would be a great well. In the deepwater GOM, in which oil is the preferable product, there is little gas infrastructure so basically gas without liquids is considered noncommercial. Associated gas is reinjected for pressure maintenance, used on platforms, or (rarely) flared. Flaring requires permits and is generally not practiced as in the past.

11.7.3 Permeability in a NGH Concentration and Its Significance for NGH Conversion and Gas Production

The most important factors for long-term production from the oceanic NGH resource are:

(1) Can dissociation be initiated and maintained through a large enough volume simply and using a low energy method that does not require the use of chemicals or application of other more energy-intensive methods?
(2) Will converted gas flow at a high enough rate for gas production to reach acceptable production of at least 1–2 million cubic feet per day?

The answer, from the empirical evidence of short term production tests and pressure core data over more than a decade appears to be yes for #1 and very likely for #2, given the right well design and production plan. Permeability is the key to the efficiency of depressurization-dissociation and natural gas production (Mahabadi and Jang 2014).

Permeability is the connected pore space in a material through which fluid flow will propagate and the rate at which it flows is a combination of the relative ease with which the fluid will pass expressed in Darcy measurement (usually as millidarcy or mD) and the viscosity of the fluid. There is a huge but scattered database for flow of gas, water and oil through a wide range of porosities in which information on low pressure and temperature water and gas will provide a good analog for NGH production for those requiring quantification. In addition to the macroscopic pore permeability, NGH may have a secondary micro-permeability.

In the first technical production test in 2002 in the permafrost NGH in the Mackenzie Delta where according to NMR logging the sediment pore space fill was 80% hydrate, 20% bound and free water. Effective permeability exceeded previous estimates (Kurihara et al. 2012; Noguchi et al. 2011, 2012). During the dissociation test, there was high delivery of fluids, with pumping lasting for over an hour without drawing out enough pore water to reduce formation pressure below NGH stability. This was a proof of concept that confirmed the NGH pay zone retained substantial permeability for long-term production (Uddin et al. 2012), regardless of some modeling assumptions about hydrate growth sealing porosity. There was excellent evidence of high permeability even at very high pore saturation of NGH. NGH was load bearing but in pore water, not coating grains. Grains water wet NGH grains.

The same high effective permeability was proven on the Alaska North Slope at the Mount Elbert well where similar fluid and pressure results were achieved at 75% hydrate, 25% water, with permeability calculated to be in the 0.12–0.17 mD range. According to Anderson et al. (2011), Hunter and Digert (2014), Moridis et al. (2010, 2011), and Wilson et al. (2011), all of the model simulations showed a remarkable agreement in the prediction of long-term production. The predicted gas rates, the cumulative produced gas, and the characteristic reservoir performance were all in good agreement with the empirical production tests.

The Igick Sikumi CH_4–CO_2 replacement test (which used two-way flow in a huff-and-puff molecular replacement experiment in which very high vapor pressure CO_2 and N_2 replaced CH_4 in permafrost NGH), also showed the same high effective permeability during the production test, even though permeability measured from cores was on the order of >0.2 mD (Anderson et al. 2014).

The question is whether the same high permeability measured in the permafrost NGH also applied to oceanic NGH, which is where it is estimated that more than 95% of the NGH resource is located. The factor of whether permeability of an oceanic NGH deposit is sufficient to support long-term production appears to have been resolved in the AT-1 test carried out in March 2013 at the Nankai deposit SW of Tokyo (JOGMEC 2013). In this test, monitor wells at ATI-MT and ATI-MC, each about 25 m distance from the wellbore ATI-P that was depressurized, indicated cooling related to dissociation within 3 days (Fujii et al. 2015). Dissociation began in the main well within ten hours (Kanno et al. 2014) and in the monitoring wells subsequently.

Results conformed with pressure-core testing (Inada and Yamamoto 2015), which showed and effective permeability of $\sim$47 mD for the sandy sediments (Kanno et al. 2015) state that effective permeability for hydrate-bearing sandy sediments was in the range of 1–100 mD, which was 2–3 orders of magnitude higher than conventional estimates. The test confirmed that permeability was high enough to map the advancing margin of the dissociation envelope that migrated outward at an average rate of between 7 and 8 m per day. A line of monitoring holes would be necessary to map progress of the dissociation front in future production tests.

Although there is not yet enough permeability data to be sure on the basis of the entire play, this would indicate effective permeability (suitable for NGH depressurization-dissociation and gas production) may be up to an order of magnitude higher than that seen in the permafrost NGH and much better than assumptions for permeability used in thermodynamic modeling of dissociation performance. Boswell (2016) notes that recent evaluation of pressure cores acquired in oceanic NGH concentrations of SE Japan suggest that in situ permeability ranging from 1 to 100 mD are common, as contrasted with permafrost NGH permeability around 0.1 mD. Thus, it appears that oceanic NGH concentrations are likely to generally have good effective permeability. This is a very positive indication for long-term NGH conversion and natural gas production.

In addition, it has been found that a heterogeneous reservoir may be more productive than homogeneous because the shale bands act as heat exchange material in the producing reservoir. In this case, which is common in marine turbidites, absolute continuity of the deposit may be worse for production than a more micro-stratigraphically complex NGH reservoir. Heterogeneity, which may be a negative in a conventional reservoir, appears to be beneficial for NGH production. Overall, the base production technology of depressurization-dissociation appears to have been proven as a concept.

Short technical production tests have shown that by dropping pressure to initiate NGH conversion, the well remains well behaved and obtains stability temperature conditions very rapidly upon dropping pressure. Equilibrium appears to be reached for all variations of depressurization that have been observed, and constant temperature production prevails, which is essential for long term production without any artificially added heat to counteract the endothermic dissociation reaction.

11.7.3.1 Does Concentrated NGH Have Micro-permeability?

In permafrost NGH deposits, a thin layer of bound water remains around grains that is substantial enough to support bacteria and allow for electron exchange (Price 2007). This bound water layer also occurs in permafrost and oceanic NGH deposits. Although the physical effect and activity of this boundary layer in a depressurizing NGH reservoir has yet to be studied, we propose that the water layer can be explained most simply as an effect of an incomplete crystallization process in which complete cells cannot be formed because of a space or packing factor.

Where a water layer can support bacterial activity, it must be many water molecules thick and conform to the concept that the grain margins of NGH crystals consist of a nexus of defects and dislocations with randomly-oriented water molecules that is not wide enough to form a new water molecule cell large enough to enclose a gas molecule. This water layer is a function of incomplete crystallization at a physical barrier that stops the extension (growth) of the NGH crystal structure.

Thus, water that remains available for crystallization within a field of NGH stability remains as liquid water. In addition to NGH-sediment grain interfaces, interfaces between different crystals of NGH, which will rarely be aligned with their crystallographic axes co-oriented, will also have a high degree of misfit. Thus, aggregates of NGH crystals may key to micro-roughness on surfaces of sediment grains while diffusion of water molecule flow paths exist along the interface. Where continuous diffusional paths in the water molecule boundary layer are suitably wide and oriented, it may contribute to effective permeability even in very highly NGH-saturated deposits.

How thick this boundary layer of disordered water molecules may be and what effect on dissociation and flow it may have remains to be determined, but the possibility exists that it could contribute positively to NGH conversion for gas production. The water layer between NGH grains and between NGH crystals and sediment particles may provide a micro-permeable path in a depressurization framework, in addition to inter-grain permeability.

11.7.4 Production-Oriented Classification of Oceanic NGH

Although a number of classes of NGH deposits have been defined based on their thermodynamic properties controlling dissociation (Moridis and Kowalski 2006), the oceanic NGH thermodynamic classes having the greatest first order economic potential are geologically very similar (Table 11.2). Only Class 1 and 2 are likely to have production potential at the present time. Class 1 NGH concentrations are typically found in a permeable sand—silt host bed that is over- and underlain by impermeable strata. In the permeable host bed below the GHSZ free gas fills pore space directly under NGH. Class 2 NGH concentrations are found in the same permeable geological situations but are underlain within the permeable horizon by pore water that usually has a high concentration of hydrate-forming gas (Table 11.2).

NGH Class 3 deposits have been modeled for its production characteristics (Pooladi-Darvish et al. 2008). Class 3 deposits have no mobile gas or water phase and may be restricted to permafrost-related occurrences where water is contained in the form of either ice or NGH. NGH in permafrost terranes can occur within ice or subjacent water zones and both the occurrence and genesis of permafrost NGH can be very different from oceanic NGH (Max et al. 2006). In rare situations, essentially dry concentrations of NGH might occur in marine sediments surrounded by impermeable sediments. Even shales that look dry, however, may have considerable water present, thus we regard the word 'dry' as not enough water to participate in a

Table 11.2 Oceanic NGH classes 1–4 established for thermodynamic modeling of production characteristics

<table>
<tr><th>Class</th><th>NGH Underlain by</th><th>Host</th><th>Bounded</th><th>Migration system</th></tr>
<tr><td>1</td><td>Gas</td><td rowspan="2">Sands, silts</td><td rowspan="2">Higher permeability sediments</td><td rowspan="2">Mobile water probably active, possibly free gas</td></tr>
<tr><td>2</td><td>Water</td></tr>
<tr><td>3</td><td>No gas or water</td><td>Probably porous beds</td><td>Uncertain</td><td>None active</td></tr>
<tr><td>4</td><td>Dispersed grains, nodules aggregates and veins</td><td rowspan="2">Muddy, fine-grained sediments, Low Permeability</td><td>Few permeability boundaries</td><td rowspan="2">Low-flow, diffusion</td></tr>
<tr><td>'5'</td><td>Vein-type</td><td>Fine grained sediments, low permeability, confined to beds</td></tr>
</table>

chemical reaction to form hydrate. We regard this type, which has yet to be observed in marine sediments, to be of little practical interest as a potential oceanic NGH resource.

Class 4 and 5 are essentially the same in that they occur in muddy sediments that have no obvious gas production methodology available. Classe 4 is geological case modeled for their thermodynamic response to dissociation to determine the energy costs of conversion of NGH to its constituent gas and water (Moridis and Kowalsky 2006). These computer intensive numerical models have also been used to estimate production profiles (Sect. 11.7.5). The best concentrations appear to be formed by focused flow of pore water mineralizing solutions into and within permeable sands and silts. The NGH classes have been used for numerical modeling of production rates, energy consumption, etc. they are also useful for describing the physical disposition of potential deposits.

Revised from Moridis and Kowalsky (2006). Class 1 and 2 of potential commercial interest, Class 4, and 5, NGH in shales currently not of commercial interest. Class 3 may be restricted to permafrost NGH. Class '5' refers to stratabound NGH veins whose potential is unassessed. A further Class, which is vein-type in permafrost terrane (Max and Johnson 2011) is not yet fully understood and there is no thermodynamic production model.

The NGH deposits having the most obvious economic potential are large concentrations of NGH formed in permeable sediments in which gas dissolved in pore fluids and/or free gas migrates along permeable strata that pass into the GHSZ. Where the fluid migration system is focused, water and possibly gas can move through porosity. This brings dissolved gas and possibly also some free gas into the GHSZ in the right form to allow spontaneous crystallization. Where a geological trap such as an anticlinal occurs, however, water may not percolate freely up-dip but gas may. In this case, water movement may be primary by diffusion of water vapor within gas-filled pores (Table 11.3; Fig. 6.3 and CD in Max et al. 2006). In an anticlinal closure, movement of gas and water diffusion may be the principal means of supply of reactants into an overlying GHSZ.

In a closed structure, NGH will concentrate up-structure as free gas migrates by buoyancy independent of pore water movement. In this case, the petroleum system would closely resemble a conventional petroleum system but up-structure water would be largely consumed in forming NGH rather than being displaced down structure by concentration of free gas. Flow of dense pore water carrying salts rejected during NGH formation may affect water circulation, especially in restricted flow situations and secondary circulation is to be anticipated. NGH is closed structures may have potential for higher NGH pore fill and longer duration stability in the event of diminished gas flux because no up-dip dissoltion is possible.

It is likely that all concentrations of NGH having production potential will have an active groundwater-fed mineralizing situation being supplied through the base of the GHSZ to stabilize and add to the NGH volumes by driving continued crystallization. There will always be water in pore space below the GHSZ in an oceanic NGH situation. Usually, even where a gas-poor situation occurs, there may be a little free gas present just below NGH. The amount and thickness of the column of

Table 11.3 Recoverable oceanic NGH deposits in bounded, permeable strata linked to subjacent source of natural gas

Physical constraints/ Character	(A) thermodynamic only (no structural trap)	(B) Structural trap
Location	Upward from base GHSZ	Potential for structure filling from base of GHSZ
Underlain by	Gas or water	Gas
Permeability bounded	By impermeable strata, rarely by faults	By impermeable strata
Plumbing system	Open	Closed
Pore water	Mobile, percolating	Static
Free gas	Mobile, Advective ascending	Advective ascending
Sequestration Potential	Removed from pore water by crystallization. Remainder of dissolved gas may vent to ocean	Most gas is trapped. Arrival of new gas dissolves and continued NGH formation. Dissolved gas retained
Relation to conventional trap	Source and migration same but mode of concentration is thermodynamic trap	Very similar
Geotechnical	Structure increases sediment strength, NGH formation increases strength	
Seismic character	Anomaly in permeable horizon usually near base GHSZ	Structure filling potential
Economic potential	May be good but substantial gas will pass through	May be excellent as gas is trapped

Open and closed refer to potential of passage through entire GHSZ

free gas associated with the NGH depends on the gas flux, water percolation-flow rates and the permeability of the mineralized zone, including NGH replacing water in pore space.

NGH is not distributed uniformly within the GHSZ. The natural system causes NGH to be concentrated within sand reservoirs, rather than dispersed through the zone, and this is the basis for its likelihood to be a significant source of natural gas. NGH occurrences range from high-grade deposits with significant concentrations of NGH in sandy strata in discrete bed-differentiated sedimentary sequences to dispersed NGH in poorly defined finer grained muddy beds. One of the general characteristics of NGH deposits is that converted gas from NGH may not be concentrated for easy recovery, especially in oceanic NGH.

In an open system where fluid and gas are migrating within a permeable bed that slopes upward and passes into a GHSZ, the geological differences between Class 1 and 2, may be more controlled by the rate of gas flux at the base of the GHSZ and the rate of gas molecule uptake into NGH within in the GHSZ that would establish upward propagating diffusional gradients. Although it is possible that large concentrations of NGH may occur well away from the base of the GHSZ, all of the large concentrations and the disposition of dispersed NGH found thusfar occur almost immediately above the GHSZ base and extend upward near the base. In practical geological terms, Class 1 and 2 deposits are virtually the same and formed in 'ascending' reservoirs by natural gas-charged fluid flow.

Where gas is in contact with overlying NGH at the base of the GHSZ (Class 1) in a very mobile water system, as at Nankai (Takahashi and Tsuji 2005; Noguchi et al. 2011) and the northern Gulf of Mexico (Boswell et al. 2012) the optimum conditions for both NGH formation and controlled NGH production conversion exist. The optimum situation for NGH conversion to natural gas is where it is concentrated in a relatively thick free-gas zone over mobile water. Where there is no free gas just below the base of the GHSZ, it can be relatively easily produced at the start of production, especially if depressurization is used or where point heating is used to initiate a gas zone, which can then be maintained by depressurization. Once depressurization production is initiated, Classes 1 and 2 occurrences will converge into a situation where free gas will tend to overlie water below NGH, but the thermal structure will be different. Water drive will displace some water but pumping for depressurization may remove more than can be replaced naturally. Once NGH begins dissociation production, gas and water will be highly intermixed within the dissociation envelope—and controlled movement and possible separation of gas and water may be part of a planned production system.

11.7.5 Estimated Production Rate Profiles

Conventional Resources: Conventional giant (3–30 Tcf) and supergiant (more than 30 Tcf) gas fields have been at the heart of global production for many decades. Not only do they contain and produce the greatest volumes, but they have the steadiest and most predictable production. This governs field management and operations cost, cash flow, and profitability over the life of the field. Large fields on land and in shallow water, for instance in and around the Persian Gulf, and deepwater wells commonly produce over 50,000 barrels per day, and can maintain that rate for decades. The production profiles (generalized graphics of (usually yearly) production rate plotted against time) of these high performers show a relatively rapid rise to a steady production rate. The production may remain at that rate for 30 years or more (Bret-Rouzaut and Favennec 2011). At the end of the period of plateau production in large fields, production drop-off often manifests as a sudden drop in production that usually marks the beginning of a steep decline (Fig. 11.3).

The elapsed time from discovery to project sanction is commonly more than five years for spar facilities and in excess of eight years for tension leg platforms (TLPs), the two platform designs most commonly used in the deepest water. For floating production, storage, and offloading facilities (FPSOs) the time to first production from discovery commonly exceeds three years. Thus, it is common for investment to support development for a decade before revenue income is achieved. Deep and ultra-deepwater costs commonly exceed \$1 billion. Shortening the

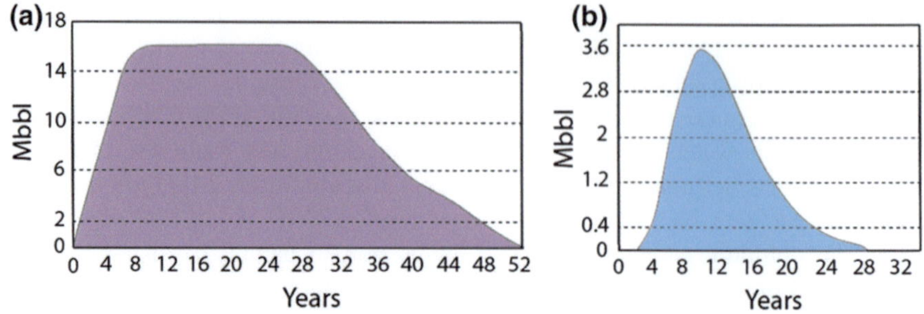

Fig. 11.3 Representative or diagrammatic production profiles redrawn after Bret-Rouzaut and Favennec (2011, p. 105). **a** Characteristic production profile for a large oilfield (~500 MMBbl). **b** Characteristic production profile for a very small oilfield (~20 MMBbl). Fields intermediate in size will vary mainly in the development of a stable production plateau, which lengthens over time with field size. Mbbl is one million barrels of oil equivalent

exploration to development cycle, during which there is no revenue, is a fundamental way of increasing commerciality by reducing the time over which investment carries all expenses. For instance, the Tubular Bells field in Mississippi Canyon offshore Louisiana, USA, remained a project under consideration for more than 11 years, but it was fast tracked from the first investment decision (FID) to production within 36 months (Barton 2015).

For gas, there have been few deepwater well completions in the past decade. There were some deepwater gas fields discovered in the late 90s that are still strong producers, but deepwater has been primarily an oil play. To our knowledge, there has been little or no exploration for deepwater gas for the past 20 years with the exception of China. Husky (2016) initiated a new Chinese gas field, the Liwan gas project 300 km southeast of Hong Kong in March 2014. Liwan consists of three natural gas fields that share a subsea production system, subsea pipeline transport, and onshore gas processing infrastructure.

The average cost of a deepwater well (GOM, East and West Africa, Brazil) appears to us to be about $300 million although remote or very difficult wells have been cost as much as $600 million. Adding to the overall E&P cost is the fact that not all developments are successful, about one in four deepwater wells are not completed or do not reach their production objectives. Each field has its particular production profile, governed by geology and by production and business factors. For instance, after CAPEX has been amortized the remaining production is literally 'money in the bank'. Production can be varied, with respect to maintaining reservoir stability, to take advantage of price swings and other market opportunities, while maintaining positive cash flow.

Smaller conventional fields have shorter lives and lower overall production volumes, but their development costs also can also be much lower. Thus, fully amortized yearly profits at full production can also be attractive, and possibly more attractive in periods of low oil prices (Sect. 11.2). Smaller conventional fields also usually show a similar increase in production to a rounded peak that marks a

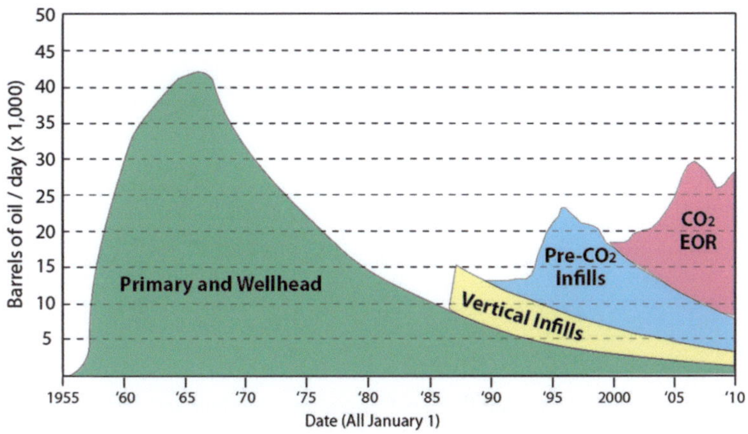

Fig. 11.4 Diagrammatic sawtooth production profile of a field undergoing secondary production (also representative of staged lowering of pressure in coalbed methane and re-fracking of a shale resource well—below). Redrawn and smoothed from a diagram of Weyburn oilfield production in western Saskatchewan, Canada attributed to Cenovus Energy (formerly Pan Canadian, EnCana). Downtime for carrying out secondary production process is not shown

maximum production. This production may last for up to as much as five years, but in contrast to the much longer production plateau of very large fields, begins a steady decline of 5–8% per year in primary production (Fig. 11.3).

The production profiles of fields in which secondary and EOR (Enhance Oil Recovery) is undertaken lengthens the life of the field, the rate of production over time, and the total volume that can be recovered from the field. Primary production usually relates to the natural flow of oil and gas from the reservoir, but in ultra-deepwater conventional deposits, pumped extraction combined with pumped boosting of the gas to a production manifold for transport is now being introduced early in the life of the field. This somewhat blurs the traditional boundary of a period of primary production being followed by secondary production using methods that stimulate further production by altering a number of factors. Chemical treatment can enhance permeability and porosity, gas injection, including injection of liquid CO_2, increases pressure and lowers viscosity of oil; fracking enhances or introduces permeability and pressure; gas injection, can substantially increase pressure; and pumped extraction can increase overall production, especially where there is a strong water drive in the reservoir. Secondary recovery methods introduce a 'sawtooth' production profile (Fig. 11.4).

Unconventional Resources: All unconventional gas resources share one key factor, that their resource is essentially stable under reservoir conditions. This requires artificial stimulation or conversion to free the gas so that it can be recovered. In the case of coalbed methane and NGH, which are both found in permeable host strata, depressurization of the reservoir frees the natural gas. In the shale resource, fracking is necessary to introduce permeability into a reservoir host that has the natural

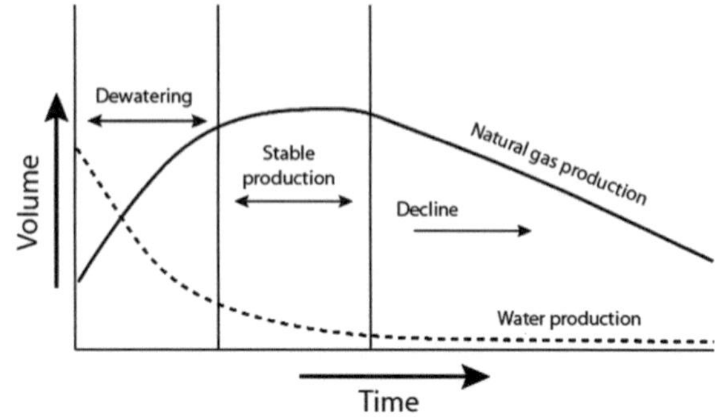

Fig. 11.5 Typical production profile for coalbed methane. Redrawn from U.S. Geological Survey (2000). Solid line—methane, dashed line—water production. Volume and time not enumerated

permeability of concrete so that the gas trapped in abundant carbonaceous-rich pore space can accumulate and flow.

Coalbed methane: As described in Chap. 10, coalbed methane begins to flow relatively soon after the pressure in its reservoir is pumped down and the adsorbed and dissolved gas coalesces as free gas and can flow from the well. Gas flow is generally related to the pressure difference that can be maintained during pump down and the formation pressure prior to pump-down, as free gas can only form when pore water is essentially supersaturated. Gas production is primarily dependent on the degree to which pressure has been lowered if pumping is staged rather than being a single, deep pumping event. Considerable variability exists because production depends on a number of commercial and technical factors that we do not discuss in detail here, but there appears to be a generally smooth and regular production profile (Fig. 11.5) for a single, strong pumping of water from the 'reservoir' (USGS 2000).

Fracking is not a feature of coalbed methane production. Simply decreasing pressure in the reservoir by pumping out water appears to be sufficient to stimulate and maintain natural gas production. In a staged production scenario, production increases when pressure is decreased and then decreases while that pressure is maintained. In a simple maintained pump down plan, staged reduction of pressure will allow for relatively continuous production. In general, the production prolife may either change as pressure is altered or an average constant production can be maintained. At some point the increased cost of pumping down the reservoir pressure and treatment or disposal of increasing amounts of produced water may render gas production uneconomic.

Shale Resources: Fracking was originally developed as a secondary recovery technique for conventional hydrocarbon deposits in the mid-20th century to increase recovery from existing fields in which production using natural pressurization had diminished to near unacceptable levels. In modified form and in horizontal wells, fracking has proven to be the enabling technology for opening the oil and gas shale to commercial exploitation. Whereas in conventional fracking the existing permeability of a reservoir is enhanced, in shale and other 'tight' plays, fractures are opened in the essentially impermeable strata so that an artificial permeability is created. The gas and liquid that was trapped in the rock will then flow.

Determining decline rates is not easy, however, because drilling companies have different approaches to production. Some want to draw down wells quickly, while others might restrict production to extend the lives of the wells or to wait until low prices rebound. The key for drillers who want to keep production totals high is for them to drill additional wells when production starts to decline significantly. But drilling a shale well, for instance, can cost from $6 to $10 million. Then, the most important question becomes, can companies afford to keep drilling such wells in the face of uneconomic gas prices?

In contrast to the relatively long and steady production of conventional wells, early shale oil and gas wells typically produce 65% of their ultimate recovery within a year. Horizontal wells drilled in the Marcellus of West Virginia and Pennsylvania, which are typical of wider shale completions since 2009, show the most rapid decline rates in the first year and 74 and 82% over three years (Hughs 2015). The problem with shale wells is that they can come on at a decent rate, but then decline rapidly (Fig. 11.6). On the other hand, if the production flow rate and gas production is high enough, this may be ideal from a project economics point of view.

Better fracking techniques over time have improved primary production of oil but because the producing reservoir (the fracked region) is the same size regardless of the rate of gas extraction, higher production results in a faster rate of production fall off. This clearly demonstrates that shale reservoirs are very small by comparison with major conventional fields and that only a high drilling and completion rate can maintain overall production. In the most recent production from the EIA example of the Permian Basin of Texas (Fig. 11.7), production rate has been halved in a little more than 7 months. From the manner in which the production profiles

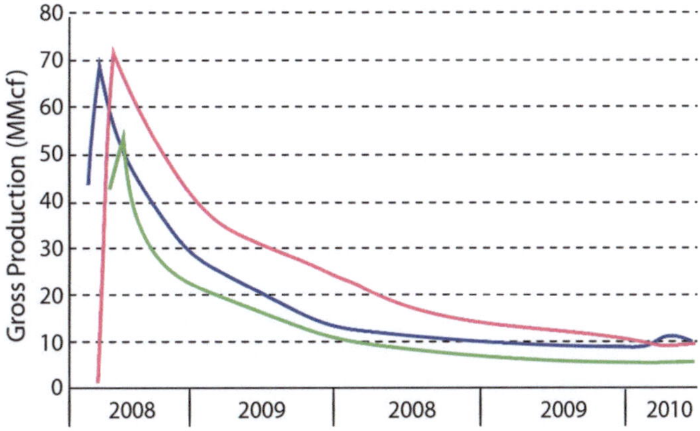

Fig. 11.6 Typical production profiles of shale gas wells showing rapid declines. Smoothed profiles for representative Fayetteville Shale wells after Mason (2011), Fig. 3

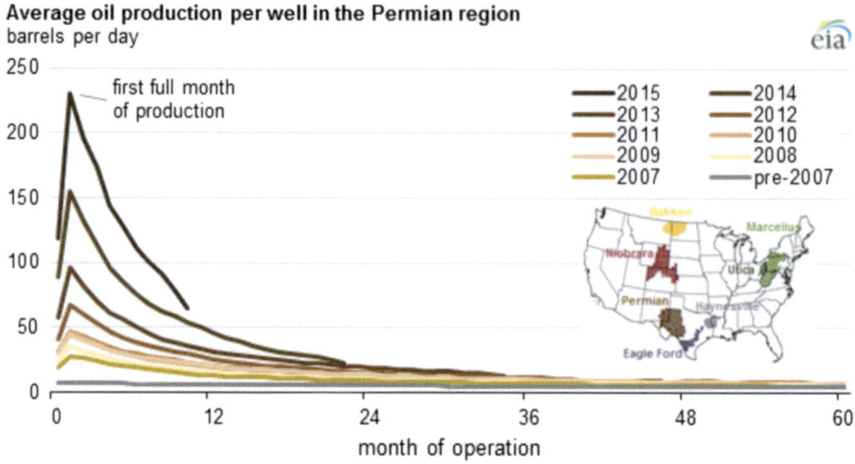

Fig. 11.7 Change in production profiles as a function of year. Data from EIA (2016b), Curtis (2016)

successively show greater slopes, it would appear that the fracked reservoir drains faster overall and reaches only residual gas flow faster. Shale play Initial production (IP) rates have evolved since 2008 when 1 MMcf/d was considered a good well. Within a short time, the IP rose to 5 MMcf/d and recently in some Dry Utica wells over 20 MMcf/d choked IPs have been achieved (Bradford 2016). This does not increase the cumulative production from a single frack of the well. Refracking will increase cumulative production but varies considerably from well to well (Kramer 2017). Production curves from different shale basins are different in detail, probably indicating local geological or fracking variations, but all shale wells show about the same high early peak rates and a rapid drop off in production (Curtis 2016). Well strategy should always consider estimated ultimate recovery (EUR) and the different means and costs of achieving maximum gas production.

Musarra and Beaubouef (2017) note that there are a number of ways to reduce CAPEX for both offshore exploration and production and off-and onshore infrastructure, which is critical for new development. When CAPEX is too high, the reaction is often to 'high-grade' or produce at the highest rates possible to maximize return over the shortest time. This has been particularly true for production from the shale resource (Fig. 11.6), in which the emphasis has been to bring volume production forward. Emeigh (2017), for instance, notes that having a production strategy that favors longer term production often requires lower CAPEX investment, especially in infrastructure (including processing). We are unaware of any studies that compare high short-term production with longer term production that in conventional resources often result in higher overall production, but it is likely that the current practices toward high short-term production waste considerable of the recoverable resource.

NGH: Large concentrations of NGH that are amenable to production using technology and methods that industry has already developed only form in the primary porosity of a permeable host strata (Chap. 2). The NGH petroleum system (Max and Johnson 2014) requires the confluence of an active groundwater system bringing mineralizing solutions of dissolved NGH-enriched groundwater within a permeable bed that passes up into the GHSZ (Chap. 2). A dissociation envelope at least 50 m across can be created around a depressurized wellbore (JOGMEC 2013), proving the effectiveness of depressurization dissociation. Further research into the extent of the envelope will allow for detailed planning for the most efficient wellbore geometry for production.

It is important to remember that whereas production data exists for commercial coalbed methane, shale gas, and conventional gas, that the modeling used to predict production rates and well performance over time can be tested against only three significant oceanic NGH production tests to date and these have been relatively short term. Modeling has come out of the laboratory where uniform quarts spheres or special laboratory sands were used in experiments. The emphasis is now to model conditions of actual pressure-cored NGH deposits so that realistic conditions will yield more realistic results. The models are increasingly based on best estimates for internal stratigraphy and sediment composition, permeability, porosity, and other factors and the results should be used only to indicate trends. In addition, whereas conventional production is based on simple extraction, dissociation conversion of NGH and the movement and separation of gas and water is inherently more complicated and to introduce constantly changing thermodynamic conditions. In addition, because conditions within the reservoir are unlikely to be static even within small regions, nonequilibrium conditions (Vafaei et al. 2012), which are always difficult to model thermodynamically, will probably persist and vary through the life of production until shut-in and the return of stable conditions. Nonetheless, modeling has become much more sophisticated for both sediment variation, heat demand and flow, material transport, and narrowing of confidence of results for both flow rates and projected production performance.

Thermodynamic modeling of Class 1 and 2 NGH deposits, which will probably host virtually all offshore NGH deposits, indicate that large concentrations of NGH will have production profiles that theoretically compare very favorably to those of large conventional fields (Fig. 11.8). Production either plateaus (a) or has an inverted broad U-shaped production peak (b). In both cases, maximum production spans about a decade. Larger NGH deposits probably will likely respond similarly to large conventional fields in that as the deposit size increases, the plateau or average period of maximum production will extend.

The TOUGH + HYDRATE thermodynamic models developed at the DOE Lawrence Berkeley laboratory (Moridis et al. 2008, 2010; Moridis and Reagan 2007a, 2007b) have evolved into an international standard for assessing controlled dissociation. The current iteration is able to input 3D data from which it can produce 2D sections showing NGH conversion progressing (Reagan et al. 2015). The model takes into account the need for heat to be provided to location of the

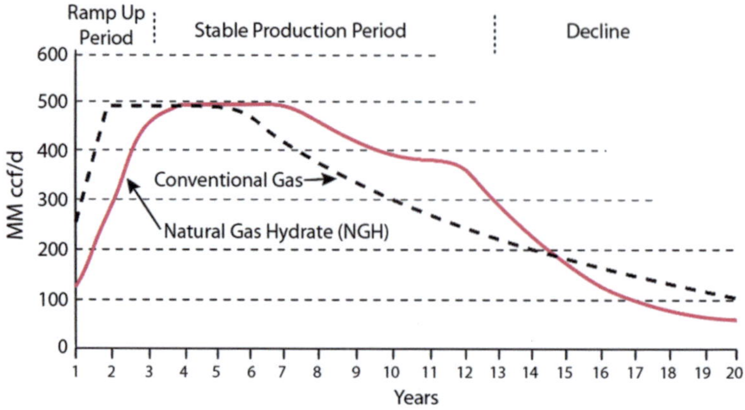

Fig. 11.8 **a** Large NGH deposit production profile, redrawn from Fig. 8 in Walsh et al. (2009) after Hancock (2008), based on Moridis and Reagan (2007b). Conditions are: depressurization only, no thermal stimulation or chemicals used. Mini-plateau at 12 years not explained but probably related to variable depressurization effect

forced dissociation from the natural environment so that the endothermic cooling of converted NGH (Chap. 2) does not drive the temperature down to the point that dissociation will cease. At the determined rates of NGH conversion and gas production, nature is able to provide the heat to allow the depressurization-dissociation reaction to run continuously.

Merey and Longinos (2018), while summarizing NGH production models used in other locations, use the HydrateResSim numerical simulator to determine production character from Class I and Class II NGH deposits that have possibly been identified in the SE Mediterranean Sea. The much warmer Mediterranean seafloor also yield promising results that suggest in only extreme thermal demand situations arising from depressurization of NGH concentrations will some heating of the product gas and water be required to reduce the possibility of ice forming within or near a production well. Where a NGH deposit has a considerable volume of lower grade or lower pore fill, all of the NGH subject to pressure drop will begin to dissociate (Chap. 5). Although we do not know the reason for the two-stage platform shown by Walsh et al. (2009), the NGH profile would be what we would expect for gas production from a bimodal grade deposit in which the low-grade fraction disassociated completely while higher grade volumes continued to dissociate.

There appears to be considerable potential for a long life for NGH wells, with production profiles similar to conventional gas wells. Modeling of gas production from a complex 3D heterogeneous NGH deposit well controlled by drilling data (Gaddipati et al. 2011; Myshakin et al. 2012) also indicates a very high potential for satisfactory production with a production plateau extending for decades. The current models of long well life, however, are mainly based on Alaska examples where the NGH is in discreet, isolated fault blocks, and with a relatively short distance of collector section within the well. Although much early modeling of production from deepwater NGH deposits has been carried out using these models, we would like to see the basic models reconsidered using deepwater physical parameters and more complex wellbores in the reservoir.

The Japanese MH-21 program has also developed a thermodynamic production model. The MH21-HYDRES model was originally developed by the University of Tokyo, and since modified and improved (Kurihara et al. 2012). The model can input 3D data and can solve five-phase and four-component problems. Its features at the time were described by Kurihara et al. (2010). The 3D approach is particularly important in dealing with small fault offsets and discontinuous or patchy NGH. Published results appear to be superficially similar to the TOUGH + HYDRATE modeling. CI STARS, which was developed by CMG limited, appears to be used much less but gives similar results.

In contrast to conventional gas wells, which are at high pressure everywhere within their reservoir at the initiation of production, NGH is present at formation pressure and not in a gaseous form. The gas in the depressurized reservoir will have a lower pressure than the original formation pressure and this will prevent the reformation of NGH, unless the pressure is allowed to rise, in which case NGH will reform rapidly. Thus, the production profile is the result of mobilizing the natural gas by converting the NGH to its constituent gas and water.

Although the production profile from an NGH deposit closely resembles that of a large conventional deposit, different physical processes are underway in each. In a conventional reservoir, the ramp-up to maximum average production is the result of the time it takes to establish flow paths to the well with the establishment of a stable hydraulic pressure gradient regime. If production of a conventional well, particularly if it is a combined oil-gas well is interrupted temporarily, restarting may take a considerable ramp-up because stable flow needs to be reestablished. The onset of decline heralds the beginning of well life without secondary recovery techniques being applied.

In a NGH well, the ramp-up time depends on the exposure of wellbore surface area of NGH to pore water media. Gas production is a function of the amount of NGH surface area that is exposed to the lower pressure gas-water mixture in the pore water. The surface area to NGH mass ratio is a critical factor in NGH conversion and gas production (Chap. 5). In general, the ramp up time to production is similar to that for coalbed methane in that depressurization is the key to production for both.

In contrast to coalbed methane, which requires deep pumping for as much as a year or more before gas production becomes significant, dissociation will initiate as soon as instability in the NGH begins. This can be achieved with a relatively small pressure drop immediately adjacent to the base of the GHSZ. As pressure is pumped down in a NGH deposit, however, gas is produced through dissociation, not exsolution of dissolved gas as with coalbed methane. Also, whereas coalbed pump-down usually appears to be single stage and deep, the NGH dissociation rate may have to be controlled by varying pressure economically to no more than that required to convert the NGH to its constituent gas and water at a desired rate. The intent is to balance the amount of gas in a reservoir with the extraction rate.

In the ramp-up initial stage of NGH production, dissociation is initiated in a particular volume from a well having an intersection roughly through the vertical height of a NGH-rich horizon. We believe that the ramp-up period to stable production (Fig. 11.7) in practice can be substantially shorter than is commonly assumed for production models. We envisage lateral and multi-lateral wells being used, in which far greater volumes of NGH-rich strata will be exposed to pressure drop than is inferred from present production modeling. From our review of the available evidence from the monitoring wells of the JOGMEC (2013) test, and from the apparent rapidity of molecular replacement during the Iġnik Sikumi test (Schoderbek et al. 2012), we interpret that permeability in a natural deposit is far better than previously assumed. This would mean that current modeling results, which are encouraging, are actually conservative with respect to the rate of NGH conversion and potential natural gas production.

The dissociation domain (which is the volume of a region around a depressurizing wellbore that is affected by the depressurization) can be crudely represented by an average radius from a production wellbore. Although a dissociation domain may have an irregular cross-section, the volume within this domain is key to well design for the reservoir. In the dissociation domain, the surface area of NGH that is exposed to pore water is the region in which NGH conversion will take place. Although it is likely that the domain will increase in size as permeability increases outward from the site of depressurization, there should be a minimum volume affected from the onset of depressurization that drives the pressure down and out of NGH stability.

If the domain is narrow and grains exposed to adjacent pore water have a similar mass to surface area ratio, decline could be very fast because once the grains cannot produce enough gas to maintain plateau production rates surface area would drop precipitously. Gas production would decline in turn. If NGH dissociation domains are large, then conversion will continue for a longer time.

We believe that it is possible to use much more complex well systems in the reservoir, having large area collector sections that can be used to initiate dissociation over a large area much faster than assumed in current models. Reservoir complexity has is yet to be fully factored into production models but the few production test results available suggest that actual performance may exceed current modeling based on simple reservoir models. In the main production period, dissociation is controlled by pressure variation for optimal production rates using an automated control system responding to sensors in the well(s). In the period of decline the amount of NGH that has not already been completely converted begins to diminish, with a consequent diminishment of NGH surface area on which dissociation can take place.

The steepness of the decline curve may be dominated by the range of sizes of NGH crystals and crystal aggregates. In the main production stage, an average or plateau rate of NGH conversion will produce gas at about the same rate if about the same surface area of NGH is dissociating, although dissociation may be retarded or accelerated by varying depressurization pressures.

Although we agree with SBC (2015) that NGH conversion modeling has become much more sophisticated, a long-term production test is required to provide real results that can be used to calibrate 3D modeling, taking into consideration not only NGH but the disposition of host sediments. This is because current modeling is assumption driven and some of the assumptions are surer than others. A production test serves the same purpose as a logged and sampled drill hole does for seismic interpretation: that of substantially lowering risk. Instead of flaring off gas from a long-term production test, it would be a preferable to capture the gas and compress it so that it can be transported, if for no other reason than to comply with the spirit of minimizing CO_2 emissions whose combustion has performed no useful work. We suggest that when a long-term production test is successful, it will be extended to transition into a production well if the economics are right. A long-term production test may well turn into the first commercial NGH activity, and perhaps it should be planned that way.

It must be remembered that the onset of parameters that separate especially conventional gas production, shale and NGH is that there is over a century of experience with conventional gas and now that there is a couple of decades of production experience with shale. As we have good evidence that early production characteristics have evolved as technology has improved, we can expect the same to apply to NGH. Presently, NGH data is very sparse, with some minor production tests and no production history. Whatever costs and production factors there may be with the first commercial NGH production, it is a certainty that with experience and evolving/improving technology, production profiles will show continuous improvement.

11.7.6 Pressure Management Summary

Management of reservoir and well pressure for NGH production will be very different from the conventional gas experience, especially in deepwater where conventional gas temperatures and pressures are very high with respect to temperature and pressure at the seafloor and sea surface. With depressurization-dissociation emerging as the primary means for NGH conversion during production, pressure within the producing deposit will have to be lower than the starting and surrounding formation pressure to maintain conversion of NGH to its constituent gas and water. Conversion requirements alter the way gas pressure is controlled in NGH wells from a conventional gas well approach (Table 11.4). Pressure control in different sections of producing wells will be critical to NGH production.

Table 11.4 Comparison of pressure within conventional and NGH production systems

DEEPWATER AND ULTRA-DEEPWATER		
Feature	Conventional	Natural gas Hydrate (NGH)
Formation pressure	Very high	Low
Reservoir pressure	Very high	Lower than formation
Pressure below wellhead	Up to reservoir pressure	Similar or lower than reservoir
Gas lift in well	No	Possibly, as part of pressure variation
Pressure gradient	Decreases from reservoir to surface except where boosted from seafloor	Variations may occur within well system and reservoir as a whole
Gas lift from seafloor	Yes, where pressure is Low. Not necessary where natural pressure is sufficient	Yes, always
Pressure control for flow assurance	No	Yes

See Chap. 2 for discussion of water depth designations

11.7.7 Infrastructure

Infrastructure is the general term for pipelines, production platforms, and all those fixed facilities that have to exist to take oil and gas from the wellhead to land and then to market. If a resource cannot be brought to market, it has no practical value. Availability of infrastructure that can be leveraged for recovery of gas from NGH (Chap. 7) is commonly considered to be a critical factor in reducing investment and decreasing development time from the decision to initiate production. Where tie-backs (connections from new wellheads) can be piggybacked (connected) with existing pipeline infrastructure, for instance, enormous savings can be made in both time and money. But there is usually little excess capacity available. Infrastructure is usually closely matched with anticipated throughput to reduce CAPEX.

In the offshore of many countries with weak regulation, natural gas has been traditionally flared because its economic value did not justify collection and transport. The economics of gas have always been more problematic than oil but since mid-2014 the low energy price structure may make collecting natural gas rather than flaring it commercially problematic for coproduced oil and gas.

Where flaring is not permitted, it is necessary to have a gas pipeline along with an oil line. In some areas in which there is considerable energy industry activity, such as the northern Gulf of Mexico (GoM), the lack of existing gas infrastructure reflects industry focus on oil and conformance with regulations that prohibit flaring. In the GoM every gas pipeline is paired with an oil pipeline because there is more gas than can be disposed of by reinjection or platform use; there is little independent infrastructure for deepwater gas. Some of the GoM oil developments were designed for very high throughput but the associated gas yield is far greater than was anticipated in planning, which may create costly production problems.

If sizable NGH deposits in the GoM were found and the economics of the deposits were opened, they would require their own pipeline infrastructure or a different solution. New infrastructure is sometimes justified. The Corrib gas field to the NW of Ireland, which was only opened in January 2016 (Sect. 10.3) is now producing through an 85+ km dedicated 20-in. pipeline to the gas terminal onshore.

There are few deepwater areas with significant natural gas infrastructure that could be piggybacked by NGH gas, and new pipelines from deepwater to a gas terminal on land may be prohibitively expensive. This calls for a reevaluation of the general means for transport of gas from wellhead to market entry point on land. Any area without existing infrastructure for natural gas may have to be treated as a remote region containing stranded gas, even an active one with deepwater oil pipelines as does the northern Gulf of Mexico. But all the means to transport stranded conventional gas to market are applicable directly to gas from NBH.

11.7.8 Solution for Stranded Gas

Stranded gas is natural gas of any type that is regarded as being too geographically isolated and too far from markets to be economic. The concept applies to any fuel but is particularly relevant to natural gas because of its comparably low energy density (at STP) and the traditional focus of the energy industry on oil. But stranded gas may be on the verge of being stranded no longer. Two factors are currently reducing the number of fields that are stranded. The first is longer and longer pipelines and improving deepwater pipeline technology. The second is new vessel technology.

Both oil and coal can be shipped around the world in their natural state and still be economic, but natural gas has to be compressed, which can be costly. Compression for long-distance transport is usually done by producing Liquefied Natural Gas (LNG). Compressed Natural Gas (CNG) is used in personal vehicles in some regions, particularly in Europe, but there are no long-distance, large CNG carriers although they are being studied due of their lower CAPEX. Transporting natural gas in the form of hydrate has been experimented with, but its transport energy density is about 2/3 that of CNG and about a quarter of LNG. Proprietary cost comparison between NGH transport and other methods are not known in detail by us.

For LNG, the principle cost is in its production and plant cost. The principle cost of CNG may be the fuel for the LNG tanker, and the main operating cost for pipelines is maintaining pressure. A first-order cost comparison between LNG and CNG is the energy required to compress the gas. Although a greater Btu content can be carried in an LNG carrier than a CNG carrier of the same size, and the transport cost of each shipment matters, the cost of compressing natural gas to CNG appears to be substantially less than in manufacturing LNG. LNG consumes about 10% of natural gas throughput or about 2,000 Btu/lb. Plant efficiency, calculated as a percent of Carnot efficiency, varies with size but 16% of inlet gas, or about 3,400 Btu/lb before any energy recovery, would not be far from a commercial average (Sinor 1991). In contrast, the typical energy requirement for compression to about 2,500 psi is on the order of 400 Btu/lb or about 2% of the energy value of the inlet gas.

Another alternative for increasing energy density for transport is to establish floating gas to liquids (GTL) facilities. Relatively small, proprietary STG+TM modular GTL plants have been developed to carry out on-site conversion of natural gas from the Marcellus and Utica shales in the NE U.S. to provide for local sources of various liquid fuels and to lower transport costs as a function of Btu/volume (Ruffman-Weiss 2016). The Btu content of virtually all synthetic hydrocarbon fuels is higher than LNG per volume (Table 7.2), but the product could be methanol or other industrial feedstock for the petrochemical industry rather than a fuel that would contribute to CO_2 emissions more than natural gas.

Units have been developed to bring natural gas from large conventional fields from remote areas or areas in which infrastructure development would be prohibitive (Chap. 8). The LNG is manufactured on floating installations in the immediate vicinity above the subsea gas fields, which may involve considerable processing of the gas + produced water and other fluids along with sand separation on the seafloor. The increased demand for gas is largely going to be met in the Asian Pacific region using FLNG vessels (Musarra 2015). On 14 May 2016 the first FLNG made the voyage from Okpo, South Korea to the Kanowit gas field 180 km off Malaysia (Offshore 11/18/16). It has been designed to not require dry docking for 20 years and has the flexibility to be redeployed to many locations to service marginal and stranded gas fields in the region (Offshore 2016). Some smaller country consortiums are specifically targeting smaller stranded deepwater gas prospects of from about 0.5 to 2 Tcf that are not of interest to major energy companies (Offshore 10/17/17).

The commerciality of FLNG depends on world supply and delivered price. For instance, in 2017 interest in widespread FLNG supply was lessened due to the abundance of LNG beginning to be exported from the United States. The LNG market is headed toward oversupply until the early 2020s, and it is likely that some other planned FLNG vessels may be delayed until the world market situation becomes clearer (Tay 2017).

Although thus far the FLNG technology has been developed entirely for conventional gas, it can be used without any alteration in process, except to the extent that much less processing may be necessary. Gas from NGH will be purer and will be associated with lower temperature, less saline, and purer water than conventional gas. This should marginally decrease the cost of processing. As with other deep-water technologies that were developed for conventional gas and oil, NGH production utilizing FLNG can piggyback on the CAPEX already spent for conventional gas production and transport to market. The optimal size and weight for

FLNG vessels are almost certainly commercial rather than technical in character (ABS 2015; White 2015).

The development of floating natural gas compression facilities, with new vessel design to accommodate stranded gas, solves the problem of monetizing the value of gas that might otherwise have to be reinjected (incidental gas produced at an oil field) or abandoned. A pertinent legal and political issue is whether governments will treat natural gas in the same way as they have with oil. Will countries let their gas resources flow to shore first, as would be the case with a land-based LNG plant without coming onshore or will an oil-related tax and fee solution be found?

11.7.9 How Soon NGH?

NGH for many years at the end of the 20th century was regarded as a resource that would be available sometime in the future (Kvenvolden 1988, 1993; Max 2003; Collett et al. 2009). The production cost of natural gas from permafrost hydrate was considered to be a slight additional cost to producing conventional natural gas, especially from permafrost NGH deposits where free gas lay below and in contact with NGH above (Max et al. 2006). The major part of the NGH resource, the 95% estimated to be in marine sediments on continental margins, appears to remain something of a mystery to petroleum geologists. This is probably because NGH is a solid crystalline mineral deposit, and is as far outside of their field of experience as is a strata-bound lead-zinc deposit.

As with other energy resources, NGH has to be evaluated as a gas resource that must compete with other gas resources in order to be commercially viable, unless there are special circumstances such as those affecting U.S. Shale oil and gas. NGH deposits will be identified (Chap. 3) and valued (Chap. 6), and then the percentage and volume estimates for the amount of gas that can be recovered will largely determine the likelihood of near-term development. The amount of gas in place is relatively unimportant unless a large percentage of it can be recovered. A case in point is shale gas, a large resource base that was without value prior to development of the fracking technology to create permeability in the beds and allows the gas to flow and be recovered. But the most important commercial factor will be how much gas can be recovered that is commercially competitive with other gas sources, both in producer and delivered gas markets.

To assess the production characteristics and economics of marine NGH, Walsh et al. (2009) used the TOUGH+HYDRATE in his reservoir simulation cost analyses to compare production from NGHs to that from a conventional gas reservoir. Overall cost estimates include a pipeline, production facility, and subsea development for both conventional and NGH production and the extra costs associated with gas production from hydrate. These extra costs include such factors as additional wells and artificial lift to manage water production. At the 50% confidence level, they determined that the additional cost associated with production from deepwater NGHs vs. conventional gas deposits would add \$3.50–\$4.00 (U.S. dollars) per Mcf. We suggest that if NGH-specific technology and practices were to be applied (Chap. 8) this 'NGH penalty' could be lowered substantially. In any case, for deep importing countries, cost competition is with imports, not other natural gas production.

Natural gas is on the road to becoming an internationally traded commodity with a common price. The market and the pricing will be similar to that of oil. Every producer will try to make the maximum profit and every importer will try and minimize their outlays. Although delivered gas will have to bear the cost of pipelines or costs of compression and transport, the general economic structure of both gas and oil is similar. Natural gas has additional costs of compression, but oil has the additional costs of refining. Natural gas does not have the early downstream processing costs of oil. All gas resources have a common base for commercialization, their fully amortized production cost, and their value at sale. Other issues, such as potential or actual environmental impact, are not yet a first order determinant.

The basic factor governing when natural gas from NGH may become an important resource will be governed more by its production cost than its delivery cost as transport becomes widely available. Locally produced NGH that could be brought ashore to market at a low cost would be more economic than higher cost delivered gas, even if its production cost were lower. Monetization or increasing the value of natural gas by shipping from a low-cost producer such as the United States and selling into a higher cost market is a first-order economic objective. Although NGH has often been characterized as a last resort future resource that will be opened only when it must be, if the cost of its exploration and production can be brought down sufficiently to make it commercially attractive in importing countries, it has the potential to become widely produced, not only by traditionally importing countries.

The dropping gas prices caused by the sudden abundance of gas since 2008 has led to the deferring of development of once promising conventional gas deposits (Max et al. 2013). An example is the Shtokman field, (one of the world's largest natural gas fields) located in the northwestern part of the South Barents Basin in the Russian sector of the Barents Sea, 600 km north of Kola Peninsula. When the price of gas rises to the point where these conventional fields that are more expensive to produce become commercial, they will be opened up, which will have the effect of lengthening the time factor for opening production in NGH if the exploration and production cost of NGH cannot be brought down.

Although we take no position on the size of the shale resource, we hold open the possibility that the shale gas skeptics may be more correct than the 'shale enthusiasts' simply to be conservative about reserves. If there is far less technically and commercially recoverable natural gas in the shale resource (with due regard to the fact that most of the gas produced to date may be incidental gas produced along with shale oil), then there may be an indigenous gas shortage in the United States within a decade. If a gas shortage develops, then the pressure to produce from NGH resources will be strengthened because of a considerably higher breakeven price. If the gas shortage does not develop, production from NGH could still be attractive on a competitive basis if the cost of NGH E&P can be brought down through better technology and practices (Chap. 8). In any case, increasing export of LNG from the United States will have the effect of increasing demand, with a consequent pressure for the increase of the price of natural gas.

Energy security will almost certainly be the greatest driver for development of the NGH resource. It cannot be a coincidence that the most aggressive NGH energy-related programs are operated by deep importing countries (India and China) that are industrializing in the pursuit of higher living standards (Chap. 1). These countries have also made commitments to reduce CO_2 emissions, which can only be done in the short term by switching from coal, and to a lesser extent from oil, to natural gas. Japan and South Korea, which are more fully developed and with high living standards, both have major NGH development programs although Japan is under less pressure after its government decided to reopen some of its nuclear electric generation plants in 2015. Even though the world price of imported natural gas is very low as supply increases, Japan is progressing its NGH project, even in the energy downturn (Cheang 2016).

We suggest that once the issue of low-cost E&P for NGH has been implemented and the first production has begun, major companies, especially those with excess offshore capabilities, will engage in serious NGH E&P so as not to be left out of the technology surge. The JOGMEC (2013) drill stem technical production test (Chap. 3) confirmed that significant continuous production was possible and revealed that sand control is a primary well survivability factor. Further production tests by Japan and China in the spring and summer of 2017 (Lees 2017) have confirmed that sanding remains a production issue, but continuous production of over a month appears to have been achieved in the Chinese test from silty turbidites whose productivity had previously been thought to be less than productive in character.

Although no major technology issue needs to be resolved before the offshore NGH resource can be opened, optimized leveraging of existing technology and the development of new technology that accomplish existing practices more efficiently and less costly can make the NGH resource much more commercially attractive. But both leverage of existing technology and development and implementation of new technology should allow for cost cutting. NGH development still faces important operational challenges including successful long-term production testing. There is a lot of natural gas but fuel switching, energy demand indicates that gas is the energy future (Islam 2015). Time will tell, but the future of NGH could be much nearer than the indefinite future.

References

ABS. (2015). *Floating offshore liquefied gas terminals. Guide for building and classing* (originally published 2010, revised 2015). American Bureau of Shipping (352 pp.).

AD. (2017). *The future of gas.* Addleshaw Goddard (28 pp.). https://www.addleshawgoddard.com/globalassets/insights/energy/special-report-the-future-of-gas.pdf. Accessed May 28, 2017.

Anderson, B., Boswell, R., Collett, T. S., Farrell, H., Ohtsuki, S. White, M., et al. (2014). Review of the findings of the Iġnick Sikumi CO_2-CH_4 gas hydrate exchange field trial. In *Proceedings of the 8th International Conference on Gas Hydrates (ICGH8-2014)*, July 28–August 1, Beijing, China, (17 pp.).

Anderson, B. J., Kurihara, M., White, M. D., Moridis, G. J., Wilson, S. J., Pooldadi-Darvish, M., et al. (2011). Regional long-term production modeling from a single well test, Mount Elbert Gas Hydrate Stratigraphic Test Well, Alaska North Slope. *Marine and Petroleum Geology, 28,* 493–501.

Anderson, D. M., Stotts, G. W. J., Mattar, L., Fekete Associates Inc., Lik, D., & Blasingame, T. A. (2006). *The 2006 SPE Annual Technical Conference and Exhibition* held in San Antonio, Texas, U.S.A., 24–27 September 2006, (15 pp.).

Barton, C. M. (2015, May 25). *Industry continues to provide solutions for deepwater production challenges* (8 pp.). Oilpro.

Barton, C., Hambling, H., Albaugh, E. K., Mahlstedt, B. & Davis, D. (2015). Deepwater solutions & records for concept selection. Going deeper with production technology. *Poster May, Offshore Magazine Inset. 1 sheet.*

Berman, A. E. (2013, September). Let's be honest about shale gas. *Houston Geological Society Bulletin* (Hardcopy handout), 33, 35.

Bishop, R. (2015). RSK (UK) Limited. Luncheon presentation to New Orleans Geological Society, 6 July 2015.

Blas, J., & Smith, G. (2016). Oil's nightmare scenario dominates Davos. Bloomberg Business. Oil's nightmare scenario dominates Davos. Bloomberg Business. Accessed February 4, 2016.

Bloomberg. (2015). Beijing shivers as natural gas import delay turns down heat. *Bloomberg News*, December 28. http://www.bloomberg.com/news/articles/2015-12-28/beijing-curtails-natural-gas-supply-as-petrochina-lng-delayed. Accessed January 11, 2016.

Bonney, J. (2015). *TOTE launches first LNG-powered container ship*. JOC.com. http://www.joc.com/maritime-news/container-lines/sea-star-line/tote-launches-first-lng-powered-container-ship_20150420.html. Accessed April 27, 2016.

Boswell, R. (2016). Emerging issues in the development of geologic models for gas hydrate numerical simulation. *Fire in the Ice, 16*(1), 19–22.

Boswell, R., & Collett, T. S. (2011). Current perspectives on gas hydrate resources. *Royal Society of Chemistry, Energy & Environmental Science, 4,* 1206–1215. https://doi.org/10.1039/c0ee00203h.

Boswell, R., Collett, T. S., Myshakin, E., Ajayi, T., & Seol, Y. (2017). The increasingly complex challenge of gas hydrate reservoir simulation. In *Proceedings of Ninth International conference on Gas Hydrates (IGCH9)*, Denver, Colrado, USA, 25–30 June 2017 (10 pp.).

Boswell, R, Collett, T. S., Frye, M., Shedd, W., McConnell, D. R., & Shelander, D. (2012) Subsurface gas hydrates in the northern Gulf of Mexico. *Marine and Petroleum Geology, 34,* 4–30.

Bradford, A. (2016). *Associated natural gas PDP warrants close watching*. OilPro. http://oilpro.com/post/22887/associated-natural-gas-pdp-warrants-close-watching?utm_source=DailyNewsletter&utm_medium=email&utm_campaign=newsletter&utm_term=2016-03-04&utm_content=Article_7_txt. Accessed April 2, 2016.

Bret-Rouzaut, N., & Favennec, J.-P. (2011). *Oil and gas exploration and production* (310 pp.). TECHNIP.

Brown, M. (2017). *China, Japan extract combustible ice from seafloor*. PennEnergy. http://www.pennenergy.com/articles/pennenergy/2017/05/fossil-fuel-china-japan-extract-combustible-ice-from-seafloor.html?cmpid=enl_pennenergy_pennenergydailypetroleum_2017-05-22&email_address=michaelmax1@mac.com&eid=288179069&bid=1760760. Accessed May 22, 2017.

Cheang, C. Y. (2016). *Japan progresses methane hydrate project, ignores industry downturn* (5 pp.). Rigzone. http://www.rigzone.com/news/oil_gas/a/145552/Japan_Progresses_Methane_Hydrate_Project_Ignores_Industry_Downturn/?all=HG2. Accessed March 8, 2017.

Chia, C., *The U.S. natural gas market—A quiet revolution?* OilPro http://oilpro.com/announcement/4964/us-natural-gas-market-quiet-revolution?utm_source=DailyNewsletter&utm_medium=email&utm_campaign=newsletter&utm_term=2017-06-28&utm_content=Article_3_txt. Accessed June 29, 2017.

Clemente, J. (2017). *The rapidly expanding global liquefied natural gas market*. Forbes. https://www.forbes.com/sites/judeclemente/2017/07/09/the-rapidly-expanding-global-liquefied-natural-gas-market/#79a944752213. Accessed July 10, 2017.

Collett, T. S., Johnson, A. H., Knapp, C. C., & Boswell, R. (2009) In Collett, T. S., Johnson, A. H., Knapp, C. C., & Boswell, R. (Eds.), *Natural gas hydrates: A review* (Vol. 89, pp. 146–219). American Association of Petroleum Geologists Memoir. ISBN 13: 918-0-89181-370-5.

Curtis, T. (2016). Unravelling the US Shale Productivity Gains. Oxford Institute for Energy Studies, OIES Paper WPM 69 (19 pp.).

Downing, B., & Livingston, D. (2015). *Sharp declines in well production typical in Ohio's Utica Shale*. http://www.ohio.com/news/local/sharp-declines-in-well-production-typical-in-ohio-s-utica-shale-1.555015. Accessed April 6, 2015.

DW. (2015). *World LNG market forecast 2016–2020*. http://www.douglas-westwood.com/wp-content/uploads/2015/10/K573-World-LNG-Market-Forecast-2016-2020-LEAFLET.pdf. Accessed January 8, 2016.

DW 11/9/15. (2015). *LNG—The coolest player in the European Gas Game*. http://www.douglas-westwood.com/dw-monday-lng-the-coolest-player-in-the-european-gas-game/. Accessed April 27, 2016.

EIA. (2016a). *Natural gas overview, data, analysis & projections*. http://www.eia.gov/dnav/ng/hist/n9140us2M.htm. Accessed May 19, 2016.

EIA. (2016b). *Initial production rates in tight oil formations continue to rise*. U.S. Energy Information Agency. https://www.eia.gov/todayinenergy/detail.cfm?id=24932. Accessed February 12, 2016.

Emeigh, G. (2017). Offshore leadership forum kicks off efforts to move industry toward long-term health. *Offshore, 77*(2), 24.

Etherington, J., Pollen, T., & Zuccolo, L. (2005). Oil and Gas Reserves Committee (OGRC) of the Society of Petroleum Engineers "Mapping" Subcommittee Final Report—December 2005. *Comparison of selected reserves and resource classifications and associated definitions*. http://www.spe.org/industry/docs/OGR_Mapping.pdf. Accessed September 21, 2015.

Fahey, M. *Oil prices and budgets: The OPEC countries most at risk. CNBC*. http://www.cnbc.com/2015/12/03/oil-prices-and-budgetsthe-opec-countries-most-at-risk.html. Accessed March 25, 2016.

Fekete. (2010). *Technical recoverability of gas hydrate in the U.S. Gulf of Mexico Type I, II, and III Reservoirs* (108 pp.). Prepared for the Bureau of Ocean energy Management. Fekete Associates Inc. Accessed August 19, 2015.

FERC. (2015). *Proposed North American LNG export terminals. Federal Energy Regulatory Commission*. http://www.ferc.gov/industries/gas/indus-act/lng/lng-proposed-export.pdf. Accessed April 22, 2016.

Freeman, L. (2017). Billions of dollars of bad oil and gas loans. *Oil & Gas Financial Journal*. Laura Freeman Consulting, Houston. http://www.ogfj.com/articles/print/volume-14/issue-9/departments/capital-perspectives/billions-of-dollars-of-bad-oil-and-gas-loans.html?cmpid=enl_ogfj_ogfj_weekly_update_2017-10-02&email_address=michaelmax1@mac.com&eid=288179069&bid=1882105. Accessed October 2, 2017.

Fujii, T., Suzuk, K., Tokujiro, T., Tamaki, M., Komatsu, Y., Konno, Y., et al. (2015). Geological setting and characterization of a methane hydrate reservoir distributed at the first offshore production test site on the Daini-Atsumi Knoll in the eastern Nankai Trough, Japan. *Marine and Petroleum Geology, 66,* 310–322.

Gaddipati, M., Myshakin, E. M., Boswell, R., & Anderson, B. J. (2011). Gas production modeling from a complex 3-D description of marine hydrate deposits. In *Proceedings of the 7th International Conference on Gas Hydrates (ICGH 2011),* Edinburgh, Scotland, United Kingdom, July 17–21.

GIIGNL. (2015). *Retail LNG Handbook Retail LNG & The Role of LNG Import Terminals Center for liquefied natural gas (CLNG)* (107 pp.). http://www.lngfacts.org. Accessed March 15, 2016.

Gronholt-Pedersen, J. (2016). *U.S. exports first shale gas as LNG tanker sails from Sabine Pass terminal.* Reuters. http://www.reuters.com/article/us-shale-export-idUSKCN0VY08B. Accessed March 12, 2016.

Hall, C. A. S., Lambert, J. G., & Balogh, S. B. (2014). EROI of different fuels and the implications for society. *Energy Policy, 64,* 141–152. http://dx.doi.org/10.1016/j.enpol.2013.05.049. Accessed January 20, 2016.

Hancock, S., (2008). Gas hydrates: Commercial sooner than expected. Presented at *CERI Natural Gas Conference*, Calgary, Alberta, February 2008. Referred to as source for figure in Walsh (2009).

Harrington, M. (2016). *Will the O&G industry's crash cascade into a 2008 styled credit contagion?* OilPro. http://oilpro.com/post/21427/solving-2016-dilemma-part-two-cascade-effect-here?utm_source=DailyNewsletter&utm_medium=email&utm_campaign=newsletter&utm_term=2016-01-11&utm_content=Feature_1_txt. Accessed January 11, 2016.

Hubbert. M. K. (1956). *Nuclear energy and fossil fuels.* Shell Development Company, Exploration and Production Research Division, Publication Number 95 (22 pp.).

Hughs, D. (2015). *Drilling deeper: A reality check on U.S. government forecasts for a lasting tight oil & shale gas boom.* Part 3: Shale gas (153 pp.). http://www.postcarbon.org/wp-content/uploads/2014/10/Drilling-Deeper_PART-3-Shale-Gas.pdf. Accessed January 23, 2016.

Hunter, R., & Digert, S. (2014). Final Technical Report September 30, 2001–March 31, 2014 Resource Characterization and Quantification of Natural Gas Hydrate and Associated Free-Gas Accumulations in the Prudhoe Bay–Kuparuk River Area on the North Slope of Alaska. Prepared for: U.S. Department of Energy National Energy Technology Laboratory (370 pp.).

Husky. (2016). *The Liwan gas project.* http://www.huskyenergy.com/operations/growthpillars/asiapacificregion/liwan.asp. Accessed April 27, 2016.

Hussein, A. A. (2015). *Analysis: Why Hubbert's peak oil theory fails.* Oilpro. http://oilpro.com/post/18246/why-hubbert-peak-oil-theory-fails?utm_source=DailyNewsletter&utm_medium=email&utm_campaign=newsletter&utm_term=2015-09-08&utm_content=Article_10_txt. Accessed September 8, 2015.

IEA. (2016). *Gas medium-term market report 2016, market analysis and forecasts to 2021* (131 pp.). International Energy Agency. ISBN (print) 978-92-64-25772-6/PDF 978-92-64-25773-3.

Inada, N., & Yamamoto, K. (2015). Data report: Hybrid Pressure Coring System tool review and summary of recovery result from gas-hydrate related coring in the Nankai Project. *Marine and Petroleum Geology, 66,* 323–345.

Islam, M. R. (2015). *Unconventional gas reservoirs evaluation, appraisal, and development* (615 pp.). Elsevier, Amsterdam. ISBN: 978-0-12-800390-9.

Itoh, S. (2013). Energy security in Northeast Asia: A pivotal moment for the U.S.-Japan Alliance. Brookings Institution, Northeast Asia Commentary Series # 67 (7 pp.).

JOC. (2015). *TOTE launches second LNG-fueled container ship.* http://www.joc.com/maritime-news/container-lines/sea-star-line/tote-launches-second-lng-fueled-container-ship_20150831.html. Accessed April 27, 2016.

JOGMEC. (2013). *News Release: Gas production from methane hydrate layers confirmed* (3 pp.). Japan Oil, Gas and Metals National Corporation, Tokyo. http://www.jogmec.go.jp/english/news/release/release0110.html. Accessed April 26, 2016.

Kanno, T., Takekoshi, M., Wang, S., Chee, M., Fukuhara, M., Osawa, Yamamoto, K., Fujii, T., Takayama, T & Suzuki, K. 2014. In-situ temperature measurement of gas hydrate dissociation during the world-first offshore production test. Paper OTC-25235-MS. In *Proceedings of the Offshore Technology Conference* held in Houston, Texas, USA, 5–8 May 2014 (5 pp.).

Kanno, Y., Yoneda, J., Egawa, Kosuke, Ito, T., Jin, Y., Kida, M., et al. (2015). Permeability of sediment cores from methane hydrate deposit in the Eastern Nankai Trough. *Marine and Petroleum Geology, 66,* 487–495.

Kramer, D. (2017). Refrackuring may not be all it's cracked up to be. *Physics Today, 70*(11), 27–29.

Kurihara, M., Sato, A., Ouchi, H., Ebinuma, T., Suzuki, K., Masuda, et al. (2010). Prediction of production test performances in Eastern Nankai trough methane hydrate reservoirs using 3D reservoir model. OTC 20737. In *Offshore Technology Conference* held in Houston, Texas, USA, 3–6 May 2010 (22 pp.).

Kurihara, M., Sato, A., Funatsu, K., Ouchi, H., Yamamoto, K., Fujii, T., et al. (2012). Analysis of 2007 and 2008 gas hydrate production tests on the Aurora/JOGMEC/NRCan Mallik 2L-38 well through numerical simulation. *Geological Survey of Canada, Bulletin 601*, 217–259. In S. R. Dallimore, K. Yamamoto, J. F. Wright, & G. Bellefleur (Eds.), *Scientific results from the JOGMEC/NRCan/Aurora Mallik 2007–2008 gas hydrate production research well program*, Mackenzie Delta, Northwest Territories, Canada.

Kvenvolden, K. A. (1988). Methane hydrate—A major reservoir of carbon in the shallow geosphere? *Chemical Geology, 71,* 41–51.

Kvenvolden, K. A. (1993). Gas hydrate as a potential energy resource—A review of their methane content. In D. G. Howell (Ed.), *The future of energy gases: U.S. geological survey professional paper* (Vol. 1570, pp. 555–561).

Lee, J. (2014). *Why LNG doesn't trade like oil.* Alaska Natural Gas Transportation Projects, Office of the Federal Coordinator. http://www.arcticgas.gov/why-lng-does-not-trade-like-oil. Accessed August 5, 2015.

Lees, G. (2017). China hypes up methane hydrates but Japan leads the way. InnovOil July 2017, 32–33.

Mahabadi, N., & Jang, J. (2014). Relative water and gas permeability for gas production from hydrate-bearing sediments. *Geochemistry, Geophysics, Geosystems, 15,* 2346–2353. https://doi.org/10.1002/2014GC005331.

Malik, N. (2017). Tanker's U-Turn shows how shale is changing world gas. Bloomberg increased market flexibility. Accessed March 9, 2017.

Mason, J. E. (2011). Well production profiles assess Fayetteville shale gas potential. *Oil & Gas Journal.* http://www.ogj.com/articles/print/volume-109/issue-14/drilling-production/well-production-profiles-assess-fayetteville-shale.html. Accessed January 27, 2016.

Max, M. D. (Ed). (2003). *Natural gas hydrate in oceanic and permafrost environments* (2nd ed., 422 pp.). London, Boston, Dordrecht: Kluwer Academic Publishers (now Springer).

Max, M. D., & Johnson, A. H. (2011). Methane hydrate/clathrate conversion. In M. R. Khan (Ed.), *Clean hydrocarbon fuel conversion technology* (pp. 413–434). Woodhead Publishing Series in Energy No. 19. Cambridge, U.K.: Woodhead Publishing Ltd. ISBN 1 84569 727 8, ISBN-13: 978 1 84569 727 3.

Max, M. D., & Johnson, A. H. (2014). Hydrate petroleum system approach to hydrate natural gas exploration. *Petroleum Geoscience, 20*(2), 187–199. Geological Society of London (Published in Online First 21 March 2014). https://doi.org/10.1144/petgeo2012-049.

Max, M. D., & Johnson, A. H. (2016). *Exploration and production of oceanic natural gas hydrate: Critical factors for commercialization* (405 pp.). Springer International Publishing AG.

Max, M. D., Johnson, A. H., & Dillon, W. P. (2006). *Economic geology of natural gas hydrate* (p. 341). Berlin, Dordrecht: Springer.

Max, M. D., Johnson, A. H., & Dillon, W. P. (2013). *SpringerBriefs in Energy: Natural gas hydrate Arctic Ocean deepwater resource potential* (113 pp.).

Merey, Ş., & Longinos, S. N. (2018). Numerical simulations of gas production from Class 1 hydrate and Class 3 hydrate in the Nile Delta of the Mediterranean Sea. *Journal of Natural Gas Science & Engineering*. https://doi.org/10.1016/j.jngse.2018.01.001.

Moridis, G. J., Collett, T. S., Boswell, R., Kurihara, M., Reagan, M. T., Koh, C., et al. (2008). Toward production from gas hydrates: Current status, assessment of resources, and simulation-based evaluation of technology and potential. In *Proceedings of the SPE Unconventional Reservoirs Conference*, Keystone, CO, U.S.A, 10–12 February 2008, (43 pp.).

Moridis, G. J., & Kowalsky, M. (2006). Gas production from unconfined Class 2 oceanic hydrate accumulations. In Max, M. D. (Ed.) (2003). *Natural gas hydrate in oceanic and permafrost environments* (2nd ed., pp. 249–266). London, Boston, Dordrecht: Kluwer Academic Publishers (now Springer).

Moridis, G. J., Reagan, M. T. (2007a). Strategies for gas production from oceanic class 3 hydrate accumulations, OTC 18865. In *2007 Offshore Technology Conference*, Houston, Texas, 30 April–3 May 2007.

Moridis, G. J., & Reagan, M. T. (2007b). Gas production from oceanic class 2 hydrate accumulations. OTC Paper 18866. In *Proceedings of the Offshore Technology Conference*, Houston TX, 30 April–3 May 2007 (33 pp.).

Moridis, G. J., Reagan, M. T. Boswell, R., Collett, T., & Zhang, K. (2010). Preliminary evaluation of the production potential of recently discovered hydrate deposits in the Gulf of Mexico. OTC 21049. In *Offshore Technology Conference* held in Houston, Texas, USA, 3–6 (27 pp.).

Moridis, G. J., Silpngarmlert, S., Reagan, M. T., Collett, T., & Zhang, K. (2011). Gas production from a cold, stratigraphically-bounded gas hydrate deposit at the Mount Elbert Gas Hydrate Stratigraphic Test Well, Alaska North Slope: Implications of uncertainties. *Marine and Petroleum Geology, 28,* 517–534.

Morrow, E., & Wallace, L. (2016). Industry looks to reduce project complexity to lower cost. *Offshore Magazine, 76*(1), 31.

Murphy, D. J., & Hall, C. A. S. (2010). Year in review—EROI or energy return on (energy) invested. *Annals of the New York Academy of Sciences, 1185,* 102–118.

Musarra, S. P. (2015). Gas demand driving mega projects off Asia/Pacific. *Offshore* 12/29/15. http://www.offshore-mag.com/articles/print/volume-75/issue-5/international-field-development/gas-demand-driving-mega-projects-off-asia-pacific.html?cmpid=EnlOSDailyDecember292015&eid=288179069&bid=1263176. Accessed December 29, 2015.

Musarra, S. P., & Beaubouef, B. (2017). Offshore leadership forum addresses capex challenges. *Offshore, 77*(2), 25–29.

Myshakin, E. M., Gaddipati, M., Rose, K., & Anderson, B. J. (2012). Numerical simulations of depressurization-induced gas production from gas hydrate reservoirs at the Walker Ridge 313 site, northern Gulf of Mexico. *Marine and Petroleum Geology, 34*, 169–185.

NSWA. (2014). *The economic impact of eliminating the percentage depletion allowance*. National Stripper Well Association. Webpage access https://nswa.us/custom/showpage.php?id=25. Accessed August 15, 2015.

Noguchi, S., Furukawa, T., Aung, T. T., & Oikawa, N. (2011). Reservoir architecture of methane hydrate bearing turbidite channels in the eastern Nankai Trough, Japan. In *Proceedings of the 7th International Conference on Gas Hydrates (ICGH 2011)*, Edinburgh, Scotland, United Kingdom, July 17–21, 2011 (9 pp.).

Noguchi, S., Fujii, T., Takayama, T., Murray, D. R., Fujii, K., Yamamoto, K., et al. (2012). Dissociation behaviour of gas hydrate throughout a production test, based on cased-hole log data, of the Aurora/JOGMEC/NRCan Mallik 2L-38 gas hydrate production research well. *Geological Survey of Canada, Bulletin, 601*, 205–216. In S. R. Dallimore, K. Yamamoto, J. F. Wright, & G. Bellefleur (Eds.), *Scientific results from the JOGMEC/NRCan/Aurora Mallik 2007–2008 Gas Hydrate Production Research Well Program*, Mackenzie Delta, Northwest Territories, Canada.

O&G. (2016). IHS: Decommissioning of aging offshore oikl and gas facilities increasing, annual spending rising to $13B by 2040. Oil&Gas Financ J. http://www.ogfj.com/articles/2016/11/ihs-decommissioning-of-aging-offshore-oil-and-gas-facilities-increasing-annual-spending-rising-to-13b-by-2040.html. Accessed December 1, 2016.

Offshore 1/18/16. (2016). *Det norske targets lower-cost offshore Norway production*. http://www.offshore-mag.com/articles/2016/01/det-norske-targets-lower-cost-offshore-norway-oil-and-gas-production.html. Accessed January 18, 2016.

Offshore 3/12/16. (2016). *LNG vessel*. http://www.harveygulf.com/green.html. Accessed April 27, 2016.

Offshore 11/18/16. (2016). World's first FLNG facility starts up. *Offshore*. http://www.offshore-mag.com/articles/2016/11/world-s-first-flng-facility-starts-up.html. Accessed November 18, 2016.

Offshore 10/17/17. (2017). *Partnership unveils stranded gas FLNG proposal*. http://www.offshore-mag.com/articles/2017/10/partnership-unveils-stranded-gas-flng-proposal.html?cmpid=enl_offshore_offshore_daily_newsletter_2017-10-18&email_address=michaelmax1@mac.com&eid=288179069&bid=1898186. Accessed October 18, 17.

Otani, A. 2/24/16. (2016). *In a first, Cheniere to export US liquefied natural gas*. CNBC. http://www.cnbc.com/2016/02/24/in-a-first-cheniere-to-export-us-liquefied-natural-gas.html.

PEES. (2016). *Vietnam's PV gas selects Honeywell to produce liquefied petroleum gas*. http://www.pennenergy.com/articles/pennenergy/2016/01/refinery-news-vietnam-s-pv-gas-selects-honeywell-to-produce-liquefied-petroleum-gas.html?cmpid=EnlDailyPetroJanuary142016&eid=288179069&bid=1277592. Accessed January 14, 2016.

Price, P. B. (2007). Microbial life in glacial ice and implications for a cold origin of life. *Federation of European Microbiological Societies Microbiology Ecology, 59,* 217–231.

Pooladi-Darvish, M., Zatspeina, O., & Hong, H. (2008). Behavior of gas production from type III hydrate reservoirs. In *Proceedings of the 6th International Conference on Gas Hydrates (ICGH 2008)*, Vancouver, British Columbia, Canada, July 6–10, 2008 (11 pp.).

Reagan, M. T., Moridis, G. J., Johnson, J. N., Pan, L., Freeman, C. M., Boyle, K. L., et al. (2015). Field-scale simulation of production from oceanic gas hydrate deposits. *Transport in Porous Media, 108,* 151–169. https://doi.org/10.1007/s11242-014-0330-7.

Reed, J. (2016). Here's One Supermajor That's NOT Planning Job Cuts. 1/20/16. http://oilpro.com/post/21661/here-one-supermajor-not-planning-job-cuts?utm_source=DailyNewsletter&utm_medium=email&utm_campaign=newsletter&utm_term=2016-01-20&utm_content=Article_6_txt. Accessed 20 Jan 2016.

Rodgers, K. J. (2017). *What is the ROI (and EROI) on subsidized renewable energy*. Natural Gas Now. http://naturalgasnow.org/roi-eroei-subsidized-renewable-energy/. Accessed June 30, 2017.

Ruffman-Weiss, L. (2016). Primus green energy methanol plant project is slated for 2017 delivery in the Marcellus Region. *GTE Ink, The Magazine of the Green Economy*. http://thegreeneconomy.com/content/launch-methanol-plant-project-announced?utm_source=ZohoCampaigns&utm_campaign=News+and+Events+from+The+Green+Economy_2016-03-14_1&utm_medium=email. Accessed March 30, 2016.

SBC. (2015). *Natural Gas Series FactBook*. Gas Hydrates SBC Energy Institute (SBC-EI) Schlumberger (99 pp.). https://www.sbc.slb.com/~/media/Files/SBC%20Energy%20Institute/SBC%20Energy%20Institute_Gas%20Hydrates_FactBook.pdf. Accessed December 22, 2015.

Scott, T. (2015). *$30 Crude leads to $6 natural gas*. Oilpro. http://oilpro.com/post/20989/30-crude-leads-to-6-natural-gas?utm_source=DailyNewsletter&utm_medium=email&utm_campaign=newsletter&utm_term=2015-12-21&utm_content=Article_14_txt. Accessed April 27, 2016.

Schoderbek, D., Martin, K., Howard, J., Silpngarmlert, S., & Hester, K. (2012). North Slope hydrate field trial: CO_2-CH_4 exchange. Paper OTC-23725, presented at *Offshore Technology Conference* (13 pp.).

Sinor, J. E. (1991). *Comparison of CNG and LNG technologies for transportation applications*. Report for National Renewable Energy Laboratory (57 pp.).

Snow, N. (2017). Trump signals intention to achieve American 'energy dominance". *Oil & Gas Journal.* http://www.ogj.com/articles/2017/06/trump-signals-intention-to-achieve-american-energy-dominance.html?cmpid=enl_ogj_ogjwashingtonpulseupdate_2017-07-07&email_address=michaelmax1@mac.com&eid=288179069&bid=1804347. Accessed July 8, 2017.

Takahashi, H., & Tsuji, Y. (2005). Multi-Well Exploration Program in 2004 for Natural Hydrate in the Nankai-Trough Offshore Japan. OTC 17162. *2005 Offshore Technology Conference* held in Houston, TX, U.S.A., 2–5 May 2005 (10 pp.).

Tay, M. (2017). *Floating Liquefied Natural Gas production bows out as U.S. exports roil market.* Oilpro. Energyworld (the Economic Times) (3 pp.). http://oilpro.com/links/37411/floating-liquefied-natural-gas-production-bows-out-us-exports-roi. Accessed March 7, 2017.

Thomas, K. (2016). *Will new Panama Canal be an LNG game-changer?* OilPro. http://oilpro.com/links/31784/new-panama-canal-lng-game-changer. Accessed May 19, 2016.

Triepke, J. 12/18/15. (2015). *Everything will bottom in 2016.* Oilpro (Lloyd's Register Energy). http://oilpro.com/post/20940/everything-bottom-2016?utm_source=Dai…campaign=newsletter&utm_term=2015-12-17&utm_content=Article_10_txt. Accessed December 18, 2015.

Tullo, A. H. & Johnson, J. (2013, March). The Gas Wars. *Chemical and Engineering News*, 9–13.

Tverbert, G. (2017). *Researchers have been underestimating the cost of wind and solar* (35 pp.). EnergyCollective. http://www.theenergycollective.com/gail-tverberg/2409208/researchers-underestimating-cost-wind-solar. Accessed July 30, 2017.

Uddin, M., Wright, J. F., Dallimore, S. R., & Coombe, D. (2012). Gas hydrate production from the Mallik reservoir: Numerical history matching and long-term production forecasting. *Geological Survey of Canada, Bulletin,* 601, 261–289. In S. R. Dallimore, K. Yamamoto, J. F. Wright, & G. Bellefleur (Eds.), *Scientific results from the JOGMEC/NRCan/Aurora Mallik 2007–2008 Gas Hydrate Production Research Well Program*, Mackenzie Delta, Northwest Territories, Canada.

USGS. (2000). Coal-bed methane: Potential and concerns. U.S. Geological Survey Fact Sheet FS-123-00 (2 pp.). Accessed November 27, 2015.

USGS. (2014). International gas hydrate research. U.S. Geological Survey slide show. http://energy.gov/sites/prod/files/2014/04/f14/International_Review_USGS_Collett%5B1%5D.pdf. Accessed January 24, 2016.

Vafaei, M. T., Kmamme, B., Chejara, A., & Jemai, K. (2012). Nonequilibrium modeling of hydrate dynamics in reservoir. Energy & fuels. *American Chemical Society, 26*, 3564–3576. https://doi.org/10.1021/ef300348r.

Walsh, M. R., Hancock, S. H., Wilson, S. J., Patil, S. L., Moridis, G. J., Boswell, R., et al. (2009). Preliminary report on the commercial viability of gas production from natural gas hydrates. *Energy Economics, 31,* 815–823.

Watts, W. (2016). U.S. oil production is declining, but how fast will it fall? Market watch 4/4/16. http://www.marketwatch.com/story/shale-oil-production-is-finally-falling-but-unclear-how-fast-2016-02-26. Accessed March 12, 2016.

White, N. (2015, October). Study provides insight into optimal weight, size for FLNG vessels. *Offshore, 58*, 60–61.

Wilson, S. J., Hunter, R. B., Collett, T. S., Hancock, S., Boswell, R., & Anderson, B. J. (2011). Alaska North Slope regional gas hydrate production modeling forecasts. *Marine and Petroleum Geology, 28,* 460–477.

Zheng, S., Cao, Y., Wu, N., & He, Y. (2011). Numerical analysis on gas production efficiency from hydrate deposits by thermal stimulation: application to the Shenhu Area. *South China Sea Energies, 4,* 294–313. https://doi.org/10.3390/en4020294.

Definitions

Commonly used in a number of chapters
1 metric ton liquefied natural gas (LNG) = 48,700 ft^3 of natural gas
1 billion cubic meters NG = ~35.3 billion cubic feet NG
Barrel (U.S.), 42 gallons (Bbl)
Billion cubic meters (Bcm)
Barrel of oil equivalent (BPE)
Capital Expenditure (CAPEX)
Exploration and Production (E&P)
Floater: Drillship or semisubmersible, or floating production, storage, processing, offloading, etc.
Gas Hydrate Stability Zone (GHSZ)
Liquefied natural gas (LNG)
Metres below present sea level (mbpsl)
Milllion Btu (of natural gas, approximately 1086 ft^3 of methane)
Million cubic feet (Mcf)
Million metric tons (MM)
Net primary production (NPP)
Natural gas hydrate (NGH)
Oil and Gas (O&G)
Parts per million (ppm)
Standard cubic meter Sm^3
Thousand cubic feet (Mft^3)
Trillion cubic feet (of natural gas) (Tcf)
Trillion cubic meters (of natural gas) (Tcm)

M. D. Max and A. H. Johnson, *Exploration and Production of Oceanic Natural Gas Hydrate*, https://doi.org/10.1007/978-3-030-00401-9

Index

M. D. Max and A. H. Johnson, *Exploration and Production of Oceanic Natural Gas Hydrate*, https://doi.org/10.1007/978-3-030-00401-9

D

MIX
Papier aus verantwortungsvollen Quellen
Paper from responsible sources
FSC® C105338

If you have any concerns about our products,
you can contact us on
ProductSafety@springernature.com

In case Publisher is established outside the EU,
the EU authorized representative is:
Springer Nature Customer Service Center GmbH
Europaplatz 3, 69115 Heidelberg, Germany

Printed by Libri Plureos GmbH
in Hamburg, Germany